Machine Learning for Signal Processing

Machine Learning for Signal Processing

Data Science, Algorithms, and Computational Statistics

Max A. Little

OXFORD
UNIVERSITY PRESS

Great Clarendon Street, Oxford, OX2 6DP,
United Kingdom

Oxford University Press is a department of the University of Oxford.
It furthers the University's objective of excellence in research, scholarship,
and education by publishing worldwide. Oxford is a registered trade mark of
Oxford University Press in the UK and in certain other countries

First published 2019
First published in paperback 2024

Published in the United States of America by Oxford University Press
198 Madison Avenue, New York, NY 10016, United States of America

British Library Cataloguing in Publication Data
Data available

Library of Congress Cataloging in Publication Data
Data available

ISBN 978–0–19–871493–4 (Hbk.)
ISBN 978–0–19–889655–5 (Pbk.)

DOI: 10.1093/oso/9780198714934.001.0001

Printed and bound by
CPI Group (UK) Ltd, Croydon, CR0 4YY

Preface

Digital signal processing (DSP) is one of the 'foundational' but somewhat invisible, engineering topics of the modern world, without which, many of the technologies we take for granted: the digital telephone, digital radio, television, CD and MP3 players, WiFi, radar, to name just a few, would not be possible. A relative newcomer by comparison, statistical machine learning is the theoretical backbone of exciting technologies that are by now starting to reach a level of ubiquity, such as automatic techniques for car registration plate recognition, speech recognition, stock market prediction, defect detection on assembly lines, robot guidance and autonomous car navigation. Statistical machine learning has origins in the recent merging of classical probability and statistics with artificial intelligence, which exploits the analogy between intelligent information processing in biological brains and sophisticated statistical modelling and inference.

DSP and statistical machine learning are of such wide importance to the knowledge economy that both have undergone rapid changes and seen radical improvements in scope and applicability. Both DSP and statistical machine learning make use of key topics in applied mathematics such as probability and statistics, algebra, calculus, graphs and networks. Therefore, intimate formal links between the two subjects exist and because of this, an emerging consensus view is that DSP and statistical machine learning should not be seen as separate subjects. The many overlaps that exist between the two subjects can be exploited to produce new digital signal processing tools of surprising utility and efficiency, and wide applicability, highly suited to the contemporary world of pervasive digital sensors and high-powered and yet cheap, computing hardware. This book gives a solid mathematical foundation to the topic of statistical machine learning for signal processing, including the contemporary concepts of the *probabilistic graphical model* (PGM) and *nonparametric Bayes*, concepts which have only more recently emerged as important for solving DSP problems.

The book is aimed at advanced undergraduates or first-year PhD students as well as researchers and practitioners. It addresses the foundational mathematical concepts, illustrated with pertinent and practical examples across a range of problems in engineering and science. The aim is to enable students with an undergraduate background in mathematics, statistics or physics, from a wide range of quantitative disciplines, to get quickly up to speed with the latest techniques and concepts in this fast-moving field. The accompanying software will enable readers to test out the techniques to their own signal analysis problems. The presentation of the mathematics is much along the lines of a standard undergraduate physics or statistics textbook, free of distracting technical complexities and jargon, while not sacrificing rigour. It would be an excellent textbook for emerging courses in machine learning for signals.

Contents

Preface   v

List of Algorithms   xiii

List of Figures   xv

1   **Mathematical foundations**   1
   1.1   Abstract algebras   1
       Groups   1
       Rings   3
   1.2   Metrics   4
   1.3   Vector spaces   5
       Linear operators   7
       Matrix algebra   7
       Square and invertible matrices   8
       Eigenvalues and eigenvectors   9
       Special matrices   10
   1.4   Probability and stochastic processes   12
       Sample spaces, events, measures and distributions   12
       Joint random variables: independence, conditionals, and marginals   14
       Bayes' rule   16
       Expectation, generating functions and characteristic functions   17
       Empirical distribution function and sample expectations   19
       Transforming random variables   20
       Multivariate Gaussian and other limiting distributions   21
       Stochastic processes   23
       Markov chains   25
   1.5   Data compression and information theory   28
       The importance of the information map   31
       Mutual information and Kullback-Leibler (K-L) divergence   32
   1.6   Graphs   34
       Special graphs   35
   1.7   Convexity   36
   1.8   Computational complexity   37
       Complexity order classes and *big-O* notation   38

Tractable versus intractable problems:
NP-completeness 38

2 Optimization 41
2.1 Preliminaries 41
Continuous differentiable problems and critical
points 41
Continuous optimization under equality constraints: La-
grange multipliers 42
Inequality constraints: duality and the Karush-Kuhn-Tucker
conditions 44
Convergence and convergence rates for iterative
methods 45
Non-differentiable continuous problems 46
Discrete (combinatorial) optimization problems 47
2.2 Analytical methods for continuous convex problems 48
L_2-norm objective functions 49
Mixed L_2-L_1 norm objective functions 50
2.3 Numerical methods for continuous convex problems 51
Iteratively reweighted least squares (IRLS) 51
Gradient descent 53
Adapting the step sizes: line search 54
Newton's method 56
Other gradient descent methods 58
2.4 Non-differentiable continuous convex problems 59
Linear programming 59
Quadratic programming 60
Subgradient methods 60
Primal-dual interior-point methods 62
Path-following methods 64
2.5 Continuous non-convex problems 65
2.6 Heuristics for discrete (combinatorial) optimization 66
Greedy search 67
(Simple) tabu search 67
Simulated annealing 68
Random restarting 69

3 Random sampling 71
3.1 Generating (uniform) random numbers 71
3.2 Sampling from continuous distributions 72
Quantile function (inverse CDF) and inverse transform
sampling 72
Random variable transformation methods 74
Rejection sampling 74
Adaptive rejection sampling (ARS) for log-concave densities 75
Special methods for particular distributions 78
3.3 Sampling from discrete distributions 79
Inverse transform sampling by sequential search 79

	Rejection sampling for discrete variables	80
	Binary search inversion for (large) finite sample spaces	81
3.4	Sampling from general multivariate distributions	81
	Ancestral sampling	82
	Gibbs sampling	83
	Metropolis-Hastings	85
	Other MCMC methods	88

4 Statistical modelling and inference — **93**

4.1	Statistical models	93
	Parametric versus nonparametric models	93
	Bayesian and non-Bayesian models	94
4.2	Optimal probability inferences	95
	Maximum likelihood and minimum K-L divergence	95
	Loss functions and empirical risk estimation	98
	Maximum a-posteriori and regularization	99
	Regularization, model complexity and data compression	101
	Cross-validation and regularization	105
	The bootstrap	107
4.3	Bayesian inference	108
4.4	Distributions associated with metrics and norms	110
	Least squares	111
	Least L_q-norms	111
	Covariance, weighted norms and Mahalanobis distance	112
4.5	The exponential family (EF)	115
	Maximum entropy distributions	115
	Sufficient statistics and canonical EFs	116
	Conjugate priors	118
	Prior and posterior predictive EFs	122
	Conjugate EF prior mixtures	123
4.6	Distributions defined through quantiles	124
4.7	Densities associated with piecewise linear loss functions	126
4.8	Nonparametric density estimation	129
4.9	Inference by sampling	130
	MCMC inference	130
	Assessing convergence in MCMC methods	130

5 Probabilistic graphical models — **133**

5.1	Statistical modelling with PGMs	133
5.2	Exploring conditional independence in PGMs	136
	Hidden versus observed variables	136
	Directed connection and separation	137
	The Markov blanket of a node	138
5.3	Inference on PGMs	139

| | | Exact inference | 140 |
| | | Approximate inference | 143 |

6 Statistical machine learning — **149**

6.1	Feature and kernel functions	149
6.2	Mixture modelling	150
	Gibbs sampling for the mixture model	150
	E-M for mixture models	152
6.3	Classification	154
	Quadratic and linear discriminant analysis (QDA and LDA)	155
	Logistic regression	156
	Support vector machines (SVM)	158
	Classification loss functions and misclassification count	161
	Which classifier to choose?	161
6.4	Regression	162
	Linear regression	162
	Bayesian and regularized linear regression	163
	Linear-in parameters regression	164
	Generalized linear models (GLMs)	165
	Nonparametric, nonlinear regression	167
	Variable selection	169
6.5	Clustering	171
	K-means and variants	171
	Soft K-means, mean shift and variants	174
	Semi-supervised clustering and classification	176
	Choosing the number of clusters	177
	Other clustering methods	178
6.6	Dimensionality reduction	178
	Principal components analysis (PCA)	179
	Probabilistic PCA (PPCA)	182
	Nonlinear dimensionality reduction	184

7 Linear-Gaussian systems and signal processing — **187**

7.1	Preliminaries	187
	Delta signals and related functions	187
	Complex numbers, the unit root and complex exponentials	189
	Marginals and conditionals of linear-Gaussian models	190
7.2	Linear, time-invariant (LTI) systems	191
	Convolution and impulse response	191
	The discrete-time Fourier transform (DTFT)	192
	Finite-length, periodic signals: the discrete Fourier transform (DFT)	198
	Continuous-time LTI systems	201
	Heisenberg uncertainty	203
	Gibb's phenomena	205
	Transfer function analysis of discrete-time LTI systems	206

	Fast Fourier transforms (FFT)	208
7.3	LTI signal processing	212
	Rational filter design: FIR, IIR filtering	212
	Digital filter recipes	220
	Fourier filtering of very long signals	222
	Kernel regression as discrete convolution	224
7.4	Exploiting statistical stability for linear-Gaussian DSP	226
	Discrete-time Gaussian processes (GPs) and DSP	226
	Nonparametric power spectral density (PSD) estimation	231
	Parametric PSD estimation	236
	Subspace analysis: using PCA in DSP	238
7.5	The Kalman filter (KF)	242
	Junction tree algorithm (JT) for KF computations	243
	Forward filtering	244
	Backward smoothing	246
	Incomplete data likelihood	247
	Viterbi decoding	247
	Baum-Welch parameter estimation	249
	Kalman filtering as signal subspace analysis	251
7.6	Time-varying linear systems	252
	Short-time Fourier transform (STFT) and perfect reconstruction	253
	Continuous-time wavelet transforms (CWT)	255
	Discretization and the discrete wavelet transform (DWT)	257
	Wavelet design	261
	Applications of the DWT	262
8	**Discrete signals: sampling, quantization and coding**	**265**
8.1	Discrete-time sampling	266
	Bandlimited sampling	267
	Uniform bandlimited sampling: Shannon-Whittaker interpolation	267
	Generalized uniform sampling	270
8.2	Quantization	273
	Rate-distortion theory	275
	Lloyd-Max and entropy-constrained quantizer design	278
	Statistical quantization and dithering	282
	Vector quantization	286
8.3	Lossy signal compression	288
	Audio companding	288
	Linear predictive coding (LPC)	289
	Transform coding	291
8.4	Compressive sensing (CS)	293
	Sparsity and incoherence	294
	Exact reconstruction by convex optimization	295
	Compressive sensing in practice	296

9 Nonlinear and non-Gaussian signal processing   **299**
  9.1   Running window filters   299
        Maximum likelihood filters   300
        Change point detection   301
  9.2   Recursive filtering   302
  9.3   Global nonlinear filtering   302
  9.4   Hidden Markov models (HMMs)   304
        Junction tree (JT) for efficient HMM computations 305
        Viterbi decoding   306
        Baum-Welch parameter estimation   306
        Model evaluation and structured data classification 309
        Viterbi parameter estimation   309
        Avoiding numerical underflow in message passing 310
  9.5   Homomorphic signal processing   311

10 Nonparametric Bayesian machine learning and signal processing   **313**
  10.1   Preliminaries   313
        Exchangeability and de Finetti's theorem   314
        Representations of stochastic processes   316
        Partitions and equivalence classes   317
  10.2   Gaussian processes (GP)   318
        From basis regression to kernel regression   318
        Distributions over function spaces: GPs   319
        Bayesian GP kernel regression   321
        GP regression and Wiener filtering   325
        Other GP-related topics   326
  10.3   Dirichlet processes (DP)   327
        The Dirichlet distribution: canonical prior for the categorical distribution 328
        Defining the Dirichlet and related processes   331
        Infinite mixture models (DPMMs)   334
        Can DP-based models actually infer the number of components? 343

Bibliography   **345**

Index   **353**

List of Algorithms

2.1   Iteratively reweighted least squares (IRLS).   52
2.2   Gradient descent.   53
2.3   Backtracking line search.   54
2.4   Golden section search.   55
2.5   Newton's method.   56
2.6   The subgradient method.   61
2.7   A primal-dual interior-point method for linear programming (LP).   64
2.8   Greedy search for discrete optimization.   67
2.9   (Simple) tabu search.   68
2.10 Simulated annealing.   69
2.11 Random restarting.   70
3.1   Newton's method for numerical inverse transform sampling.   73
3.2   Adaptive rejection sampling (ARS) for log-concave densities.   76
3.3   Sequential search inversion (simplified version).   80
3.4   Binary (subdivision) search sampling.   81
3.5   Gibb's Markov-Chain Monte Carlo (MCMC) sampling. . .   83
3.6   Markov-Chain Monte Carlo (MCMC) Metropolis-Hastings (MH) sampling.   86
5.1   The junction tree algorithm for (semi-ring) marginalization inference of a single variable clique.   142
5.2   Iterative conditional modes (ICM).   144
5.3   *Expectation-maximization* (E-M).   146
6.1   Expectation-maximization (E-M) for general i.i.d. mixture models.   153
6.2   The K-means algorithm.   172
7.1   Recursive, Cooley-Tukey, radix-2, decimation in time, fast Fourier transform (FFT) algorithm.   210
7.2   Overlap-add FIR convolution.   224
8.1   The Lloyd-Max algorithm for fixed-rate quantizer design.   279
8.2   The K-means algorithm for fixed-rate quantizer design.   . 280
8.3   Iterative variable-rate entropy-constrained quantizer design. 281
9.1   *Baum-Welch* expectation-maximization (E-M) for hidden Markov models (HMMs).   308
9.2   *Viterbi training* for hidden Markov models (HMMs). . . .   310
10.1 Gaussian process (GP) *informative vector machine* (IVM) regression.   325
10.2 Dirichlet process mixture model (DPMM) Gibbs sampler.   337

10.3 Dirichlet process means (DP-means) algorithm. 340
10.4 Maximum a-posteriori Dirichlet process mixture collapsed
      (MAP-DP) algorithm for conjugate exponential family
      distributions. 342

List of Figures

1.1	Mapping between abstract groups	2
1.2	Rectangle symmetry group	2
1.3	Group Cayley tables	3
1.4	Metric 2D circles for various distance metrics	4
1.5	Important concepts in 2D vector spaces	5
1.6	Linear operator flow diagram	7
1.7	Invertible and non-invertible square matrices	8
1.8	Diagonalizing a matrix	9
1.9	Distributions for discrete and continuous random variables	13
1.10	Empirical cumulative distribution and density functions	20
1.11	The 2D multivariate Gausian PDF	21
1.12	Markov and non-Markov chains	25
1.13	Shannon information map	28
1.14	Undirected and directed graphs	33
1.15	Convex functions	35
1.16	Convexity, smoothness, non-differentiability	37
2.1	Lagrange multipliers in constrained optimization	42
2.2	Objective functions with differentiable and non-differentiable points	46
2.3	Analytical shrinkage	50
2.4	Iteratively reweighted least squares (IRLS)	52
2.5	Gradient descent with constant and line search step sizes	54
2.6	Constant step size, backtracking and golden section search	55
2.7	Convergence of Newton's method with and without line search	57
2.8	Piecewise linear objective function (1D)	61
2.9	Convergence of the subgradient method	62
2.10	Convergence of primal-dual interior point	64
2.11	Regularization path of total variation denoising	65
3.1	Rejection sampling	75
3.2	Adaptive rejection sampling	77
3.3	Ancestral sampling	82
3.4	Gibb's sampling for the Gaussian mixture model	83
3.5	Metropolis sampling for the Gaussian mixture model	85
3.6	Slice sampling	90
4.1	Overfitting and underfitting: polynomial models	103

4.2   Minimum description length (MDL)   104
4.3   Cross-validation for model complexity selection   105
4.4   Quantile matching for distribution fitting   125
4.5   Convergence properties of Metropolis-Hastings (MH) sam-
      pling   131

5.1   Simple 5-node probabilistic graphical model (PGM)   133
5.2   Notation for repeated connectivity in graphical models   134
5.3   Some complex graphical models   135
5.4   Three basic probabilistic graphical model patterns   137
5.5   D-connectivity in graphical models   138
5.6   The Markov blanket of a node   139
5.7   Junction trees   143
5.8   Convergence of iterative conditional modes (ICM)   145

6.1   Gaussian mixture model for Gamma ray intensity strati-
      fication   151
6.2   Convergence of expectation-maximization (E-M) for the
      Gaussian mixture model   152
6.3   Classifier decision boundaries   160
6.4   Loss functions for classification   161
6.5   Bayesian linear regression   163
6.6   Linear-in-parameters regression   165
6.7   Logistic regression   166
6.8   Nonparametric kernel regression   167
6.9   Overfitted linear regression on random data: justifying
      variable selection   169
6.10 Lasso regression   170
6.11 Mean-shift clustering of human movement data   176
6.12 Semi-supervised versus unsupervised clustering   178
6.13 Dimensionality reduction   179
6.14 Principal components analysis (PCA): traffic count signals 181
6.15 Kernel principal components analysis (KPCA): human
      movement data   184
6.16 Locally linear embedding (LLE): chirp signals   185

7.1   The sinc function   189
7.2   Heisenberg uncertainty   204
7.3   Gibb's phenomena   206
7.4   Transfer function analysis   208
7.5   Transfer function of the simple FIR low-pass filter   214
7.6   Transfer function of the truncated ideal low-pass FIR filter 216
7.7   The bilinear transform   220
7.8   Computational complexity of long-duration FIR imple-
      mentations   224
7.9   Impulse response and transfer function of discretized ker-
      nel regression   225
7.10 Periodogram power spectral density estimator   233

7.11 Various nonparametric power spectral density estimators   234
7.12 Linear predictive power spectral density estimators   236
7.13 Regularized linear prediction analysis model selection   238
7.14 Wiener filtering of human walking signals   239
7.15 Sinusoidal subspace principal components analysis (PCA)
     filtering   241
7.16 Subspace MUSIC frequency estimation   242
7.17 Typical 2D Kalman filter trajectories   243
7.18 Kalman filter smoothing by Tikhonov regularization   252
7.19 Short-time Fourier transform analysis   254
7.20 Time-frequency tiling: Fourier versus wavelet analysis   256
7.21 A selection of wavelet functions   261
7.22 Vanishing moments of the Daubechies wavelet   262
7.23 Discrete wavelet transform wavelet shrinkage   263

8.1 Digital MEMS sensors   265
8.2 Discrete-time sampling hardware block diagram   267
8.3 Non-uniform and uniform and bandlimited sampling   268
8.4 Shannon-Whittaker uniform sampling in the frequency
     domain   269
8.5 Digital Gamma ray intensity signal from a drill hole   271
8.6 Quadratic B-spline uniform sampling   272
8.7 Rate-distortion curves for an i.i.d. Gaussian source   277
8.8 Lloyd-Max scalar quantization   280
8.9 Entropy-constrained scalar quantization   282
8.10 Shannon-Whittaker reconstruction of density functions   283
8.11 Quantization and dithering   285
8.12 Scalar vesus vector quantization   286
8.13 Companding quantization   287
8.14 Linear predictive coding for data compression   289
8.15 Transform coding: bit count versus coefficient variance   291
8.16 Transform coding of musical audio signals   292
8.17 Compressive sensing: discrete cosine transform basis   295
8.18 Random demodulation compressive sensing   297

9.1 Quantile running filter for exoplanetary light curves   301
9.2 First-order log-normal recursive filter: daily rainfall signals 302
9.3 Total variation denoising (TVD) of power spectral density
     time series   303
9.4 Bilateral filtering of human walking data   304
9.5 Hidden Markov modelling (HMM) of power spectral time
     series data   306
9.6 Cepstral analysis of voice signals   312

10.1 Infinite exchangeability in probabilistic graphical model
     form   315
10.2 Linking kernel and regularized basis function regression   319

10.3 Random function draws from the linear basis regression
      model   321
10.4 Gaussian process (GP) function draws with various co-
      variance functions   322
10.5 Bayesian GP regression in action   323
10.6 Gaussian process (GP) regression for classification   326
10.7 Draws from the Dirichlet distribution for $K = 3$ 327
10.8 The Dirichlet process (DP)   331
10.9 Random draws from a Dirichlet process (DP)   332
10.10 Comparing partition sizes of Dirichlet versus Pitman-Yor
      processes   334
10.11 Parametric versus nonparametric estimation of linear pre-
      dictive coding (LPC) coefficients   339
10.12 DP-means for signal noise reduction: exoplanet light curves 341
10.13 Probabilistic graphical model (PGM) for the Bayesian
      mixture model   342
10.14 Performance of MAP-DP versus DP-means nonparamet-
      ric clustering   344

Mathematical foundations

1

Statistical machine learning and signal processing are topics in applied mathematics, which are based upon many abstract mathematical concepts. Defining these concepts clearly is the most important first step in this book. The purpose of this chapter is to introduce these foundational mathematical concepts. It also justifies the statement that much of the art of statistical machine learning as applied to signal processing, lies in the choice of convenient mathematical models that happen to be useful in practice. Convenient in this context means that the algebraic consequences of the choice of mathematical modeling assumptions are in some sense manageable. The seeds of this manageability are the elementary mathematical concepts upon which the subject is built.

1.1   Abstract algebras

We will take the simple view in this book that mathematics is based on logic applied to *sets*: a set is an unordered collection of objects, often *real numbers* such as the set $\{\pi, 1, e\}$ (which has three *elements*), or the set of all real numbers $\mathbb{R}$ (with an infinite number of elements). From this modest origin it is a remarkable fact that we can build the entirety of the mathematical methods we need. We first start by reviewing some elementary principles of (abstract) *algebras*.

Groups

An *algebra* is a structure that defines the rules of what happens when pairs of elements of a set are acted upon by *operations*. A kind of algebra known as a *group* $(+, \mathbb{R})$ is the usual notion of addition with pairs of real numbers. It is a group because it has an *identity*, the number zero (when zero is added to any number it remains unchanged, i.e. $a + 0 = 0 + a = a$, and every element in the set has an *inverse* (for any number a, there is an inverse $-a$ which means that $a + (-a) = 0$. Finally, the operation is *associative*, which is to say that when operating on three or more numbers, addition does not depend on the order in which the numbers are added (i.e. $a + (b + c) = (a + b) + c)$. Addition also has the intuitive property that $a + b = b + a$, i.e. it does not matter if the numbers are swapped: the operator is called *commutative*, and the group is then called an *Abelian group*. Mirroring addition is multiplication acting on the set of real numbers with zero removed $(\times, \mathbb{R} - \{0\})$, which is also an Abelian group. The identity element is 1, and the inverses

Machine Learning for Signal Processing: Data Science, Algorithms, and Computational Statistics. Max A. Little.
© Max A. Little 2019. Published in 2019 by Oxford University Press. DOI: 10.1093/oso/9780198714934.001.0001

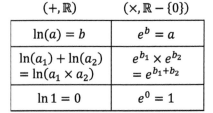

$(+, \mathbb{R})$	$(\times, \mathbb{R} - \{0\})$
$\ln(a) = b$	$e^b = a$
$\ln(a_1) + \ln(a_2)$ $= \ln(a_1 \times a_2)$	$e^{b_1} \times e^{b_2}$ $= e^{b_1 + b_2}$
$\ln 1 = 0$	$e^0 = 1$

Fig. 1.1: Illustrating abstract groups and mapping between them. Shown are the two continuous groups of real numbers under addition is (left column) and multiplication (right column), with identities 0 and 1 respectively. The homomorphism of exponentiation maps addition onto multiplication (left to right column), and the inverse, the logarithm, maps multiplication back onto addition (right to left column). Therefore, these two groups are homomorphic.

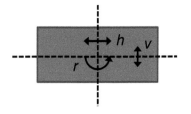

Fig. 1.2: The group of symmetries of the rectangle, $V_4 = (\circ, \{e, h, v, r\})$. It consists of horizontal and vertical flips, and a rotation of 180° about the centre. This group is isomorphic to the group $M_8 = (\times_8, \{1, 3, 5, 7\})$ (see Figure 1.3).

are the reciprocals of each number. Multiplication is also associative, and commutative. Note that we cannot include zero because this would require the inclusion of the inverse of zero $1/0$, which does not exist (Figure 1.1).

Groups are naturally associated with *symmetries*. For example, the set of rigid geometric transformations of a rectangle that leave the rectangle unchanged in the same position, form a group together with compositions of these transformations (there are flips along the horizontal and vertical midlines, one clockwise rotation through 180° about the centre, and the identity transformation that does nothing). This group can be denoted as $V_4 = (\circ, \{e, h, v, r\})$, where e is the identity, h the horizontal flip, v the vertical flip, and r the rotation, with the composition operation $\circ$. For the rectangle, we can see that $h \circ v = r$, i.e. a horizontal followed by a vertical flip corresponds to a 180° rotation (Figure 1.2).

Very often, the fact that we are able to make some convenient algebraic calculations in statistical machine learning and signal processing, can be traced to the existence of one or more symmetry groups that arise due to the choice of mathematical assumptions, and we will encounter many examples of this phenomena in later chapters, which often lead to significant computational efficiencies. A striking example of the consequences of groups in classical algebra is the explanation for why there are no solutions that can be written in terms of addition, multiplication and roots, to the general polynomial equation $\sum_{i=0}^{N} a_i x^i = 0$ when $N \geq 5$. This fact has many practical consequences, for example, it is possible to find the eigenvalues of a general matrix of size $N \times N$ using simple analytical calculations when $N < 5$ (although the analytical calculations do become prohibitively complex), but there is no possibility of using similar analytical techniques when $N \geq 5$, and one must resort to numerical methods, and these methods sometimes cannot guarantee to find all solutions!

Many simple groups with the same number of elements are *isomorphic* to each other, that is, there is a unique function that maps the elements of one group to the elements of the other, such that the operations can be applied consistently to the mapped elements. Intuitively then, the identity in one group is mapped to that of the other group. For example, the rotation group V_4 above is isomorphic to the group $M_8 = (\times_8, \{1, 3, 5, 7\})$, where $\times_8$ indicates multiplication modulo 8 (that is, taking the remainder of the multiplication on division by 8, see Figure 1.3).

Whilst two groups might not be isomorphic, they are sometimes *homomorphic*: there is a function between one group and the other that maps each element in the first group to one or more elements in the second group, but the mapping is still consistent under the second operation. A very important example is the exponential map, $\exp(x)$, that converts addition over the set of real numbers, to multiplication over the set of positive real numbers: $e^{a+b} = e^a e^b$; a powerful variant of this map is widely used in statistical inference to simplify and stabilize calculations involving the probabilities of independent statistical events, by converting them into calculations with their associated information content. This is the negative logarithmic map $-\ln(x)$, that converts probabilities under multiplication, into entropies under addition. This map is very widely used in statistical inference as we shall see.

For a more detailed but accessible background to group theory, read Humphreys (1996).

Rings

Whilst groups deal with one operation on a set of numbers, *rings* are a slightly more complex structure that often arises when two operations are applied to the same set. The most immediately tangible example is the operations of addition and multiplication on the set of integers $\mathbb{Z}$ (the positive and negative whole numbers with zero). Using the definition above, the set of integers under addition form an Abelian group, whereas under multiplication the integers form a simple structure known as a *monoid* – a group without inverses. Multiplication with the integers is associative, and there is an identity (the positive number one), but the multiplicative inverses are not integers (they are fractions such as $1/2$, $-1/5$ etc.) Finally, in combination, integer multiplication *distributes* over integer addition: $a \times (b + c) = a \times b + a \times c = (b + c) \times a$. These properties define a ring: it has one operation that together with the set forms an Abelian group, and another operation that, with the set, forms a monoid, and the second operation distributes over the first. As with integers, the set of real numbers under the usual addition and multiplication also has the structure of a ring. Another very important example is the set of *square matrices* all of size $N \times N$ with real elements under normal matrix addition and multiplication. Here the multiplicative identity element is the *identity matrix* of size $N \times N$, and the additive identity element is the same size square matrix with all zero elements.

Rings are powerful structures that can lead to very substantial computational savings for many statistical machine learning and signal processing problems. For example, if we remove the condition that the additive operation must have inverses, then we have a pair of monoids that are distributive. This structure is known as a *semiring* or *semifield* and it turns out that the existence of this structure in many machine learning and signal processing problems makes these otherwise computationally intractable problems feasible. For example, the classical *Viterbi algorithm* for determining the most likely sequence of hidden states in a *Hidden Markov Model* (HMM) is an application of the *max-sum semifield* on the *dynamic Bayesian network* that defines the stochastic dependencies in the model.

Both Dummit and Foote (2004) and Rotman (2000) contain detailed introductions to abstract algebra including groups and rings.

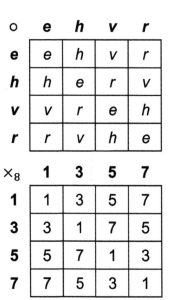

Fig. 1.3: The table for the symmetry group $V_4 = (\circ, \{e, h, v, r\})$ (top), and the group $M_8 = (\times_8, \{1, 3, 5, 7\})$ (bottom), showing the isomorphism between them obtained by mapping $e \mapsto 1$, $h \mapsto 3$, $v \mapsto 5$ and $r \mapsto 7$.

1.2   Metrics

Distance is a fundamental concept in mathematics. Distance functions play a key role in machine learning and signal processing, particularly as measures of similarity between objects, for example, digital signals encoded as items of a set. We will also see that a statistical model often implies the use of a particular measure of distance, and this measure determines the properties of statistical inferences that can be made.

A *geometry* is obtained by attaching a notion of distance to a set: it becomes a *metric space*. A *metric* takes two points in the set and returns a single (usually real) value representing the distance between them. A metric must have the following properties to satisfy intuitive notions of distance:

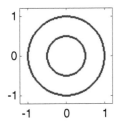

(1) Non-negativity: $d(x, y) \geq 0$,

(2) Symmetry: $d(x, y) = d(y, x)$,

(3) Coincidence: $d(x, x) = 0$, and

(4) Triangle inequality: $d(x, z) \leq d(x, y) + d(y, z)$.

Respectively, these requirements are that (1) distance cannot be negative, (2) the distance going from x to y is the same as that from y to x, (3) only points lying on top of each other have zero distance between them, and (4) the length of any one side of a triangle defined by three points cannot be greater than the sum of the length of the other two sides. For example, the *Euclidean metric* on a D-dimensional set is:

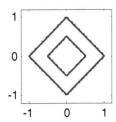

$$d(\boldsymbol{x}, \boldsymbol{y}) = \sqrt{\sum_{i=1}^{D} (x_i - y_i)^2} \tag{1.1}$$

This represents the notion of distance that we experience in everyday geometry. The defining properties of distance lead to a vast range of possible geometries, for example, the *city-block geometry* is defined by the absolute distance metric:

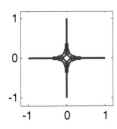

$$d(\boldsymbol{x}, \boldsymbol{y}) = \sum_{i=1}^{D} |x_i - y_i| \tag{1.2}$$

City-block distance is so named because it measures distances on a grid parallel to the co-ordinate axes. Distance need not take on real values, for example, the *discrete metric* is defined as $d(x, y) = 0$ if $x = y$ and $d(x, y) = 1$ otherwise. Another very important metric is the *Mahalanobis distance*:

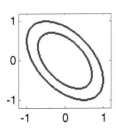

$$d(\boldsymbol{x}, \boldsymbol{y}) = \sqrt{(\boldsymbol{x} - \boldsymbol{y})^T \boldsymbol{\Sigma}^{-1} (\boldsymbol{x} - \boldsymbol{y})} \tag{1.3}$$

Fig. 1.4: Metric 2D circles $d(x, 0) = c$ for various distance metrics. From top to bottom, Euclidean distance, absolute distance, the distance $d(\boldsymbol{x}, \boldsymbol{y}) = \left(\sum_{i=1}^{D} |x_i - y_i|^{0.3} \right)^{0.3^{-1}}$, and the Mahalanobis distance for $\Sigma_{11} = \Sigma_{22} = 1.0$, $\Sigma_{12} = \Sigma_{21} = -0.5$. The contours are $c = 1$ (red lines) and $c = 0.5$ (blue lines).

This distance may not be axis-aligned: it corresponds to the finding the Euclidean distance after applying an arbitrary stretch or compression along each axis followed by an arbitrary D-dimensional rotation (indeed if $\boldsymbol{\Sigma} = \boldsymbol{I}$, the identity matrix, this is identical to the Euclidean distance).

Figure 1.4 shows plots of 2D *circles* $d(\boldsymbol{x}, \boldsymbol{0}) = c$ for various metrics, in particular $c = 1$ which is known as the *unit circle* in that metric space.

For further reading, metric spaces are introduced in Sutherland (2009) in the context of real analysis and topology.

1.3   Vector spaces

A *space* is just the name given to a set endowed with some additional mathematical structure. A (real) *vector space* is the key structure of linear algebra that is a central topic in most of classical signal processing — all digital signals are vectors, for example. The definition of a (finite) vector space begins with an ordered set (often written as a column) of N real numbers called a *vector*, and a single real number called a *scalar*. To that vector we attach the addition operation which is both associative and commutative, that simply adds every corresponding element of the numbers in each vector together, written as $\boldsymbol{v} + \boldsymbol{u}$. The identity for this operation is the vector with N zeros, $\boldsymbol{0}$. Additionally, we define a scalar multiplication operation that multiplies each element of the vector with a scalar, λ. Using the scalar multiplication by $\lambda = -1$, we can then form inverses of any vector. Scalar multiplication should not matter in which order two scalar multiplications occur, e.g. $\lambda(\mu\boldsymbol{v}) = (\lambda\mu)\boldsymbol{v} = (\mu\lambda)\boldsymbol{v} = \mu(\lambda\boldsymbol{v})$. We also require that scalar multiplication distributes over vector addition, $\lambda(\boldsymbol{v} + \boldsymbol{u}) = \lambda\boldsymbol{v} + \lambda\boldsymbol{u}$, and scalar addition distributes over scalar multiplication, $(\lambda + \mu)\boldsymbol{v} = \lambda\boldsymbol{v} + \mu\boldsymbol{v}$.

Every vector space has at least one *basis* for the space: this is a set of linearly independent vectors, such that every vector in the vector space can be written as a unique linear combination of these basis vectors (Figure 1.5). Since our vectors have N entries, there are always N vectors in the basis. Thus, N is the *dimension* of the vector space. The simplest basis is the so-called *standard basis*, consisting of the N vectors $\boldsymbol{e}_1 = (1, 0, \ldots, 0)^T$, $\boldsymbol{e}_2 = (0, 1, \ldots, 0)^T$ etc. It is easy to see that a vector $\boldsymbol{v} = (v_1, v_2, \ldots, v_N)^T$ can be expressed in terms of this basis as $\boldsymbol{v} = v_1\boldsymbol{e}_1 + v_2\boldsymbol{e}_2 + \cdots + v_N\boldsymbol{e}_N$.

By attaching a *norm* to a vector space (see below), we can measure the length of any vector, the vector space is then referred to as a *normed space*. To satisfy intuitive notions of length, a norm $V(\boldsymbol{u})$ must have the following properties:

(1) Non-negativity: $V(\boldsymbol{u}) \geq 0$,
(2) Positive scalability: $V(\alpha\boldsymbol{u}) = |\alpha| V(\boldsymbol{u})$,
(3) Separation: If $V(\boldsymbol{u}) = 0$ then $\boldsymbol{u} = \boldsymbol{0}$, and
(4) Triangle inequality: $V(\boldsymbol{u} + \boldsymbol{v}) \leq V(\boldsymbol{u}) + V(\boldsymbol{v})$.

Often, the notation $\|\boldsymbol{u}\|$ is used. Probably most familiar is the Euclidean norm $\|\boldsymbol{u}\|_2 = \sqrt{\sum_{i=1}^N u_i^2}$, but another norm that gets heavy use in statistical machine learning is the L_p-norm:

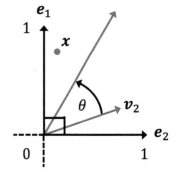

Fig. 1.5: **Important concepts in 2D vector spaces. The standard basis** (e_1, e_2) **is aligned with the axes. Two other vectors** (v_1, v_2) **can also be used as a basis for the space, all that is required is they are linearly independent of each other (in the 2D case, they are not simply scalar multiples of each other). Then** x **can be represented in either basis. The dot product between two vectors is proportional to the cosine of the angle between them:** $\cos(\theta) \propto \langle v_1, v_2 \rangle$. **Because** (e_1, e_2) **are at mutual right angles, they have zero dot product and therefore the basis is orthogonal. Additionally, it is an orthonormal basis because they are unit norm (length)** $(\|e_1\| = \|e_2\| = 1)$. **The other basis vectors are neither orthogonal nor unit norm.**

$$\|\boldsymbol{u}\|_p = \left(\sum_{i=1}^{N} |u_i|^p\right)^{\frac{1}{p}} \tag{1.4}$$

of which the Euclidean ($p = 2$) and city-block ($p = 1$) norms are special cases. Also of importance is the max-norm $\|\boldsymbol{u}\|_\infty = \max_{i=1...N} |u_i|$, which is just the length of the largest co-ordinate.

There are several ways in which a product of vectors can be formed. The most important in our context is the *inner product* between two vectors:

$$\alpha = \langle \boldsymbol{u}, \boldsymbol{v} \rangle = \sum_{i=1}^{N} u_i v_i \tag{1.5}$$

This is sometimes also described as the *dot product* $\boldsymbol{u} \cdot \boldsymbol{v}$. For complex vectors, this is defined as:

$$\langle \boldsymbol{u}, \boldsymbol{v} \rangle = \sum_{i=1}^{N} u_i \bar{v}_i \tag{1.6}$$

where $\bar{a}$ is the complex conjugate of $a \in \mathbb{C}$.

We will see later that the dot product plays a central role in the statistical notion of correlation. When two vectors have zero inner product, they are said to be *orthogonal*; geometrically they meet at a right-angle. This also has a statistical interpretation: for certain random variables, orthogonality implies statistical independence. Thus, orthogonality leads to significant simplifications in common calculations in classical DSP.

A special and very useful kind of basis is an *orthogonal basis* where the inner product between every pair of distinct basis vectors is zero:

$$\langle \boldsymbol{v}_i, \boldsymbol{v}_j \rangle = 0 \qquad \text{for all } i \neq j, \, i, j = 1, 2 \ldots N \tag{1.7}$$

In addition, the basis is *orthonormal* if every basis vector has unit norm $\|\boldsymbol{v}_i\| = 1$ – the standard basis is orthonormal, for example (Figure 1.5). Orthogonality/orthonormality dramatically simplifies many calculations over vector spaces, partly because it is straightforward to find the N scalar coefficients a_i of an arbitrary vector $\boldsymbol{u}$ in this basis using the inner product:

$$a_i = \frac{\langle \boldsymbol{u}, \boldsymbol{v}_i \rangle}{\|\boldsymbol{v}_i\|^2} \tag{1.8}$$

which simplifies to $a_i = \langle \boldsymbol{u}, \boldsymbol{v}_i \rangle$ in the orthonormal case. Orthonormal bases are the backbone of many methods in DSP and machine learning.

We can express the Euclidean norm using the inner product: $\|\boldsymbol{u}\|_2 = \sqrt{\boldsymbol{u} \cdot \boldsymbol{u}}$. An inner product satisfies the following properties:

(1) Non-negativity: $\boldsymbol{u} \cdot \boldsymbol{v} \geq 0$,

(2) Symmetry: $\boldsymbol{u} \cdot \boldsymbol{v} = \boldsymbol{v} \cdot \boldsymbol{u}$, and

(3) Linearity: $(\alpha\boldsymbol{u})\cdot\boldsymbol{v}=\alpha\left(\boldsymbol{u}\cdot\boldsymbol{v}\right)$.

There is an intuitive connection between distance and length: assuming that the metric is homogeneous $d\left(\alpha\boldsymbol{u},\alpha\boldsymbol{v}\right)=|\alpha|\,d\left(\boldsymbol{u},\boldsymbol{v}\right)$ and translation invariant $d\left(\boldsymbol{u},\boldsymbol{v}\right)=d\left(\boldsymbol{u}+\boldsymbol{a},\boldsymbol{v}+\boldsymbol{a}\right)$, a norm can be defined as the distance to the origin $\|\boldsymbol{u}\|=d\left(\boldsymbol{0},\boldsymbol{u}\right)$. A commonly occurring example of this the so-called *squared L_2 weighted norm* $\|\boldsymbol{u}\|_{\mathbf{A}}^{2}=\boldsymbol{u}^{T}\mathbf{A}\boldsymbol{u}$ which is just the squared Mahalanobis distance $d\left(\boldsymbol{0},\boldsymbol{u}\right)^{2}$ discussed earlier with $\boldsymbol{\Sigma}^{-1}=\mathbf{A}$.

On the other hand, there is one sense in which every norm *induces* an associated metric with the construction $d\left(\boldsymbol{u},\boldsymbol{v}\right)=\|\boldsymbol{u}-\boldsymbol{v}\|$. This construction enjoys extensive use in machine learning and statistical DSP to quantify the "discrepancy" or "error" between two signals. In fact, since norms are *convex* (discussed later), it follows that metrics constructed this way from norms, are also convex, a fact of crucial importance in practice.

A final product we will have need for in later chapters is the *elementwise product* $\boldsymbol{w}=\boldsymbol{u}\circ\boldsymbol{v}$ which is obtained by multiplying each element in the vector together $w_n=u_nv_n$.

Linear operators

A linear operator or *map* acts on vectors to create other vectors, and while doing so, preserve the operations of vector addition and scalar multiplication. They are homomorphisms between vector spaces. Linear operators are fundamental to classical digital signal processing and statistics, and so find heavy use in machine learning. Linear operators L have the *linear combination* property:

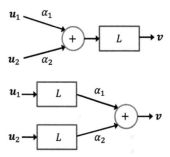

$$L\left[\alpha_1\boldsymbol{u}_1+\alpha_2\boldsymbol{u}_2+\cdots+\alpha_N\boldsymbol{u}_N\right]= \tag{1.9}$$
$$\alpha_1L\left[\boldsymbol{u}_1\right]+\alpha_2L\left[\boldsymbol{u}_2\right]+\cdots+\alpha_NL\left[\boldsymbol{u}_N\right]$$

What this says is that the operator commutes with scalar multiplication and vector addition: we get the same result if we first scale, then add the vectors, and then apply the operator to the result, or, apply the operator to each vector, then scale them, and them add up the results (Figure 1.6).

Matrices (which we discuss next), differentiation and integration, and expectation in probability are all examples of linear operators. The linearity of integration and differentiation are standard rules which can be derived from the basic definitions. Linear maps in two-dimensional space have a nice geometric interpretation: straight lines in the vector space are mapped onto other straight lines (or onto a point if they are *degenerate* maps). This idea extends to higher dimensional vector spaces in the natural way.

Matrix algebra

When vectors are 'stacked together' they form a powerful structure

Fig. 1.6: A 'flow diagram' depicting linear operators. All linear operators share the property that the operator L applied to the scaled sum of (two or more) vectors $\alpha_1u_1+\alpha_2u_2$ (top panel), is the same as the scaled sum of the same operator applied to each of these vectors first (bottom panel). In other words, it does not matter whether the operator is applied before or after the scaled sum.

which is a central topic of much of signal processing, statistics and machine learning: *matrix algebra*. A *matrix* is a 'rectangular' array of $N \times M$ elements, for example, the 3×2 matrix $\mathbf{A}$ is:

$$\mathbf{A} = \begin{pmatrix} a_{11} & a_{12} \\ a_{21} & a_{22} \\ a_{31} & a_{32} \end{pmatrix} \tag{1.10}$$

This can be seen to be two length three vectors stacked side-by-side. The elements of a matrix are often written using the subscript notation a_{ij} where $i = 1, 2, \ldots, N$ and $j = 1, 2, \ldots, M$. Matrix addition of two matrices, is commutative: $\mathbf{C} = \mathbf{A} + \mathbf{B} = \mathbf{B} + \mathbf{A}$, where the addition is element-by-element i.e. $c_{ij} = a_{ij} + b_{ij}$.

As with vectors, there are many possible ways in which matrix multiplication could be defined: the one most commonly encountered is the row-by-column inner product. For two matrices $\mathbf{A}$ of size $N \times M$ and $\mathbf{B}$ of size $M \times P$, the product $\mathbf{C} = \mathbf{A} \times \mathbf{B}$ is a new matrix of size $N \times P$ defined as:

$$c_{ij} = \sum_{k=1}^{M} a_{ik} b_{kj} \qquad i = 1, 2, \ldots, N, \, j = 1, 2, \ldots, P \tag{1.11}$$

This can be seen to be the matrix of all possible inner products of each row of $\mathbf{A}$ by each column of $\mathbf{B}$. Note that the number of columns of the left hand matrix must match the number of rows of the right hand one. Matrix multiplication is associative, it distributes over matrix addition, and it is compatible with scalar multiplication: $\alpha \mathbf{A} = \mathbf{B}$ simply gives the new matrix with entries $b_{ij} = \alpha a_{ij}$, i.e. it is just columnwise application of vector scalar multiplication. Matrix multiplication is, however, *non-commutative*: it is not true in general that $\mathbf{A} \times \mathbf{B}$ gives the same result as $\mathbf{B} \times \mathbf{A}$.

A useful matrix operator is the *transpose* that swaps rows with columns; if $\mathbf{A}$ is an $N \times M$ matrix then $\mathbf{A}^T = \mathbf{B}$ is the $M \times N$ matrix $b_{ji} = a_{ij}$. Some of the properties of the transpose are: it is self-inverse $\left(\mathbf{A}^T\right)^T = \mathbf{A}$; respects addition $(\mathbf{A} + \mathbf{B})^T = \mathbf{A}^T + \mathbf{B}^T$; and it reverses the order of factors in multiplication $(\mathbf{AB})^T = \mathbf{B}^T \mathbf{A}^T$.

Square and invertible matrices

So far, we have not discussed how to solve matrix equations. The case of addition is easy because we can use scalar multiplication to form the negative of a matrix, i.e. given $\mathbf{C} = \mathbf{A} + \mathbf{B}$, finding $\mathbf{B}$ requires us to calculate $\mathbf{B} = \mathbf{C} - \mathbf{A} = \mathbf{C} + (-1)\mathbf{A}$. In the case of multiplication, we need to find the "reciprocal" of a matrix, e.g. to solve $\mathbf{C} = \mathbf{AB}$ for $\mathbf{B}$ we would naturally calculate $\mathbf{A}^{-1}\mathbf{C} = \mathbf{A}^{-1}\mathbf{AB} = \mathbf{B}$ by the usual algebraic rules. However, things become more complicated because $\mathbf{A}^{-1}$ does not exist in general. We will discuss the conditions under which a matrix does have a multiplicative inverse next.

All *square matrices* of size $N \times N$ can be summed or multiplied to-

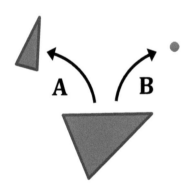

Fig. 1.7: A depiction of the geometric effect of invertible and non-invertible square matrices. The invertible square matrix A maps the triangle at the bottom to the 'thinner' triangle (for example, by transforming the vector for each vertex). It scales the area of the triangle by the determinant $|\mathbf{A}| \neq 0$. However, the non-invertible square matrix B collapses the triangle onto a single point with no area because $|\mathbf{B}| = 0$. Therefore, $\mathbf{A}^{-1}$ is well-defined, but $\mathbf{B}^{-1}$ is not.

gether in any order. A square matrix $\mathbf{A}$ with all zero elements except for the *main diagonal*, i.e. $a_{ij} = 0$ unless $i = j$ is called a *diagonal matrix*. A special diagonal matrix, $\mathbf{I}$, is the *identity matrix* where the main diagonal entries $a_{ii} = 1$. Then, if the equality $\mathbf{AB} = \mathbf{I} = \mathbf{BA}$ holds, the matrix $\mathbf{B}$ must be well-defined (and unique) and it is the inverse of $\mathbf{A}$, i.e. $\mathbf{B} = \mathbf{A}^{-1}$. We then say that $\mathbf{A}$ is *invertible;* if it is not invertible then it is *degenerate* or *singular*.

The $N \times N$ identity matrix is denoted $\mathbf{I}_N$ or simply $\mathbf{I}$ when the context is clear and the size can be omitted.

There are many equivalent conditions for matrix invertibility, for example, the only solution to the equation $\mathbf{Ax} = \mathbf{0}$ is the vector $\mathbf{x} = \mathbf{0}$ or the columns of $\mathbf{A}$ are linearly independent. But one particularly important way to test the invertibility of a matrix is to calculate the *determinant* $|\mathbf{A}|$: if the matrix is singular, the determinant is zero. It follows that all invertible matrices have $|\mathbf{A}| \neq 0$. The determinant calculation is quite elaborate for a general square matrix, formulas exist but geometric intuition helps to understand these calculations: when a linear map defined by a matrix acts on a geometric object in vector space with a certain volume, the determinant is the *scaling factor* of the mapping. Volumes under the action of the map are scaled by the magnitude of the determinant. If the determinant is negative, the *orientation* of any geometric object is reversed. Therefore, invertible transformations are those that do not collapse the volume of any object in the vector space to zero (Figure 1.7).

Another matrix operator which finds significant use is the *trace* $tr(\mathbf{A})$ of a square matrix: this is just the sum of the diagonals, e.g. $\mathrm{tr}(\mathbf{A}) = \sum_{i=1}^{N} a_{ii}$. The trace is invariant to addition $\mathrm{tr}(\mathbf{A} + \mathbf{B}) = \mathrm{tr}(\mathbf{A}) + \mathrm{tr}(\mathbf{B})$, transpose $\mathrm{tr}(\mathbf{A}^T) = \mathrm{tr}(\mathbf{A})$ and multiplication $\mathrm{tr}(\mathbf{AB}) = \mathrm{tr}(\mathbf{BA})$. With products of three or more matrices the trace is invariant to cyclic permutations, with three matrices: $\mathrm{tr}(\mathbf{ABC}) = \mathrm{tr}(\mathbf{CAB}) = \mathrm{tr}(\mathbf{BCA})$.

Eigenvalues and eigenvectors

A ubiquitous computation that arises in connection with algebraic problems in vector spaces is the *eigenvalue problem* for the given $N \times N$ square matrix $\mathbf{A}$:

$$\mathbf{Av} = \lambda \mathbf{v} \qquad (1.12)$$

Any non-zero $N \times 1$ vector $\mathbf{v}$ which solves this equation is known as an *eigenvector* of $\mathbf{A}$, and the scalar value λ is known as the associated *eigenvalue*. Eigenvectors are not unique: they can be multiplied by any non-zero scalar and still remain eigenvectors with the same eigenvalues. Thus, often the unit length eigenvectors are sought as the solutions to (1.12).

It should be noted that (1.12) arises for vector spaces in general e.g. linear operators. An important example occurs in the vector space of functions $f(x)$ with differential the operator $L = \frac{d}{dx}$. Here, the corresponding eigenvalue problem is the differential equation $L[f(x)] = \lambda f(x)$, for which the solution is $f(x) = ae^{\lambda x}$ for any (non-zero) scalar value a. This is known as an *eigenfunction* of the differential operator

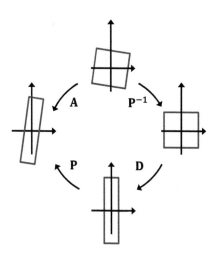

Fig. 1.8: An example of diagonalizing a matrix. The diagonalizable square matrix A has diagonal matrix D containing the eigenvalues, and transformation matrix P containing the eigenbasis, so $A = PDP^{-1}$. A maps the rotated square (top), to the rectangle in the same orientation (at left). This is equivalent to first 'unrotating' the square (the effect of P^{-1}) such that it is aligned with the co-ordinate axes, then stretching/compressing the square along each axis (the effect of D), and finally rotating back to the original orientation (the effect of P).

L.

If they exist, the eigenvectors and eigenvalues of a square matrix $\mathbf{A}$ can be found by obtaining all scalar values λ such that $|(\mathbf{A} - \lambda\mathbf{I})| = 0$. This holds because $A\boldsymbol{v} - \lambda\boldsymbol{v} = \mathbf{0}$ if and only if $|(\mathbf{A} - \lambda\mathbf{I})| = 0$. Expanding out this determinant equation leads to an N-th order polynomial equation in λ, namely $a_N\lambda^N + a_{N-1}\lambda^{N-1} + \cdots + a_0 = 0$, and the roots of this equation are the eigenvalues.

This polynomial is known as the *characteristic polynomial* for $\mathbf{A}$ and determines the existence of a set of eigenvectors that is also a basis for the space, in the following way. The fundamental theorem of algebra states that this polynomial has exactly N roots, but some may be *repeated* (i.e. occur more than once). If there are no repeated roots of the characteristic polynomial, then the eigenvalues are all distinct, and so there are N eigenvectors which are all linearly independent. This means that they form a basis for the vector space, which is the *eigenbasis* for the matrix.

Not all matrices have an eigenbasis. However matrices that do are also *diagonalizable*, that is, they have the same geometric effect as a diagonal matrix, but in a different basis other than the standard one. This basis can be found by solving the eigenvalue problem. Placing all the eigenvectors into the columns of a matrix $\mathbf{P}$ and all the corresponding eigenvalues into a diagonal matrix $\mathbf{D}$, then the matrix can be rewritten:

$$\mathbf{A} = \mathbf{PDP}^{-1} \tag{1.13}$$

See Figure 1.8. A diagonal matrix simply scales all the coordinates of the space by a different, fixed amount. They are very simple to deal with, and have important applications in signal processing and machine learning. For example, the Gaussian distribution over multiple variables, one of the most important distributions in practical applications, encodes the probabilistic relationship between each variable in the problem with the *covariance matrix*. By diagonalizing this matrix, one can find a linear mapping which makes all the variables statistically independent of each other: this dramatically simplifies many subsequent calculations.

Despite the central importance of the eigenvectors and eigenvalues a linear problem, it is generally not possible to find all the eigenvalues by analytical calculation. Therefore one generally turns to iterative numerical algorithms to obtain an answer to a certain precision.

Special matrices

Beyond what has been already discussed, there is not that much more to be said about general matrices which have $N \times M$ degrees of freedom. Special matrices with fewer degrees of freedom have very interesting properties and occur frequently in practice.

Some of the most interesting special matrices are *symmetric matrices* with real entries – *self-transpose* and so square by definition, i.e. $\mathbf{A}^T = \mathbf{A}$. These matrices are always diagonalizable, and have an orthogonal eigenbasis. The eigenvalues are always real. If the inverse exists, it is also symmetric. A symmetric matrix has $\frac{1}{2}N(N+1)$ unique entries, on the order of half the N^2 entries of an arbitrary square matrix.

Positive-definite matrices are a special kind of symmetric matrix for which $\boldsymbol{v}^T\mathbf{A}\boldsymbol{v} > 0$ for any non-zero vector $\boldsymbol{v}$. All the eigenvalues are positive. Take any real, invertible matrix $\mathbf{B}$ (so that $\mathbf{B}\boldsymbol{v} \neq \mathbf{0}$ for all such $\boldsymbol{v}$) and let $\mathbf{A} = \mathbf{B}^T\mathbf{B}$, then $\boldsymbol{v}^T\mathbf{B}^T\mathbf{B}\boldsymbol{v} = (\mathbf{B}\boldsymbol{v})^T(\mathbf{B}\boldsymbol{v}) = \|\mathbf{B}\boldsymbol{v}\|_2^2 > 0$ making $\mathbf{A}$ positive-definite. As will be described in the next section, these kinds of matrices are very important in machine learning and signal processing because the covariance matrix of a set of random variables is positive-definite for exactly this reason.

Orthonormal matrices have all columns which are vectors that form an orthonormal basis for the space. The determinant of these matrices is either $+1$ or -1. Like symmetric matrices they are always diagonalizable, although the eigenvectors are generally complex with modulus 1. An orthonormal matrix is always invertible, the inverse is also orthonormal and equal to the transpose, $\mathbf{A}^T = \mathbf{A}^{-1}$. The subset with determinant $+1$, correspond to *rotations* in the vector space.

For *upper (lower) triangular matrices*, the diagonal and the entries above (below) the diagonal are non-zero, the rest zero. These matrices often occur when solving matrix problems such as $\mathbf{A}\boldsymbol{v} = \boldsymbol{b}$, because the matrix equation $\mathbf{L}\boldsymbol{v} = \boldsymbol{b}$ is simple to solve by *forward substitution* if $\mathbf{L}$ is lower-triangular. Forward substitution is a straightforward sequential procedure which first obtains v_1 in terms of b_1 and l_{11}, then v_2 in terms of b_1, l_{21} and l_{22} etc. The same holds for upper triangular matrices and *backward substitution*. Because of the simplicity of these substitution procedures, there exist methods for decomposing a matrix into a product of upper or lower triangular matrices and a companion matrix.

Toeplitz matrices are matrices with $2N - 1$ degrees of freedom that have constant diagonals, that is, the elements of $\mathbf{A}$ have entries $a_{ij} = c_{i-j}$. All *discrete convolutions* can be represented as Toeplitz matrices, and as we will discuss later, this makes them of fundamental importance in DSP. Because of the reduced degrees of freedom and special structure of the matrix, a Toeplitz matrix problem $\mathbf{A}\boldsymbol{x} = \boldsymbol{b}$ is computationally easier to solve than a general matrix problem: a method known as the Levinson recursion dramatically reduces the number of arithmetic operations needed.

Circulant matrices are Toeplitz matrices where each row is obtained from the row above by rotating it one element to the right. With only N degrees of freedom they are highly structured and can be understood as discrete circular convolutions. The eigenbasis which diagonalizes the matrix is the *discrete Fourier basis* which is one of the cornerstones of classical DSP. It follows that any circulant matrix problem can be very efficiently solved using the *fast Fourier transform* (FFT).

Dummit and Foote (2004) contains an in-depth exposition of vector spaces from an abstract point of view, whereas Kaye and Wilson (1998) is an accessible and more concrete introduction.

1.4   Probability and stochastic processes

Probability is a formalization of the intuitive notion of uncertainty. Statistics is built on probability. Therefore, statistical DSP and machine learning has, at it's root, the quantitative manipulation of uncertainties. *Probability theory* contains the axiomatic foundation of uncertainty.

Sample spaces, events, measures and distributions

We start with a set of elements, say, Ω, which are known as *outcomes*. This is what we get as the result of a measurement or experiment. The set Ω is known as the *sample space* or *universe*. For example, the die has six possible outcomes, so the sample space is $\Omega = \{1, 2, 3, 4, 5, 6\}$. Given these outcomes, we want to quantify the probability of certain *events* occurring, for example, that we get a six or an even number in any throw. These events form an abstract $\sigma-algebra$, $\mathcal{F}$, which is, all the sets of subsets of outcomes that can be constructed by applying the elementary set operations of complement and (countable) unions to a selection of the elements in 2^{Ω} (the set of all subsets of Ω). The elements of $\mathcal{F}$ are the events. For example, in the coin toss, there are two possible outcomes, heads and tails, so $\Omega = \{H, T\}$. A set of events that are of interest make up a$\sigma-$algebra $\mathcal{F} = \{\varnothing, \{H\}, \{T\}, \Omega\}$, so that we can calculate the probability of heads or tails, none, or heads *or* tails occurring (N.B. the last two events are in some senses 'obvious' — the first is impossible and the second inevitable — so they require no calculation to evaluate, but we will see that to do probability calculus we always need the empty set and the set of all outcomes).

Given the pair $\Omega, \mathcal{F}$ we want to assign *probabilities* to events, which are real numbers lying between 0 and 1. An event with probability 0 is impossible and will never occur, whereas if the event has probability 1 then it is certain to occur. A mapping that determines the probability of any event is known as a *measure function* $\mu : \mathcal{F} \to \mathbb{R}$. An example would be the measure function for the fair coin toss which is $\mu(\{\varnothing\}) = 0$, $\mu(\{H, T\}) = 1$, $\mu(\{H\}) = \mu(\{T\}) = \frac{1}{2}$. A measure satisfies the following rules:

(1) Non-negativity: $\mu(A) \geq 0$ for all $A \in \mathcal{F}$,

(2) Unit measure: $\mu(\Omega) = 1$ and,

(3) Disjoint additivity: $\mu\left(\bigcup_{i=1}^{\infty} A_i\right) = \sum_{i=1}^{\infty} \mu(A_i)$ if the events $A_i \in \mathcal{F}$ do not overlap with each other (that is, they are mutually disjoint and so contain no elements from the sample space in common).

We mainly use the notation $P(A)$ for the probability (measure) of event A. We can derive some important consequences of these rules. For example, if one event is wholly contained inside another, it must have smaller probability: if $A \subseteq B$ then $P(A) \leq P(B)$ with equality if $A = B$. Similarly, the probability of the event not occurring is one minus the probability of that event: $P(\bar{A}) = 1 - P(A)$.

Of great importance to statistics is the sample space of real numbers. A useful σ-algebra is the *Borel algebra* formed from all possible (open) intervals of the real line. With this algebra we can assign probabilities to ranges of real numbers, e.g. $P([a,b])$ for the real numbers $a \leq b$. An important consequence of the axioms is that $P(\{a\}) = 0$, i.e. point set events have zero probability. This differs from discrete (countable) sample spaces where the probability of any single element from the sample space can be non-zero.

Given a set of all possible events it is often natural to associate numerical 'labels' to each event. This is extremely useful because then we can perform meaningful numerical computations on the events. *Random variables* are functions that map the outcomes to numerical values, for example the random variable that maps coin tosses into the set $\{0,1\}$, $X(\{T\}) = 0$ and $X(\{H\}) = 1$. *Cumulative distribution functions* (CDFs) are measures as defined above, but where the events are selected through the random variable. For example, the CDF of the (fair) coin toss as described above would be:

$$P(\{A \in \{H,T\} : X(A) \leq x\}) = \begin{cases} \frac{1}{2} & \text{for } x = 0 \\ 1 & \text{for } x = 1 \end{cases} \quad (1.14)$$

This is a special case of the *Bernoulli distribution* (see below). Two common shorthand ways of writing the CDF are $F_X(x)$ and $P(X \leq x)$.

When the sample space is the real line, the random variable is *continuous*. The associated probability of an event, in this case, a (half open) interval of the real line is:

$$P((a,b]) = P(\{A \in \mathbb{R} : a < X(A) \leq b\}) = F_X(b) - F_X(a) \quad (1.15)$$

Often we can also define a distribution through a *probability density function* (PDF) $f_X(x)$:

$$P(A) = \int_A f_X(x)\,dx \quad (1.16)$$

where $A \in \mathcal{F}$, and in practice statistical DSP and machine learning is most often (though not exclusively) concerned with $\mathcal{F}$ being the set of all open intervals of the real line (the Borel algebra), or some subset of the real line such as $[0,\infty)$. To satisfy the unit measure requirement, we must have that $\int_\mathbb{R} f_X(x)\,dx = 1$ (for the case of the whole real line). In the discrete case, the equivalent is the *probability mass function* (PMF) that assigns a probability measure to each separate outcome. To simplify the notation, we often drop the random variable subscript when the context is clear, writing e.g. $F(x)$, $f(x)$.

We can deduce certain properties of CDFs. Firstly, they must be non-decreasing, because the associated PMF/PDFs must be non-negative. Secondly, if X is defined on the range $[a,b]$, we must have that $F_X(a) = 0$ and $F_X(b) = 1$ (in the commonly occurring case where either a or b are infinite, then we would have, e.g. $\lim_{x\to-\infty} F_X(x) = 0$ and/or

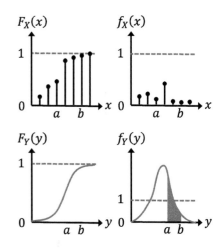

Fig. 1.9: Distribution functions and probabilities for ranges of discrete (top) and continuous (bottom) random variables X and Y respectively. Cumulative distribution functions (CDF) F_X and F_Y are shown on the left, and the associated probability mass (PMF) and probability density functions (PDF) f_X and f_Y on the right. For discrete X defined on the integers, the probability of the event A such that $a \leq X(A) \leq b$ is $\sum_{x=a}^{b} f_X(x)$ which is the same as $F_X(b) - F_X(a-1)$. For the continuous Y defined on the reals, the probability of the event $A = (a,b]$ is the given area under the curve of f_Y, i.e. $P(A) = \int_A f_Y(y)\,dy$. This is just $F_Y(b) - F_Y(a)$ by the fundamental theorem of calculus.

$\lim_{x\to\infty} F_X(x) = 1$). An important distinction to make here between discrete and continuous random variables, is that the PDF can have $f(x) > 1$ for some x in the range of the random variable, whereas PMFs must have $0 \le f(x) \le 1$ for all values of x in range. In the case of PMFs this is necessary to satisfy the unit measure property. These concepts are illustrated in Figure 1.9.

An elementary example of a PMF is the fair coin for which:

$$f(x) = \frac{1}{2} \quad \text{for } x \in \{0, 1\} \tag{1.17}$$

To satisfy unit measure, we must have $\sum_{a \in X(\Omega)} f(a) = 1$. The measure of an event is similarly:

$$P(A) = \sum_{a \in X(A)} f(a) \tag{1.18}$$

Some ubiquitous PMFs include the Bernoulli distribution which represents the binary outcome:

$$f(x) = \begin{cases} 1 - p & \text{for } x = 0 \\ p & \text{for } x = 1 \end{cases} \tag{1.19}$$

A compact representation is $f(x) = (1 - p)^{1-x} p^x$. A very important continuous distribution is the *Gaussian distribution*, whose density function is:

$$f(x; \mu, \sigma) = \frac{1}{\sqrt{2\pi\sigma^2}} \exp\left(-\frac{(x - \mu)^2}{\sigma^2}\right) \tag{1.20}$$

The semicolon is used to separate the random variable from the adjustable (non-random) parameters that determine the form of the precise distribution of X. When the parameters are considered random variables, the bar notation $f(x \mid \mu, \sigma)$ is used instead, indicating that X depends, in a consistent probabilistic sense, on the value of the parameters. This latter situation occurs in the *Bayesian framework* as we will discuss later.

Joint random variables: independence, conditionals, and marginals

Often we are interested in the probability of multiple simultaneous events occurring. A consistent way to construct an underlying sample space is to form the set of all possible combinations of events. This is known as the *product sample space*. For example, the product sample space of two coin tosses is the set $\Omega = \{(H, H), (H, T), (T, H), (T, T)\}$ with σ−algebra $\mathcal{F} = \{\varnothing, \{(H, H)\}, \{(H, T)\}, \{(T, H)\}, \{(T, T)\}, \Omega\}$. As with single outcomes, we want to define a probability measure so that we can evaluate the probability of any joint outcome. This measure is known as the *joint CDF*:

$$F_{XY}(x,y) = P(X \leq x \text{ and } Y \leq y) \tag{1.21}$$

In words, the joint CDF is the probability that the pair of random variables X, Y simultaneously take on values that are equal to x, y at the most. For the case of continuous random variables, each defined on the whole real line, this probability is a multiple integration:

$$F_{XY}(x,y) = \int_{-\infty}^{y} \int_{-\infty}^{x} f(u,v) \, du \, dv \tag{1.22}$$

where $f(u,v)$ is the *joint PDF*. The sample space is now the plane $\mathbb{R}^2$, and so in order to satisfy the unit measure axiom, it must be that $\int_{\mathbb{R}} \int_{\mathbb{R}} f(u,v) \, du \, dv = 1$. The probability of any region A of $\mathbb{R}^2$ is then the multiple integral over that region: $P(A) = \int_A f(u,v) \, du \, dv$.

The corresponding discrete case has that $P(A \times B) = \sum_{a \in X(A)} \sum_{b \in Y(B)} f(a,b)$ for any product of events where $A \in \Omega_X$, $B \in \Omega_Y$, and Ω_X, Ω_Y are the sample spaces of X and Y respectively, and $f(a,b)$ is the *joint PMF*. The joint PMF must sum to one over the whole product sample space: $\sum_{a \in X(\Omega_X)} \sum_{b \in Y(\Omega_Y)} f(a,b) = 1$.

More general joint events over N variables are defined similarly and associated with multiple CDFs, PDFs and PMFs, e.g. $f_{X_1 X_2 \dots X_N}(x_1, x_2 \dots x_N)$ and, when the context is clear from the arguments of the function, we drop the subscript in the name of the function for notational simplicity. This naturally allows us to define distribution functions over vectors of random variables, e.g. $f(\boldsymbol{x})$ for $\boldsymbol{X} = (X_1, X_2 \dots X_N)^T$ where typically, each element of the vector comes from the same sample space.

Given the joint PMF/PDF, we can always 'remove' one or more of the variables in the joint set by *integrating out* this variable, e.g.:

$$f(x_1, x_3, \dots, x_N) = \int_{\mathbb{R}} f(x_1, x_2, x_3, \dots, x_N) \, dx_2 \tag{1.23}$$

This computation is known as *marginalization*.

When considering joint events, we can perform calculations about the *conditional probability* of one event occurring, when another has already occurred (or is otherwise fixed). This conditional probability is written using the bar notation $P(X = x \mid Y = y)$: described as the 'probability that the random variable $X = x$, given that $Y = y$'. For PMFs and PDFs we will shorten this to $f(x \mid y)$. This probability can be calculated from the joint and single distributions of the conditioning variable:

$$f(x \mid y) = \frac{f(x,y)}{f(y)} \tag{1.24}$$

In effect, the conditional PMF/PDF is what we obtain from restricting the joint sample space to the set for which $Y = y$, and calculating the measure of the intersection of the joint sample space for any chosen x. The division by $f(y)$ ensures that the conditional distribution is itself a normalized measure on this restricted sample space, as we can show by marginalizing out X from the right hand side of the above equation.

If the distribution of X does not depend upon Y, we say that X is *independent of* Y. In this case $f(x|y) = f(x)$. This implies that $f(x,y) = f(x)f(y)$, i.e. the joint distribution over X,Y factorizes into a product of the marginal distributions over X,Y. Independence is a central topic in statistical DSP and machine learning because whenever two or more variables are independent, this can lead to very significant simplifications that in some cases, make the difference between whether a problem is tractable at all. In fact, it is widely recognized these days that the main goal of statistical machine learning is to find *good factorizations* of the joint distribution over all the random variables of a problem.

Bayes' rule

If we have the distribution function of a random variable conditioned on another, is it possible to swap the role of conditioned and conditioning variables? The answer is yes: provided that we have all the marginal distributions. This leads us into the territory of *Bayesian reasoning*. The calculus is straightforward, but the consequences are of profound importance to statistical DSP and machine learning. We will illustrate the concepts using continuous random variables, but the principles are general and apply to random variables over any sample space. Suppose we have two random variables X,Y and we know the conditional distribution of X given Y, then the conditional distribution of Y on X is:

$$f(y|x) = \frac{f(x|y)f(y)}{f(x)} \qquad (1.25)$$

This is known as *Bayes' rule*. In the Bayesian formalism, $f(x|y)$ is known as the *likelihood*, $f(y)$ is known as the *prior*, $f(x)$ is the *evidence* and $f(y|x)$ is the *posterior*.

Often, we do not know the distribution over X; but since the numerator in Bayes' rule is the joint probability of X and Y, this can be obtained by marginalizing out Y from the numerator:

$$f(y|x) = \frac{f(x|y)f(y)}{\int_{\mathbb{R}} f(x|y)f(y)\,dy} \qquad (1.26)$$

This form of Bayes' rule is ubiquitous because it allows calculation of the posterior knowing only the likelihood and the prior.

Unfortunately, one of the hardest and most computationally intractable problems in applying the Bayesian formalism arises when attempting to evaluate integrals over many variables to calculate the posterior in (1.26). Fortunately however, there are common situations in which it is not necessary to know the evidence probability. A third restatement of Bayes' rule makes it clear that the evidence probability can be considered a 'normalizer' for the posterior, ensuring that the posterior satisfies the unit measure property:

$$f\left(y|\,x\right) \propto f\left(x|\,y\right) f\left(y\right) \tag{1.27}$$

This form is very commonly encountered in many statistical inference problems in machine learning. For example, when we wish to know the value of a parameter or random variable given some data which maximizes the posterior, and the evidence probability is independent of this variable or parameter, then we can exclude the evidence probability from the calculations.

Expectation, generating functions and characteristic functions

There are many ways of *summarizing* the distribution of a random variable. Of particular importance are *measures of central tendency* such as the *mean* and *median*. The mean of a (continuous) random variable X is the sum over all possible outcomes weighted by the probability of that outcome:

$$E\left[X\right] = \int_{\Omega} x f\left(x\right) dx \tag{1.28}$$

Where not obvious from the context, we write $E_X\left[X\right]$ to indicate that this integral is with respect to the random variable X. In the case of discrete variables this is $E\left[X\right] = \sum_{a \in X(\Omega)} x f\left(x\right)$. As discussed earlier, expectation is a linear operator, i.e. $E\left[\sum_i a_i X_i\right] = \sum_i a_i E\left[X_i\right]$ for arbitrary constants a_i. A constant is invariant under expectation: $E\left[a\right] = a$. The mean is also known as the *expected value*, and the integral is called the *expectation*. The expectation plays a central role in probability and statistics, and can in fact be used to construct an entirely different axiomatic view on probability. The expectation with respect to an arbitrary transformation of a random variable, $g\left(X\right)$ is:

$$E\left[g\left(X\right)\right] = \int_{\Omega} g\left(x\right) f\left(x\right) dx \tag{1.29}$$

Using this we can define a hierarchy of summaries of the distribution of a random variable, known as the *k-th moments*:

$$E\left[X^k\right] = \int_{\Omega} x^k f\left(x\right) dx \tag{1.30}$$

From the unit measure property of probability it can be seen that the zeroth moment $E\left[X^0\right] = 1$. The first moment coincides with the mean. *Central moments* are those defined "around" the mean:

$$\mu_k = E\left[\left(X - E\left[X\right]\right)^k\right] = \int_{\Omega} \left(x - \mu\right)^k f\left(x\right) dx \tag{1.31}$$

where μ is the mean. A very import central moment is the variance, $\text{var}\left[X\right] = \mu_2$, which is a *measure of spread* (about the mean) of the distribution. The *standard deviation* is the square root of this $\text{std}\left[X\right] = \sqrt{\mu_2}$. Higher order central moments such as skewness (μ_3) and kurtosis (μ_4) measure aspects such as the asymmetry and sharpness of distribution, respectively.

For joint distributions with joint density function $f(x,y)$, the expectation is:

$$E\left[g\left(X,Y\right)\right] = \int_{\Omega_Y}\int_{\Omega_X} g\left(x,y\right)f\left(x,y\right)dx\,dy \qquad (1.32)$$

From this, we can derive the *joint moments*:

$$E\left[X^j Y^k\right] = \int_{\Omega_Y}\int_{\Omega_X} x^j y^k f\left(x,y\right)dx\,dy \qquad (1.33)$$

An important special case is the *joint second central moment*, known as the *covariance*:

$$
\begin{aligned}
\mathrm{cov}\left[X,Y\right] &= E\left[\left(X - E\left[X\right]\right)\left(Y - E\left[Y\right]\right)\right] \qquad (1.34)\\
&= \int_{\Omega_Y}\int_{\Omega_X}\left(x - \mu_X\right)\left(y - \mu_Y\right)f\left(x,y\right)dx\,dy
\end{aligned}
$$

where μ_X, μ_Y are the means of X, Y respectively.

Sometimes, the hierarchy of moments of a distribution serve to define the distribution uniquely. A very important kind of expectation is the *moment generating function* (MGF), for discrete variables:

$$M\left(s\right) = E\left[\exp\left(sX\right)\right] = \sum_{x\in X(\Omega)}\exp\left(sx\right)f\left(x\right)dx \qquad (1.35)$$

The real variable s becomes the new independent variable replacing the discrete variable x. When the sum (1.35) converges absolutely, then the MGF exists and can be used to find all the moments for the distribution of X:

$$E\left[X^k\right] = \frac{d^k M}{dt^k}\left(0\right) \qquad (1.36)$$

This can be shown to follow from the series expansion of the exponential function. Using the Bernoulli example above, the MGF is $M\left(s\right) = 1 - p + p\exp\left(s\right)$. Often, the distribution of a random variable has a simple form under the MGF that makes the task of manipulating random variables relatively easy. For example, given a linear combination of independent random variables:

$$X_N = \sum_{n=1}^{N} a_n X_n \qquad (1.37)$$

it is not a trivial matter to calculate the distribution of X_N. However, the MGF of the sum is just:

$$M_{X_N}\left(s\right) = \prod_{n=1}^{N} M_{X_n}\left(a_n s\right) \qquad (1.38)$$

from which the distribution of the sum can sometimes be recognized immediately. As an example, the MGF for an (unweighted) sum of N i.i.d. Bernoulli random variables with parameter p, is:

$$M_{X_N}(s) = (1 - p + p \exp(s))^N \tag{1.39}$$

which is just the MGF of the *binomial distribution*.

A similar expectation is the *characteristic function* (CF), for continuous variables:

$$\psi(s) = E\left[\exp(isX)\right] = \int_\Omega \exp(isx) f(x) \, dx \tag{1.40}$$

where $i = \sqrt{-1}$. This can be understood as the Fourier transform of the density function. An advantage over the MGF is that the CF always exists. It can therefore be used as an alternative way to define a distribution, a fact which is necessary for some well-known distributions such as the *Levy* or *alpha-stable* distributions. Well-known properties of Fourier transforms make it easy to use the CF to manipulate random variables. For example, given a random variable X with CF $\psi_X(s)$, the random variable $Y = X + m$ where m is a constant is:

$$\psi_Y(s) = \psi_X(s) \exp(ism) \tag{1.41}$$

From this, given that the CF of the standard normal Gaussian with mean zero and unit variance, is $\exp\left(-\frac{1}{2}s^2\right)$, the shifted random variable Y has CF $\psi_Y(s) = \exp\left(ism - \frac{1}{2}s^2\right)$. Another property, similar to the MGF, is the linear combination property:

$$\psi_{X_N}(s) = \prod_{i=1}^N \psi_{X_i}(a_i s) \tag{1.42}$$

We can use this to show that for a linear combination (1.37) of independent Gaussian random variables with mean μ_n and variance σ_n^2, the CF of the sum is:

$$\psi_{X_N}(s) = \exp\left(is \sum_{n=1}^N a_n \mu_n - \frac{1}{2}s^2 \sum_{n=1}^N a_n^2 \sigma_n^2\right) \tag{1.43}$$

which can be recognized as another Gaussian with mean $\sum_{n=1}^N a_n \mu_n$ and variance $\sum_{n=1}^N a_n^2 \sigma_n^2$. This shows that the Gaussian is *invariant* to linear transformations, a property known as *(statistical) stability*, which is of fundamental importance in classical statistical DSP.

Empirical distribution function and sample expectations

If we start with a PDF or PMF, then the specific values of the parameters of these functions determine the mathematical form of the distribution. However, often we want some given data to "speak for itself" and determine a distribution function directly. An important, and simple way to do this, is using the *empirical cumulative distribution function*, or ECDF:

$$F_N(x) = \frac{1}{N} \sum_{n=1}^{N} \mathbf{1}\,[x_n \leq x] \tag{1.44}$$

where $\mathbf{1}\,[.]$ takes a logical condition as argument, and is 1 if the condition is true, 0 if false. Thus, the ECDF counts the number of data points that are equal to or below the value of the variable x. It looks like a staircase, jumping up one count at the value of each data point. The ECDF estimates the CDF of the distribution of the data, and it can be shown that, in a specific, probabilistic sense, this estimator converges on the true CDF given an infinite amount of data. By differentiating (1.44), the associated PMF (PDF) is a sum of Kronecker (Dirac) delta functions:

$$f_N(x) = \frac{1}{N} \sum_{n=1}^{N} \delta\,[x_n - x] \tag{1.45}$$

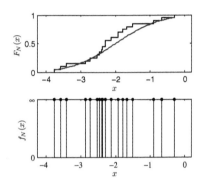

See Figure 1.10. The simplicity of this estimator makes it very useful in practice. For example, the expectation with respect to the function $g(X)$ of the ECDF for a continuous random variable is:

$$
\begin{aligned}
E\,[g(X)] &= \int_{\Omega} g(x)\, f_N(x)\, dx \tag{1.46} \\
&= \int_{\Omega} g(x)\, \frac{1}{N} \sum_{n=1}^{N} \delta\,[x_n - x]\, dx \\
&= \frac{1}{N} \sum_{n=1}^{N} \int_{\Omega} g(x)\, \delta\,[x_n - x]\, dx = \frac{1}{N} \sum_{n=1}^{N} g(x_n)
\end{aligned}
$$

Fig. 1.10: Empirical cumulative distribution function (ECDF) and density functions (EPDF) based on a sample of size $N = 20$ from a Gaussian random variable with $\mu = -2$ and $\sigma = 1.0$. For the estimated ECDF (top), the black 'steps' occur at the value of each sample, whereas the blue curve is the theoretical CDF for the Gaussian. The EPDF (bottom) consists of an infinitely tall Dirac 'spike' occurring at each sample.

using the *sift* property of the delta function, $\int f(x)\, \delta\,[x - a]\, dx = f(a)$. Therefore, the expectation of a random variable can be estimated from the average of the expectation function applied to the data. These estimates are known as *sample expectations*, the most well-known of which is the *sample mean* $\mu = E\,[X] = \frac{1}{N} \sum_{n=1}^{N} x_n$.

Transforming random variables

We often need to apply some kind of transformation to a random variable. What happens to the distribution over this random variable after transformation? Often, it is straightforward to compute the CDF, PMF or PDF of the transformed variable. In general for a random variable X with CDF $F_X(x)$, if the transformation $Y = g(X)$ is invertible, then $F_Y(y) = F_X(g^{-1}(y))$ if g^{-1} is increasing, or $F_Y(y) = 1 - F_X(g^{-1}(y))$ if g^{-1} is decreasing. For the corresponding PDFs, $f_Y(y) = f_X(g^{-1}(y)) \left| \frac{dg^{-1}}{dy}(y) \right|$. This calculus can be extended to the case where g is generally non-invertible except on any number of isolated points. An example of this process is the transformation $g(X) = X^2$, which converts the Gaussian into the *chi-squared distribution*.

Another ubiquitous example is the effect of a linear transformation, i.e. $Y = \sigma X + \mu$, for arbitrary scalar values μ, σ, for which $f_Y(y) = \frac{1}{|\sigma|} f_X\left(\frac{y-\mu}{\sigma}\right)$. This can be used to prove that, for example, a great many distributions are invariant under scaling by a positive real value (*exponential* and *gamma* distributions are important examples), or are contained in a *location-scale family,* that is, each member of the family of distributions can be obtained from any other in the same family by an appropriate translation and scaling (this holds for the Gaussian and *logistic* distributions).

Multivariate Gaussian and other limiting distributions

As discussed earlier in this section, invariance to linear transformations (stability) is one of the hallmarks of the Gaussian distribution. Another is the *central limit theorem*: for an infinite sequence of random variables with finite mean and variance which are all independent of each other, the distribution of the sum of this sequence of random variables will tend to the Gaussian. A simple proof using CFs exists. In many contexts this theorem is given as a justification for choosing the Gaussian distribution as a model for some given data.

These desirable properties of a single Gaussian random variable carry over to vectors with D elements; over which the multivariate Gaussian distribution is:

$$f(\boldsymbol{x}; \boldsymbol{\mu}, \boldsymbol{\Sigma}) = \frac{1}{\sqrt{(2\pi)^D |\boldsymbol{\Sigma}|}} \exp\left(-\frac{1}{2}(\boldsymbol{x} - \boldsymbol{\mu})^T \boldsymbol{\Sigma}^{-1} (\boldsymbol{x} - \boldsymbol{\mu})\right) \quad (1.47)$$

where $\boldsymbol{x} = (x_1, x_2, \ldots, x_D)^T$, the mean vector $\boldsymbol{\mu} = (\mu_1, \mu_2, \ldots, \mu_D)^T$ and $\boldsymbol{\Sigma}$ is the *covariance matrix*. Equal probability density contours of the multivariate Gaussian are, in general, (hyper)-ellipses in D dimensions. The maximum probability density occurs at $\boldsymbol{x} = \boldsymbol{\mu}$. The positive-definite covariance matrix can be decomposed into a rotation and expansion or contraction in each axis, around the point $\boldsymbol{\mu}$. Another very special and important property of the multivariate Gaussian is that all marginals are also (multivariate) Gaussians. Similarly, this means that conditioning one multivariate Gaussians on another, gives another multivariate Gaussian. All these properties are depicted in Figure 1.11 and described algebraically below.

It is simple to extend the statistical stability property of the univariate Gaussian to multiple dimensions to show that this property also applies to the multivariate normal. Firstly, we need the notion of the CF for joint random variables $\boldsymbol{X} \in \mathbb{R}^D$, which is $\psi_{\boldsymbol{X}}(\boldsymbol{s}) = E\left[\exp\left(i\boldsymbol{s}^T\boldsymbol{X}\right)\right]$ for the variable $\boldsymbol{s} \in \mathbb{C}^D$. Now, consider an i.i.d. vector of standard normal univariate RVs, $\boldsymbol{X} \in \mathbb{R}^D$. The joint CF will be $\psi_{\boldsymbol{X}}(\boldsymbol{s}) =$

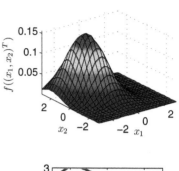

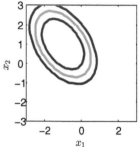

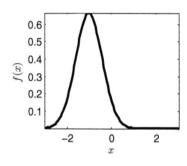

Fig. 1.11: An example of the multivariate $(D = 2)$ Gaussian with PDF $f\left((x_1, x_2)^T; \boldsymbol{\mu}, \boldsymbol{\Sigma}\right)$ (top). Height is probability density value, and the other two axes are x_1, x_2. The contours of constant probability are ellipses (middle), here shown for probability density values of 0.03 (blue), 0.05 (cyan) and 0.08 (black). The maximum probability density coincides with the mean (here $\boldsymbol{\mu} = (-1, 1)^T$). The marginals are all single-variable Gaussians, an example PDF of one of the marginals with mean $\mu = -1$ is shown at bottom.

$\prod_{i=1}^{D} \exp\left(-\frac{1}{2}s_i^2\right) = \exp\left(-\frac{1}{2}s^T s\right)$. This is the CF of the *standard multivariate Gaussian* which has zero mean vector and identity covariance, $\boldsymbol{X} \sim \mathcal{N}(\mathbf{0}, \mathbf{I})$. What happens if we apply the (affine) transformation $\boldsymbol{Y} = \mathbf{A}\boldsymbol{X} + \boldsymbol{b}$ with the (full-rank) matrix $\mathbf{A} \in \mathbb{R}^{D \times D}$ and $\boldsymbol{b} \in \mathbb{R}^D$? The effect of this transformation on an arbitrary CF is:

$$
\begin{aligned}
\psi_{\boldsymbol{Y}}(\boldsymbol{s}) &= E_{\boldsymbol{X}}\left[\exp\left(i\boldsymbol{s}^T \mathbf{A}\boldsymbol{X}\right)\right] \\
&= \exp\left(i\boldsymbol{s}^T \boldsymbol{b}\right) E_{\boldsymbol{X}}\left[\exp\left(i\left(\mathbf{A}^T \boldsymbol{s}\right)^T \boldsymbol{X}\right)\right] \qquad (1.48) \\
&= \exp\left(i\boldsymbol{s}^T \boldsymbol{b}\right) \psi_{\boldsymbol{X}}\left(\mathbf{A}^T \boldsymbol{s}\right)
\end{aligned}
$$

Inserting the CF of the standard multivariate normal, we get:

$$
\begin{aligned}
\psi_{\boldsymbol{Y}}(\boldsymbol{s}) &= \exp\left(i\boldsymbol{s}^T \boldsymbol{b}\right) \exp\left(-\frac{1}{2}\left(\mathbf{A}^T \boldsymbol{s}\right)^T \left(\mathbf{A}^T \boldsymbol{s}\right)\right) \\
&= \exp\left(i\boldsymbol{s}^T \boldsymbol{b} - \frac{1}{2}\boldsymbol{s}^T \left(\mathbf{A}\mathbf{A}^T\right)\boldsymbol{s}\right) \qquad (1.49)
\end{aligned}
$$

which is the CF of another multivariate normal, from which we can say that $\boldsymbol{Y}$ is multivariate normal with mean $\boldsymbol{\mu} = \boldsymbol{b}$ and covariance $\boldsymbol{\Sigma} = \mathbf{A}\mathbf{A}^T$, or $\boldsymbol{Y} \sim \mathcal{N}(\boldsymbol{\mu}, \boldsymbol{\Sigma})$. It is now straightforward to predict that application of another affine transformation $\boldsymbol{Z} = \mathbf{B}\boldsymbol{Y} + \boldsymbol{c}$ leads to another multivariate Gaussian:

$$
\begin{aligned}
\psi_{\boldsymbol{Z}}(\boldsymbol{s}) &= \exp\left(i\boldsymbol{s}^T \boldsymbol{c}\right) \exp\left(i\left(\mathbf{B}^T \boldsymbol{s}\right)^T \boldsymbol{\mu} - \frac{1}{2}\left(\mathbf{B}^T \boldsymbol{s}\right)^T \boldsymbol{\Sigma}\left(\mathbf{B}^T \boldsymbol{s}\right)\right) \\
&= \exp\left(i\boldsymbol{s}^T \left(\mathbf{B}\boldsymbol{\mu} + \boldsymbol{c}\right) - \frac{1}{2}\boldsymbol{s}^T \left(\mathbf{B}\boldsymbol{\Sigma}\mathbf{B}^T\right)\boldsymbol{s}\right) \qquad (1.50)
\end{aligned}
$$

which has mean $\mathbf{B}\boldsymbol{\mu} + \boldsymbol{c}$ and covariance $\mathbf{B}\boldsymbol{\Sigma}\mathbf{B}^T$. This construction can also be used to show that all marginals are also multivariate normal, if we consider the orthogonal projection, with a set of dimension indices P, e.g. projection down onto $P = \{2, 6, 7\}$. Then the projection matrix $\mathbf{P}$ has entries $b_{P,P} = 1$ and the rest of the entries are zero. Then we have the following CF:

$$
\begin{aligned}
\psi_{\boldsymbol{Z}}(\boldsymbol{s}) &= \exp\left(i\boldsymbol{s}^T (\mathbf{P}\boldsymbol{\mu}) - \frac{1}{2}\boldsymbol{s}^T \left(\mathbf{P}\boldsymbol{\Sigma}\mathbf{P}^T\right)\boldsymbol{s}\right) \\
&= \exp\left(i\boldsymbol{s}_P^T \boldsymbol{\mu}_P - \frac{1}{2}\boldsymbol{s}_P^T \boldsymbol{\Sigma}_P \boldsymbol{s}_P\right) \qquad (1.51)
\end{aligned}
$$

where $\boldsymbol{a}_P$ indicates a vector with the elements in P deleted, and $\mathbf{A}_p$ indicates a matrix with the rows and columns in P deleted. This is yet another multivariate Gaussian. Similar arguments, can be used to show that the conditional distributions are also multivariate normal. Partition $\boldsymbol{X}$ into two sub-vectors $\boldsymbol{X}_P$ and $\boldsymbol{X}_{\bar{P}}$ where $\bar{P} = \{1, 2, \ldots D\} \backslash P$ such that $P \cup \bar{P} = \{1, 2, \ldots, D\}$. Then $f(\boldsymbol{x}_P | \boldsymbol{x}_{\bar{P}}) = \mathcal{N}(\boldsymbol{x}_P; \boldsymbol{m}, \mathbf{S})$ where

$m = \mu_P - \Sigma_{P\bar{P}}\Sigma_{\bar{P}\bar{P}}^{-1}(x_{\bar{P}} - \mu_{\bar{P}})$ and $\mathbf{S} = \Sigma_{PP} - \Sigma_{P\bar{P}}\Sigma_{\bar{P}\bar{P}}^{-1}\Sigma_{\bar{P}P}$. For detailed proofs, see Murphy (2012, section 4.3.4).

Although the Gaussian is special, it is not the only distribution with the statistical stability property: another important example is the α-*stable* distribution (which includes the Gaussian as a special case); in fact, a generalization of the central limit theorem states that the distribution of the sum of independent distributions with infinite variance tends towards the α-*stable* distribution. Yet another broad class of distributions that are also invariant to linear transformations are the *elliptical distributions* whose densities, when they exist are defined in terms of a function of the Mahalanobis distance from the mean, $d(x,\mu) = \sqrt{(x-\mu)^T \Sigma^{-1}(x-\mu)}$. These useful distributions also have elliptically-distributed marginals, and the multivariate Gaussian is a special case.

Just as the central limit theorem is often given as a justification for choosing the Gaussian, the *extreme value theorem* is often a justification for choosing one of the *extreme value distributions*. Consider an infinite sequence of identically distributed but independent random variables, the maximum of this sequence is either *Frechet, Weibull* or *Gumbel* distributed, regardless of the distribution of the random variables in the sequence.

Stochastic processes

Stochastic processes are key objects in statistical signal processing and machine learning. In essence, a stochastic process is just a collection of random variables X_t on the same sample space Ω (also known as the *state space*), where the index t comes from an arbitrary set T which may be finite or infinite in size, uncountable or countable. When the index set is finite such as $T = \{1, 2 \ldots N\}$, then we may consider the collection to coincide with a vector of random variables.

In this book the index is nearly always synonymous with *(relative) time*, and therefore plays a crucial role since nearly all signals are time-based. When the index is a real number i.e $T = \mathbb{R}$, the collection is known as a *continuous-time stochastic process*. Countable collections are known as *discrete-time stochastic processes*. Although continuous-time processes are important for theoretical purposes, all recorded signals we capture from the real world are discrete-time and finite. These signals must be stored digitally so that signal processing computations may be performed on them. A signal is typically *sampled* from the real world at uniform intervals in time, and each sample takes up a finite number of digital bits.

This latter constraint influences the choice of sample space for each of the random variables in the process. A finite number of bits can only encode a finite range of discrete values, so, being faithful to the digital representation might suggest $\Omega = \{0, 1, 2 \ldots K - 1\}$ where $K = 2^B$ for B bits, but this digital representation can also be arranged to encode real numbers with a finite precision, a common example being the *32 bit floating point* representation, and this might be more faithful to how we model the real-world process we are sampling (see Chapter 8 for details). Therefore, it is often mathematically realistic and convenient to work with stochastic processes on the sample space of real numbers $\Omega = \mathbb{R}$. Nevertheless, the choice of sample space depends crucially upon the real-world interpretation of the recorded signal, and we should not forget that the actual computations may only ever amount to an approximation of the mathematical model upon which they are based.

If each member of the collection of variables is independent of every other and each one has the same distribution, then to characterize the distributional properties of the process, all that matters is the distribution of every X_t, say $f(x)$ for the PDF over the real state space. Simple processes with this property are said to be *independent and identically distributed* (i.i.d.) and they are of crucial importance in many applications in this book, because the joint distribution over the entire process factorizes into a product of the individual distributions which leads to considerable computational simplifications in statistical inference problems.

However, much more interesting signals are those for which the stochastic process is time-dependent, where each of the X_t is not, in general, independent. The distributional properties of any process can be analyzed by consideration of the joint distributions of all finite-length collections of the constituent random variables, known as the *finite-dimensional distributions* (f.d.d.s) of the process. For the real state space the f.d.d.s are defined by a vector of N time indices $\boldsymbol{t} = (t_1, t_2 \ldots t_N)^T$, where the vector $(X_{t_1}, X_{t_2} \ldots X_{t_N})^T$ has the PDF $f_{\boldsymbol{t}}(\boldsymbol{x})$ for $\boldsymbol{x} = (x_1, x_2 \ldots x_N)^T$. It is these f.d.d.s which encode the dependence structure between the random variables in the process.

The f.d.d.s encode much more besides dependence structure. Statistical DSP and machine learning make heavy use of this construct. For example, *Gaussian processes* which are fundamental to ubiquitous topics such as *Kalman filtering* in signal processing and *nonparametric Bayes' inference* in machine learning are processes for which the f.d.d.s are all multivariate Gaussians. Another example is the *Dirichlet process* used in nonparametric Bayes' which has *Dirichlet distributions* as f.d.d.s. (see Chapter 10).

Strongly stationary processes are special processes in which the f.d.d.s are invariant to time translation, i.e. $(X_{t_1}, X_{t_2} \ldots X_{t_N})^T$ has the same f.d.d. as $(X_{t_1+\tau}, X_{t_2+\tau} \ldots X_{t_N+\tau})^T$, for $\tau > 0$. This says that the local distributional properties will be the same at all times. This is yet another mathematical simplification and it is used extensively. A less restrictive notion of stationarity occurs when the first and second joint moments are invariant to time delays: $\operatorname{cov}[X_t, X_s] = \operatorname{cov}[X_{t+\tau}, X_{s+\tau}]$ and $E[X_t] = E[X_s]$ for all t, s and $\tau > 0$. This is known as *weak stationarity*. Strong stationarity implies weak stationarity, but the implication does not necessarily work the other way except in special cases (stationary Gaussian processes are one example for which this is true). The temporal covariance (*autocovariance*) of a weakly stationary process depends only on time translation τ: $\mathrm{cov}\,[X_t, X_{t+\tau}] = \mathrm{cov}\,[X_0, X_\tau]$ (see Chapter 7).

Markov chains

Another broad class of simple non-i.i.d. stochastic processes are those for which the dependence in time has a finite effect, known as *Markov chains*. Given a discrete-time process with index $t = \mathbb{Z}$ over a discrete state space, such a process satisfies the *Markov property*:

$$
\begin{aligned}
f\left(X_{t+1} = x_{t+1} \,|\, X_t = x_t, X_{t-1} = x_{t-1} \ldots X_1 = x_1\right) \\
= f\left(X_{t+1} = x_{t+1} \,|\, X_t = x_t\right) \qquad (1.52)
\end{aligned}
$$

for all $t \geq 1$ and all x_t. This conditional probability is called the *transition distribution*.

What this says is that the probability of the random variable at any time t depends only upon the previous time index, or stated another way, the process is independent given knowledge of the random variable at the previous time index. The Markov property leads to considerable simplifications which allow broad computational savings for inference in hidden Markov models, for example. (Note that a chain which depends on $M \geq 1$ previous time indices is known as a *M-th order Markov chain*, but in this book we will generally only consider the case where $M = 1$, because any higher-order chain can always be rewritten as a 1st-order chain by the choice of an appropriate state space). Markov chains are illustrated in Figure 1.12.

A further simplification occurs if the temporal dependency is time translation invariant:

$$
f\left(X_{t+1} = x_{t+1} \,|\, X_t = x_t\right) = f\left(X_2 = x_2 \,|\, X_1 = x_1\right) \qquad (1.53)
$$

for all t. Such Markov processes are strongly stationary. Now, using the basic properties of probability we can find the probability distribution over the state at time $t + 1$:

$$
\begin{aligned}
\sum_{x'} f\left(X_{t+1} = x_t \,|\, X_t = x'\right) f\left(X_t = x'\right) &= \sum_{x'} f\left(X_{t+1} = x_{t+1}, X_t = x'\right) \\
&= f\left(X_{t+1} = x_{t+1}\right) \qquad (1.54)
\end{aligned}
$$

In words: the unconditional probability over the state at time $t + 1$ is obtained by marginalizing out the probability over time t from the joint distribution of the state at time $t + 1$ and t. But, by definition of the transition distribution, the joint distribution is just the product of the transition distribution times the unconditional distribution at time t.

For a chain over a finite state space of size K, we can always re-label each state with a natural number $\Omega = \{1, 2 \ldots K\}$. This allows us to construct a very computationally convenient notation for any finite

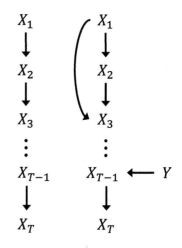

Fig. 1.12: **Markov (left) and non-Markov (right) chains. For a Markov chain, the distribution of the random variable at time t, X_t, depends only upon the value of the previous random variable X_{t-1}; this is true for the chain on the left. The chain on the right is non-Markov because X_3 depends upon X_1, and X_{T-1} depends upon an 'external' random variable Y.**

discrete-state (strongly stationary) chain:

$$p_{ij} = f(X_2 = i \mid X_1 = j) \tag{1.55}$$

The $K \times K$ matrix $\mathbf{P}$ is called the *transition matrix* of the chain. In order to satisfy the axioms of probability, this matrix must have non-negative entries, and it must be the case that $\sum_i p_{ij} = 1$ for all $j = 1, 2 \ldots K$. These properties make it a (*left*) *stochastic matrix*. This matrix represents the conditional probabilities of the chain taking one 'step' from t to $t+1$ and furthermore, $\mathbf{P}^s$ is the transition matrix taking the state of the chain multiple steps from t to index $t + s$, as we now show. Letting the K-element vector $\boldsymbol{p}^t$ contain the PMF for X_t, i.e. $p_i^t = f(X_t = i)$, then equation (1.54) can be written in component form:

$$p_i^{t+1} = \sum_j p_{ij} p_j^t = (\mathbf{P}\boldsymbol{p}^t)_i \tag{1.56}$$

or in matrix-vector form:

$$\boldsymbol{p}^{t+1} = \mathbf{P}\boldsymbol{p}^t \tag{1.57}$$

In other words, the PMF at time index $t + 1$ can be found by multiplying the PMF at time index t by the transition matrix. Now, equation (1.57) is just a *linear recurrence system* whose solution is simply $\boldsymbol{p}^{t+1} = \mathbf{P}^t \boldsymbol{p}^1$, or more generally, $\boldsymbol{p}^{t+s} = \mathbf{P}^s \boldsymbol{p}^t$. This illustrates how the distributional properties of stationary Markov chains over discrete state spaces can be found using tools from matrix algebra.

We will generally be very interested in *stationary distributions* $\boldsymbol{q}$ of strongly stationary chains, which are invariant under application of the transition matrix, so that $\mathbf{P}\boldsymbol{q} = \boldsymbol{q}$. These occur as limiting PMFs, i.e. $\boldsymbol{p}^t \to \boldsymbol{q}$ as $t \to \infty$ from some starting PMF $\boldsymbol{p}^1$. Finding a stationary distribution amounts to solving an eigenvector problem with eigenvalue 1: the corresponding eigenvector will have strictly positive entries and sum to 1.

Whether there are multiple stationary distributions, just one, or none, is a key question to answer of any chain. If there is one, and only one, limiting distribution, it is known as the *equilibrium distribution* $\boldsymbol{\mu}$, and $\boldsymbol{p}^t \to \boldsymbol{\mu}$ from *all possible* starting PMFs, effectively "forgetting" the initial state $\boldsymbol{p}^1$. This property is crucial to many applications in machine learning. Two important concepts which help to establish this property for a chain are *irreducibility*: that is, whether it is possible to get from any state to any other state, taking a finite number of transitions, and *aperiodicity*: the chain does not get stuck in a repeating "cycle" of states.

Formally, for an irreducible chain, the transition matrix for $s > 0$ steps must be non-zero for all $i, j \in K$:

$$(\mathbf{P}^s)_{ij} > 0 \tag{1.58}$$

There are several equivalent ways to define aperiodicity, here is one. For all $i \in K$:

$$\gcd\{s : (\mathbf{P}^s)_{ii} > 0\} = 1 \tag{1.59}$$

where gcd is the greatest common divisor. This definition implies that the chain returns to the same state i at irregular times.

Another very important class of chains are those which satisfy the *detailed balance* condition, in component form:

$$p_{ij}q_j = p_{ji}q_i \tag{1.60}$$

for some for some distribution over states $\mathbf{q}$ with $q_i = f(X = i)$. Expanding this out, we get:

$$
\begin{aligned}
f(X = i \,|\, X' = j) \, f(X' = j) &= f(X = j \,|\, X' = i) \, f(X' = i) \\
f(X = i, X' = j) &= f(X = j, X' = i)
\end{aligned} \tag{1.61}
$$

which implies that $p_{ij} = p_{ji}$, in other words, these chains have symmetric transition matrices, $\mathbf{P} = \mathbf{P}^T$, which are *doubly stochastic* in that both the columns and rows are normalized. Interestingly, the distribution $\mathbf{q}$ is also a stationary distribution of the chain because:

$$\sum_j p_{ij} q_j = \sum_j p_{ji} q_i = q_i \sum_j p_{ji} = q_i \tag{1.62}$$

This symmetry also implies that such chains are *reversible* if in the stationary state, i.e. $\mathbf{p}^t = \mathbf{q}$:

$$
\begin{aligned}
f(X_{t+1} = i \,|\, X_t = j) \, f(X_t = j) &= f(X_{t+1} = j \,|\, X_t = i) \, f(X_t = i) \\
f(X_{t+1} = i, X_t = j) &= f(X_t = i, X_{t+1} = j)
\end{aligned} \tag{1.63}
$$

So, the roles of X_{t+1} and X_t in the joint distribution of states at adjacent indices can be swapped, and this means that the f.d.d.s, and hence all the distributional properties of the chain, are invariant to time reversal.

In the case of continuous state spaces where $\Omega = \mathbb{R}$, the probability of each time index is a PDF, $f(X_t = x_t)$, and the transition probabilities are conditional PDFs, $f(X_{t+1} = x_{t+1} \,|\, X_t = x_t)$. For example, consider the following linear stochastic recurrence relation:

$$X_{t+1} = aX_t + \epsilon_t \tag{1.64}$$

with $X_1 = x_1, a$ is an arbitrary real scalar, and ϵ_t are zero-mean, i.i.d. Gaussian random variables with standard deviation σ. This is a discrete-time Markov process with the following transition PDF:

$$f(X_{t+1} = x_t \,|\, X_t = x_t) = \mathcal{N}\left(x_{t+1}; ax_t, \sigma^2\right) \tag{1.65}$$

i.e. a Gaussian with mean aX_{t-1} and standard deviation σ. This important example of a Gaussian Markov chain is known as an *AR(1)* or *first-order autoregressive* process; more general $AR(q)$ *processes* defined by chains of higher order, are ubiquitous in signal processing and time series analysis and will be discussed in detail in Caper 7.

Although we can no longer use finite matrix algebra in order to predict probabilistic properties of chains with continuous state spaces, much of the theory developed above for finite state chains applies with the obvious modifications. In the case of continuous sample space, the equivalent to equation (1.54) for the PDF of the next time index is:

$$f\left(X_{t+1} = x_{t+1}\right) = \int f\left(X_{t+1} = x_{t+1} \mid X_t = x'\right) f\left(X_t = x'\right) dx' \quad (1.66)$$

so that, if it exists, an invariant PDF satisfies:

$$f\left(X = x\right) = \int f\left(X_{t+1} = x \mid X_t = x'\right) f\left(X = x'\right) dx' \quad (1.67)$$

Detailed balance requires:

$$f\left(X = x \mid X' = x'\right) f\left(X' = x'\right) = f\left(X = x' \mid X' = x\right) f\left(X' = x\right) \quad (1.68)$$

Indeed, for the AR(1) case described above, it can be shown that the chain is reversible with Gaussian stationary distribution $f\left(x\right) = \mathcal{N}\left(x; 0, \sigma^2/\left(1 - a^2\right)\right)$.

For further reading, Grimmett and Stirzaker (2001) is a very comprehensive introduction to probability and stochastic processes, whereas Kemeny and Snell (1976) contains highly readable coverage of the basics of finite, stationary Markov chains.

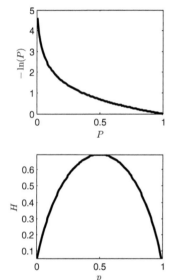

Fig. 1.13: The (Shannon) information map (top) and entropy (bottom). The information map $-\ln\left(P\right)$ maps probabilities in the range $[0, 1]$ to entropies over $[+\infty, 0]$. Entropy is the expected information of a random variable $H\left[X\right] = E\left[-\ln f\left(X\right)\right]$, which is shown here for the Bernoulli random variable with PMF $f\left(x\right) = p^x\left(1 - p\right)^{1-x}$ for $x \in \{0, 1\}$, as a function of p. Since the Bernoulli distribution is uniform for $p = \frac{1}{2}$ ($f\left(x\right) = \frac{1}{2}$), it follows that the entropy is maximized at that value, which is clear from the plot.

1.5   Data compression and information theory

Much of mathematical and statistical modelling of the world can be considered a form of *data compression* — representing measured physical phenomena in a form that is more compact than the 'size' of the recorded signal obtained from the real world. There are many open questions about exactly how to measure the size of a mathematical model and a set of measured data. For example, the *Kolmogorov complexity*, $K\left[S\right]$ is the size (number of instructions or number of symbols) of the smallest computer program that can reproduce the sequence of symbols S making up a digital representation of the measurement. This measure of size of a data set has the desirable property that it is invariant to the choice computer programming language used (up to a constant).

For instance, the algorithmic complexity of the sequence $S_1 = (1, 3, 1, 3, 1, 3, 1, 3, 1, 3, 1, 3, 1, 3, 1, 3)$ is (at most) just the length of the program which outputs the sub-sequence $(1, 3)$ nine times.

Whereas, the Kolmogorov complexity of the sequence $S_2 = (8, 2, 7, 8, 2, 1, 8, 2, 6, 8, 2, 8, 8, 2, 3, 8, 2, 9)$, which has the same length as S_1, does not, apparently, have such an obviously simple program which can reproduce it (although we can certainly imagine simple schemes which could encode this sequence more compactly than the length of S_2). Hence, we can hypothesize that $K[S_1] < K[S_2]$ i.e. that the complexity of S_2 is greater than of S_1.

We can always say that an upper bound (but not a 'tight' bound) on the complexity is $K[S] \leq |S| + C$, i.e. the length of S plus some (small) constant C, because a program which simply has a table from which it retrieves the items from the sequence in the correct order, can reproduce any sequence. This upper bound program has to contain the entire sequence (hence the $|S|$ part), and the program which retrieves the sequence in order (the C part). We can also exhibit programs which are *too simple*: these are short but do not exactly reproduce the sequence (for example, for S_2 above, the program which outputs the sub-sequence $(8, 2, 4)$ six times has complexity less than $|S|$, but it gets every third element of the sequence wrong). These 'lossy' programs can serve as lower bounds. The actual $K[S]$ lies between those bounds somewhere. The main difficulty is that finding $K[S]$ is *non-computable*, that is, we can prove that it is impossible, in general, to write a program which, given any sequence, can output the Kolmogorov complexity of that sequence (Cover and Thomas, 2006, page 163). This means that for any sequence and a program which reproduces it, we can only ever be sure that we have an upper bound for $K[S]$.

A related quantity which *is* computable, applicable to random variables, is the *Shannon information map* $-\ln P$ for the probability value P. This is the foundation of the (somewhat poorly-named) discipline of *information theory*. Since $P \in [0, 1]$, the corresponding information ranges over $[+\infty, 0]$ (Figure 1.13). Entropy measures the average information contained in the distribution of a random variable. The *Shannon entropy* of a discrete random variable X, defined on the sample space Ω with PMF or PDF $f(x)$is:

$$H[X] = E[-\ln f(X)] \qquad (1.69)$$

For discrete random variables this is $H[X] = -\sum_{x \in X(\Omega)} f(x) \ln f(x)$, whereas for continuous variables this becomes $H[X] = -\int_\Omega f(x) \ln f(x) \, dx$. In the continuous case, $H[X]$ is known as the *differential entropy*. It is important to note that while this is indeed an expectation of a function of the random variable, the expectation function is the distribution function of the variable (Figure 1.13).

To date there are some reasonable contraversies about the unique axiomatic foundation of information theory, but the following axioms are sufficient and necessary to define Shannon entropy:

(1) Permutation invariance: The events can be relabeled without changing $H[X]$,

(2) Finite maximum: Of all distributions on a fixed sample space,

$H[X]$ is maximized by the uniform distribution (for the discrete case), or the Gaussian distribution (for the continuous case),

(3) Non-informative new impossible events: Extending a random variable by including additional events that have zero probability measure does not change $H[X]$,

(4) Additivity: $H[X,Y] = H[X] + E_X[H[Y|X]]$, in words: the entropy of a joint distribution is the sum of the entropy of X, plus the average entropy of Y conditioned on X.

Note that axiom 4 above implies that if X, Y are independent, then $H[X,Y] = H[X] + H[Y]$, i.e. the entropy of two independent random variables is the sum of the entropies of the individual variables. Axiom 4 can, for example, be can be replaced by this weaker statement, in which case the Shannon entropy is not the only entropy that satisfies all the axioms. However, the Shannon entropy is the only entropy that behaves in a consistent manner for conditional probabilities:

$$P(X|Y) = \frac{P(X,Y)}{P(Y)} \implies H[X|Y] = H[X,Y] - H[Y] \qquad (1.70)$$

This is intuitive: it states that the information in X given Y, is what we get by subtracting the information in Y from the joint information contained in X and Y simultaneously. For this reason we will generally work with the Shannon entropy in this book, so that by referring to 'entropy' we mean Shannon entropy.

For random processes, Kolmogorov complexity and entropy are intimately related. Consider a discrete-time, i.i.d. process X_n on the discrete, finite set Ω. Also, denote by $K[S|N]$ the *conditional* Kolmogorov complexity where the length of the sequence N is assumed known to the program. We can show that as $N \to \infty$ Cover and Thomas (2006, page 154):

$$E[K[(X_1, \ldots, X_N)|N]] \to N H_2[X] \qquad (1.71)$$

where $H_2[X]$ is the entropy with the logarithm taken to base 2 in order to measure it in *bits*. In other words, the expected (conditional) Kolmogorov complexity of an i.i.d. sequence, converges to the *total entropy* of the sequence. This remarkable connection arises because Shannon entropy can also be defined using a *universally good coding scheme* for all i.i.d. random sequences (for example, *Huffman coding* which creates a uniquely decodable sequence of bits), that is, a way of constructing *compressed* representations of realizations of any random variable. Shannon's famous *source-coding theorem* shows that the number of bits in that (universal) compressed representation lies in between $N H_2[X]$ and $N H_2[X] + N$ (Cover and Thomas, 2006, page 88).

Another way of expressing this compression is that given $|\Omega| \leq 2^L$, then the integer$L$ is the number of bits needed to represent all members the sample space Ω. So, it would take $N \times L$ bits to represent the sequence S in an uncompressed binary representation. By contrast, using entropy coding, then since $H_2[X] \leq \log_2 |\Omega|$ and provided the distribution of X is non-uniform, we can achieve a reduction in the number of bits required to represent S without loss. The entropy tells us exactly how much compression we can achieve on average in the best case.

The importance of the information map

Why is information theory and data compression so important to machine learning and statistical signal processing? The most fundamental is that, in logarithmic base 2, the information map $-\log_2 f(\boldsymbol{x})$ of a distribution over a random sequence $\boldsymbol{X} = (X_1, \ldots, X_N)$, quantifies the expected length of the compressed sequence, as follows. For an i.i.d. sequence of length N, over the discrete set with $M = |\Omega|$ elements, then if each outcome has probability p_i, $i \in \{1, \ldots M\}$, then outcome i will be expected to occur $N p_i$ times. The distribution of the sequence is:

$$f(\boldsymbol{x}) = \prod_{n=1}^{N} p_{x_n} \approx \prod_{i=1}^{M} p_i^{N p_i} \tag{1.72}$$

Now, applying the information map:

$$\begin{aligned} -\log_2 f(\boldsymbol{x}) &\approx -\sum_{i=1}^{M} N p_i \log_2 p_i = -N \sum_{i=1}^{M} p_i \log_2 p_i \\ &= N H[X] \end{aligned}$$

which approximately coincides with both the total entropy, and the expected Kolmogorov complexity, asymptotically. The exact nature of this approximation is made rigorous through the concepts of *typical sets* and the *asymptotic equipartition principle* (Cover and Thomas, 2006, chapter 3). So, the information map gets us directly from the probabilistic to the data compression modeling points of view in machine learning.

A key algebraic observation here is that applying the information map $-\ln P(X)$ has the effect of converting multiplicative probability calculations into additive information calculations. For conditional probabilities:

$$-\ln P(X|Y) = -\ln \left(\frac{P(X,Y)}{P(Y)} \right) = -\ln P(X,Y) + \ln P(Y) \tag{1.73}$$

and for independent random variables:

$$-\ln P(X,Y) = -\ln (P(X) P(Y)) = -\ln P(X) - \ln P(Y) \tag{1.74}$$

This mapping has computational advantages. Products of many probabilities can become unmanageably small and we run into difficulties when doing computations with finite numerical precision, whereas, information is additive on the scale $\mathbb{R}^+$ and the corresponding information calculations are more stable.

The information map also has algebraic advantages. For example, a ubiquitous problem involves maximizing a joint probability with respect to some parameter, where the joint distribution factorizes due to independence. In which case, the joint distribution becomes a sum of negative log probabilities. Minimizing this negative log probability is equivalent to maximizing the joint probability, and this minimization can usually be performed by setting the derivative with respect to the parameter, of this sum of negative log probabilities, to zero. But, since differentiation is linear, the condition that holds at the maximum joint probability is that the sum of the derivatives of each negative log probability is zero. In this form, the maximum probability problem is generally much easier to solve.

Yet another advantage is that a vast number of important distributions, including the Gaussian and Bernoulli distributions, belong to the so-called *exponential family* (see Section 4.5), that can be written in the form $f(\boldsymbol{x};\boldsymbol{p}) = \exp(\boldsymbol{p}\cdot\boldsymbol{g}(\boldsymbol{x}) - a(\boldsymbol{p}) + h(\boldsymbol{x}))$. Applying the information map to this family, we get $-\ln f(\boldsymbol{x};\boldsymbol{p}) = -\boldsymbol{p}\cdot\boldsymbol{g}(\boldsymbol{x}) + a(\boldsymbol{p}) - h(\boldsymbol{x})$ which stabilizes and simplifies all the probability calculations using these distributions.

In abstract terms, probability densities form a semifield on $[0, +\infty]$ with the operations of addition and multiplication. The additive and multiplicative identities are the usual 0 and 1. After applying the negative log transformation, multiplication is converted to addition with the identity 0. The corresponding operation to addition is the *(negative) log-exp-sum* operation $-\ln(e^{-x} + e^{-y})$, with identity $+\infty$. This reveals the main disadvantage of the information map: addition of probabilities is no simpler, and perhaps somewhat more complex, under the negative log transformation. As an illustration, consider marginalization $P(X) = \sum_{y\in Y(\Omega_Y)} P(X, Y = y)$ where applying the information map just leads to $-\ln P(X) = -\ln\sum_{y\in Y(\Omega_Y)} P(X, Y = y)$, and no further simplification is possible. We will see in later chapters a number of effective computational tricks for dealing with this problem.

Mutual information and Kullback-Leibler (K-L) divergence

While the joint entropy $H[X, Y]$ characterizes the total amount of information contained in a joint distribution, the *mutual information* quantifies the amount of information *shared between* random two variables, X, Y with distribution functions f_X, f_Y and joint distribution function f_{XY} is:

$$I[X, Y] = E_{XY}\left[-\ln\left(\frac{f_X(X) f_Y(Y)}{f_{XY}(X, Y)}\right)\right] \qquad (1.75)$$

where the expectation is with respect to the joint distribution. For discrete sample spaces this is:

$$I\left[X,Y\right] = -\sum_{x\in X\left(\Omega_X\right)}\sum_{y\in Y\left(\Omega_Y\right)} f_{XY}\left(x,y\right)\ln\left(\frac{f_X\left(x\right)f_Y\left(y\right)}{f_{XY}\left(x,y\right)}\right) \quad (1.76)$$

and for the continuous case:

$$I\left[X,Y\right] = -\int_{\mathbb{R}}\int_{\mathbb{R}} f_{XY}\left(x,y\right)\ln\left(\frac{f_X\left(x\right)f_Y\left(y\right)}{f_{XY}\left(x,y\right)}\right) dx\,dy \quad (1.77)$$

Using the properties of the logarithm and the linearity of expectation, it is straightforward to show that $I\left[X,Y\right] = H\left[X\right] + H\left[Y\right] - H\left[X,Y\right]$, i.e. mutual information is the difference between the sum of the marginal entropies and the joint entropy. What are the properties of this quantity? The most important ones are listed below:

(1) Independence: $I\left[X,Y\right] = 0$ if and only if X is independent of Y,

(2) Non-negativity: $I\left[X,Y\right] \geq 0$, with equality for X,Y independent,

(3) Symmetry: $I\left[X,Y\right] = I\left[Y,X\right]$,

(4) Self-information: $I\left[X,X\right] = H\left[X\right]$.

The first three properties have the 'flavour' of a metric. Contrast this with covariance: this is zero for independent variables but if two random variables have zero covariance, this does not imply that they are independent (except in special cases, for example, when they are both Gaussian). Therefore, mutual information is often used as a general measure of dependence between variables.

The information map can also be used to quantify the 'dissimilarity' between the distributions of two random variables. For example, given the two random variables X,Y with PDFs/PMFs f_X, f_Y, the *Kullback-Leibler (K-L) divergence* is defined as:

$$D\left[X,Y\right] = E_X\left[-\ln\left(\frac{f_Y\left(X\right)}{f_X\left(X\right)}\right)\right] \quad (1.78)$$

This is also known as the *information gain* or *relative entropy*. It can also be expressed as the difference in average information contained in the two distributions, with respect to the distribution of X:

$$
\begin{aligned}
D\left[X,Y\right] &= E_X\left[-\ln f_Y\left(X\right)\right] - E_X\left[-\ln f_X\left(X\right)\right] \quad (1.79)\\
&= H_X\left[Y\right] - H\left[X\right]
\end{aligned}
$$

where $H_X\left[Y\right]$ is known as the *cross-entropy* of Y with X. So, we can see from this that if X,Y have the same distribution, then $D\left[X,Y\right] = 0$. Furthermore, K-L divergence is non-negative $D\left[X,Y\right] \geq 0$. Does this make it a metric? Unfortunately not, because it is not symmetric and does not satisfy the triangle inequality. Nonetheless, the K-L divergence

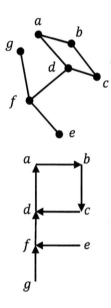

Fig. 1.14: Undirected (top) and directed (bottom) graphs. Both graphs share the same underlying undirected graph (an illustration of the fact that geometry is not important, only connectivity matters). They also share the cycle *abcda*, but the cycle *adcba* does not exist in the directed graph. Removing the vertex *b* from both graphs removes the cycles, turning them both into trees, the bottom graph becomes a directed acyclic graph (DAG).

is of great importance to statistical inference in the many signal processing problems in this book.

The classic textbook by Cover and Thomas (2006) contains an excellent in-depth treatment of the material in this section.

1.6 Graphs

One of the central themes of this book is the idea that we want to find good mathematical models for signals which represent the relationships between variables and parameters in the system that generated those signals. For example, if one random variable in the signal X_1 depends upon another X_2, which in turn depends upon another X_3, we could have a graph which somehow represents that relationship $X_3 \rightarrow X_2 \rightarrow X_1$ in a convenient form. *Graphs* (also known as *networks*) are the one of the most general mathematical structures that can be used to represent these kind of relationships (among many other things).

A graph G is a mathematical object composed of a finite set of *vertices* (*nodes*) V and *edges* E. Vertices are connected by edges. If there is an explicit directionality to the edge, then the graph is said to be *directed*, if not, it is *undirected* (Figure 1.14). Edges may have *weights*, these are usually real values – a graph without weights is called *unweighted*. The *size* of the graph is simply the number of edges $N = |E|$.

We can bring this object into a form where we can do computations on it with a representation called the *adjacency matrix*. Typically, we will only consider graphs that have either no edge, or one edge between nodes. In this case, for an unweighted graph, this is a matrix with either ones or zeros; weights on edges can be placed in this matrix instead. By construction, this matrix is symmetric for undirected graphs, and asymmetric for directed graphs.

The number of edges connected to any node is called the (node) *degree*. Clearly, for an unweighted, undirected graph, this is the row (and column) sum corresponding to that node. For directed graphs, there are two quantities: the *in-degree* and *out-degree* corresponding to the number of edges entering the node and leaving it, respectively. These are the appropriate row and column sums for the asymmetric adjacency matrix.

A *walk* (or *path*) in a graph is a traversal between two nodes, through a sequence of edges which are all connected. A *cycle* is a path that is periodic, that is, the sequence of edges and vertices repeats after a certain number of steps. A *loop* is a self-edge, that is, an edge connecting a node to itself. (However, note that in some texts a loop is synonymous with a cycle – we will be explicit wherever necessary.)

The *distance* between two nodes in a graph is the number of edges in the shortest possible walk connecting them. Even though two nodes might be connected, in a directed graph, it may not be possible to find a path between them that respects the direction of all the edges in that path. Undirected graphs do not have this restriction, obviously. The *diameter* of a graph is the largest distance in the graph, that is, the length of the longest path between any two nodes.

Special graphs

Because the foundational concepts above are so simple, we end up with a vast array of possible graphs and it becomes hard to say anything meaningful or useful about them in general. But, by fixing a particular property or set of properties of interest, one can define classes of graphs with very important mathematical properties. For example, the particular vertex-edge connectivity of a graph is one of the most important features (Figure 1.14).

A *connected* graph is one where all nodes can be reached by paths from all other nodes, i.e. there are no *isolated nodes*. We will generally only be concerned with connected graphs, because, if they are not connected then we can simply pick out all the isolated nodes and study these as a separate graph of smaller size.

A *fully-connected* graph is one in which all nodes are connected to every other node and there are no loops. The unweighted adjacency matrix for a fully-connected graph contains all ones, except the diagonal which is all zeros.

Fully connected graphs are in some senses uninteresting, because they do not contain any 'structure' we can exploit. Indeed, one of the most interesting kinds of directed network is the *directed acyclic graph* (DAG). A DAG is a connected graph with no directed cycles (note that a cycle in an undirected graph could disappear when directionality is introduced). This graph arises naturally in the case of chains of probabilistic conditioning between random variables, known in the machine learning literature as *probabilistic graphical models* (PGMs). Indeed, PGMs are one of the key concepts in this book.

Other important graphs are those which have no loops or cycles, known as *trees*, by obvious analogy to the visual appearance (Figure 1.14). Trees are simple and have some very important properties: any two nodes are connected by a unique path; if any edge is removed, the tree splits into more than one piece and is no longer connected; a tree has $N - 1$ edges $|V| = |E| - 1$; and a loop or cycle will be formed if any new edge is added that joins two existing nodes. These properties create very considerable computational simplifications which make statistical inference on a DAG tree easy in many practical situations. The classical example is the HMM discussed earlier.

Bipartite graphs are connected graphs that split the nodes into two disjoint sets, V_1, V_2. There are no edges connecting nodes in the same set – edges only connect nodes in V_1 to those in V_2 and vice-versa. Important examples in machine learning include the undirected bipartite graphs underlying *conditional random fields* (CRFs).

For a more in-depth treatment of the topic in isolation, Bollobás (1998) is recommended, whereas Henle (1994) contains an accessible introduction to graphs as topological objects.

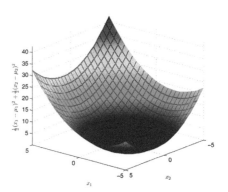

Fig. 1.15: **A typical (convex)** *objective function* in two variables $F(x) = \frac{1}{2}(x_1 - \mu_1)^2 + \frac{1}{2}(x_2 - \mu_2)^2$ with $\mu_1 = -2$, $\mu_2 = 1$. **The colour indicates the value of** F. **Often, the goal will be to minimize this function with respect to the parameter vector** x, **in this case the optimal values are simply** $\hat{x} = (\mu_1, \mu_2)^T$.

1.7   Convexity

Statistical decision-making has at heart the topic of *optimization*: finding the 'best' solution, usually a compromise between conflicting requirements. Statistical inference involves finding the optimum value of some variable or parameter under given constraints; this will be posed as a problem of maximizing the joint probability over the variables and parameters in the problem, or minimizing the corresponding negative log, K-L divergence, or other quantity of the probability or PDF. Thus, in this book, optimization problems can generally be posed as a problem in maximization/minimization of an *objective function* $F : \mathbb{R}^D \to \mathbb{R}$:

$$\hat{\boldsymbol{x}} = \arg\min_{\boldsymbol{x}} F(\boldsymbol{x}) \tag{1.80}$$

See Figure 1.15. When such an optimization problem is *concave* (maximization) or *convex* (minimization), any *locally optimal solution* is the one that solves the problem. Note that we can always convert a concave problem into a convex problem by replacing F with $-F$ and minimizing rather than maximizing.

Convex optimization problems have the very important consequence that if an optimal solution exists, it can be found by a *local search*. This often translates into the intuitive idea that by choosing a starting point somewhere and revising this in the 'downhill' direction, we will eventually arrive at the optimal solution. The inevitable complication to this simple picture is that for downhill search to work, it must be possible to define a gradient ∇F – special techniques are needed when the gradient is not well defined at every point (Figure 1.16).

A convex function is one for which the *(zeroth order) convexity condition* holds for all $\boldsymbol{x}, \boldsymbol{y}$ in the domain of F and $0 \leq \theta \leq 1$, known as *Jensen's inequality*:

$$F(\theta\boldsymbol{x} + (1-\theta)\boldsymbol{y}) \leq \theta F(\boldsymbol{x}) + (1-\theta)F(\boldsymbol{y}) \tag{1.81}$$

This condition means that the line joining any two points $(\boldsymbol{x}, F(\boldsymbol{x}))$ and $(\boldsymbol{y}, F(\boldsymbol{y}))$ is never below the graph of F (Figure 1.16).

Many important functions used in statistical machine learning and signal processing are convex. Here is a list that forms a useful starting point:

- Negative logarithm: $-\ln(x)$,
- Exponential: $\exp(ax)$ for any $a \in \mathbb{R}$,
- Absolute power: $|x|^p$ for $p \geq 1$,
- Negative entropy: $\sum_{i=1}^{N}[x_i \ln(x_i)]$,
- Norms (including the max-norm),
- Log-sum-exp: $\ln\left(\sum_{i=1}^{N}\exp(x_i)\right)$.

For differentiable functions F, convexity can be characterized using calculus. The *first-order convexity condition* for all $\boldsymbol{x}, \boldsymbol{y}$ in the domain of F states that the function is convex if and only if:

$$F\left(\boldsymbol{y}\right) \geq F\left(\boldsymbol{x}\right) + \nabla F\left(\boldsymbol{x}\right)^{T}\left(\boldsymbol{y}-\boldsymbol{x}\right) \qquad (1.82)$$

The right-hand side of this equation is the tangent line at $\boldsymbol{x}$, i.e. the line with the same gradient as F at $\boldsymbol{x}$. An explanation of this condition is that the tangent line at every point is below or touches the function at that point (Figure 1.16). Consider the situation where the gradient vanishes in each dimension $\nabla F\left(\boldsymbol{x}\right) = \boldsymbol{0}$, then from (1.82), $F\left(\boldsymbol{y}\right) \geq F\left(\boldsymbol{x}\right)$ for all $\boldsymbol{y}$ in the domain of F, which means that F is indeed minimized when the gradient in each dimension vanishes.

Again for twice-differentiable functions, the *second-order convexity condition* holds that a function F is convex if and only if, for all $\boldsymbol{x}$ in the domain of F:

$$\nabla^{2} F\left(\boldsymbol{x}\right) \geq \boldsymbol{0} \qquad (1.83)$$

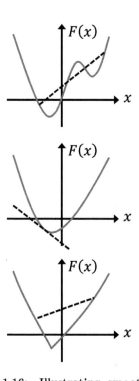

where the inequality can be interpreted elementwise. This simply states that the second derivative is everywhere non-negative, i.e. the function has non-negative curvature in each dimension (Figure 1.16). From the basic list of convex functions above, we can form combinations that are also convex. We discuss some very important examples next.

Non-negative linear combinations. If a set of functions $F_i\left(\boldsymbol{x}\right)$, $i = 1, 2 \ldots N$ is convex, then $\sum_{i=1}^{N} a_i F_i\left(\boldsymbol{x}\right) + b$ is also convex for $a_i \geq 0$ and $b \in \mathbb{R}$. This is known as a *convex combination*. This is the typical form of composite objective function that occurs in *maximum a-posteriori* (*MAP*) inference, where some of the a_i are known as *regularization* parameters, and the b arises due to the normalization of the posterior.

Maximum of convex functions. The maximum at a given point $\boldsymbol{x}$ of a set of N convex functions $F_i\left(\boldsymbol{x}\right), i = 1, 2 \ldots N$, is convex. For example, this book makes frequent use of *(convex) piecewise-linear functions* $F\left(\boldsymbol{x}\right) = \max\left\{\boldsymbol{a}_1^T \boldsymbol{x} + b_1, \boldsymbol{a}_2^T \boldsymbol{x} + b_2 \ldots \boldsymbol{a}_N^T \boldsymbol{x} + b_N\right\}$ where they form the functional basis of piecewise-linear PDFs.

Marginalization of concave functions. If $f\left(x, y\right)$ is concave in x for every $y \in \mathbb{R}$, then $\int_{\mathbb{R}} f\left(x, y\right) dy$ is concave in x. It turns out that many important joint distribution functions are concave in at least one of the random variables over which it is defined; therefore marginalizing out the other variable preserves concavity (this is true for the multivariate Gaussian for example).

Readers wanting a comprehensive introduction to convex analysis will benefit greatly from reading Boyd and Vandenberghe (2004) which also discusses numerous practical applications.

Fig. 1.16: Illustrating smooth but non-convex (top), smooth and convex (middle), and non-differentiable convex functions (bottom). Non-convex functions violate the *zeroth-order convexity condition*: the line joining every pair of points on the graph of the function must always be above the graph. Differentiable convex functions have tangent lines that are below the graph of the function, except where they touch the graph (the *first-order condition*). They also have non-negative second derivatives in each dimension (the *second-order condition*). Non-differentiable convex functions satisfy the zeroth-order condition, but because the gradient is not defined everywhere, it is not possible to minimize such functions by picking a starting point and refining this in the downhill direction (as it is with smooth, convex functions).

1.8 Computational complexity

Most of the theory underlying machine learning and DSP is aimed at constructing algorithms which can be programmed for computer hardware of some kind. It is always the case that there are finite resources available to implement these algorithms in practice. For this reason, a very important consideration is the 'efficiency' of such algorithms. An algorithm that makes more efficient use of resources is likely to be preferred in practice over another, less efficient alternative. It is useful to be able to quantify the efficiency of algorithms in order to predict which ones will be practical, and also how to modify an existing algorithm to make it more efficient. The quantification of the efficiency of algorithms is a topic in computer science known as *computational complexity* theory.

Complexity order classes and *big-O* notation

The two efficiency issues that come up most often are *time* and *space* efficiency: time refers to the number of fundamental (irreducible) computational steps that must be taken in order to execute an algorithm, without regard to the specific hardware. Space in this context, means the number of fundamentally irreducible data structures that must be held in computer memory in order to perform the algorithm. We can usually expect that time and space requirements depend critically upon the size N of the input data for the algorithm, for example, a long digital signal will take more time to process and require more memory to store than a short signal.

Often, what is of interest is not the very specific amount of time or space required in any one instance, more, we are interested to know how these computational resource requirements *scale* with changing input size, relative to competing algorithms. For example, consider Algorithm 1 which requires $f(N) = N^2 + 5N$ fundamental steps, this has roughly the same resource requirements as Algorithm 2 needing $g(N) = \frac{1}{2}N^2$ steps. The equivalence of these two algorithms in this aspect, is represented using *big-O* notation: both algorithms are said to have time complexity of order $O(N^2)$ because the N^2 term dominates as $N \to \infty$ in both f and g. Although we can show that $f(N) > g(N)$ for all N, in the limit $\frac{f(N)}{g(N)} \to 2$ as $N \to \infty$, therefore Algorithm 1 'only' requires twice the number of steps as Algorithm 2. Contrast this with Algorithm 3 taking $h(N) = 2^N + 6N^2$ steps: the ratios $\frac{h(N)}{f(N)}$ and $\frac{h(N)}{g(N)}$ grow without bound with increasing N. Thus, for large inputs, the running time of Algorithms 1 & 2 will be of similar magnitude, but Algorithm 3 can be expected to take orders of magnitude more time.

Given these equivalences in order, we can identify a hierarchy of computational complexity and efficiency (Table 1.1). The most efficient algorithms will be completely, or almost entirely insensitive to N, whereas, the least efficient will have rapidly increasing complexity with increasing N.

Order	Name
$O(1)$	constant
$O(\log N)$	logarithmic
$O(N)$	linear
$O(N \log N)$	log-linear
$O(N^k),\ k > 1$	polynomial
$O(k^N),\ k > 1$	exponential
$O(N!)$	factorial

Table 1.1: Commonly encountered computational complexity order classes, in *big-O* notation. Most efficient classes are at the top, least efficient towards the bottom. The polynomial classes and better are typically considered to be 'tractable' in practice.

Tractable versus intractable problems: NP-completeness

Of course, what order class counts as efficient in an absolute sense is debatable, but exponential time complexity can entail computations which would take astronomically large time periods to solve problems for even modest N. As an example, consider that an order $O(2^N)$ algorithm for an input of length $N = 1000$ would require $2^{1000} \approx 10^{300}$ steps, and even with computer processing speeds on the order of 10^9 calculations per second, the algorithm would take approximately 3×10^{283} years to complete! Now consider that digital signals can easily reach lengths of $N = 10^6$ or greater, and we can appreciate that exponential time algorithms are going to be useless in practice. An alternative algorithm of polynomial order $O\left(N^2\right)$ would take $1000^2 = 10^6$ steps which is solvable in a fraction of a second on ubiquitous computer hardware available today. This radical contrast between polynomial and exponential order complexity classes is a good motivation for taking the consensus view that algorithms in order classes at most as bad as polynomial are tractable, and those with exponential order and worse, are intractable.

So, faced with a particular mathematical problem, we would want to find the most efficient algorithm to solve it. Typically, the most obvious solution involves some kind of *brute-force computation* which simply solves or evaluates a constituent series of mathematical expressions in the problem exhaustively, and this solution will not be very efficient. But, by exploiting special features in the problem (such as underlying group structure) we can often find an algorithm with improved efficiency. A well-known example is the computation of the *discrete Fourier transform*: the naive algorithm which simply evaluates the summation in the problem term-by-term has time complexity $O\left(N^2\right)$, but an algorithm known as the *Fast Fourier Transform* can solve the problem exactly in $O\left(N \log N\right)$ instead. With ingenuity then, some efficient algorithms have been devised for common computational problems encountered in machine learning and DSP.

This raises the question of whether there exists highly efficient algorithms for *every* mathematical problem. Unfortunately, the answer appears to be "no". Remarkably, there are some problems for which there is no known, efficient algorithm, and in general, we cannot expect to have algorithms with anything better than exponential time complexity. These algorithms are all, essentially, equivalent. For technical reasons, the details of which are not important, these problems are known as *non-polynomial time complete (NP-complete)* in the literature on algorithm complexity. They include some very important problems in statistical machine learning and DSP, such as *feature selection, k-means clustering* and *sparse signal recovery*. In these cases, as we will see later, the best we can hope for are tractable algorithms which find good approximate solutions.

Computational complexity and NP-completeness, and the related concept of NP-hardness are discussed in depth in the very readable textbook by Dasgupta *et al.* (2008).

Optimization

2

Decision-making under uncertainty is a central topic of this book. A common scenario is the following: data is recorded from some (digital) sensor device, and we know (or assume) that there is some "underlying" signal contained in this data, which is obscured by noise. The goal is to extract this signal, but the noise causes this task to be impossible: we can never know the actual underlying signal. We must make mathematical assumptions that make this task possible at all. Uncertainty is formalized through the mathematical machinery of probability (see Section 1.4), and decisions are made that find the optimal choices under these assumptions. This chapter explores the main methods by which these optimal choices are made in DSP and machine learning.

2.1 Preliminaries

As discussed in Section 1.7, optimization problems are typically posed in the standard form of minimizing some (generalized) *objective function* $F : X \to \mathbb{R}$ with respect to the parameter vector $x \in X$. When $X = \mathbb{R}^D$, this is known as a *continuous problem*. By contrast, where X consists of a finite set, or is countably infinite, the problem is *discrete* or *combinatorial*. How we attempt to solve this problem will depend critically upon the properties of F: in particular, whether it is continuous or discrete, whether it has one or more points that are 'locally' optimal (in a sense to be made explicit later), and if F is continuous, how many derivatives exist.

Continuous differentiable problems and critical points

One of the most important classes of problems are continuous problems for which F can be differentiated at least twice. For these problems we can state a necessary criteria for a point $\hat{x}$ to be a *local minimum* of F: the gradient must be zero in each of the D directions, i.e. $\nabla F(\hat{x}) = \mathbf{0}$. By local, we mean that $F(\hat{x}) < F(\hat{x} + \Delta\hat{x})$ for any sufficiently small displacement vector $\|\Delta\hat{x}\| > 0$ (note that, we could have an infinite number of isolated points for which this is true). These are known as *critical points*. All this follows from basic calculus.

However, the gradient condition does not suffice: if, in addition, the matrix of second partial derivatives, known as the *Hessian matrix* $\nabla^2 F(\hat{x})$, is positive definite, then $\hat{x}$ is indeed a minimum, and the func-

Machine Learning for Signal Processing: Data Science, Algorithms, and Computational Statistics. Max A. Little.
© Max A. Little 2019. Published in 2019 by Oxford University Press. DOI: 10.1093/oso/9780198714934.001.0001

tion is increasing in each dimension, moving in any direction away from the critical point. Otherwise, it could be a *local maximum* or a *saddle-point*, which increases in some dimensions and decreases in others. In one dimension, $D = 1$, for a critical point to be a minimum, the second derivative must be positive at that point. Functions F for which there is only one minima across the whole range of the function, are convex; indeed the Hessian is positive definite everywhere, as stated in Section 1.7.

Most practical continuous problems cannot be solved analytically. Even so, if in addition to being everywhere differentiable, the problem is convex, there is a simple procedure which is guaranteed, in a finite number of computational steps, to find a solution with a certain required level of accuracy. Starting with a 'guess' x_n at iteration n, and taking an appropriate 'step' in the direction of steepest descent $-\nabla F(x_n)$, we will have an improved estimate x_{n+1}. Another, similar step arrives at an even better estimate, and this can be repeated indefinitely. This is known as the *method of gradient descent*. This fortunate situation, where a simple greedy algorithm can work for a large class of practical problems, is perhaps the most attractive feature of convex, differentiable problems, and is explored in detail in later sections. However, we will see that such a simple idea often performs very poorly in practice and will introduce some of the most effective remedies for this limitation.

Continuous optimization under equality constraints: Lagrange multipliers

Sometimes, we are also faced with a set of N *(equality) constraint functions* $G_i : \mathbb{R}^D \to \mathbb{R}$:

$$\begin{aligned} \text{minimize} \quad & F(x) \\ \text{subject to} \quad & G_i(x) = 0, \ i = 1, 2 \ldots N \end{aligned} \tag{2.1}$$

Problems such as (2.1) are known as *(continuous) constrained optimization* problems. Constraints introduce some additional complications: in general, some (sub)-regions of $\mathbb{R}^D$ will be 'forbidden' because within those regions, the constraint equations do not hold. The regions of $\mathbb{R}^D$ where the constraints do hold are called the *feasible region* or *feasible set*.

The goal of (2.1) is to find the point $\hat{x}$ within the feasible region where F is minimal. For problems where F is differentiable, there is a simple solution known as the *method of Lagrange multipliers*, that allows us to find this $\hat{x}$. A basic geometric intuition can be understood by example (Figure 2.1). Here, we wish to minimize the function $F(x) = x_1^2 + 2\left(x_2 - \frac{1}{2}\right)^2 - 2$, subject to the single constraint $G(x) = x_1^2 + x_2^2 - 1 = 0$. This constraint fixes the unit circle as the feasible region. Shown in the figure are *level curves*, that is, sets of x for which $F(x) = c$, some constant. Also shown are the directions of steepest descent $-\nabla F(x)$ at

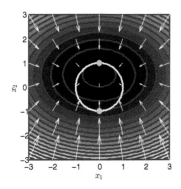

Fig. 2.1: Illustrating the method of Lagrange multipliers for equality constrained continuous optimization. The convex objective function F with contours $F(x) = c$ (blue) is shown as a grey colour gradient from black (smaller values of F) to light grey (larger values). Arrows show the (downhill) gradient $-\nabla F$, these are normal vectors perpendicular to the contours. The white circle is the constraint which constitutes the feasible set. The two yellow dots are the only points where the gradients $-\nabla F$ are parallel to the normal vectors to the feasible set. At these points, it is not possible to decrease F by (infinitesimal) movements along the feasible set, therefore they represent potential solutions to the optimization problem (in this case, the point $(0, 1)$ corresponds to the smallest value of F).

a selection of points. The critical observation is that the level curves of F are locally parallel to the feasible set at only two points $(0, \pm 1)^T$, and the objective function F is smaller at $(0, 1)^T$ than at $(0, -1)^T$. So, the solution to the constrained minimization problem is $(0, 1)^T$.

To understand why this is true, we can observe that one way to find a solution, is to start anywhere on the feasible set and try to minimize F by local movements along the set, in the direction with the largest decrease in F. When we have reached one of the points $(0, \pm 1)^T$, a local movement leads to no significant change in F, because local movements along the feasible set are also local movements along a level set (which is, by definition, constant).

This is the geometric intuition, the corresponding algebraic formulation goes as follows. The vector of steepest descent $-\nabla F(x)$ is the normal vector to the level set at x, which is perpendicular to the level set there. Similarly, the normal vector to the feasible set is perpendicular to the feasible set at x. Since the level set and the feasible set are parallel, it follows that their normal vectors are also parallel. However, they will generally have different magnitudes, hence we can write:

$$- \nabla F(\hat{x}) = \nu \nabla G(\hat{x}) \tag{2.2}$$

with an arbitrary scalar value $\nu > 0$. Therefore, we are implying that optimal solutions must satisfy both (2.2) and $G(\hat{x}) = 0$, and a simple way to encode these requirements this is to introduce the *Lagrangian*:

$$L(x, \nu) = F(x) + \nu G(x) \tag{2.3}$$

for which $\nabla_x L(x, \nu) = 0$ recovers $\nabla F(x) + \nu \nabla G(x) = 0$ (the same as (2.2)), and $\nabla_\nu L(x, \nu) = 0$ is simply the constraint $G(x) = 0$. The additional degree of freedom, ν, is called the *Lagrange multiplier* for the problem.

Returning to our example above, we have $L(x, \nu) = x_1^2 + 2\left(x_2 - \frac{1}{2}\right)^2 - 2 + \nu\left(x_1^2 + x_2^2 - 1\right)$ so that $\nabla_x L(x, \nu) = (2x_1(1 + \nu), (4 + 2\nu)x_2 - 2)^T$. Assuming generally $x_1 \neq 0$, we get $\nu = -1$ and $x_2 = \frac{1}{2+\nu} = 1$. Finally, applying the constraint equation, we discover that $x_1 = 0$, which agrees with the geometric solution.

When there are more constraints, we treat these in essentially the same way by forming the general Lagrangian:

$$L(x, \nu) = F(x) + \sum_{i=1}^{N} \nu_i G_i(x) \tag{2.4}$$

with the length-N vector of Lagrange multipliers ν. Finding the optimal solution requires solving the $D + N$ equations $\nabla_x L(x, \nu) = 0$ and $\nabla_\lambda L(x, \nu) = 0$. It is usually the case that we cannot solve these equations analytically, and we will need to use some kind of numerical procedure.

Inequality constraints: duality and the Karush-Kuhn-Tucker conditions

In many situations, in addition to equality constraints, we have a set of M *inequality constraint functions* $H_j : \mathbb{R}^D \to \mathbb{R}$:

$$\begin{aligned} \text{minimize} \quad & F(\boldsymbol{x}) && (2.5) \\ \text{subject to} \quad & G_i(\boldsymbol{x}) = 0, \; i = 1, 2 \ldots N \\ & H_j(\boldsymbol{x}) \leq 0, \; j = 1, 2 \ldots M \end{aligned}$$

and the associated Lagrangian is:

$$L(\boldsymbol{x}, \boldsymbol{\nu}, \boldsymbol{\lambda}) = F(\boldsymbol{x}) + \sum_{i=1}^{N} \nu_i G_i(\boldsymbol{x}) + \sum_{j=1}^{M} \lambda_j H_j(\boldsymbol{x}) \qquad (2.6)$$

with the additional M-length vector of Lagrange multipliers $\boldsymbol{\lambda}$. The vectors $(\boldsymbol{\nu}, \boldsymbol{\lambda})$ are called *dual variables*.

The inequality multipliers are subject to two restrictions. Firstly, the *complementary slackness* condition must hold at the optimal solution:

$$\lambda_j H_j(\hat{\boldsymbol{x}}) = 0, \; j = 1, 2 \ldots M$$

i.e. for each of the inequality constraints, either the multiplier is zero, or the constraint condition is zero (or both). To see why, consider the case when $H_j(\hat{\boldsymbol{x}}) < 0$; then the location of $\hat{\boldsymbol{x}}$ is the same as when the inequality is ignored, and so it can be "removed" from the problem by setting $\lambda_j = 0$. On the other hand, if the optimal point lies on the boundary of an inequality constraint so that $H_j(\hat{\boldsymbol{x}}) = 0$, then the associated multiplier λ_j can be non-zero. Secondly, the inequality multipliers must be non-negative, $\boldsymbol{\lambda} \geq 0$, for the following reason. Consider that for the feasible set where the inequality constraints are satisfied as equalities, the vector of steepest descent in F must be parallel with, but point in the opposite direction to, the normal to this set. This is precisely the situation as in (2.2), so we must have $\boldsymbol{\lambda} \geq 0$.

It is straightforward to show (Boyd and Vandenberghe, 2004, pp. 223) that if $\boldsymbol{x}$ satisfies the constraints in problem (2.6), and since $\boldsymbol{\lambda} > 0$, the *dual (objective) function* $\Lambda(\boldsymbol{\nu}, \boldsymbol{\lambda}) = \inf_{\boldsymbol{x}} L(\boldsymbol{x}, \boldsymbol{\nu}, \boldsymbol{\lambda})$ is a lower bound on the optimal value $F(\hat{\boldsymbol{x}})$ for (2.6). So, the best lower bound is obtained by solving the concave (*Lagrangian*) *dual problem*:

$$\begin{aligned} \text{maximize} \quad & \Lambda(\boldsymbol{\nu}, \boldsymbol{\lambda}) && (2.7) \\ \text{subject to} \quad & \lambda_i > 0, \; i = 1, 2 \ldots N \end{aligned}$$

with respect to the multiplier vectors $(\boldsymbol{\nu}, \boldsymbol{\lambda})$. In this context, (2.6) is known as the *primal problem*. We can view points $(\boldsymbol{x}, \boldsymbol{\nu}, \boldsymbol{\lambda})$ in $\mathbb{R}^{D+N+M}$ as the new variables for the combined "primal-dual" optimization problem.

The quantity $F(\boldsymbol{x}) - \Lambda(\boldsymbol{\nu}, \boldsymbol{\lambda}) \geq 0$, i.e. the difference between the primal and dual objectives, is known as the *duality gap*. It can be shown that, for convex primal problems, if the primal constraints are satisfied strictly, then $F(\hat{\boldsymbol{x}}) = \Lambda\left(\hat{\boldsymbol{\nu}}, \hat{\boldsymbol{\lambda}}\right)$ so that the duality gap is zero (Boyd and Vandenberghe, 2004, pp. 226). This desirable situation is called *strong duality*, and it provides a convenient strategy for checking the *convergence* of a numerical procedure for solving convex optimization problems, which we discuss later.

It is useful to collect together all the conditions which must be true for the primal-dual variables $(\boldsymbol{x}, \boldsymbol{\nu}, \boldsymbol{\lambda})$ to be optimal. Assuming that the primal problem is convex and strong duality holds, then the *Karush-Kuhn-Tucker* (KKT) conditions apply at the primal-dual optimal point $\left(\hat{\boldsymbol{x}}, \hat{\boldsymbol{\nu}}, \hat{\boldsymbol{\lambda}}\right)$:

$$G_i(\hat{\boldsymbol{x}}) = 0, \; i = 1, 2 \ldots N$$
$$H_j(\hat{\boldsymbol{x}}) \leq 0, \; j = 1, 2 \ldots M$$
$$\hat{\lambda}_j \geq 0, \; j = 1, 2 \ldots M$$
$$\nabla F(\hat{\boldsymbol{x}}) + \sum_{i=1}^{N} \hat{\nu}_i \nabla G_i(\hat{\boldsymbol{x}}) + \sum_{j=1}^{M} \hat{\lambda}_j \nabla H_j(\hat{\boldsymbol{x}}) = 0 \qquad (2.8)$$
$$\hat{\lambda}_j H_j(\hat{\boldsymbol{x}}) = 0, \; j = 1, 2 \ldots M$$

The first three conditions are just a restatement of the primal-dual equality and inequality constraints. Since $\hat{\lambda}_j > 0$, the Lagrangian at $L\left(\boldsymbol{x}, \hat{\boldsymbol{\nu}}, \hat{\boldsymbol{\lambda}}\right)$ must be convex with respect to $\boldsymbol{x}$, so it follows that gradient of the Lagrangian with respect to the primal variables must be zero at the optimal point – this gives the fourth condition. The fifth condition is complementary slackness described earlier.

Convergence and convergence rates for iterative methods

Most numerical procedures for solving optimization problems are *iterative*, that is, they produce a sequence of 'guesses' $\boldsymbol{x}_0, \boldsymbol{x}_1 \ldots \boldsymbol{x}_n$. We would, at least, want these guesses to converge on fixed point $\boldsymbol{x}^\star$, i.e. $\boldsymbol{x}_{n+1} = \boldsymbol{x}_n = \boldsymbol{x}^\star$ (or more practically, $\boldsymbol{x}_{n+1} \approx \boldsymbol{x}_n$ for n sufficiently large). Correspondingly, we want at least $|F(\boldsymbol{x}_n) - F(\boldsymbol{x}^\star)| < \epsilon$ for some small *tolerance* $\epsilon > 0$ for n sufficiently large. Ideally also, we want this fixed point to be the solution to the optimization problem, i.e. $\boldsymbol{x}_n \to \boldsymbol{x}^\star = \hat{\boldsymbol{x}}$ as n increases. And, of course, we want this convergence to be as fast as possible. Given two convergent algorithms, how do we decide which converges faster? The *convergence rate* is one approach to answering this question.

Consider an iterative algorithm which converges on the fixed point $\boldsymbol{x}^\star$, then, for $q \geq 1$, this algorithm *converges with order q at rate* $\rho \in [0, 1]$ when:

$$\lim_{n\to\infty} \frac{|F(\boldsymbol{x}_{n+1}) - F(\boldsymbol{x}^{\star})|}{|F(\boldsymbol{x}_n) - F(\boldsymbol{x}^{\star})|^q} = \rho \qquad (2.9)$$

Writing $\Delta F_n = |F(\boldsymbol{x}_n) - F(\boldsymbol{x}^{\star})|$ we have that $\Delta F_{n+1} \to \rho \Delta F_n^q$ and so $\Delta F_n \to \Delta F_0^{q^n} R$ where $R = \left(\rho^{q^n-1}\right)\rho^{1-q}$. Therefore, $\ln \Delta F_n \to q^n (\ln \Delta F_0 + \ln \rho)$, and in the limit $q \to 1$, we have $\ln \Delta F_n \to \ln \Delta F_0 + n \ln \rho$. For this reason, this limiting special case is known as *linear convergence* (it will be a line on a linear-log plot of n versus ΔF_n) and additionally, iterations with $\rho = 0$ are said to be *superlinearly* convergent. The other extreme, $\rho = 1$ is called *sublinear* convergence. The special case $q = 2$ is known as *quadratic* convergence . A key point is that with linear convergence $q = 1$, it does not matter how large ΔF_0 is from the optimal solution, the iterates get closer to optimal at the same exponential rate, whereas, for other values of q, the algorithm should have $\Delta F_0 < \rho^{-1}$ otherwise many initial iterates will be spent 'warming up' where the linear algorithm has already made progress.

Non-differentiable continuous problems

Continuous optimization problems for which derivatives exist everywhere have the local and global optimality criteria described in the previous section. However, there are many practical problems in statistical machine learning for which the derivatives are not defined for some points in the domain of F. Then, there is no guarantee that necessary condition $\nabla F = \mathbf{0}$ can be obtained. Perhaps the simplest practical example is the absolute value function $F(x) = |x|$ for which the gradient at $x = 0$ is not defined. Also, most non-differentiable problems are not solvable analytically. Solving these problems requires some special mathematical concepts and techniques which we introduce next.

Consider a continuous objective function $F : \mathbb{R} \to \mathbb{R}$. While the gradient might not exist at an isolated point x, a *subderivative* at that point always exists. This is any number $m \in \mathbb{R}$ such that:

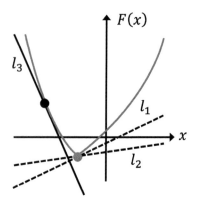

Fig. 2.2: An objective function $F(x)$ defined on $\mathbb{R}$ with both differentiable (black) and non-differentiable (blue) points. For the non-differentiable point, two 'subtangent lines' l_1, l_2 are shown. By contrast, the differentiable point has only one tangent line l_3, whose gradient is the derivative at that point. The non-differentiable point is also a local minima of the function, because there is a horizontal subtangent line with zero gradient going through that point.

$$F(x + \Delta x) \geq m\Delta x + F(x) \qquad (2.10)$$

for all $\Delta x \in \mathbb{R}$. Each value of m is the gradient of a 'subtangent line' that is nowhere above $F(x + \Delta x)$, see Figure 2.2.

The set of all such gradients m is called the *subdifferential* at x, denoted $\partial F(x)$. It is a closed interval lying between the left-sided and right-sided derivatives, $[F'_-(x), F'_+(x)]$. This concept generalizes the notion of differential because, if the function is differentiable at x, then the left and right derivatives are equal so that the subdifferential consists of only one number, which is the ordinary gradient $F'(x)$. Finally, when $0 \in \partial F(x)$, then the point x is a local optima of F and if, in addition, F is convex, then this point is the global minima. These facts about subdifferentials suggest various practical methods for solving convex, non-differentiable continuous problems which will be explored in later sections.

To illustrate, the function $F(x) = |x| + (1/2)x^2$ has subdifferential:

$$\partial F(x) = \begin{cases} x+1, & x > 0 \\ [-1,1], & x = 0 \\ x-1, & x < 0 \end{cases} \qquad (2.11)$$

from which we can see that $x = 0$ is a local optima, and since one can show that the function is convex, this point is also the global minima.

The notion of subderivative and subdifferential can be extended to multiple dimensions. For a function $F : \mathbb{R}^D \to \mathbb{R}$, a *subgradient* of F at $\boldsymbol{x}$ is any vector $\boldsymbol{u}$ such that:

$$F(\boldsymbol{x} + \Delta\boldsymbol{x}) \geq \langle \boldsymbol{u}, \Delta\boldsymbol{x} \rangle + F(\boldsymbol{x}) \qquad (2.12)$$

for all $\Delta\boldsymbol{x} \in \mathbb{R}^D$. Analogous to the $D = 1$ case, this is simply the gradient of any 'subtangent hyperplane' touching, or lying below $F(\boldsymbol{x} + \Delta\boldsymbol{x})$. The subdifferential $\partial F(\boldsymbol{x})$ is just the set of all such subgradients, and it follows that $\boldsymbol{x}$ is a local optima when $\boldsymbol{0} \in \partial F(\boldsymbol{x})$.

Discrete (combinatorial) optimization problems

When the objective function is defined on a finite, or countably infinite, set of values or objects, X, then the optimization is discrete or combinatorial. As an example, the problem of *variable selection* in regression is commonly encountered in statistical machine learning; here, we are given a large number of possible variables to include in a model, and the goal is to find the small set of variables that are 'best' in some sense appropriate to the context. Contrasted with continuous problems, such objective functions lack the same basic notions of continuity and differentiability critical for determining the existence of, and finding, local and/or global optima. Because of this, discrete problems require entirely different mathematical methods to continuous problems.

Clearly, it would be of tremendous value if there were one, unified mathematical approach to solving all such discrete problems. Considering the special case when X is a finite set, an obviously universal solution to a (minimization) problem is *exhaustive (brute-force) enumeration*: compute the value of F for every $x \in X$ and simply pick the value $\hat{x}$ for which $F(\hat{x}) < F(x)$ for all $x \neq \hat{x}$. This is known as *exhaustive search*. Assuming there is such a unique optimal solution, it is obvious that this procedure is guaranteed to find it. It would of course also allow us to discover if there is no such unique optimal solution.

Unfortunately, exhaustive search like this is not a practical solution to most optimization problems because of *combinatorial explosion*. Consider, for example, the problem of variable selection from a set of D variables mentioned earlier. We can define an objective on the set $X = \{0,1\}^D$, so each $x \in X$ is just a string of D ones and zeros, a one in position $1 \leq d \leq D$ indicating that variable number d is to be selected. We generally want to consider any possible combination of variables, so, we are not restricted to selecting a certain number less than D. Given this, the size of X is 2^D. Therefore, the exhaustive search, which must check every possible combination in turn, would require $2^4 = 16$ steps if $D = 4$, but $2^{32} > 4 \times 10^9$ if $D = 32$. In other words, while exhaustive search is a universal solution for finite problems, it can have exponential time complexity and so, in general, it is practically useless for all but the smallest finite domains X. It is, of course, intractable for infinite domains anyway.

Often, a discrete problem will have some special features that make it a lot more tractable than this. For example, if we restrict the variable selection problem to a choice of only a fixed number of variables $d < D$ in each subset, there are $D!/(d!\,(D-d)!)$ possible subsets. Then, an exhaustive search becomes more tractable, since the number of subsets approaches $D^d/d!$ as $D \to \infty$, which is polynomial in D.

In the quest for unified approaches to discrete optimization problems, one encounters a large class of seemingly quite disparate problems (such as the famous *traveling salesman problem* – find the cycle consisting of all vertices in an edge-weighted graph with the smallest weight sum) that can be formulated as *integer linear programming* (ILP) problems. An ILP requires minimizing an integer linear combination $\boldsymbol{c}^T \boldsymbol{x}$ of the non-negative integer parameter vector $\boldsymbol{x} \in \mathbb{Z}^D$, subject to an entirely integer matrix system constraint. Thus, if a discrete machine learning or DSP problem can be reformulated as an ILP, then we can turn to one of the many known algorithms which can find exact solutions, such as *cutting-plane* and *enumerative methods*. It is disappointing to discover, however, that ILP is NP-complete, so in general, it will have worst case exponential time complexity, making it prohibitive for typical problems encountered in machine learning and DSP.

The inevitable conclusion one draws from the state-of-the-art theory of discrete optimization is that, unless there are special features of a problem that mitigate the combinatorial explosion, we are forced to accept approximate solutions. Some of the key ideas here are known as *heuristics*: optimization strategies that are not necessarily guaranteed to find the optimal solution, but can find one that is good enough. Important, and often fairly successful strategies include *greedy search*, *tabu search*, *simulated annealing* and others, these will be the focus of later sections.

2.2   Analytical methods for continuous convex problems

An analytical method is one for which the optimum can be obtained by a prescribed sequence of computational steps, derived by algebraic manipulation of the optimization problem. Analytical methods are unfortunately incredibly rare, because most of the practical problems we describe in this book are not amenable to the kind of algebraic manipulation which makes such solutions possible.

L_2-norm objective functions

There is an extremely important and ubiquitous example of objective functions with analytical minimizers: *L_2-norm objective functions* having the general form:

$$F(\boldsymbol{x}) = \frac{1}{2} \|\mathbf{A}\boldsymbol{x} - \boldsymbol{b}\|_2^2 \qquad (2.13)$$

where $\mathbf{A} \in \mathbb{R}^{M \times D}$, $\boldsymbol{b} \in \mathbb{R}^M$ and the parameters $\boldsymbol{x} \in \mathbb{R}^D$ which we will alter to minimize F. These are known as *linear least-squares* problems, and their solution is particularly simple. The first obvious remark to make is that this objective function is convex in $\boldsymbol{x}$ – this follows because it is a positive quadratic polynomial in each dimension.

Since the norm in (2.13) is squared, we can expand out the norm to get $F(\boldsymbol{x}) = (1/2)(\mathbf{A}\boldsymbol{x} - \boldsymbol{b})^T (\mathbf{A}\boldsymbol{x} - \boldsymbol{b})$, which is then further expanded to:

$$F(\boldsymbol{x}) = \frac{1}{2}\boldsymbol{x}^T \mathbf{A}^T \mathbf{A}\boldsymbol{x} - \boldsymbol{x}^T \mathbf{A}^T \boldsymbol{b} + \frac{1}{2}\boldsymbol{b}^T \boldsymbol{b} \qquad (2.14)$$

and by differentiating with respect to the parameters and setting to zero, leads to:

$$\frac{\partial F}{\partial \boldsymbol{x}}(\boldsymbol{x}) = \mathbf{A}^T \mathbf{A}\boldsymbol{x} - \mathbf{A}^T \boldsymbol{b} = \boldsymbol{0} \qquad (2.15)$$

or $\mathbf{A}^T \mathbf{A}\boldsymbol{x} = \mathbf{A}^T \boldsymbol{b}$, so that the solution to the optimization problem is:

$$\hat{\boldsymbol{x}} = \left(\mathbf{A}^T \mathbf{A}\right)^{-1} \mathbf{A}^T \boldsymbol{b} \qquad (2.16)$$

This solution is so ubiquitous in statistics that it has it's own name: *the normal equation*, and it shows that solving a least squares problem requires only standard matrix arithmetic, a fact of considerable computational convenience. We will find this equation cropping in a large number of distinct contexts throughout this book. It is of fundamental importance in *classical linear* DSP, and gets heavy use in statistical machine learning as well.

Some special cases are particularly valuable. If $\mathbf{A}$ is orthonormal, then since $\mathbf{A}^T \mathbf{A} = \mathbf{I}$, then (2.16) simplifies to $\hat{\boldsymbol{x}} = \mathbf{A}^T \boldsymbol{b}$. The geometric viewpoint here is instructive: since $\mathbf{A}$ is an orthonormal basis, the linear least squares problem in this basis is solved by mapping $\boldsymbol{b}$ into this new basis. The inverse mapping is obtained by left multiplying with $\mathbf{A}$ to get $\boldsymbol{\zeta} = \mathbf{A}\hat{\boldsymbol{x}}$, in other words, by projecting back into the original basis. Later, we will see that the *discrete Fourier transform* is an instance of this orthonormal basis mapping, with $\hat{\boldsymbol{x}}$ known as the *frequency domain* signal, and $\boldsymbol{\zeta}$ the *time domain* signal, and $\mathbf{A}$ is known as the *discrete Fourier matrix*.

We will often encounter so-called *L_2-norm regularization* (also known as *Tikhonov regularization* or *ridge regression*), where the objective function is of the form $F(\boldsymbol{x}) = (1/2)\|\mathbf{A}\boldsymbol{x} - \boldsymbol{b}\|_2^2 + \frac{\gamma}{2}\|\boldsymbol{x}\|_2^2$, which is similar to least squares but including a *regularization* term $\gamma \geq 0$ which forces the entries of $\boldsymbol{x}$ go to zero as $\gamma \to \infty$. This can be rewritten as:

$$F(\boldsymbol{x}) = \frac{1}{2}\|\mathbf{A}\boldsymbol{x} - \boldsymbol{b}\|_2^2 + \frac{1}{2}\|\boldsymbol{\Gamma}\boldsymbol{x}\|_2^2 \tag{2.17}$$

with the solution $\hat{\boldsymbol{x}} = \left(\mathbf{A}^T\mathbf{A} + \boldsymbol{\Gamma}^T\boldsymbol{\Gamma}\right)^{-1}\mathbf{A}^T\boldsymbol{b}$, which is only a trivial modification to the normal equation above. Examining the special case $\boldsymbol{\Gamma} = \gamma\mathbf{I}$ is instructive: with $\gamma = 0$, we recover the normal equation (2.16). In the limit as $\gamma \to \infty$, then $\boldsymbol{x} \to \gamma^{-2}\mathbf{A}^T\boldsymbol{b}$ which is $\mathbf{0}$.

A similar form of normal equation also occurs for the extremely useful case of *weighted least squares problems* where $F(\boldsymbol{x}) = (1/2)\|\mathbf{A}\boldsymbol{x} - \boldsymbol{b}\|_{\mathbf{W}}^2$ where $\mathbf{W}$ is a positive-definite weight matrix. In this case, the normal equation is just $\hat{\boldsymbol{x}} = \left(\mathbf{A}^T\mathbf{W}\mathbf{A}\right)^{-1}\mathbf{A}^T\mathbf{W}\boldsymbol{b}$.

The matrix inversion and matrix products make the complexity of these optimization solutions $O\left(N^3\right)$.

Mixed L_2-L_1 norm objective functions

There is another very interesting special case for which an analytical solution is possible in a limited context: so-called *mixed L_2-L_1-norm* problems in the form:

$$F(\boldsymbol{x}) = \frac{1}{2}\|\mathbf{A}\boldsymbol{x} - \boldsymbol{b}\|_2^2 + \gamma\|\boldsymbol{x}\|_1 \tag{2.18}$$

This is convex in $\boldsymbol{x}$ because it is a convex combination of two convex norms. These are known as *Lasso regression* problems. When $\mathbf{A}$ is orthonormal, it can be shown that the optimal value for the minimization problem is:

$$\hat{x}_i = \text{sgn}(\tilde{x}_i)\left[|\tilde{x}_i| - \gamma\right]_+ \tag{2.19}$$

for all $i = 1, 2 \ldots D$, where $[x]_+ = x$ if $x \geq 0$ and zero otherwise, and $\tilde{x} = \mathbf{A}^T\boldsymbol{b}$ (which is just the normal equation solution when $\gamma = 0$ described earlier). It is very interesting to compare this solution to that of the ridge solution (2.17) when $\mathbf{A}$ is orthonormal:

$$\hat{x}_i = \frac{1}{1 + \gamma}\tilde{x}_i \tag{2.20}$$

In both cases, increasing γ pushes each $\hat{x}_i$ towards zero, but unlike (2.20) for which $\hat{x}_i \to 0$ as $\gamma \to \infty$, the Lasso solution (2.19) has $\hat{x}_i = 0$ for $\gamma \geq |\tilde{x}_i|$. So, the Lasso solution goes to zero for some finite value of the penalty parameter γ (Figure 2.3). This interesting 'shrinkage' of the solution to zero for finite γ has gained a lot of attention in statistical machine learning and DSP because it has some important statistical properties that will be explored later.

Since we need to perform matrix inversion and matrix products, the solution is has computational complexity $O\left(N^3\right)$. Unfortunately, when $\mathbf{A}$ is not orthonormal the Lasso analytical solution (2.19) no longer holds and we have to resort to numerical techniques to solve mixed norm problems like (2.18).

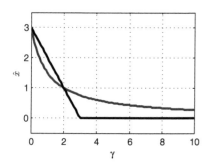

Fig. 2.3: Two important kinds of analytical 'shrinkage' arising from the use of penalized objective functions: L_2-norm (ridge) penalty (blue) and L_1-norm (Lasso) penalty (black). The initial least squares solution (for regularization parameter $\gamma = 0$) is $\hat{x} = 3.0$, and as γ increases, this goes to zero. For the ridge penalty, we have that $\hat{x} \to 0$ in the limit, whereas for the Lasso penalty, $\hat{x} = 0$ for $\gamma \geq \tilde{x}$, where $\tilde{x}$ is the non-penalized least squares solution.

2.3 Numerical methods for continuous convex problems

We have seen earlier that objective functions using the L_2-norm are somewhat unique in that the optimal solution is always computable analytically, and we discussed another limited set of circumstances in which an analytical solution was possible. Are there any other kind of (convex) objective functions with analytical solutions? Indeed, one-dimensional problems $D = 1$ are particularly amenable to this but in practice we encounter high-dimensional, nonlinear problems for which solving the zero derivative condition for x is essentially intractable. In this section, we explore some of the many numerical methods that have been devised to find approximate (and in fact, often extremely accurate) solutions to continuous convex optimization problems for which no analytical solution is tractable.

Iteratively reweighted least squares (IRLS)

Our first example is a class of problems which are, in some sense, just a small 'departure' from L_2-norm problems, where we can arrange to use a linear least squares solution as part of solving these nonlinear problems. We will introduce an *error* or *loss function f* which is (strictly) convex and at least twice differentiable, and symmetric about 0 (i.e. $f(e) = f(-e)$) with $f(0) = 0$. We will require that f is unbounded (so that $f(e) \to \infty$ as $|e| \to \infty$). We also define the *weight function* $w(e) = f'(e)/e$, and will want to have chosen f such that $w(e) < \infty$ (it is bounded) and $w(e)$ is not increasing. Using this error function, we will assemble a nonlinear objective function:

$$F(\boldsymbol{\beta}) = \sum_{i=1}^{N} f\left(\sum_{j=1}^{D} X_{ij}\beta_j - y_i\right) \tag{2.21}$$

and set up the optimization problem of minimizing F with respect to the parameter vector $\boldsymbol{\beta} \in \mathbb{R}^D$. This can be understood as a form of *fitting* or *regression*, where we want to minimize the total error between the predicted values of 'output' vector $\boldsymbol{y} \in \mathbb{R}^N$ and the linear sum of each parameter β_j and the 'input' vectors in the rows of the matrix $\mathbf{X}$. Each contribution to the error term is through the loss function. In the special case that $f(e) = e^2$, we have the linear least squares solution discussed earlier. The objective function (2.21) is also convex in $\boldsymbol{\beta}$ and differentiable (because it is a convex combination of convex functions).

To find the optimal value of F we need the derivative with respect to each parameter β_k:

$$\frac{\partial F}{\partial \beta_k}(\beta_k) = \sum_{i=1}^{N} X_{ik} f'\left(\sum_{j=1}^{D} X_{ij}\beta_j - y_i\right) \tag{2.22}$$

By denoting the error terms $e_i = \sum_{j=1}^{D} X_{ij}\beta_j - y_i$, and writing $w_i = w(e_i)$, then we can express the above as:

$$\frac{\partial F}{\partial \beta_k}(\beta_k) = \sum_{i=1}^{N} X_{ij} w_i \left(\sum_{j=1}^{D} X_{ij}\beta_j - y_i \right) \tag{2.23}$$

In matrix form, at the global minimum we have:

$$\mathbf{X}^T \mathbf{W}(\mathbf{X}\boldsymbol{\beta} - \boldsymbol{y}) = \mathbf{0} \tag{2.24}$$

with the diagonal weight matrix $W_{ii} = w_i$. This is just the (rearranged) normal equations for the weighted least squares problem introduced earlier, which we can readily solve as $\hat{\boldsymbol{\beta}} = (\mathbf{X}^T \mathbf{W} \mathbf{X})^{-1} \mathbf{X}^T \mathbf{W}\boldsymbol{y}$. However, notice that the weights depend upon the error terms e_i, which depend upon the estimated parameters, which of course are unknown at this point. A proposed solution to this impasse is to start with some initial, good guesses for the parameters β_j^0, find the errors e_i, compute the weights w_i, and then solve the weighted least squares problem to get new estimates β_j^1. This method is called *iteratively reweighted least squares* (IRLS), described next.

Algorithm 2.1 Iteratively reweighted least squares (IRLS).

(1) *Initialization.* Choose some good estimates for $\boldsymbol{\beta}_0$, set iteration number $n = 0$, set tolerance $\epsilon > 0$,

(2) *Compute weights.* Using $\boldsymbol{\beta}_n$, find error terms e_i from which we can compute the weights w_i and construct weight matrix $\mathbf{W}$,

(3) *Solve weighted least squares.* Obtain new parameter estimates $\boldsymbol{\beta}_{n+1} = (\mathbf{X}^T \mathbf{W} \mathbf{X})^{-1} \mathbf{X}^T \mathbf{W}\boldsymbol{y}$, compute new objective function value $F(\boldsymbol{\beta}_{n+1})$, calculate objective function improvement $\Delta F = |F(\boldsymbol{\beta}_{n+1}) - F(\boldsymbol{\beta}_n)|$ and if $\Delta F < \epsilon$, exit with solution $\boldsymbol{\beta}_n$,

(4) *Iteration.* Update $n \leftarrow n + 1$, go back to step 2.

By placing the earlier conditions on the function f (and assuming that $\mathbf{X}$ is full rank), it has been shown that this iterative method converges to the optimal solution regardless of the choice of $\boldsymbol{\beta}_0$ (Byrd and Payne, 1979, Theorem 3). Clearly though, choosing good starting values for the parameters will speed up this convergence.

As an example of IRLS, consider the *logistic* error function $f(e) = \ln \cosh(e)$ with $f'(e) = \tanh(e)$ and weight function $w(e) = \tanh(e)/e$ (Figure 2.4). This demonstrates the good convergence of the IRLS algorithm, reaching a solution optimal to a tolerance of $\epsilon = 10^{-8}$ within just eight iterations.

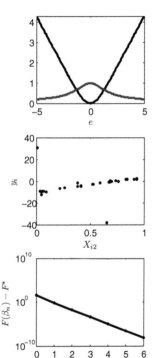

Fig. 2.4: Illustrating iteratively reweighted least squares (IRLS). Error function $f(e)$ (black, top panel) and weight function $w(e) = f'(e)e^{-1}$ (blue, top panel). The error function is almost linear with increasing e, by contrast with the square error function e^2. An example regression set up (middle) with data drawn from the linear model $y_i = \beta_0 + \beta_1 X_{i2} + \epsilon_i$, with $X_{i1} = 1$. The random data ϵ_i is i.i.d. inverse Gaussian, which has a far higher probability of extremely large 'outliers' which confounds any least squares fit of the regression parameters β. Starting from the least squares fit with objective function $F(\beta^0)$, the iterations converge at a linear rate ($F^\star$ is the optimal value of F given the specific data).

It can be shown that the convergence rate is linear (Byrd and Payne, 1979, Theorem 6), so to obtain a solution with tolerance ϵ will require worst case $O\left(-\ln \epsilon\right)$ iterations. Since the computational complexity of the weighted least squares minimization is $O\left(N^3\right)$ (this is due to the complexity of the matrix multiplication and inversion), this gives a total complexity of $O\left(-N^3 \ln \epsilon\right)$.

IRLS is ubiquitous and gets heavy use in statistical machine learning, wherever the dependency upon the input variables $\mathbf{X}$ is linear but the error function is not one that leads to an analytical solution. In particular, IRLS is commonly used to find parameters of so-called *generalized linear models* which are a class of *regression problems* applicable to a broad and very useful class of distributions for the error terms e_i. We will see later that IRLS can be understood as a modified example of *Newton's method*.

Gradient descent

As discussed earlier, an obvious approach to optimizing a continuous, convex problem is to find an accurate numerical solution to the equation $\nabla F = \mathbf{0}$. The intuitive method of gradient descent was invoked, described next.

Algorithm 2.2 Gradient descent.

(1) *Initialization.* Choose some good estimates for $\boldsymbol{x}_0$, a convergence tolerance ϵ, a set of positive step sizes $\alpha_n \in \mathbb{R}$, and set iteration number $n = 0$,

(2) *Select step size.* Choose step size α_n,

(3) *Gradient descent step.* Obtain new parameter estimates $\boldsymbol{x}_{n+1} = \boldsymbol{x}_n - \alpha_n \nabla F\left(\boldsymbol{x}_n\right)$, compute new objective function value $F\left(\boldsymbol{x}_{n+1}\right)$, calculate objective function improvement $\Delta F = \left|F\left(\boldsymbol{x}_{n+1}\right) - F\left(\boldsymbol{x}_n\right)\right|$ and if $\Delta F < \epsilon$, exit with solution $\boldsymbol{x}_{n+1}$,

(4) *Iteration.* Update $n \leftarrow n + 1$, go back to step 2.

Assuming the α_n are chosen appropriately in step 2, this algorithm obtains a decreasing sequence of iterates $F\left(\boldsymbol{x}_0\right) > F\left(\boldsymbol{x}_1\right) > \cdots > F\left(\boldsymbol{x}_n\right)$. The immediate question then becomes one of choosing step sizes such that convergence to the global minimum is guaranteed, preferably converging as fast as possible.

Perhaps the simplest choice is the *constant step size* $\alpha_n = \alpha$. Obviously, if α is too large, it could happen that $F\left(\boldsymbol{x}_{n+1}\right) \geq F\left(\boldsymbol{x}_n\right)$ for some n, so, can we guarantee convergence? The answer is yes, but under certain conditions, namely (Bertsekas, 1995, Proposition 1.2.3):

$$\|\nabla F\left(\boldsymbol{x}\right) - \nabla F\left(\boldsymbol{y}\right)\| \leq \lambda \|\boldsymbol{x} - \boldsymbol{y}\| \tag{2.25}$$

for all $\boldsymbol{x}, \boldsymbol{y} \in \mathbb{R}^D$ and some $\lambda > 0$, and the step size must satisfy $\nu \leq \alpha \leq (2 - \nu)/\lambda$ for some $\nu > 0$. This implies that the step size must decrease with the maximum slope of the gradient of the objective function, roughly, for objective functions with "large" curvature at some point in their domain, we must have correspondingly small step sizes. Finding λ and ν to ensure convergence may of course be very hard in practice, which is one good motivation for using somewhat more sophisticated approaches which adapt the step size on the fly.

Adapting the step sizes: line search

Consider the objective function value at each new step in the gradient descent Algorithm 2.2 as a function of the step size $\Phi(\alpha) = F(\boldsymbol{x}_n - \alpha \nabla F(\boldsymbol{x}_n))$. The the optimal step sizes are:

$$\hat{\alpha}_n = \arg \min_{\alpha} \Phi(\alpha) \qquad (2.26)$$

such that $\hat{\alpha}_n > 0$, which is a one-dimensional optimization problem along the ray $\boldsymbol{x}_n - \alpha \nabla F(\boldsymbol{x}_n)$, $\alpha > 0$ in $\mathbb{R}^D$. An algorithm that can solve this minimization problem is known as an *exact line search* method. In most cases, exact line search is not possible because minimizing $\Phi(\alpha)$ is unlikely to be analytically tractable. So, numerical methods are usually used instead. Here is a very simple, approximate line search algorithm which can be used in step 2 of Algorithm 2.2.

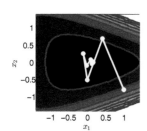

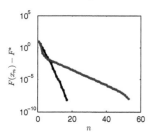

Fig. 2.5: Comparing gradient descent with constant step size and backtracking line search. The iterates $\boldsymbol{x}_n$ of gradient descent with backtracking (top panel, white) are superimposed over the contours of the objective function $F(\boldsymbol{x}) = \exp\left(x_1 + \frac{5}{2}x_2 - \frac{3}{5}\right) + \exp\left(x - \frac{5}{2}x_2 - \frac{3}{5}\right) + \exp\left(\frac{2}{5} - x_1\right)$, showing convergence from the starting point $\boldsymbol{x}_0 = (1.0, -0.8)^T$. The rate of convergence of both algorithms is linear (bottom), but using backtracking (black) requires half the number of steps to reach convergence of $\epsilon = 10^{-8}$ than with constant step size (blue). Backtracking parameters are $c = 10^{-4}$, $d = 0.8$.

Algorithm 2.3 Backtracking line search.

(1) *Initialization.* Set $a > 0$, $c \in (0, 1/2)$ and $d \in (0, 1)$, initialize with $\alpha = a$,

(2) *Backtracking.* If $\Phi(\alpha) \leq F(\boldsymbol{x}_n) + \alpha c \|\nabla F(\boldsymbol{x}_n)\|_2^2$, exit with solution α, otherwise update $\alpha \leftarrow d\alpha$,

(3) *Iteration.* Go back to step 2.

The upper bound in step 2, $F(\boldsymbol{x}_n) + \alpha c \|\nabla F(\boldsymbol{x}_n)\|_2^2$, is known as the *Armijo bound*. For $c = 1$, this is simply the first order Taylor expansion at $\boldsymbol{x}_n$ which is below the convex curve F, and for $0 < c < 1/2$ there is always an α small for the condition to hold. Algorithm 2.3 is guaranteed to find such a step size. One can show that gradient descent with backtracking converges on a minimizer for F, and the worst-case rate of convergence is linear (Boyd and Vandenberghe, 2004, Page 468). Figure 2.5 shows the improvement obtainable with simple backtracking (convergence rate $\rho \approx 0.26$) against constant step size (rate $\rho \approx 0.75$). Typical values for backtracking are $a = 1$, $c \geq 10^{-4}$ and $d \leq 0.8$.

Backtracking does not solve problem (2.26), therefore the step size at each iteration is sub-optimal, implying that the convergence rate of gradient descent could be improved with a better line search algorithm.

Of course, any decrease in the complexity of gradient descent will come at the cost of increased complexity line search. We will next describe a straightforward one-dimensional search algorithm known as *golden section search*. This is a *bracketing* method: we maintain an interval which always encloses the optimum step size, $\alpha_n \in [\alpha_l, \alpha_u]$, and successively shrink this interval in proportion to the *golden ratio* $\phi = \left(\sqrt{5} - 1\right)/2$, such that an approximation to $\hat{\alpha}_n$ is found to a desired tolerance. The major advantage of this algorithm is that it does not require knowledge of the directional gradient $\frac{\partial \Phi}{\partial a}(\alpha)$. Here is the algorithm in full.

Algorithm 2.4 Golden section search.

(1) *Find initial bracket.* Set a tolerance $\epsilon > 0$, and initial bracket search parameters $d > 1$, $\Delta\alpha > 0$. Search for a subsequence beginning at j such that $\Phi\left(\alpha_j^\star\right) > \Phi\left(\alpha_{j+1}^\star\right)$ and $\Phi\left(\alpha_{j+1}^\star\right) < \Phi\left(\alpha_{j+2}^\star\right)$, in the set $\alpha_k^\star = \pm d^k \Delta\alpha$, $k = 1, 2 \ldots$. The initial bracket $[\alpha_l, \alpha_u]$ will be defined by $\alpha_l = \alpha_j^\star$ and $\alpha_u = \alpha_{j+2}^\star$,

(2) *Initialization.* Set $r = \phi(\alpha_u - \alpha_l)$, $\alpha_1 = \alpha_l + r$, $\alpha_2 = \alpha_u - r$, iteration number $m = 0$, and calculate initial estimate $\hat{\alpha}_0 = (\alpha_u + \alpha_l)/2$ and objective function value $\Phi(\hat{\alpha}_0)$,

(3) *Subdivision.* If $\Phi(\alpha_1) < \Phi(\alpha_2)$ update $\alpha_l \leftarrow \alpha_2$, $\alpha_2 \leftarrow \alpha_1$ and $\alpha_1 \leftarrow \alpha_l + \phi(\alpha_u - \alpha_l)$. Otherwise for $\Phi(\alpha_1) \geq \Phi(\alpha_2)$ update $\alpha_u \leftarrow \alpha_1$, $\alpha_1 \leftarrow \alpha_2$ and $\alpha_2 \leftarrow \alpha_u - \phi(\alpha_u - \alpha_l)$,

(4) *Convergence test.* Form new estimate of solution $\hat{\alpha}_{m+1} = (\alpha_u + \alpha_l)/2$, compute new objective function value $\Phi(\hat{\alpha}_{m+1})$, calculate objective function improvement $\Delta\Phi = |\Phi(\hat{\alpha}_{m+1}) - \Phi(\hat{\alpha}_m)|$ and if $\Delta\Phi < \epsilon$, exit with solution $\hat{\alpha}_{m+1}$,

(5) *Iteration.* Update $m \leftarrow m + 1$, go back to step 3.

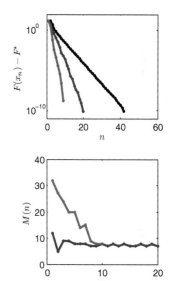

Fig. 2.6: Contrasting the performance of different algorithms for step size selection in the gradient descent problem of Figure 2.5 (top): constant step size (black), backtracking line search (blue), and golden section search (red). All algorithms show linear convergence rate, but golden section search requires the fewest iterations. However, taking into account $M(n)$, the number of iterations of the step size selection algorithms (bottom), we can see that golden section search requires many more iterations per gradient step than the other algorithms (note that constant step size has $M(n) = 1$).

The initial bracketing simply searches for increasingly large intervals such that the minimum, as indicated by a set of three points, is guaranteed to be within one of these intervals eventually. Note that it is possible for this search to return $\alpha_l < 0$, but the interval refinement should eventually return $\hat{\alpha}_m > 0$.

Gradient descent with exact line search such as this, converges on a minimizer for F, and the worst-case rate of convergence is linear (Boyd and Vandenberghe, 2004, page 467). Performance comparisons of constant step size, backtracking and golden section search are drawn in Figure 2.6. An important conclusion to draw from these comparisons is that, although adaptive step size methods will generally converge in fewer iterations of the gradient descent Algorithm 2.2 than with fixed step size, when the number of iterations of the adaptive step size selection algorithms are factored in, the real performance gain of more complex algorithms such as golden section search, is unclear. This is indeed one of the reasons why backtracking is so popular – it has a good compromise between gradient descent convergence and backtracking iterations. Smart choices of initialization of adaptive step size algorithms can help reduce the number of iterations (Nocedal and Wright, 2006, pages 59-60).

Newton's method

We have seen that gradient descent, while straightforward, has (worst case) linear convergence rate. Can we improve upon this rate? In this section we will describe *Newton's method*, an example of an algorithm with generally quadratic convergence.

Locally, approximate the objective function with a quadratic Taylor expansion:

$$
\begin{aligned}
F\left(\boldsymbol{x}\right) \approx\ & F\left(\boldsymbol{x}_n\right) + \left(\boldsymbol{x} - \boldsymbol{x}_n\right)^T \nabla F\left(\boldsymbol{x}_n\right) + \\
& \frac{1}{2}\left(\boldsymbol{x} - \boldsymbol{x}_n\right)^T \nabla^2 F\left(\boldsymbol{x}_n\right)\left(\boldsymbol{x} - \boldsymbol{x}_n\right)
\end{aligned} \tag{2.27}
$$

where $\nabla^2 F\left(\boldsymbol{x}_n\right)$ is the Hessian matrix. Taking the derivative of this, we get:

$$
\nabla F\left(\boldsymbol{x}\right) = \nabla F\left(\boldsymbol{x}_n\right) + \nabla^2 F\left(\boldsymbol{x}_n\right)\left(\boldsymbol{x} - \boldsymbol{x}_n\right) \tag{2.28}
$$

The minimum of this occurs where $\boldsymbol{x} - \boldsymbol{x}_n = -\left[\nabla^2 F\left(\boldsymbol{x}_n\right)\right]^{-1}\nabla F\left(\boldsymbol{x}_n\right)$. This immediately suggests a simple iterative method, as described next.

Algorithm 2.5 Newton's method.

(1) *Initialization.* Choose some good estimates for $\boldsymbol{x}_0$, a tolerance $\epsilon > 0$, and set iteration number $n \leftarrow 0$,

(2) *Select step size.* Choose a step size α_n,

(3) *Minimize local quadratic approximation.* Obtain new parameter estimates $\boldsymbol{x}_{n+1} = \boldsymbol{x}_n - \alpha_n \left[\nabla^2 F\left(\boldsymbol{x}_n\right)\right]^{-1}\nabla F\left(\boldsymbol{x}_n\right)$, compute new objective function value $F\left(\boldsymbol{x}_{n+1}\right)$, calculate objective function improvement $\Delta F = |F\left(\boldsymbol{x}_{n+1}\right) - F\left(\boldsymbol{x}_n\right)|$ and if $\Delta F < \epsilon$, exit with solution $\boldsymbol{x}_{n+1}$,

(4) *Iteration.* Set $n \leftarrow n + 1$, go back to step 2.

A step size $\alpha_n > 0$ is included to allow adaptation using line search, but (pure) Newton's method has $\alpha_n = 1$. Note that the (scaled) *Newton step* $\Delta\boldsymbol{x}_n = -\alpha_n \left[\nabla^2 F\left(\boldsymbol{x}_n\right)\right]^{-1}\nabla F\left(\boldsymbol{x}_n\right)$ is a descent direction because, assuming the Hessian is positive definite, we have $\nabla F\left(\boldsymbol{x}_n\right)^T \Delta\boldsymbol{x}_n =$

$-\alpha_n \nabla F\left(\boldsymbol{x}_n\right)^T \left[\nabla^2 F\left(\boldsymbol{x}_n\right)\right]^{-1} \nabla F\left(\boldsymbol{x}_n\right)$ which must be negative. Therefore, the iterations will have the descent property $F\left(\boldsymbol{x}_0\right) > F\left(\boldsymbol{x}_1\right) > \cdots > F\left(\boldsymbol{x}_n\right)$.

The choice of step sizes is of course important. There are many ways to select this, for example, by using some kind of line search algorithm as discussed earlier. Note that for the backtracking line search, the Armijo bound becomes $F\left(\boldsymbol{x}_n\right) + \alpha c \nabla F\left(\boldsymbol{x}_n\right)^T \left[\nabla^2 F\left(\boldsymbol{x}_n\right)\right]^{-1} \nabla F\left(\boldsymbol{x}_n\right)$.

As one might anticipate, for such a simple algorithm, there are situations when it fails. Firstly, notice that we have to compute the inverse of the Hessian matrix and this may not always work. The algorithm relies upon the local curvature being positive definite, and while this must hold at the global minimum, far from this the curvature may not be positive definite. However, assuming that the following conditions hold: F is twice differentiable, $\nabla^2 F\left(\boldsymbol{x}_n\right)$ is positive definite and Lipschitz continuous (i.e. there exists an $L > 0$ such that $\left\|\nabla^2 F\left(\boldsymbol{x}_n\right) - \nabla^2 F\left(\boldsymbol{x}_n\right)\right\| \leq L \left\|\boldsymbol{x} - \boldsymbol{y}\right\|$ for all $\boldsymbol{x}, \boldsymbol{y} \in \mathbb{R}^D$), then Newton's method converges to the minimum and does so quadratically (Nocedal and Wright, 2006, Theorem 3.5), which is one of the reasons for the popularity of the method. With backtracking or other line search, Newton's method can be even more effective than this in practice (Figure 2.7).

Newton's method forms the basis of a very large number of optimization algorithms. For example, IRLS introduced earlier is a modified Newton's method, as we now show. We can write the gradient of the IRLS objective function as $\nabla F = \mathbf{X}^T \boldsymbol{d}$, where $\boldsymbol{d} = \left[f'\left(e_1\right), f'\left(e_1\right) \ldots f'\left(e_N\right)\right]^T$, and the Hessian as $\nabla^2 F = \mathbf{X}^T \mathbf{D} \mathbf{X}$, where $\mathbf{D}$ is the diagonal matrix defined by $D_{ii} = f''\left(e_i\right)$. The Newton step with step sizes $\alpha_n = 1$ is:

$$\boldsymbol{\beta}_{n+1} = \boldsymbol{\beta}_n - \left(\mathbf{X}^T \mathbf{D} \mathbf{X}\right)^{-1} \mathbf{X}^T \boldsymbol{d} \qquad (2.29)$$

We can approximate $f''\left(e\right) \approx \left(f'\left(e\right) - f'\left(0\right)\right)/\left(e - 0\right)$ since $f'\left(0\right) = 0$ because the error function is convex, symmetric about 0, and differentiable, this approximation is therefore $f'\left(e\right)/e = w\left(e\right)$. So, the weight function is an approximation to the second derivative of the error function. It follows that $\mathbf{D} \approx \mathbf{W}$, the IRLS weight matrix. With this, we can introduce a new iteration:

$$\boldsymbol{\beta}_{n+1} = \boldsymbol{\beta}_n - \left(\mathbf{X}^T \mathbf{W} \mathbf{X}\right)^{-1} \mathbf{X}^T \mathbf{W} \boldsymbol{e} \qquad (2.30)$$

where $\boldsymbol{e} = \left[e_1, e_2 \ldots e_N\right]^T$. Finally, writing $\boldsymbol{\beta}_n = \left(\mathbf{X}^T \mathbf{W} \mathbf{X}\right)^{-1} \mathbf{X}^T \mathbf{W} \mathbf{X} \boldsymbol{\beta}_n$ and using $\boldsymbol{e} = \mathbf{X} \boldsymbol{\beta}_n - \boldsymbol{y}$, this becomes:

$$\begin{aligned} \boldsymbol{\beta}_{n+1} &= \left(\mathbf{X}^T \mathbf{W} \mathbf{X}\right)^{-1} \mathbf{X}^T \mathbf{W} \left(\mathbf{X} \boldsymbol{\beta}_n - \boldsymbol{e}\right) \qquad (2.31) \\ &= \left(\mathbf{X}^T \mathbf{W} \mathbf{X}\right)^{-1} \mathbf{X}^T \mathbf{W} \boldsymbol{y} \end{aligned}$$

which is step 3 in Algorithm 2.1.

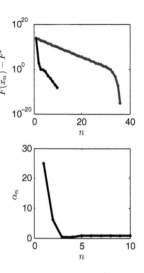

Fig. 2.7: Rate of convergence (top) of Newton's method with and without line search, for the problem of Figure 2.5 with initial parameter estimates $\boldsymbol{x}_0 = \left(6.0, 10.4\right)^T$ very far from the minimum at $\left(\left(1 - \ln 2\right)/2, 0\right)^T$. Constant step size $\alpha_n = 1$ (blue), backtracking line search (black), backtracking parameters $c = 10^{-5}$, $d = 0.5$ and $a = 100$. The constant step size shows quadratic convergence rate with the typical very rapid convergence towards the end, reaching the tolerance $\epsilon = 10^{-8}$ in 37 iterations, but with line search, convergence is achieved in just 11 iterations. This happens because backtracking initially selects very large step sizes for the first few iterations, converging on $\alpha_n = 1$ (bottom).

Other gradient descent methods

In this section we will discuss other gradient descent methods which have shown considerable promise in practical applications, including quasi-Newton and conjugate gradients.

Newton's method is attractive because of the simplicity of the iteration and quadratic convergence, but it suffers from the need to compute and invert the Hessian at each step, an $O\left(N^3\right)$ operation. This may be a critical computational drawback. *Quasi-Newton methods*, as with gradient descent, require evaluations of the function F and the gradient vector ∇F, but not the Hessian, hence they require only $O\left(N^2\right)$ matrix-vector multiplications at each step. They are based upon the idea that the Hessian, or preferably the inverse Hessian, can be approximated and updated from the gradient. One of the most successful quasi-Newton methods, the *BFGS (Broyden–Fletcher–Goldfarb–Shanno) method*, keeps track of an approximation, $\mathbf{H}_n$, to the inverse Hessian. This makes it simple to apply each Newton step, because it avoids a matrix inversion: $\boldsymbol{x}_{n+1} = \boldsymbol{x}_n + \alpha_n \boldsymbol{\Delta}_n$, with the BFGS step $\boldsymbol{\Delta}_n = -\mathbf{H}_n \nabla F\left(\boldsymbol{x}_n\right)$. Convergence to the minimum of F is assured (Powell, 1976) if the step lengths α_n satisfy the *Wolfe conditions*:

$$
\begin{aligned}
F\left(\boldsymbol{x}_{n+1}\right) &\leq F\left(\boldsymbol{x}_n\right) + c\alpha_n \nabla F\left(\boldsymbol{x}_n\right)^T \boldsymbol{\Delta}_n & (2.32)\\
\nabla F\left(\boldsymbol{x}_{n+1}\right)^T \boldsymbol{\Delta}_n &\geq d\,\nabla F\left(\boldsymbol{x}_n\right)^T \boldsymbol{\Delta}_n & (2.33)
\end{aligned}
$$

and $0 < c < 1/2$, and $c < d < 1$. Note that the condition (2.32) is the Armijo bound for the case of BFGS method. Of course, one has to weigh the costs of line search against the cost of inverting the Hessian, at each step.

Conjugate gradient methods attempt to solve one of the major problems with gradient descent methods: that a step can 'undo' progress made in some direction in a previous step. This happens because although each step starts off in the steepest descent direction $\boldsymbol{u} = -\nabla F\left(\boldsymbol{x}\right)$, at the end of the line search is a new direction $\boldsymbol{v}$ which is perpendicular to $\boldsymbol{u}$. Hence, gradient descent can slowly 'zig-zag' to convergence in cases where the objective function has large gradients in some dimensions, but not others. We want to choose the next direction $\boldsymbol{v}$ to avoid this 'undoing' problem. It can be shown that the *conjugate direction* $\boldsymbol{u}^T \nabla^2 F\left(\boldsymbol{x}\right) \boldsymbol{v}$ has this property. Conjugate gradient methods perturb the direction at each step, $\boldsymbol{v} = -\nabla F\left(\boldsymbol{x}\right) + \gamma \boldsymbol{\Delta}$, in such a way that the direction of the next step is always conjugate to the previous direction. The direction updates are simple and fast, of order $O\left(N\right)$. Note that for $\gamma = 0$, the gradient descent algorithm is recovered.

Assuming there is a constant δ such that $\|\nabla F\left(\boldsymbol{x}\right)\| \leq \delta$ for all $\boldsymbol{x}$ inside the bounded level set $\{\boldsymbol{x} : F\left(\boldsymbol{x}\right) \leq F\left(\boldsymbol{x}_0\right)\}$, it can be shown that such conjugate gradient iterations eventually converge to the minimum of F (Nocedal and Wright, 2006, Theorem 5.7).

2.4 Non-differentiable continuous convex problems

The techniques presented above assume that the objective function is everywhere differentiable. This is a significant restriction because many problems in statistical machine learning and nonlinear DSP can be formulated as convex objective functions which are non-differentiable, at one or more points in the domain of the function. This section introduces some of the more useful classes of algorithms for minimizing such non-differentiable objective functions.

Linear programming

While there is a vast universe of possible non-differentiable optimization problems, there is a sufficiently broad class of problems which can be formulated in a standard way, known as *linear programs* (LPs). For example, most convex *piecewise-linear* objectives (Figure 2.8) can be written in LP form. An LP is a constrained optimization problem which can be expressed in the following (standard) form (Boyd and Vandenberghe, 2004, page 147):

$$\text{minimize} \quad \boldsymbol{c}^T \boldsymbol{x} \qquad (2.34)$$
$$\text{subject to} \quad \mathbf{A}\boldsymbol{x} = \boldsymbol{b}, \, \boldsymbol{x} \geq 0$$

where $\boldsymbol{c} \in \mathbb{R}^D$, $\mathbf{A} \in \mathbb{R}^{N \times D}$ and $\boldsymbol{b} \in \mathbb{R}^N$. The non-negativity constraint on $\boldsymbol{x}$ is applied elementwise. As an example, the L_1-norm regression problem discussed earlier can be written in this form:

$$\text{minimize} \quad \mathbf{1}^T (\boldsymbol{u} + \boldsymbol{v}) \qquad (2.35)$$
$$\text{subject to} \quad \mathbf{A}\boldsymbol{x} + \boldsymbol{u} - \boldsymbol{v} = \boldsymbol{b}, \, \boldsymbol{u}, \boldsymbol{v} \geq 0$$

To understand how this works, consider that the equation $\mathbf{A}\boldsymbol{x} = \boldsymbol{b}$ is overdetermined, that is, there are fewer unknown $\boldsymbol{x}$'s than data points. This means that in general, $\mathbf{A}\boldsymbol{x} \approx \boldsymbol{b}$ which means that there is a *residual* $\mathbf{A}\boldsymbol{x} + \boldsymbol{r} = \boldsymbol{b}$. Now, each element of this residual can clearly be either positive or negative. However, we have $\boldsymbol{u}, \boldsymbol{v} \geq 0$, so, if we substitute $\boldsymbol{r} = \boldsymbol{u} - \boldsymbol{v}$, then if any element of the residual is negative, we must have $u_i > v_i$, and $v_i \geq u_i$ otherwise. At the same time, we are minimizing the sum of these non-negative quantities. Therefore, the only way this sum can be minimized, is to have $u_i = 0$, $v_i = -r_i$ if $r_i < 0$ and $v_i = 0$, $u_i = r_i$ if $r_i \geq 0$. This means that the minimized objective function is indeed the sum of the absolute value of the residuals corresponding to the L_1-norm regression estimate. The advantage we get by re-formulating non-differentiable problems this way is that there is a very extensive literature on the topic of minimizing LPs, and a large number of algorithms which can be used to minimize them.

There are some crucial things to know about LPs. Firstly, solutions always lie on the geometrically *isolated* vertices of a *convex polytope* in $\mathbb{R}^D$. As a result, LPs are essentially *combinatorial* in nature, which means that solutions can be found by methods which enumerate some or all of these vertices in turn. One such method is the *simplex method*, discovered in the 1940s. It is technically quite involved and despite often having very good performance in practice, can be shown to have worst-case exponential-time complexity, and so we will not describe it in this book (see Nocedal and Wright, 2006, chapter 13, for a very thorough description). In more recent years, it has been shown that methods which use both primal and dual variables, known as *primal-dual interior-point* methods, have, in theory, polynomial-time complexity. We will discuss these methods in some detail later.

Quadratic programming

In machine learning we will frequently encounter a convex objective function which is quadratic in $\boldsymbol{x}$, the vast majority of which can be formulated as a *quadratic program* (QP):

$$\text{minimize} \quad \frac{1}{2}\boldsymbol{x}^T \mathbf{Q}\boldsymbol{x} + \boldsymbol{c}^T \boldsymbol{x} \tag{2.36}$$
$$\text{subject to} \quad \mathbf{A}\boldsymbol{x} = \boldsymbol{b}, \ \boldsymbol{x} \geq 0$$

where $\boldsymbol{c} \in \mathbb{R}^D$, $\mathbf{A} \in \mathbb{R}^{N \times D}$ and $\boldsymbol{b} \in \mathbb{R}^N$. The symmetric matrix $\mathbf{Q} \in \mathbb{R}^{D \times D}$ must be positive definite for this objective to be (strictly) convex. QPs are obviously a generalization of LPs as can be seen if $\mathbf{Q}$ is the zero matrix.

Important examples of QPs include the mixed L_2-L_1 Lasso regression problem discussed earlier: for the general design matrix $\mathbf{A}$ there is no analytical solution (as there is in the orthogonal case), so we can use methods designed to solve QP problems numerically. One way of reformulating the Lasso problem (2.18) as a QP is:

$$\text{minimize} \quad \frac{1}{2}\boldsymbol{u}^T \boldsymbol{u} + \gamma \mathbf{1}^T \left(\boldsymbol{w}^+ + \boldsymbol{w}^-\right) \tag{2.37}$$
$$\text{subject to} \quad \boldsymbol{u} = \mathbf{A}\left(\boldsymbol{w}^+ - \boldsymbol{w}^-\right) - \boldsymbol{b}, \ \left(\boldsymbol{w}^+, \boldsymbol{w}^-\right) \geq 0$$

where $\boldsymbol{u} \in \mathbb{R}^N$ and, as earlier, we use the non-negative variables $\boldsymbol{w}^+ \in \mathbb{R}^D, \boldsymbol{w}^- \in \mathbb{R}^D$ so that $w_i^+ + w_i^- = |w_i|$ and $w_i^+ - w_i^- = w_i$. The solutions are $\boldsymbol{x} = \boldsymbol{w}^+ - \boldsymbol{w}^-$. Other important QP problems in machine learning include the *support vector machine classifier* which we will discuss later.

Subgradient methods

Earlier (Section 2.1) we introduced the notion of subderivative, which will play an important role in the development of numerical methods for non-differentiable problems. We will next introduce the *subgradient method*, described below.

Algorithm 2.6 The subgradient method.

(1) *Initialization.* Choose some good estimates for x_0, set iteration number $n \leftarrow 0$, set $x^\star = \varnothing$ and $F^\star = +\infty$, choose the maximum number of iterations N,

(2) *Take subgradient step.* Obtain new parameter estimates $\boldsymbol{x}_{n+1} = \boldsymbol{x}_n - \alpha_n \boldsymbol{\delta}_n$, where $\boldsymbol{\delta}_n \in \partial F(\boldsymbol{x}_n)$ is any subgradient at $\boldsymbol{x}_n$,

(3) *Maintain best solution.* If $F(\boldsymbol{x}_n) < F^\star$ then update $F^\star \leftarrow F(\boldsymbol{x}_n)$ and $\boldsymbol{x}^\star = \boldsymbol{x}_n$,

(4) *Iteration.* Set $n \leftarrow n+1$, if $n < N$ go back to step 2, otherwise exit with solution $\boldsymbol{x}^\star$.

The method has, of course, superficial similarity with gradient descent (Algorithm 2.2), but with important differences: the gradient at each step is chosen as one possible subgradient. The iteration is also not a descent update because it is possible that $F(\boldsymbol{x}_{n+1}) > F(\boldsymbol{x}_n)$ for some n. Therefore, we need to keep track of the optimum solution. Another important distinction is the lack of any formal stopping criteria, the iteration simply continues until the maximum number of iterations is reached.

Another key difference is that the step sizes $\alpha_n > 0$ are chosen in advance of running the algorithm, rather than chosen dynamically during the iteration, and these step sizes set the convergence properties of the algorithm. Here is a general statement about convergence of the subgradient method: assuming the norm of all subgradients are less than or equal to some constant G (so that $\|\partial F(\boldsymbol{x})\| \leq G$, for all $\boldsymbol{x} \in \mathbb{R}^D$), and that the iteration starts at a distance of at most R away from the optimal value $\hat{\boldsymbol{x}}$ (i.e. $\|\boldsymbol{x}_0 - \hat{\boldsymbol{x}}\| \leq R$), then Nesterov (2004, Theorem 3.2.2):

$$F^\star - \hat{F} \leq \frac{R^2 + G^2 \sum_{n=1}^{N} \alpha_n^2}{2 \sum_{n=1}^{N} \alpha_n} \tag{2.38}$$

Therefore, if the right hand size goes to zero as $N \to \infty$, the iteration converges. Some choices of step size which guarantee convergence are:

$$\alpha_n = \frac{a}{n} \tag{2.39}$$

$$\alpha_n = \frac{a}{\sqrt{n}} \tag{2.40}$$

$$\alpha_n = \frac{a}{n \|\boldsymbol{\delta}_n\|} \tag{2.41}$$

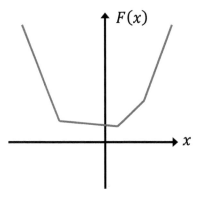

Fig. 2.8: A 1D *piecewise linear* objective function $F(x)$ defined on $\mathbb{R}$. This is constructed from a finite set of linear segments joined together at points (where the gradient does not exist).

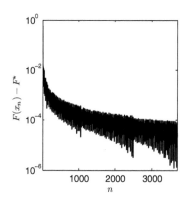

$$F(x_n) - F^*$$

Fig. 2.9: Convergence of the subgradient method for the L_1-norm regression problem with objective function $F(x) = \|\mathbf{A}x - b\|_1$ ($\mathbf{A} \in \mathbb{R}^{100 \times 3}$ and $b \in \mathbb{R}^{100}$, with the elements for A and b drawn at random from the unit normal distribution). The step sizes follow the diminishing step-length sequence $0.1/(n\|\delta_n\|)$.

for some small value of $a > 0$.

Non-differentiable problems include *piecewise linear* (PWL) objectives which are just a set of intersecting lines, and intersecting (hyper)-planes in higher dimensions (Figure 2.8). A ubiquitous PWL objective involves the L_1-norm $F(x) = \|g(x)\|_1$ for an arbitrary function g, for example the linear example $g(x) = \mathbf{A}x - b$, for the arbitrary matrix $\mathbf{A} \in \mathbb{R}^{N \times D}$ and $b \in \mathbb{R}^N$. This is known as the L_1-*norm regression problem*, or *median quantile regression*. Note that for the L_2-norm this is just the classical least-squares regression problem. A subgradient is $\mathbf{A}^T \mathrm{sgn}(\mathbf{A}x - b)$. Figure 2.9 demonstrates the convergence of the subgradient method for this problem. We can see that it is painfully slow, taking nearly 4,000 iterations to reach a tolerance of $\epsilon = 10^{-6}$.

Primal-dual interior-point methods

When a non-differentiable problem can be formulated as a constrained optimization problem (such as a linear or quadratic program), an alternative to exploiting subgradient information is to produce a sequence of iterations on both the primal and dual variables. Algorithms which use this strategy are known as *primal-dual interior-point* methods. Here we will illustrate this for the LP problem. It relies upon the concepts of convex duality and the application of Newton's method to the KKT conditions described earlier.

Given the primal LP (2.34), the dual problem is another LP:

$$\text{minimize} \quad b^T \nu \tag{2.42}$$
$$\text{subject to} \quad \mathbf{A}^T \nu - \lambda = -c, \ \lambda \geq 0$$

with $\nu \in \mathbb{R}^N$ and $\lambda \in \mathbb{R}^D$. Note that we have converted this dual to a convex problem by changing the sign of ν. The primal-dual KKT conditions in the LP case are:

$$
\begin{aligned}
\mathbf{A}\hat{x} - b &= 0 \\
\hat{x} &\geq 0 \\
\hat{\lambda} &\geq 0 \\
\mathbf{A}^T \hat{\nu} - \hat{\lambda} + c &= 0 \\
\hat{\lambda}_j \hat{x}_j &= 0, \ j = 1, 2 \ldots D
\end{aligned}
\tag{2.43}
$$

We can cast these conditions as a problem of finding the roots of the following nonlinear equation, subject to the constraints $x, \lambda \geq 0$:

$$
\begin{bmatrix} \mathbf{A}^T \nu - \lambda + c \\ \mathbf{A}x - b \\ \mathbf{\Lambda X 1} \end{bmatrix} = \begin{bmatrix} 0 \\ 0 \\ 0 \end{bmatrix}
\tag{2.44}
$$

The matrices $\mathbf{X}, \mathbf{\Lambda}$ are diagonal with $X_{ii} = x_i$, $\Lambda_{ii} = \lambda_i$, and $\mathbf{1}$ is a $D \times 1$ vector of ones. Primal-dual methods attempt to find these roots using some kind of iterative method. The origin of the "interior-point" name is that these methods satisfy the constraints strictly, i.e. $\boldsymbol{x}, \boldsymbol{\lambda} > 0$. One approach to solving these equations is to use Newton's method, which leads to the following linear system to solve for the descent direction $(\Delta \boldsymbol{x}, \Delta \boldsymbol{\nu}, \Delta \boldsymbol{\lambda})$ (Boyd and Vandenberghe, 2004, page 244):

$$
\begin{bmatrix} \mathbf{0} & \mathbf{A}^T & \mathbf{I} \\ \mathbf{A} & \mathbf{0} & \mathbf{0} \\ \mathbf{\Lambda} & \mathbf{0} & \mathbf{X} \end{bmatrix} \begin{bmatrix} \Delta \boldsymbol{x} \\ \Delta \boldsymbol{\nu} \\ \Delta \boldsymbol{\lambda} \end{bmatrix} = \begin{bmatrix} \boldsymbol{e}_{\mathrm{d}} \\ \boldsymbol{e}_{\mathrm{p}} \\ \boldsymbol{e}_{\mathrm{c}} \end{bmatrix} \tag{2.45}
$$

The vector $\boldsymbol{e}_{\mathrm{d}} = -\mathbf{A}^T \boldsymbol{\nu} + \boldsymbol{\lambda} - \boldsymbol{c}$ is the *dual residual*, $\boldsymbol{e}_{\mathrm{p}} = -\mathbf{A}\boldsymbol{x} + \boldsymbol{b}$ is the *primal residual* vector, and $\boldsymbol{e}_{\mathrm{c}} = -\mathbf{\Lambda}\mathbf{X}\mathbf{1}$ is the *complementary slackness residual*. The iterates are updated using a step size chosen such that the iterates always satisfy the constraints $\boldsymbol{x}, \boldsymbol{\lambda} > 0$. One practical formula which achieves this is:

$$
\alpha_{\mathrm{p}} = \min \left(1, \eta \min_{i: \Delta x_i < 0} \left[-\frac{x_i}{\Delta x_i} \right] \right) \tag{2.46}
$$

$$
\alpha_{\mathrm{d}} = \min \left(1, \eta \min_{j: \Delta \lambda_j < 0} \left[-\frac{\lambda_j}{\Delta \lambda_j} \right] \right)
$$

The parameter $\eta \in [0.9, 1.0)$ is chosen to ensure that the step sizes always lead to a descent direction which satisfies the constraints strictly.

Note that the iterates are not guaranteed to lie on the feasible set for the primal-dual problem, since the dual and primal residuals are only zero when the optimal solution is found on the boundary of the feasible set. Therefore, one useful measure of convergence of the method is the average disagreement from complementary slackness, $\mu = \left(\boldsymbol{x}^T \boldsymbol{\lambda} \right) / D$, which is known as the *duality measure*. Typically, the full Newton step length violates the constraints, so, a more tractable aim is to seek to lower the duality measure. Most implementations use a modified complementary slackness residual $\boldsymbol{e}_{\mathrm{c}} = -\mathbf{\Lambda}\mathbf{X}\mathbf{1} + \sigma \mu \mathbf{1}$, where $\sigma \in (0, 1)$ is a small constant chosen to set the extent to which the duality measure should be reduced on each iteration. A practical primal-dual algorithm for the LP problem is given in Algorithm 2.7.

Useful insight into the behaviour of this algorithm can be had through an exploration of the concept of the *central path*. This is a set of primal-dual points $(\boldsymbol{x}, \boldsymbol{\nu}, \boldsymbol{\lambda})$ which satisfies the LP KKT conditions strictly, but with a modified complementary slackness condition $\lambda_j x_j = \tau$, $j = 1, 2 \ldots M$ where $\tau > 0$. As the *path parameter* $\tau \to 0$, the KKT conditions approach the LP KKT conditions (2.43) that hold at the optimum point $\left(\hat{\boldsymbol{x}}, \hat{\boldsymbol{\nu}}, \hat{\boldsymbol{\lambda}} \right)$. The central path parametrized by τ acts as a "guide" which leads the iterates towards the optimal LP solution, which always lie on the boundary of the feasible set. The parameter σ is sometimes known as a *centering parameter* because when $\sigma = 1$, the Newton step takes the iterates towards the central path rather than directly towards the solution when $\sigma = 0$.

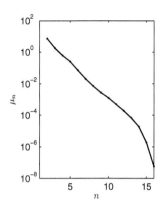

Fig. 2.10: **Illustrating convergence of the primal-dual interior-point LP method applied to the** L_1**-norm regression problem with objective function** $F(x) = \|Ax - b\|_1$ **(A** $\in \mathbb{R}^{201 \times 13}$ **and** $b \in \mathbb{R}^{201}$**, with the elements for A and b drawn at random from the unit normal distribution). The duality measure tolerance** $\epsilon = 10^{-6}$**, reduction factor** $\sigma = 10^{-6}$**, and step length parameter** $\eta = 0.98$**.**

Algorithm 2.7 A primal-dual interior-point method for linear programming (LP).

---

(1) *Initialization.* Set duality measure $\epsilon > 0$, choose good estimates for the primal-dual variables (x_0, ν_0, λ_0) which satisfy the constraints $x_0, \lambda_0 > 0$; choose a duality measure reduction factor $\sigma \in (0, 1)$ and step length parameter $\eta \in [0.9, 1.0)$; set iteration number $n \leftarrow 0$,

(2) *Find Newton direction.* Solve (2.45) to find the descent direction $(\Delta x_n, \Delta \nu_n, \Delta \lambda_n)$, and obtain feasible primal and dual step lengths $\alpha_{\mathrm{p}}, \alpha_{\mathrm{d}}$ using (2.46),

(3) *Take descent step.* Obtain new primal-dual variable estimates $x_{n+1} = x_n + \alpha_{\mathrm{p}} \Delta x_n$, $\nu_{n+1} = \nu_n + \alpha_{\mathrm{d}} \Delta \nu_n$ and $\lambda_{n+1} = \lambda_n + \alpha_{\mathrm{d}} \Delta \lambda_n$, compute new duality measure μ_{n+1}, and if $\mu_{n+1} < \epsilon$, exit with primal-dual solution $(x_{n+1}, \nu_{n+1}, \lambda_{n+1})$,

(4) *Iteration.* Set $n \leftarrow n + 1$ and go back to step 2.

Generally speaking, primal-dual interior-point methods have more complex iterations than the simple subgradient method, largely because they require the solution of a system of linear equations in $2D + N$ variables. However, to reach duality measure accuracy of $\mu \leq \epsilon$, algorithms such as 2.7 (known as *short-step* interior-point methods) require $O\left(-\sqrt{D} \ln(\epsilon)\right)$ iterations in the worst case (Gondzio, 2012), which is usually an order of magnitude less iterations than subgradient methods. Figure 2.10 demonstrates the L_1-norm regression problem solved in 16 iterations for a duality measure below $\epsilon = 10^{-6}$.

In practice, the theoretical worst case number of iterations is improved upon substantially, typical values are 20-50 iterations for $\epsilon = 10^{-6}$. However, because primal-dual methods maintain an estimate of both the primal and dual variables, they have space complexity requirements of the order $O(D + N + M)$ which is always larger than the $O(D)$ subgradient method. These trade-offs need to be explored in each DSP application separately.

Path-following methods

Earlier, we discussed the fact that there very few known optimization problems which can be solved analytically. However, in many cases in which there is a single parameter γ upon which the solution depends (often a *regularization parameter* such as γ in the Lasso problem (2.18)), we can obtain an exact, or nearly exact, solution for some initial value $\gamma_0 < \gamma$ of this parameter, call it $\hat{x}(\gamma_0)$. Then, we can use this to find a curve in $\mathbb{R}^D$, $\hat{x}(\tilde{\gamma})$ parameterized by $\gamma_0 \leq \tilde{\gamma} \leq \gamma$, where $\hat{x}(\gamma)$ is the desired solution. This curve is known as the *regularization path*. We will illustrate the method on an example where exact solutions can be obtained in this way, for the problem of *total variation denoising* (TVD):

$$F\left(\boldsymbol{x}\right) = \frac{1}{2}\sum_{n=1}^{N}\left(x_n - y_n\right)^2 + \gamma\sum_{n=1}^{N-1}\left|x_{n+1} - x_n\right| \qquad (2.47)$$

applied to vectors $\boldsymbol{x}, \boldsymbol{y} \in \mathbb{R}^N$.

This is a mixed L_2-L_1-norm problem involving a regularization term which is the sum of absolute differences of adjacent elements of $\boldsymbol{x}$ (Mallat, 2009, page 728). Thus, the regularization term is non-negative, and the only non-zero contributions to this term occur where $x_{n+1} \neq x_n$, i.e. adjacent values of x_n are different. This means that if $x_n = c$ for some constant c, then the regularization term is zero. It can be shown that the regularization paths of this problem consist of linear segments which are joined together at a set of L gradient change points of the regularization parameter $0 = \gamma_0 < \gamma_1 < \cdots < \gamma_L$ (Tibshirani and Taylor, 2011):

$$\hat{\boldsymbol{x}}\left(\tilde{\gamma}\right) = \hat{\boldsymbol{x}}\left(\gamma_i\right) + \left(\tilde{\gamma} - \gamma_i\right)\boldsymbol{\delta}_i \qquad (2.48)$$

for $\gamma_i \leq \tilde{\gamma} \leq \gamma_{i+1}$, $i = 0, 1 \ldots L - 1$, and $\boldsymbol{\delta}_j \in \mathbb{R}^N$ are the path gradient vectors for each segment. Each γ_{i+1} is found from knowledge of $\hat{\boldsymbol{x}}\left(\gamma_i\right)$ and the gradients $\boldsymbol{\delta}_i$ (Tibshirani and Taylor, 2011). One interesting property of the regularization paths for this TVD problem are that, as γ increases, adjacent values become "fused" together and share the same path for all larger values of γ. Typical paths are shown in Figure 2.11.

Similar exact path formulas have been discovered for many useful methods in machine learning, including Lasso regression and the support vector machine, and in principle, piecewise linear paths exist for a surprisingly wide range of problems with convex objective functions (Hastie *et al.*, 2009, page 89). The main disadvantage of these path following techniques is high computational complexity. For example, the TVD problem above requires $O\left(N^2\right)$ operations to compute the full path (Tibshirani and Taylor, 2011). Of course, if it is known that only the partial path is required, this would save on computational effort.

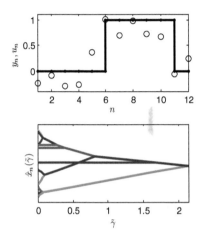

Fig. 2.11: The regularization path of the total variation denoising problem (2.47), applied to input $y = u + e$, where u is a piecewise constant, and each e_n is i.i.d. Gaussian with mean zero and standard deviation $\sigma = 0.25$. The bottom panel shows the piecewise linear regularization paths, that is, values of the optimal solution $\hat{x}\left(\tilde{\gamma}\right)$ to (2.47) for each value of $\tilde{\gamma}$, obtained using (2.48).

2.5 Continuous non-convex problems

For problems which are continuous but non-convex, there is no longer any guarantee that the local descent methods described earlier can find the global minimum, that is, the value $\hat{\boldsymbol{x}}$ such that $F\left(\hat{\boldsymbol{x}}\right) \leq F\left(\boldsymbol{x}\right)$ for all $\boldsymbol{x} \in \mathbb{R}^D$. This is the topic of *global optimization* (Horst *et al.*, 2000). We can certainly envisage situations in which such global, continuous optimization problems can be solved 'easily', for example, if it known that there are M isolated local optima within a bounded region, we can proceed to convergence using a descent method starting from a set of initial guesses $\boldsymbol{x}_0$ within that region. We have an exact criteria for deciding when to stop the search: when we have found exactly M different local minima. The global minimum is one of these local minima.

Unfortunately, unless we have specific information about the location of $\hat{x}$, such continuous non-convex problems are generally NP-complete: we cannot know whether a local optima is the global one, so that convergence properties of local optimization algorithms tell us nothing about the global optimum. The focus of continuous non-convex optimization over a bounded region usually therefore shifts to performing an exhaustive search, using various tricks to increase the efficiency of this search. One important class of methods are *branch-and-bound* algorithms. These operate on the premise that, if the minimum value of a region can be bounded above and below, then it is possible to avoid searching entire sub-regions on the basis that the lower bound in that sub-region is larger than the upper bounds of all other sub-regions (Horst *et al.*, 2000). By splitting (called *branching*) the search region up into smaller and smaller regions, it is possible to efficiently narrow down the sub-region in which the global optimum lies, closing the gap between the upper and lower bounds on $F(\hat{x})$.

If an approximate solution can be tolerated, much as with non-convex combinatorial problems, we can apply *heuristics* such as *random restarting* coupled with searches for local minima, described in the next section.

2.6 Heuristics for discrete (combinatorial) optimization

In this section we explore in depth a selection of efficient methods for approximately solving discrete minimization problems. An important concept frequently used in this section is the (*combinatorial*) *neighbourhood* $\mathcal{N}(x)$ of a point $x \in X$. These are points that are in some way 'close' to x – close in the sense that it is easy to find these points by some trivial change to the structure of x (they will usually be in the feasible set for constrained problems). An important special case is the neighbourhood defined using a distance function d:

$$\mathcal{N}_r(x) = \{y \in X : d(x,y) \leq r\} \qquad (2.49)$$

where $d : X \times X \to \mathbb{R}^+$ is a metric on X. An example of a neighbourhood in the variable selection problem introduced earlier where $X = \{0,1\}^D$, would be all those binary strings which differ in at most r places from x. So, the neighbourhood $\mathcal{N}_1(x)$ contains all those subsets of variables that include one variable not in x, or drop one variable from x. In this case the metric is the *Hamming distance*:

$$d(x,y) = \sum_{i=1}^{D} \mathbf{1}[x_i \neq y_i] \qquad (2.50)$$

which simply counts the number of digits in x differing from y (this is in fact a convex combination of discrete metrics for each dimension).

Greedy search

This is one of the simplest heuristics. Often known as *iterative improvement* or *greedy search*, the method repeatedly tries to find better solutions by replacing the current best guess with a better candidate from the neighbourhood of the current best guess. This algorithm is described below.

Algorithm 2.8 Greedy search for discrete optimization.

(1) *Initialization.* Choose a candidate solution x_0, set iteration number $n \leftarrow 0$,

(2) *Greedy improvement.* Set x_{n+1} as the best improved point from $\mathcal{N}(x_n)$, if no improvement can be found, exit with the solution x_n.

(3) *Iteration.* Set $n \leftarrow n + 1$, go back to step 2.

The best candidate chosen in step 2 is typically either the first point $y \in \mathcal{N}(x_n)$ encountered for which $F(y) < F(x_n)$ (*first improvement*), or it is the best point in $\mathcal{N}(x_n)$, i.e. the point y such that $F(y) < F(z)$ for all $z \in \mathcal{N}(x_n)$, where $z \neq y$ (*best improvement*). If the neighbourhood set is large, first improvement may be more computationally efficient than best improvement, but may require an increase in the number of iterations needed to reach a local minima. An important property of the sequence of candidate solutions is that $F(x_0) > F(x_1) > \cdots > F(x_n)$, i.e. greedy search is guaranteed to decrease the objective function at each iteration. Crucially though, because greedy search terminates when no further local improvement can be found, it can easily get trapped in local minima.

(Simple) tabu search

This heuristic attempts to avoid the local optima trap by giving the search the opportunity to 'escape' from such doldrums. It does this by (a) maintaining a list of recently visited candidate solutions, (b) preventing the search from re-selecting these previous candidates, and (c) allowing the search to go 'uphill' as well as 'downhill'. This procedure, known as *tabu search*, is described in Algorithm 2.9 below.

In step 2, the most improved point can be defined as the point $y \in \mathcal{A}$ for which $F(y) < F(z)$ for all $z \in \mathcal{A}$, where $z \neq y$. Note that this does not necessarily mean that $F(x_{n+1}) < F(x_n)$, because we have explicitly excluded the point x_n from the allowable set $\mathcal{A}$. This is why tabu search can sometimes increase the objective function with an iteration and so escape from a local optima. It is also why we need to maintain a record of the optimal point discovered so far. Also, because it has short-term memory of recently visited candidate points, it can also avoid cycling through a sequence of points smaller than length L. In this way, it can prevent some of the same, sub-optimal solutions from being visited repeatedly.

Algorithm 2.9 (Simple) tabu search.

(1) *Initialization.* Initialize the *tabu list* $\mathcal{T} = X^L$ with all empty entries. Choose a maximum number of iterations, N, and the *tabu list* length L. Choose a candidate solution x_0, set iteration number $n \leftarrow 0$, and set x_0 as the first item in $\mathcal{T}$. Also set $x^\star = \varnothing$ and $F^\star = +\infty$,

(2) *Allowed neighbourhood improvement.* Set $\mathcal{A} = \mathcal{N}(x_n) - \mathcal{T}$ (the neighbourhood of x_n *excluding* points in the tabu list). Set x_{n+1} as the point from $\mathcal{A}$ with the lowest objective function value. If instead $\mathcal{A} = \varnothing$, exit with solution $x^\star$,

(3) *Maintain best solution.* If $F(x_{n+1}) < F^\star$, set $F^\star \leftarrow F(x_{n+1})$ and $x^\star = x_{n+1}$,

(4) *Tabu list update.* Shift every previously visited point in $\mathcal{T}$ back by one place, dropping the last (oldest) point from $\mathcal{T}$, and putting x_n into the first (most recent) position,

(5) *Iteration.* While $n < N$ update $n \leftarrow n + 1$, go back to step 2, otherwise exit with solution $x^\star$.

Of course, it is still possible for the algorithm to get trapped in a cycle longer than L. A variation on the algorithm is to have *infinite memory*: that is, to increase the length of the tabu list with each iteration hence storing all previously visited points. This would, of course, prevent all possible cycling but it may also increase the chance that a local region is exhaustively explored and the algorithm terminates because it cannot escape that region. Thus, tabu search, although simple, requires the choice of two parameters N and L, and there is no optimal choice which works in every case.

Simulated annealing

Annealing is a process of slowly cooling an amorphous crystalline solid such as glass, to strengthen it by removing defects which have structural weaknesses. It is basically greedy search which seeks neighbourhood improvements, but with the adaptation that it will accept increases in the objective function with a certain probability. The probability is analogous to the annealing temperature, and as with annealing, this probability decreases over time. As a greedy search, it can get trapped in local optima, but if the temperature is sufficiently high, there is a chance that it will be able to escape.

Algorithm 2.10 Simulated annealing.

---
(1) *Initialization.* Choose a candidate solution x_0, set iteration number $n = 0$, set $F^\star = \infty$, set a maximum number of iterations, N, and choose a positive *annealing schedule* parameter $k \in \mathbb{R}$,

(2) *Neighbourhood search rejection/acceptance.* Choose any y from $\mathcal{N}(x_n)$. If $F(y) < F(x_n)$, set $x_{n+1} \leftarrow y$, otherwise, choose uniform random number q from $[0,1]$, and if $q > \exp\left(-\left[F(y) - F(x_n)\right]\frac{n}{kN}\right)$, set $x_{n+1} \leftarrow y$,

(3) *Maintain best solution.* If $F(x_{n+1}) < F^\star$, set $F^\star \leftarrow F(x_{n+1})$ and $x^\star = x_{n+1}$,

(4) *Iteration.* While $n < N$ update $n \leftarrow n + 1$, go back to step 2, otherwise exit with solution $x^\star$.

The probability function here is inspired by the *Boltzman distribution* from thermodynamics. The worse the new choice from the neighbour, the smaller the probability of acceptance. The temperature $T = \frac{N}{n}$ decreases as $n \to N$. If k is large, the schedule will be slow, and for k small, the temperature decreases fast. It is clear to see the main limitation of this technique is that there are no obvious ways to set k nor the number of iterations N. If the cooling occurs too quickly, there is high risk of the solution getting trapped in local optima. On the other hand if the cooling occurs too slowly, the algorithm loses computational efficiency. Nonetheless, simulated annealing has demonstrated quite remarkable success across a range of hard discrete problems. We will also see in Chapter 3 that this algorithm can be directly applied to solve problems of sampling from an otherwise intractable multivariate distribution.

Random restarting

However smart an approximation method it will always be the case that after the approximation algorithm terminates, we will not know if we have found the optimal solution or if there is a better one to be found. *Random restarting* (Algorithm 2.11) is a ubiquitous strategy which can be used to generate initial candidates for methods which require them, and to find potentially better approximations. The idea is to set up several random starting points and to try each one in turn, keeping track of the best solution we have found so far.

In this way, we give the approximation algorithm several opportunities to find a better solution. Of course, there is no guarantee that random restarting will actually find a better solution, but at least it will give us increased confidence that $x^\star$ is indeed trustworthy.

Algorithm 2.11 Random restarting.

(1) *Initialization.* Set $F^\star = +\infty$, set $x^\star = \varnothing$, set iteration number $m = 0$, choose maximum number of iterations M,

(2) *Random restart.* Choose random candidate x_0 and run approximation algorithm until termination to get approximation algorithm output x_n,

(3) *Maintain best solution.* If $F(x_n) < F^\star$ then update $F^\star \leftarrow F(x_n)$ and $x^\star = x_n$,

(4) *Iteration.* Set $m \leftarrow m + 1$, if $m > M$ the maximum number of iterations, then terminate with solution $x^\star$, otherwise go back to step 2.

Random sampling

3

This chapter provides an overview of generating samples from random variables with a given (joint) distribution, and using these samples to find quantities of interest from digital signals. This task plays a fundamental role in many problems in statistical machine learning and DSP. For example, effectively simulating the behaviour of the statistical model offers a viable alternative to optimization problems arising from some models for signals with large numbers of variables.

3.1 Generating (uniform) random numbers

Starting from first principles, let us assume we want to program a computer to simulate the throw of a die: that is, to generate Bernoulli random numbers with parameter $p = 0.5$. Is this possible? Actually the answer is "no", because computers are (assuming they are functioning correctly) entirely *deterministic*, that is, given the same inputs, a program will always produce the same outputs. This is incompatible with the fact that each random outcome is generally different, and we cannot predict in advance which outcome will occur. However, computers excel at producing so-called *pseudo-random* numbers: that is, numbers which are superficially random, but contain a 'hidden pattern'. This pattern is typically in the form of repetition which recurs after an extraordinarily long sequence of outcomes.

The art and science of generating pseudo-random numbers usually revolves around picking a generator which has an astronomically long sequence of outcomes. Typical generators use a *nonlinear recurrence relation* which shows behaviour which is almost indistinguishable from *chaotic dynamics*. A chaotic sequence never repeats, but it remains bounded. A chaotic recurrence can be strikingly simple. For example, the recurrence $x_{n+1} = x_n^2$ with $x_0 = \exp(i\pi\theta)$ for $n = 0, 1, 2 \ldots$ where θ is any value whose binary digit sequence does not terminate and does not repeat. This chaotic behaviour occurs because the solution to the recurrence relation is $x_n = \exp(i\pi 2^n \theta)$. This is obviously bounded because it is constrained to the unit circle in the complex plane, but the quantity $2^n \theta$ shifts the bits in θ one place to the left on each iteration, discarding the most significant bit. So, if the binary digit sequence of x_0 does not terminate or repeat, neither will the sequence x_n. Transcendental numbers such as $\theta = \pi$ or $\theta = e$ have this non-terminating property

Machine Learning for Signal Processing: Data Science, Algorithms, and Computational Statistics. Max A. Little.
© Max A. Little 2019. Published in 2019 by Oxford University Press. DOI: 10.1093/oso/9780198714934.001.0001

and therefore generate chaotic sequences in the above recurrence.

Unfortunately, in practice, because hardware floating-point real values lose precision on each iteration, pseudo-random number generators based on such chaotic recurrence relations are not useful. Instead, random number generators often use modular integer arithmetic recurrence relations which do not lose precision. These sequences might be called "pseudo-chaotic" since although they are bounded by the modulus, they do not produce infinite sequences, i.e. they are actually periodic but with (hopefully) extremely long period. Then, the art of good pseudo-random number generator design comes down to choosing a recurrence relation (or other algorithm) which produces sufficiently long sequences of numbers before repeating.

Simple, well-used examples of pseudo-random recurrences are *linear congruential* generators $x_{n+1} = bx_n + c \pmod{m}$ with each $x_n \in \mathbb{N}$, and the values $a, b, m \in \mathbb{N}$ are chosen carefully such that the sequence $x_0, x_1, \ldots$ has length m (Press, 1992, page 276), regardless of the choice of starting value x_0. Each value x_n will lie in the range $[0, m-1]$. Although very simple, linear congruential recurrences can have flaws such as serial correlations between values. It is a delicate matter to create generators which remove flaws such as this; for a selection of well-tested algorithms with good numerical and randomness properties, see Press (1992, Chapter 7.1).

3.2 Sampling from continuous distributions

An extremely common problem in machine learning and DSP is to generate continuous random numbers with a given CDF, PDF, or other description of the measure of the random variable. For a univariate continuous random variable where the PDF is uniform in the range $[u_l, u_h]$, the solution is trivial: simply create a sequence of pseudo-random numbers $x_0, x_1, x_2 \ldots$ using the generator described above, and set $u_n = u_l + (u_h - u_l)\, x_n/m$ where m is the modulus. Outside of this special situation, however, things get considerably more complex. There are few, if any, generally-applicable algorithms. This means that, the type of description we have for the random variable often determines the best (or only practical) approach available.

Quantile function (inverse CDF) and inverse transform sampling

One special case in which a general algorithm does exist is the situation where the distribution of the random variable is described using the inverse of the CDF, $p = F(x)$, the so-called *quantile function*:

$$Q(p) = F^{-1}(p) \tag{3.1}$$

This is defined on the set of probabilities $[0, 1]$. With this function in hand, it is straightforward to generate random numbers from this random variable by creating a sequence of uniform numbers in the range $[0, 1]$, call it $u_0, u_1, u_2 \ldots$ and setting $x_n = Q(u_n)$. This method is known as *inverse transform sampling*. Given its simplicity, it is worth investigating the practical scope of the method. It is usually preferable if Q has a simple closed form, this makes the necessary function evaluations straightforward. What are the properties of the quantile function? If the range of the random variable is $[x_l, x_h]$, then we must have $\lim_{p \to 0} Q(p) = x_l$ and $\lim_{p \to 0} Q(p) = x_h$, and because F is non-decreasing, so is Q. Quite a wide range of PDFs which occur in statistical DSP and machine learning have closed form quantile functions, these include the *exponential, Laplace* (double exponential), *logistic, Weibull* and *Gumbel* distributions. Furthermore, a simple and quite broad 'calculus' of quantile functions, including linear combinations, products, reciprocals, powers, logarithms allow one to construct a wide range of distributions, all of which are amenable to the inverse transform sampling method (Gilchrist, 2000). These are discussed in detail in Section 4.6.

An obvious next question which arises: if we only have the CDF, can we invert the CDF to use this sampling method? The answer is clearly "yes", but with the proviso that one must turn to numerical methods to solve the equation $F(x^\star) - p = 0$ for $x^\star$, the quantile associated with p. As an important example, if we have an analytical form for the CDF, it is usually not difficult to differentiate this analytically and so find the associated PDF f. This information is needed for Newton's method (which is the *root-finding* counterpart to the same algorithm introduced in Section 2.3). For each sample required, we first need to generate an associated probability value p. Now, given a tolerance value $\epsilon > 0$:

Algorithm 3.1 Newton's method for numerical inverse transform sampling.

---

(1) *Initialization.* Choose a good starting point quantile value x_0, set iteration number $n = 0$,

(2) *Update quantile estimate.* Update $x_{n+1} = x_n - (F(x_n) - p)/f(x_n)$,

(3) *Check convergence.* Calculate quantile value error $\Delta F = |F(x_{n+1}) - p|$ and if $\Delta F < \epsilon$, exit with solution x_n,

(4) *Iteration.* Update $n \leftarrow n + 1$, go back to step 2.

A straightforward way of understanding convergence of this method is to treat the Newton quantile update step as a *fixed-point iteration* $x_{n+1} = g(x_n)$ where $g(x) = x - (F(x) - p)/f(x)$. Then, the iteration converges on the associated quantile to p provided:

$$\left| (F(x_n) - p) f'(x_n) [f(x_n)]^{-2} \right| < 1 \qquad (3.2)$$

for all $n = 0, 1, 2 \ldots$. Thus, as long as both CDF and PDF are continuously differentiable, and x_0 is sufficiently close to $x^\star$ (i.e. $x_0 \in [x^\star - \delta, x^\star + \delta]$, where δ must be chosen to ensure that the condition above is satisfied), using this method we can sample from pretty much any (univariate) distribution where both the CDF and PDF are available.

The main limitation is that for all iterations we must have $f(x_n) \neq 0$, and in practice we must have $f(x_n)$ sufficiently large to avoid numerical overflow. This can often be hard to ensure. Similarly, it is not clear how to choose δ in general, often we need to know something about the particular distribution to guarantee a convergent choice (for example, whether it is unimodal). Nonetheless, when Newton's method does converge, it can be shown to have quadratic convergence rate (Nocedal and Wright, 2006, Theorem 11.2), and in practice it typically takes at most 10 iterations to reach a solution to more precision than is available on the computational hardware. Similarly, many of these shortcomings of the basic Newton's method above can be mitigated by relatively sophisticated adaptations (for example, by including an update step size parameter and line search, see Section 2.3).

Of course, we must not forget that this numerical method must be performed for each sample required, which is a cost which must be factored in when attempting to generate large random sample sequences. Devroye (1986, pages 31-35) contains a nice description of other numerical inverse transform methods based upon the *bisection* and *secant* root-finding methods, both of which only require the CDF.

Random variable transformation methods

As alluded to earlier, there are many distributions whose quantile function is obtained by some simple combination of closed-form transformations (Gilchrist, 2000). For these distributions the inverse transform sampling method is highly efficient. Table 4.4 lists a very important set of distributions in *location-scale-shape* form, which can all be written as:

$$\left(\frac{X - \mu}{\sigma} \right)^\kappa = Q(U) \qquad (3.3)$$

where U is a uniform random variable in the range $[0, 1]$. Inverting this gives the very simple random number generator $x = \mu + \sigma Q(u)^{1/\kappa}$.

Rejection sampling

The previous section demonstrated a general approach to sampling from a distribution when we have a closed form for the quantile function or CDF in hand, or can construct one out of simple analytical transformations. Very often, however, we only have the PDF and it is usually not possible to integrate this to find the associated CDF, and even more difficult to invert this to find the quantile function. With only the PDF available, we can instead use *rejection sampling*: first, we generate a *proposal* sample vector $\tilde{x}$ from a *proposal distribution* g. The proposal distribution must satisfy $f(x) \le cg(x)$ for the entire sample space, where the *rejection constant* $c \ge 1$. Next, we generate a sample u from the uniform distribution on $[0, 1]$. Finally, if $u \times c \times g(\tilde{x})/f(\tilde{x}) < 1$, we accept $\tilde{x}$ as a sample from f, otherwise we reject it and repeat from the first step (Devroye, 1986, page 42).

The principle underlying this very simple algorithm is the following theorem: if the random vector (X, U) over the joint sample space $\mathbb{R}^D \times [0, 1]$ is uniformly distributed on the following set:

$$A = \{(x, u) : x \in \mathbb{R}^D, 0 < u < bf(x)\} \tag{3.4}$$

where $b > 0$, then X has the density $f(x)$ (Devroye, 1986, Theorem 3.1). The set A is the *acceptance region*, that is, the region of the joint sample space where the proposal samples are accepted by the algorithm. It is the region under the graph f (up to a scale factor, see Figure 3.1).

Rejection sampling, like inverse transform sampling, is valuable for the broad scope of applicability. The main pitfall is that unless the proposal distribution closely follows the target distribution, there will be a lot of wasted samples (see Figure 3.1 where the uniform distribution as a proposal for the univariate Gaussian over a finite interval leads to 83% of the samples being rejected). Indeed, a simple argument shows that the expected fraction of rejections is $1 - c^{-1}$, so it is important to keep the rejection constant as small as possible (Devroye, 1986, page 42). Thus, the efficiency of the method depends crucially upon the choice of proposal distribution.

Typically, we would want to choose a proposal distribution which has a simple random sample generator, for example, by the use of inverse transform sampling. For example, for a unimodal univariate density, the proposal distribution needs to have larger probability in the extremes, and a 'sharper' peak than the target density. This is true of the Laplace distribution as a proposal for the normal distribution. The Laplace distribution has quantile function $Q(p) = \mu + \sigma\,\mathrm{sgn}\left(\frac{1}{2} - p\right) \ln\left(1 - 2\left|p - \frac{1}{2}\right|\right)$ for the location and scale parameters (μ, σ), and we can show that the optimal rejection constant is $c = \sqrt{2e/\pi}$, giving an expected rejection fraction of around 24% (Devroye, 1986, page 44).

Adaptive rejection sampling (ARS) for log-concave densities

Rejection sampling is popular, because, provided one can find a good proposal density, it is perhaps the most widely applicable method barring inverse transform sampling. However, some samples must always be discarded by the process which raises the question about whether anything can be 'learned' from failure to accept a sample. *Adaptive rejection sampling* (ARS), described in Algorithm 3.2, tries to do just that for *log-concave* densities, that is, densities for which $\ln f(x)$ is a concave

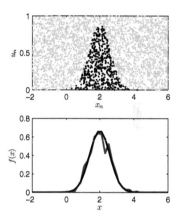

Fig. 3.1: Illustrating rejection sampling from the Gaussian density (truncated to the range $[-2, 6]$). Proposal samples are generated from the uniform density on $[-2, 6]$, and the constant $c = 6$. The top panel shows the proposal samples $(\tilde{x}_n, u_n)$ (grey) and the accepted samples (black). The bottom panel shows the estimated PDF of the accepted samples x_n (blue), and the theoretical PDF for comparison (black). In this set up, $N = 3000$ samples were generated, of which 83% were rejected.

function of x.

Algorithm 3.2 Adaptive rejection sampling (ARS) for log-concave densities.

---

(1) *Initialization.* Choose a set of (sorted) initial tangent points x_i^t for $i = 1, 2 \ldots T$. Calculate the gradients and intercepts of the tangent lines to the log density. Using these, find the points of intersection x^i of each tangent line, augmented with the sample space end points for the target density. Calculate the CDF of the piecewise exponential density at each augmented intersection point. Choose the number of samples N, and set output sample count $n = 0$,

(2) *Sample from proposal distribution.* Generate a proposal sample $\tilde{x}$ from the piecewise exponential density defined by the current tangent line intersection set x^i,

(3) *Acceptance/rejection step.* Generate a uniform random number u in the range $[0, 1]$. If $u \times g^i(\tilde{x})/f(\tilde{x}) < 1$, where g^i is the log density of the current piecewise linear upper bound, accept the proposed sample i.e. update $n \leftarrow n + 1$, and set $x_n = \tilde{x}$. Otherwise, reject the proposed sample and include it in the list of tangent points (e.g. update $T \leftarrow T + 1$ and set $x_T^t = \tilde{x}$, then re-sort the tangent point list). Update the gradients and intercepts of the tangent lines. Using these tangents, update the points of intersection x^i, augmented with the sample space end points for the target density. Finally, update the CDF of the piecewise exponential density at each augmented intersection point,

(4) *Iteration.* When $n = N$, exit, otherwise, go back to step 2.

ARS is very useful in practice because quite a large class of densities occurring in applications are log-concave, including the normal, Laplace and logistic distributions. Still others are log-concave for certain parameter ranges, such as the gamma and Weibull distributions. Additionally, many of these are exponential family densities so that the logarithm of their density functions has a quite simple analytical form. Another useful property is that products of log-concave densities are themselves log-concave: this situation occurs often in applications where a joint density is formed as the product of two or more densities.

The key idea behind ARS is that the proposal distribution g is a *piecewise exponential* density, for which $\ln g(x)$ is a simple piecewise linear (PWL) function. This PWL function is chosen such that each linear segment is tangential to the logarithm of the target density, $\ln f(x)$, at a point x^t. For log-concave densities, this construction ensures that the piecewise exponential is always an upper bound for the target density. Proposal samples $\tilde{x}$ are generated from the piecewise exponential density defined by these tangent lines, which are easy to generate because the CDF and quantile functions for the (truncated) exponential density are available in closed form such that we can use inversion sampling. If the sample is rejected, it becomes a new tangent point x^t, refining the piecewise exponential density to more closely "hug" the target density.

The performance of the algorithm for the gamma density with log-concave choice of parameters, is shown in Figure 3.2. Some details are required to sample from the piecewise exponential proposal distribution. We need the intersection points of the tangent lines augmented by the end points of the sample space for the target density, and the CDF of the piecewise exponential density at each of these intersection points. The CDF information is used to pick a linear segment from which to generate the proposal sample.

To find the intersection points, we need the gradients and intercepts of the tangent lines to the log density, e.g. $a_i = (\ln f)'(x_i^t)$ and $b_i = \ln f(x_i^t)$ for each $i = 1, 2 \ldots T$. The points of intersection are then just $x_i^i = (b_{i+1} - b_i)/(a_i - a_{i+1})$. This list needs to be augmented with the end points of the target density's sample space, (i.e. $x_0^i = 0$ and $x_{T+1}^i = \infty$ for the case of target random variables over the positive half real line).

The CDF of the piecewise exponential density at each augmented intersection point is $G^t(x_i^i) \propto [\exp(a_i x_i^i + b_i) - \exp(a_i x_{i+1}^i + b_i)]/a_i$ for $i = 0, 1, 2 \ldots T+1$, and these are then normalized by dividing through by $Z = \sum_{i=0}^{T+1} G^t(x_i^i)$. The linear segment is chosen by generating a uniform random number u in $[0, 1]$ and selecting the first segment k such that $u > G^t(x_k^i)$. Finally, the quantile function for the density of this segment in isolation is $Q^k(p) = \ln(p \exp(a_k x_{k+1}^i) + (1 - p) \exp(a_k x_k^i))/a_k$. So, given a uniform random number u in $[0, 1]$, the proposal sample from that segment is generated using inverse transform sampling, $\tilde{x} = Q^k(u)$.

An obvious comment here is that rejection of a sample entails the following: (1) updating the tangent list, (2) re-calculating the gradients and intercepts (which involves evaluating both the log of the target density and the gradient of the log of the target), (3) updating the intercept points, and finally (4) updating the CDF of the proposal density at each intersection point. So, rejection could be quite expensive computationally and at least in the initial phase before the proposal density becomes close to the target density, there will be a lot of these rejections and sampling will be slowed by this. Nonetheless, although we can always observe a rejection for any sufficiently large interval of number of required samples $[N_1, N_2]$, the cumulative number of rejections is no larger than N, and empirically at least, it appears that the cumulative number of rejections grows approximately logarithmically with N (see Figure 3.2). Therefore an estimate of the complexity is $O(\log N)$, so that if large sequences of random numbers are required the algorithm might be highly efficient.

A word or two is required about the choice of initial tangent points.

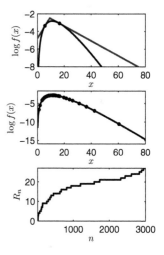

Fig. 3.2: Illustrating adaptive rejection sampling (ARS) for the gamma distribution with parameter values that ensure the density log-concave (shape $k = 3.8$ and scale $\theta = 3.95$). The target log-density function (black, top panel) is enclosed by initial tangent line segments (blue) defined by the choice of two initial tangent points (black dots) which are chosen to surround the mode. After generating $N = 3000$ samples using the ARS algorithm, the initial tangent point list is extended to include all the rejected samples (black dots, middle panel). These can be seen to be concentrated around the highest probability part of the density. The cumulative number of rejections R_n as a function of the number of accepted samples n (bottom panel) increases rapidly at first, but slows as the piecewise linear function defined by the tangent points becomes a good approximation to the target log-density.

Since the density is log-concave, it will have a mode, and if this mode is not at one of the end points of the sample space, one good choice of initial tangent points are two points either side of the mode. If the mode is at one of the end points then the density will be monotonic, hence an initial tangent point can be anywhere in the sample space.

The basic ARS algorithm presented above has been extended in numerous ways. For example, a lower bound can be constructed from *secants* (that is, lines lying underneath the function) and used to perform an additional *squeezing step* which reduces the number of evaluations of the target log-density function, which may be computationally expensive (Görür and Teh, 2011). In a similar way, the inclusion of lower bounds can remove the restriction of log-concavity if the density function is decomposed into a sum of concave and convex parts, although this decomposition requires knowledge of the inflection points of the density which may be difficult to find in general (Görür and Teh, 2011).

Special methods for particular distributions

All the above algorithms make some fairly general and often quite weak assumptions about the form or existence of the PDF or CDF for the random variable. If we have additional information about the distribution of a particular random variable of interest, we can often create special methods which are highly efficient. These are useful because in applications we often deal with very special kinds of distributions. In this section we will address a few of these.

A distribution of fundamental importance is the univariate normal distribution, $\mathcal{N}(\mu, \sigma)$. To address this by inverse transform sampling, we would need the quantile function which is $Q(p) = \mu + \sigma\Phi^{-1}(p)$, where $\Phi^{-1}(p) = \sqrt{2}\mathrm{erf}^{-1}(2p - 1)$. The function erf^{-1} is the inverse of the *error function* $f(x) = \mathrm{erf}(x)$ which is defined as the definite integral $(2/\sqrt{\pi})\int_0^x \exp(-x')\,dx'$. There is no closed form for this integral so it would not seem to be efficient to evaluate it numerically and certainly even harder to invert it in order to use inverse transform sampling. One approach is to directly approximate the inverse error function using Taylor series and this can be used to good effect, although of course the accuracy is entirely dependent upon the accuracy of the Taylor series approximation.

An alternative is to use the *Box-Muller algorithm* which is exact. First, sample two random numbers u, v uniformly on $[0, 1]$. Then, the transformed numbers $x = \sqrt{-2\ln u}\cos(2\pi v)$ and $y = \sqrt{-2\ln u}\sin(2\pi v)$ will both be $\mathcal{N}(0, 1)$-distributed (the normal with $\mu = 0$ and $\sigma = 1$, known as the *standard normal*). If the trigonometric functions are expensive, the *Marsaglia polar method* can be used instead: setting $s = u^2 + v^2$ and $r = \sqrt{(-2\ln s)/s}$, then $x = ur$ and $y = vr$ are also standard normal. To generate univariate random numbers x' with mean μ and standard deviation σ, simply set $x' = \mu + \sigma x$ (this follows because of the form of the quantile function for the univariate normal). Of course, if we only require one random number, we will have wasted one uniform variable but usually we require a stream of normal random variables, so that these two algorithms are probably the method of choice in most situations.

Given a sequence of univariate normal random variables, we can use these to generate multivariate random vectors with arbitrary covariance matrix $\boldsymbol{\Sigma}$ and mean vector $\boldsymbol{\mu}$. The approach exploits the statistical stability property of the Gaussian, that is, that linear transformations of normal random vectors are themselves normal. Consider a random vector $\boldsymbol{X} \in \mathbb{R}^N$ where each random variable X_n for $n = 1, 2 \dots N$ is standard univariate normal, then because each element of the vector is independent of the rest, the covariance matrix of the random vector is just the $N \times N$ identity matrix $\mathbf{I}$. Now, it can be shown that for the arbitrary matrix $\mathbf{A}$, the matrix-vector product $\mathbf{A}\boldsymbol{X}$ has covariance matrix $\mathbf{A}\mathbf{A}^T$. So, if we have $\mathbf{A}\mathbf{A}^T = \boldsymbol{\Sigma}$ for our desired covariance matrix $\boldsymbol{\Sigma}$, then the random vector $\mathbf{A}\boldsymbol{X}$ will have the desired covariance matrix. One possible matrix $\mathbf{A}$ is the *Cholesky decomposition* which is unique and straightforward to compute for positive definite covariance matrices. Finally, the transformed vector $\boldsymbol{\mu} + \mathbf{A}\boldsymbol{X}$ will have mean vector $\boldsymbol{\mu}$.

Devroye (1986, in particular Chapter IX) contains an excellent and detailed description of efficient samplers for important distributions such as the gamma and *Student's t*.

3.3 Sampling from discrete distributions

In this section we will examine how to sample from PMFs or CDFs defined over discrete sample spaces. We will see that there are many similarities with sampling from continuous sample spaces, but the different probabilistic principles at work lead to quite different algorithms. We will address these principles while discussing several of the most popular and generally applicable methods.

Inverse transform sampling by sequential search

Perhaps the most obvious method is basically the same as inverse transform sampling for continuous variables. The quantile function for a discrete random variable X with PMF $f(x)$, can be defined as giving the value of x for a probability $p \in [0, 1]$ such that:

$$\sum_{i<x} f(X = i) < p \leq \sum_{i \leq x} f(X = i) \tag{3.5}$$

Therefore, there are many values of p for each x and the quantile function is "step-like", as is the CDF. In terms of the CDF F, this is just the condition that $F(X = x - 1) < p \leq F(X = x)$. This observation leads to an algorithm which can generate random numbers from the PMF $f(x)$ simply by systematically searching for the quantile:

Algorithm 3.3 Sequential search inversion (simplified version).

---

(1) *Initialization.* Generate a sample u uniformly at random from the range $[0, 1]$, set $x = 0$,

(2) *Check quantile obtained.* If $u > f(x)$, exit with solution x, otherwise update $u \leftarrow u - f(x)$ and $x \leftarrow x + 1$,

(3) *Iteration.* Go back to step 2.

Sequential search inversion has the great advantage that it is generically applicable to *any* PMF (note that this algorithm requires the that sample space for X is the set of non-negative integers $\Omega = \{0, 1, 2 \ldots\}$ which is not usually a problematic restriction because most discrete PMFs are defined that way, and if not, they can often be rewritten so that this is true).

However, we need to appreciate the computational complexity of this algorithm. Clearly, the number of iterations depends upon the value of the quantile x, and is therefore itself a random number, which could in principle be infinite! However, the expected number of iterations is $E[X] + 1$ (Devroye, 1986, page 85). This is clearly going to be a problem for any distribution with infinite mean. This occurs in some rather innocuous-looking cases (consider the simple example $f(x) = 6 \big/ \left(\pi^2 (x + 1)^2 \right)$ given in Devroye, 1986, page 114). Another issue is that if the quantile x is large, we may run into accumulated numerical errors due to computational precision limitations when updating u in step 3.

The computational complexity of sequential search can be reduced if we know some special information about the PMF. Devroye (1986, Section III.2) gives examples, such as the case of unimodal distributions where the mode occurs at a large value of x, so that substantial improvements can be obtained by re-ordering the search (from incrementing x to decrementing it instead).

Rejection sampling for discrete variables

Although sequential search inversion is simple and broadly applicable, we can often do better, particularly in the case of random variables with large average values, by appealing to the same rejection principle as for continuous variables (see Section 3.2). If we have a proposal PMF $g(x)$ from which it is easy to sample, and $f(x) \leq cg(x)$ for the rejection constant $c \geq 1$, then we can accept a proposed sample $\tilde{x}$ distributed according to $g(x)$ whenever $u \times c \times g(\tilde{x}) \leq f(\tilde{x})$, for some u generated uniformly from the range $[0, 1]$. As with continuous random variables, efficiency requires finding a good proposal PMF which makes c as small as possible.

Binary search inversion for (large) finite sample spaces

One of the main distinguishing features separating discrete from continuous variables is that there are situations in which the sample space is not only countable, it is finite with say, $|\Omega| = K$ unique outcomes. In the situation where K is very small, then sequential inverse transform search will probably suffice. But, when K is large, sequential inversion can be very significantly improved by searching among the cumulative probabilities $F(x)$, $x = 1, 2 \ldots K$ using a well-known method called *binary (subdivision) search*, described below.

Algorithm 3.4 Binary (subdivision) search sampling.

---

(1) *Initialization.* Generate a number u uniformly at random from the range $[0, 1]$. From the set of sample outcomes $\Omega = \{1, 2 \ldots K\}$, choose the search set $S = \Omega$, and select x as the maximum median value of S,

(2) *Check quantile obtained or subdivide.* If $F(x - 1) < u \leq F(x)$, exit with sample x. Otherwise: split S into two halves at x, S_l and S_h, with S_h containing x. If $u \leq F(x)$, set $S \leftarrow S_l$, otherwise set $S \leftarrow S_h$, and set x as the maximum median value of S,

(3) *Iteration.* Go back to step 2.

Here, the "maximum median" is the middle value if the set has an odd number of elements, or it is the largest of the two middle values if the number of elements is even.

The algorithm works because the CDF is non-decreasing, and at each step, the search set is halved in size, thus the solution is contained in the set S of size approximately $2^{\log_2 K - n}$, where $n = 0, 1, 2 \ldots$ is the iteration number. The expected number of iterations is logarithmic in the size of the sample space, $O(\ln K)$. This is independent of x, which is very important if K is large, because the number of evaluations of the CDF required increases only slowly with the size of the sample space. This usually makes it much more efficient than the sequential search method.

3.4 Sampling from general multivariate distributions

So far, the discussion has concentrated largely on efficient sampling methods for univariate models. However, in statistical machine learning and DSP we typically have multivariate models and are interested in generating samples from the joint density $\boldsymbol{X} = (X_1, X_2 \ldots X_D)$. In the case where the multivariate model simply consists of a set of i.i.d. variables then, generating a joint sample is simply a matter of calling the same univariate generator D times, one for each X_i, $i = 1, 2 \ldots D$.

However, we will often have a set of random variables which are dependent upon each other in arbitrary ways (as described by an associated *probabilistic graphical model*) and have several different kinds of distributions, or at least, the same distribution family but with different parameters. In this case, sampling from the joint density becomes a lot more complex. Nonetheless, the univariate techniques described above often form the basis of the more complicated schemes described next.

Ancestral sampling

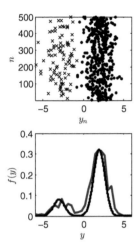

Fig. 3.3: Illustrating ancestral sampling for the two-component Gaussian mixture model with Y variable parameters $\mu_1 = -3$, $\mu_2 = 2$ and $\sigma = 1.0$. The distribution of X is Bernoulli with parameter $p = 0.2$. Samples from the conditional density $f(y|\tilde{x}_n) = N(y; \mu_{\tilde{x}_n}, \sigma)$ where $\tilde{x}_n = 1$ (black 'x's, top panel) and where $\tilde{x}_n = 2$ (black dots, top panel) are plotted against index number n (top panel). The known unconditional density of Y (black, bottom panel) is contrasted with the histogram estimate (blue line, bottom panel) after generating $N = 500$ joint samples. The influence of the unbalanced Bernoulli distribution ($p \neq 0.5$) on the distribution of Y can be seen in the relatively small number of $\tilde{x}_n = 1$ samples, which causes the mode of the density of Y at $\mu = -3$ to be smaller than that at $\mu = 2$.

Suppose we have two univariate random variables X, Y and that Y depends upon X. Suppose also that we have a procedure to generate samples from $f(x)$ and the conditional distribution $f(y|x)$. Then, since $f(x, y) = f(y|x) f(x)$, a simple approach to generating samples from the joint distribution $f(x, y)$ is to first generate a sample $\tilde{x}$ from $f(x)$, then generate a sample from the conditional of Y given X, with $X = \tilde{x}$, i.e. generate from $f(y|\tilde{x})$ to get $(\tilde{x}, \tilde{y})$. This procedure is known as *ancestral sampling* (ancestral, because in the parlance of probabilistic graphical models, X is called a *parent* of Y, so, we sample variables in order of parents first).

Ancestral sampling is a very general technique. There is no restriction upon the sample spaces of the variables, for example, we can have X discrete and Y continuous and vice versa and there is also no restriction on the number of random variables or the length of the chain of dependencies, provided only that the dependencies are not circular, that is, we do not have a variable depending upon itself through a chain of dependencies upon other variables (fortunately, this latter condition is satisfied by design for many useful probabilistic graphical models such as Bayes' nets).

Here is an example of ancestral sampling from a *mixture density* (Figure 3.3). In this simple mixture model, the discrete random variable $X \in \{1, 2\}$ and is Bernoulli distributed with parameter p being the probability that $X = 1$, i.e. $f(X = 1) = p$. The continuous random variable Y has at least one parameter which depends upon the value of X, in this case we will choose this to be Gaussian, so $f(y|x) = \mathcal{N}(y; \mu_x, \sigma)$. Then the unconditional distribution of Y (obtained by marginalizing out X from the joint distribution) is:

$$f(y) = p\mathcal{N}(y; \mu_1, \sigma) + (1 - p)\mathcal{N}(y; \mu_2, \sigma) \tag{3.6}$$

which is a weighted sum (*mixture*) of two distributions. Ancestral sampling involves first drawing a value for X, say $\tilde{x}$ (e.g. using the discrete inversion method), and then drawing a sample $\tilde{y} \sim \mathcal{N}(\mu_{\tilde{x}}, \sigma)$ (using a normal random generator).

Gibbs sampling

In the approaches to sampling described above, independent samples are generated without using the previously generated values. *Markov-Chain Monte Carlo* (MCMC) methods exploit a particularly simple idea that if the stationary distribution of a Markov chain can be engineered to be the same as the joint distribution in which we are interested, then we can simulate from the chain in order to generate the required samples. The chain can therefore be seen as an "update procedure" on current samples.

The *Gibbs sampler* or *heat bath* algorithm is perhaps one of the simplest MCMC samplers (Robert and Casella, 1999, Chapter 7). It requires that the conditional distribution of each variable dependent upon all the others is available, and that we can sample from these conditional distributions. The algorithm is described next.

Algorithm 3.5 Gibb's Markov-Chain Monte Carlo (MCMC) sampling.

---

(1) *Initialization.* Choose a random starting point for each x_i^0, $i = 1, 2 \ldots D$, choose the number of iterations N and set iteration number $n = 0$,

(2) *Sequential sampling from all conditionals.* Compute each conditional distribution $f(x_i | x_1, x_2 \ldots x_{i-1}, x_{i+1} \ldots x_D)$ and draw the following sequence of samples:

$$x_1^{n+1} \sim f(X_1 = x_1 | X_2 = x_2^n, X_3 = x_3^n, \ldots, X_D = x_D^n)$$
$$x_2^{n+1} \sim f(X_2 = x_2 | X_1 = x_1^{n+1}, X_3 = x_3^n, \ldots, X_D = x_D^n)$$
$$\vdots$$
$$x_D^{n+1} \sim f(X_D = x_D | X_1 = x_1^{n+1}, \ldots, X_{D-1} = x_{D-1}^{n+1})$$

(3) *Iteration.* While $n < N$ update $n \leftarrow n + 1$, go back to step 2, otherwise exit.

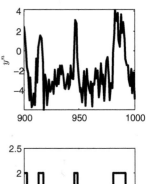

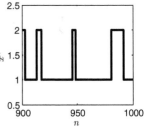

Fig. 3.4: **A subset of one run of 1000 iterations of Markov chain Monte Carlo (MCMC) Gibb's sampling applied to the two-component Gaussian mixture model with Y variable parameters $\mu_1 = -3$, $\mu_2 = 2$, $\sigma = 1.5$, and X variable (Bernoulli) parameter $p = 0.5$. Sequence correlation in the y_n (top panel) and x_n (bottom panel) samples can be seen. This particular run of the chain starting with $x^0 = 1$, produced 45% $X = 1$ and 55% $X = 2$ samples over the 1000 iterations.**

The chain uses a sequential product of transition densities, so that the chain's overall transition PDF is:

$$
\begin{aligned}
f(\boldsymbol{x} | \boldsymbol{x}') = \ & f\left(X_1^{t+1} = x_1 \,\middle|\, X_2^t = x_2', X_3^t = x_3', \ldots, X_D^t = x_D'\right) \times \\
& f\left(X_2^{t+1} = x_2 \,\middle|\, X_1^t = x_1, X_3^t = x_3', \ldots, X_D^t = x_D'\right) \times \\
& \quad \vdots \qquad\qquad\qquad\qquad\qquad\qquad\qquad\qquad (3.7) \\
& f\left(X_D^{t+1} = x_D \,\middle|\, X_1^t = x_1, X_2^t = x_2, \ldots, X_{D-1}^t = x_{D-1}\right)
\end{aligned}
$$

It is a straightforward algebraic exercise to show that this transition density leaves the joint distribution $f(\boldsymbol{X} = \boldsymbol{x})$ invariant (Robert and Casella, 1999, Theorem 7.1.9). As an example, we will apply Gibbs iterations to the simple Gaussian mixture of the previous section, with $p = \frac{1}{2}$. An example run of the chain with $\sigma = 1.5$ is shown in Figure 3.4. The two conditional distributions are:

$$f(y|x) = \mathcal{N}(\mu_x, \sigma) \tag{3.8}$$

$$f(x|y) = \frac{\mathcal{N}(y; \mu_x, \sigma)}{\mathcal{N}(y; \mu_1, \sigma) + \mathcal{N}(y; \mu_2, \sigma)} \tag{3.9}$$

One important point to draw out is that the order in which the variables in the algorithm are updated does not matter, because the overall transition PDF is unaffected by re-ordering of the conditionals (Chib and Greenberg, 1995). We could, for example, choose a randomized update order on each iteration, giving rise to an algorithm known as *random scan Gibbs*. We could just as well choose to update the variables simultaneously in groups or blocks, known as *block Gibbs*. The sequential Algorithm 3.5 is sometimes known as *systematic scan Gibbs*.

Knowing that the chain has the desired joint distribution as a stationary distribution, it remains to check the conditions for *convergence*. By convergence, we mean that $|f_n(\boldsymbol{x}) - f(\boldsymbol{x})| \to 0$ as $n \to \infty$, where $f_n(\boldsymbol{x})$ is the current joint distribution at iteration n, and $|g(\boldsymbol{x})| = \int |g(\boldsymbol{x})| \, d\boldsymbol{x}$ is the (functional) L_1-norm on the space of distribution functions.

It makes intuitive sense that if the *support* of the joint probability space of the multivariate distribution (that is, the region of the state space where the probability density is non-zero) is not *connected* (that is, it is in several separate pieces) and transitions between these pieces is not made possible by the transition distribution, then the Gibbs sampler can get "stuck" in one of these pieces if started there.

Beyond this, however, it is a somewhat subtle matter to devise both sufficient and necessary conditions for convergence that are universally applicable. However, Roberts and Smith (1994) prove the following useful convergence criteria. In the case of discrete joint PMFs $f(\boldsymbol{x})$, if the Gibbs transition matrix $f(\boldsymbol{x}|\boldsymbol{x}')$ is irreducible and $f(\boldsymbol{x})$ is non-zero for all states, then the Gibbs sampler will converge (Roberts and Smith, 1994, Lemma 1). For continuous PDFs, if the support of the desired target PDF $f(\boldsymbol{x})$ is connected, then it suffices to check two regularity conditions which merely ensure that the required conditionals for the Gibbs iteration exist and that the target density has probability mass in the domain of interest Roberts and Smith (1994, Theorem 2). The weak nature of these criteria give insight into the very broad applicability of the Gibbs iteration for practical sampling problems in machine learning and DSP, and accordingly, the method is extremely popular.

The major limitations of Gibbs sampling arise due to the difficulty in determining the *rate* at which the chain converges on the target joint distribution. For example, it is possible for a Gibbs sampler applied to a particular target distribution to be provably convergent, and yet only explore a tiny subset of the joint sample space in any "reasonable" number of iterations, because although the convergence conditions are satisfied, the joint sample space is effectively disconnected by regions of very low probability.

This "pathological" situation can occur even in quite innocuous looking models: in the Gaussian mixture model example of (3.9) shown in Figure 3.4, simply setting $\sigma = 0.5$ causes runs of 1000 iterations to become stuck in the state where $x^n = x^0$. In this model, what is happening is that as σ becomes small, the conditional $f(x|y)$ becomes close to either 1 or 0, depending upon whether y is closer in value to μ_1 or μ_2. So, we will have $x^{n+1} \to x^n$ and y^n will be only ever be generated from the initial starting component in the mixture, and the iteration becomes effectively stuck.

The issue of convergence rates is therefore of critical importance to practical Gibbs sampling and is addressed in the next chapter.

Metropolis-Hastings

Gibbs sampling above is attractive when the conditional distributions are available. In this section we will look at the very general and much more broadly applicable *Metropolis-Hastings* (MH) algorithm of which the Gibbs sampler, and other related methods, are special cases. The basic idea underlying MH is to use a (multivariate) *proposal transition distribution* from which it is easy to sample, and construct a corresponding stationary distribution which is the desired target distribution. MH can therefore be understood to be a synthesis of rejection sampling (described earlier) and MCMC.

Let us suppose that we have a proposal transition distribution $g\left(\boldsymbol{X}^{t+1} = \boldsymbol{x} \mid \boldsymbol{X}^t = \boldsymbol{x}'\right)$ from which it is easy to sample. We will use detailed balance to construct a chain which is guaranteed to have the target distribution $f(\boldsymbol{X} = \boldsymbol{x})$ as stationary distribution. Detailed balance for the proposal distribution is $g(\boldsymbol{x}'|\boldsymbol{x}) h(\boldsymbol{x}) = g(\boldsymbol{x}|\boldsymbol{x}') h(\boldsymbol{x}')$ but generally h and f are not the same. However, we can assume without loss of generality that:

$$g(\boldsymbol{x}'|\boldsymbol{x}) f(\boldsymbol{x}) > g(\boldsymbol{x}|\boldsymbol{x}') f(\boldsymbol{x}') \tag{3.10}$$

and ensure that the above is satisfied as an equality by setting:

$$g(\boldsymbol{x}'|\boldsymbol{x}) f(\boldsymbol{x}) \alpha(\boldsymbol{x}', \boldsymbol{x}) = g(\boldsymbol{x}|\boldsymbol{x}') f(\boldsymbol{x}') \tag{3.11}$$

where the *acceptance probability* is:

$$\alpha(\boldsymbol{x}', \boldsymbol{x}) = \min\left[1, \frac{g(\boldsymbol{x}|\boldsymbol{x}') f(\boldsymbol{x}')}{g(\boldsymbol{x}'|\boldsymbol{x}) f(\boldsymbol{x})}\right] \tag{3.12}$$

This leads to the MH algorithm described below.

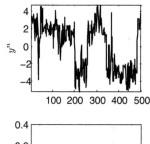

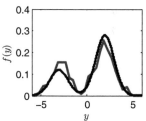

Fig. 3.5: A run of 500 iterations of Markov chain Monte Carlo (MCMC) Metropolis sampling applied to the two-component Gaussian mixture model with Y variable parameters $\mu_1 = -3$, $\mu_2 = 2$, $\sigma = 1.0$, and X variable (Bernoulli) parameter $p = 0.3$. The proposal distribution is a single Gaussian $g(x|x') = \mathcal{N}(x; x', \tau)$ with standard deviation $\tau = 1.5$. Sequence correlation and repeated samples due to rejection can be seen (top panel). The histogram estimated distribution of Y (bottom panel) is shown in blue, the analytical target density in black. Starting with $y^0 = 0$, this run delivered an acceptance rate of **67%**.

Algorithm 3.6 Markov-Chain Monte Carlo (MCMC) Metropolis-Hastings (MH) sampling.

---

(1) *Initialization.* Choose a random starting point x^0, number of iterations N, and set iteration number $n = 0$,

(2) *Generate proposal sample.* Draw $\tilde{x} \sim g(x|x^n)$,

(3) *Acceptance/rejection.*   With   probability $\alpha(\tilde{x}, x^n)$ $=$ $\min[1, (g(x^n|\tilde{x}) f(\tilde{x}))/(g(\tilde{x}|x^n) f(x^n))]$ choose $x^{n+1} \leftarrow \tilde{x}$ (accept proposal), otherwise set $x^{n+1} \leftarrow x^n$ (reject proposal),

(4) *Iteration.* While $n < N$ update $n \leftarrow n+1$, go back to step 2, otherwise exit.

Several simplifications arise in practice. If the proposal transition distribution is symmetric so that $g(x|x') = g(x'|x)$, then the acceptance probability (3.12) becomes:

$$\alpha(x', x) = \min\left[1, \frac{f(x')}{f(x)}\right] \tag{3.13}$$

This is known as *random-walk MH* or simply the *Metropolis* algorithm, and we demonstrate this simplification applied to the Gaussian mixture model (Figure 3.5). The intuition here is straightforward: if the proposed sample has higher target probability density than the previous sample, it is accepted with probability 1. If has lower probability, it is accepted with probability that is equal to the ratio of the target density at the proposed sample to the density at the previous sample.

Another simplification occurs when the proposal transition distribution is independent of the previous iteration, i.e. when $g(x|x') = g(x)$, which leads to the so-called *independence sampler*:

$$\alpha(x', x) = \min\left[1, \frac{g(x) f(x')}{g(x') f(x)}\right] \tag{3.14}$$

One interesting and important consequence of the construction of the MH chain is that since α is a ratio of probability densities, we can often omit normalization factors which can lead to considerable computational simplifications. For example, assume that $f(x) = Z^{-1}\phi(x)$ where Z is a normalization constant $Z = \int \phi(x) \, dx$, then clearly this can be avoided in the evaluation of α because $f(x')/f(x) = \phi(x')/\phi(x)$. The same simplification applies to the proposal distribution, of course.

It can also be shown (Robert and Casella, 1999, Theorem 7.1.17) that the Gibbs sampler discussed in the previous section is a special case of MH, where the proposal distribution is broken down into a composition of separate Gibbs conditional probabilities, one for each variable $i = 1, 2 \ldots D$. This also demonstrates that the acceptance probability $\alpha = 1$, i.e we always accept the Gibbs proposal sample for x^{n+1}.

In Chapter 2 we introduced the method of simulated annealing for discrete optimization. Here we show how this algorithm is also a special case of the Metropolis algorithm with a non-stationary transition distribution which changes according the iteration number n:

$$\alpha\left(\boldsymbol{x}', \boldsymbol{x}\right) = \min\left[1, \left(\frac{f\left(\boldsymbol{x}'\right)}{f\left(\boldsymbol{x}\right)}\right)^{\frac{n}{kN}}\right] \tag{3.15}$$

where N is the number of iterations of the algorithm and the real value $k > 0$ is the annealing schedule. From this, we can see that whenever $f\left(\boldsymbol{x}'\right) > f\left(\boldsymbol{x}\right)$, we accept the proposal sample. Otherwise, we accept proposals with probability equal to the density ratio $f\left(\boldsymbol{x}'\right) f\left(\boldsymbol{x}\right)^{-1} < 1$ which decreases with the iteration count, reaching a minimum at $n = N$. So, the probability of accepting the new proposal decreases with each iteration, at a rate determined by k.

To make the connection with the minimization of an objective function $F\left(x\right)$ for x chosen from a discrete set X, define a PDF $f\left(x\right) \propto \exp\left(-F\left(x\right)\right)$. We get:

$$\left(\frac{f\left(x'\right)}{f\left(x\right)}\right)^{\frac{n}{kN}} = \exp\left(-\left[F\left(x'\right) - F\left(x\right)\right]\frac{n}{kN}\right) \tag{3.16}$$

which is the same as the acceptance probability in the discrete optimization case. The unifying principle for both discrete simulated annealing and Metropolis annealing is that they are based around proposed deviations from the neighbourhood of the current sample, whose acceptance is a function of the improvement in the objective function (optimization) or increase in probability density (Metropolis sampling). In Metropolis sampling, the deviations are explicitly random and determined by the proposal transition distribution. All of the same considerations about the trade-off between the speed of cooling and the computational efficiency in terms of the number of rejections required to reach convergence apply, with the exception that here, we are interested in convergence of the distribution of $\boldsymbol{x}^n$ to the target density f rather than finding the global minima of the objective function F.

It is necessary to check the conditions under which the chain is guaranteed to converge on the target distribution. Robert and Casella (1999, Corollary 6.2.6) show that sufficient conditions for convergence under the L_1-norm to the target density are that it is possible for rejections to occur (which establishes the aperiodicity of the MH chain), and the proposal distribution is positive on the support of f (implying that the MH chain is irreducible). While these conditions are only sufficient they are quite weak and usually satisfied in practice.

As with rejection sampling, for computational efficiency reasons it is clearly of importance to pick a proposal transition distribution which leads to low rejection rates. However, as this is an MCMC method, there is the additional consideration that all regions of the joint sample space are effectively connected, so that the chain does not get "stuck".

There is, as a result, a tension between maximizing acceptance probability and minimizing serial dependence between samples produced by the algorithm.

To give a flavour of the issues involved, which are inevitably complex, we will look at the Metropolis case with symmetric (multivariate) Gaussian proposal distribution g. The effective "spread" of g is determined by the eigenvalues of the covariance matrix. If these are large, then the algorithm can take large steps away from the current x^n. This is good because it reduces the serial dependence between proposed samples by encouraging exploration of the state space, but it is bad because the probability of acceptance is also reduced. On the other hand, if the covariance eigenvalues are small, then the probability of acceptance is increased, but x^{n+1} will be close to x^n, which again increases the dependence between samples. So, the practicality of Metropolis sampling depends crucially upon a "well-tuned" choice of proposal distribution to find the optimal balance between acceptance probability and serial dependence, and unfortunately this choice depends quite crucially on the specifics of the target density (Robert and Casella, 1999, Section 6.4).

Other MCMC methods

The success of all MCMC methods depends upon obtaining iterations which can efficiently converge on the target distribution. This should ideally be achieved while minimizing the number of rejections, the sequence dependence between samples, and the computational effort. In this section we will provide a brief exposition of some more sophisticated MCMC methods than the classical Metropolis-Hastings algorithm and derivatives such as Gibbs sampling.

One of the major limitations of Metropolis sampling is that, if the search for the target distribution is to have low rejection probability, each proposal step must be very small, making exploration of the full state space very inefficient. The *Hybrid* or *Hamiltonian MCMC* (HMC) method, applicable to continuous joint target PDFs, improves the efficiency of the Metropolis algorithm by allowing large steps with a low rejection rate. This is achieved by building in artificial "momentum variables" $p(t)$ into the exploration of the state space. It is based upon exploitation of the concept of *Hamiltonian mechanics* from elementary physics. Hamiltonian mechanics describes the decomposition of the total energy of a closed system into sums of component energy terms, in this case *potential* and *kinetic* energies $H(x, p) = K(p) + V(x)$, and the resulting pairs of differential *equations of motion* which describe the time evolution $x(t), p(t)$ of the dynamics of this system, the *Hamiltonian dynamics*.

HMC methods construct a *canonical joint density* proportional to $\exp(-H(x, p))$ and by associating the target density $f(x)$ with the potential energy term, we have $f(x) f(p) \propto \exp(-K(p)) \exp(-V(x))$. This means that the joint density $f(x, p) = f(x) f(p)$ factorizes so that the required samples from $f(x)$ can be obtained simply by discarding the artificial momentum samples $\boldsymbol{p}$.

The Hamiltonian equations of motion have three critical properties (by construction): (a) they are reversible in time, (b) the quantity H is an *invariant* of these dynamics, in other words, it remains constant as $\boldsymbol{x}(t), \boldsymbol{p}(t)$ change over time, and (c) the dynamics preserves volume elements in the joint state space $(\boldsymbol{x}(t), \boldsymbol{p}(t))$.

HMC methods exploit these properties in the following way. Time reversibility of the Hamiltonian dynamics (indirectly) ensures that the canonical joint density $f(\boldsymbol{x}, \boldsymbol{p})$ is the invariant density of the sampler. By generating proposal samples using a numerical integration method for the Hamiltonian dynamics which is itself reversible (such as the *Störmer-Verlet* or *leapfrog* finite difference method, see Hairer *et al.*, 2006), detailed balance of the chain is satisfied. Exact invariance of H in principle ensures that proposal samples are drawn exactly from the canonical joint density, but some (small) errors are introduced by the numerical integrator so that a Metropolis acceptance step is required. However, because of volume preservation, the acceptance probability takes on a particularly simple form:

$$\alpha(\boldsymbol{x}', \boldsymbol{p}', \boldsymbol{x}, \boldsymbol{p}) = \min\left[1, \exp\left(-H(\boldsymbol{x}', \boldsymbol{p}') + H(\boldsymbol{x}, \boldsymbol{p})\right)\right] \qquad (3.17)$$

The key to the efficiency of MHC methods lies in the fact that the proposal samples $(\boldsymbol{x}', \boldsymbol{p}') = (\tilde{\boldsymbol{x}}, \tilde{\boldsymbol{p}})$ generated by the numerical method can be quite distant in state space from the current samples $\boldsymbol{x}^n, \boldsymbol{p}^n$, but because H is almost exactly preserved over that distance, we have $\alpha \approx 1$ for most proposals. Therefore, both the rejection rates and sample sequence autocorrelation can be much lower than with the simple Metropolis or Gibbs samplers described above.

We have to choose a distribution for the auxiliary momentum variable $\boldsymbol{p}$. For computational convenience, the multivariate Gaussian with independent dimension variables is often used, which radically simplifies the numerical integration scheme.

The success of this algorithm depends quite crucially upon "good" choices of the properties of the numerical integration algorithm used to make each proposal trajectory. Ideally we would want these proposal trajectories to be as long as possible to ensure efficient exploration of the joint sample space of the target density, but we also want to to minimize the error in the preservation of the Hamiltonian (and to guarantee stability of the numerical integration scheme), so that the acceptance probability remains high. However, a warning is in order that if the trajectory is too long, it can start to retrace itself thus *reducing* the distance between proposal samples, defeating the purpose. Unfortunately, choosing the numerical integration parameters turns out to be a difficult problem which generally requires experimentation for particular target densities f (although see Hoffman and Gelman, 2014 for one ingenious approach to addressing these choices).

The HMC method is an example of an *auxiliary variable method* since the momentum variable is introduced to solve the problem but does not actually form part of the solution. Another method which introduces an additional variable which is not part of the solution is *slice sampling*. This technique has much in common with rejection sampling, but it is an MCMC method which therefore produces dependent sample sequences. As with rejection sampling, it is based upon the concept that draws from the density function $f(\boldsymbol{x})$ can be obtained by drawing uniformly at random from the volume under the curve $(\boldsymbol{x}, u)$ where $0 \leq u \leq p(\boldsymbol{x})$. The auxiliary u samples can then be dropped. Note that it suffices to draw uniformly from the volume $0 \leq u \leq f'(\boldsymbol{x})$ where $f'(\boldsymbol{x})$ is unnormalized. Effectively, we want to sample from the following joint distribution over $\boldsymbol{x}$ and the auxiliary variable u:

$$f(\boldsymbol{x}, u) = \begin{cases} 1 & 0 \leq u \leq f(\boldsymbol{x}) \\ 0 & \text{otherwise} \end{cases} \tag{3.18}$$

so that by discarding values of u we get $\int_0^\infty f(\boldsymbol{x}, u)\, du = \int_0^{f(\boldsymbol{x})} 1\, du + \int_{f(\boldsymbol{x})}^\infty 0\, du$ which gives $f(\boldsymbol{x})$ as required.

Slice sampling uses the (block) Gibbs sampler to alternately draw from $f(u|\boldsymbol{x})$ and $f(\boldsymbol{x}|u)$. To do this, we need the first conditional distribution which is obtained from (3.18) as $f(u|\boldsymbol{x}) = f(\boldsymbol{x})^{-1}$ for $0 \leq u \leq f(\boldsymbol{x})$ and 0 otherwise. This means that the associated conditional CDF $F(u|\boldsymbol{x}) = \int_0^u f(u'|\boldsymbol{x})\, du'$ is 1 when $u \geq f(\boldsymbol{x})$ and $u f(\boldsymbol{x})^{-1}$ otherwise. So, the quantile function is $F^{-1}(p) = p f(\boldsymbol{x})$. Sampling u^{n+1} is now easily achieved using inverse transform sampling by generating p uniformly in $[0, 1]$ and multiplying this by $f(\boldsymbol{x}^n)$.

Next we have to sample from $f(\boldsymbol{x}|u)$. Because the joint PDF (3.18) is constant, the conditional CDF will be a non-decreasing piecewise linear function and hence readily invertible, so the quantile function exists and can in principle be used to sample $\boldsymbol{x}^{n+1}$. We can follow the same logic as above and this shows that we must sample uniformly from the region $S = \{\boldsymbol{x} : u < f(\boldsymbol{x})\}$ which is a "slice" through the joint PDF with $u = u^{n+1}$ fixed (hence the name of the method). So, to generate $\boldsymbol{x}^{n+1}$ we need to sample uniformly from this region S.

The concept is simple but surprisingly tricky to implement in practice. In particular, since we need to know the boundary values of $\boldsymbol{x}$ for which $f(\boldsymbol{x}) = u^{n+1}$, this in turn requires knowing the inverse PDF for $\boldsymbol{x}$. This boundary could have an arbitrarily complex shape. In the univariate case, it might be practical to use numerical inversion methods, but S may not be in one piece: there will be several disconnected intervals for multimodal densities, for example. This means that there are several branches to the inverse PDF and we would need to search numerically for each boundary point. Also, we must bear in mind that we are wasting one random sample per draw. Furthermore, as this is an MCMC algorithm the samples are not i.i.d. So, it is not clear that this method will outperform adaptive rejection sampling, for example.

Several general slice sampling schemes have been proposed. One of the most popular schemes is to first find an enclosing (hyper)-rectangle

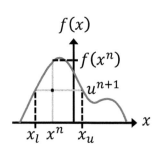

Fig. 3.6: An illustration of one step of auxiliary variable MCMC slice sampling applied to a univariate, multimodal density function $f(x)$. **Given an initial sample** x^n, **a new auxiliary sample** u^{n+1} **is drawn uniformly at random from the interval** $[0, f(x^n)]$ **(vertical grey line). This defines another interval at height** u^{n+1} **bounded by** x_l, x_u **which are the boundaries of the "slice"** $S = \{x : u^{n+1} < f(x^n)\}$. **The next sample for** x^{n+1} **is generated uniformly at random from this interval** $[x_l, x_u]$.

for the region, and then contract this using bounding random proposal samples until a sample within the region S is found (Neal, 2003). This approach can be augmented with various tricks depending upon the availability of additional information about the density function (such as whether it is unimodal or multimodal etc.)

Statistical modelling and inference

4

The modern view of statistical machine learning and signal processing is that the central task is one of finding good probabilistic models for the joint distribution over all the variables in the problem. We can then make 'queries' of this model, also known as *inferences*, to determine optimal parameter values or signals. Hence, the importance of statistical methods to this book cannot be overstated. This chapter is an in-depth exploration of what this probabilistic modeling entails, the origins of the concepts involved, how to perform inferences and how to test the quality of a model produced this way.

4.1 Statistical models

In this book, we define a *statistical model* as simply the joint distribution function defined over the vector $\boldsymbol{X} = (X_1, X_2 \ldots X_N)^T$ of random variables in the problem, e.g. $f(\boldsymbol{x})$, for $\boldsymbol{x} \in \Omega_{\boldsymbol{X}}$. The joint probability space $\Omega_{\boldsymbol{X}} = \Omega_{X_1} \times \Omega_{X_2} \times \cdots \times \Omega_{X_N}$ can be (and often is) a combination of discrete and continuous sets.

Parametric versus nonparametric models

These models are said to fall into one of two broad classes: *parametric* and *nonparametric*. Although the distinction is not entirely sharp, parametric models are so-called because they involve a vector of *parameters* $\boldsymbol{P} = (P_1, P_2 \ldots P_K)$ with K elements, which may themselves be random variables over the joint space $\Omega_{\boldsymbol{P}}$, or may just take on fixed values. The form and shape of the joint distribution function will often be heavily influenced by these parameters, so it is usually important to write $f(\boldsymbol{x}; \boldsymbol{p})$ to emphasize this influence. In the case where the parameters are random variables, we usually care centrally about the conditional distribution $f(\boldsymbol{x}|\boldsymbol{p})$. The goal of parametric statistical modeling then nearly always revolves around making statistical inferences about the optimal values of this vector of parameters, given a set of data for $\boldsymbol{X}$.

Parametric models often make quite strong assumptions about the form of the distribution function $f(\boldsymbol{x}; \boldsymbol{p})$ and the relationship between variables in the problem, for example, the multivariate Gaussian distribution is a classic example with the parameters $(\boldsymbol{\mu}, \boldsymbol{\Sigma})$. As such, they usually excel in situations where the data is sparse or of poor quality. Of

Machine Learning for Signal Processing: Data Science, Algorithms, and Computational Statistics. Max A. Little.
© Max A. Little 2019. Published in 2019 by Oxford University Press. DOI: 10.1093/oso/9780198714934.001.0001

course, the disadvantage to such an approach is that, if the actual distribution of the data varies substantially from the parametric model, the result of fitting such a model to the data can be meaningless or useless in practice. Parametric models often make heavy reliance on mathematical convenience, and this can sometimes be at odds with reality as captured in the data.

For these situations, a nonparametric model is often far better suited. The form of nonparametric models are often constrained by some generic mathematical principles, such as 'smoothness'. A nonparametric model might have only one parameter whereas a parametric model might have hundreds. An example of a nonparametric model is the *kernel density estimate*, which might have only one *bandwidth* parameter σ which controls the smoothness of the fitted distribution function. The inevitable caution about the use of nonparametric models in situations where a small amount of data is available is that, since the scope of these models can be very broad, it is easy to get an unrealistic estimate of the shape of the distribution function. Often, the nonparametric distribution obtained is merely a reflection of the particular data at hand, which represents the idiosyncrasies of a particular realization of the random variables in the problem.

In both cases, there are tools available to assess the suitability of the fit of the model, whether parametric or nonparametric, to the given data.

Bayesian and non-Bayesian models

In the description of parametric models above, we alluded to the idea that the parameter vector $\boldsymbol{P}$ is very often treated as a random vector, suggesting that we care about the conditional distribution of $\boldsymbol{X}$ given $\boldsymbol{P}$. This also means that, somewhere in the background, is the joint distribution of the variables and the parameters $f(\boldsymbol{x}, \boldsymbol{p})$, from which both sets of marginals $f(\boldsymbol{p}), f(\boldsymbol{x})$ and the conditional $f(\boldsymbol{p}|\boldsymbol{x})$ can be obtained by integration and/or summation (or combinations of the two). In fact, the so-called *Bayesian* modeling approach does not actually distinguish between 'parameters' and 'variables' in the problem because both are treated on an equal probabilistic footing, and all the marginals and conditionals can be obtained from the joint distribution over $\boldsymbol{X}, \boldsymbol{P}$ anyway.

There are multiple advantages to this Bayesian approach. For example, it is a mathematically consistent treatment of the way in which dependence between variables or parameters in the model can be manipulated, mostly using Bayes' rule, and it therefore unifies many apparently disparate inferential procedures under one mathematical framework. The main disadvantage is that Bayesian models, which demand distribution functions for all the variables and parameters in the problem, can quickly become unwieldy and over-complex, mostly because the number of degrees of freedom in the model quickly escalates. Contrasted with non-Bayesian models, therefore, the Bayesian approach is usually more 'principled' and transparent, but it is often outperformed, in terms of computational complexity and practical utility, by non-Bayesian methods.

The student of statistics should be aware that there is often vehement disagreement among dedicated Bayesian versus non-Bayesian practitioners, and there are also disagreements among Bayesian practitioners about the epistemic meaning of probabilities (for example, there are at least two competing views on this point, such as whether they encode personal 'degrees of belief' of the practitioner, or follow as a formalized extension of logical reasoning to reasoning about uncertainty). The issues are subtle and thorny, but in the final analysis, it has to be said that these disagreements are nearly all about justifying standard Bayesian probability calculus on some more general principles. As such, they do not affect the validity of Bayes' rule nor the general calculus of probabilities introduced in Chapter 1. Furthermore, under the quite basic mathematical assumption of exchangeability, the Bayesian approach is justified on purely mathematical grounds; this is discussed in more detail in Section 10.1.

So, we take the 'working' view in this book that we desire the best mathematical model for the problem at hand, using the standard calculus of probabilities. In this context, we want a useful probabilistic model which has computationally tractable inference procedures. Furthermore, it is important to appreciate that there are many very useful modeling approaches and inferential procedures which might properly be understood as 'hybrid' Bayesian/non-Bayesian methods. And indeed, nearly all Bayesian models will have *hyperparameters* that are not explicitly modeled as random variables, in other words, we will nearly always have a model in the form $f(\boldsymbol{x}|\boldsymbol{p};\boldsymbol{q})$ where the set of variables $\boldsymbol{P}$ are random, but the set of hyperparameters $\boldsymbol{Q}$ are fixed.

4.2   Optimal probability inferences

Assume we have a good distributional model $f(\boldsymbol{x};\boldsymbol{p})$ (a non-Bayesian model), $f(\boldsymbol{x}|\boldsymbol{p};\boldsymbol{q})$ (a conditional model) or perhaps $f(\boldsymbol{x},\boldsymbol{p};\boldsymbol{q})$ (a fully Bayesian model), for our data. We will typically want to extract useful information from this model. For example, we may be interested in the values of the parameters $\boldsymbol{p}$. The distribution will typically depend in a continuous fashion on these parameters, so, we will somehow want to summarize the effect of these parameters on the distribution. The simplest summary is the joint value of all the parameters which *maximizes* f: such inferences are *optimal probability* calculations, the focus of this section, which play a central role in many forms of statistical inference.

Maximum likelihood and minimum K-L divergence

In this book, we will define the *likelihood* of a model as the distribution of the data for given values of parameters of the model. Simply because there are many such distributions, there are many kinds of likelihood.

For example, this might refer to the unconditional distribution $f\left(\boldsymbol{x};\boldsymbol{p}\right)$ or the conditional distribution $f\left(\boldsymbol{x}|\boldsymbol{p}\right)$ in a Bayesian context. The *maximum likelihood* (ML) procedure maximizes the value of this distribution with respect to the parameters, e.g.:

$$\hat{\boldsymbol{p}} = \arg\max_{\boldsymbol{p}} f\left(\boldsymbol{x};\boldsymbol{p}\right) \tag{4.1}$$

This is a problem for which the optimization techniques of Chapter 2 are useful. As an example, consider the (very common) problem of maximizing the joint probability over a set of N i.i.d. data points x_n, with respect to the parameter p. The likelihood is $f\left(\boldsymbol{x};p\right) = \prod_{n=1}^{N} f\left(x_n;p\right)$ and the maximum likelihood problem for estimating the parameter p is:

$$\hat{p} = \arg\max_{p} \prod_{n=1}^{N} f\left(x_n;p\right) \tag{4.2}$$

We can often simplify this considerably by minimizing the negative logarithm instead:

$$
\begin{aligned}
\hat{p} &= \arg\min_{p} \left(-\ln \prod_{n=1}^{N} f\left(x_n;p\right)\right) \\
&= \arg\min_{p} \left(-\sum_{n=1}^{N} \ln f\left(x_n;p\right)\right)
\end{aligned}
\tag{4.3}
$$

The quantity being minimized here is known as the *negative log-likelihood* (NLL), and the quantity $\hat{p}$ is known as the *maximum likelihood estimator* (MLE) for p.

Importantly, we can ground this MLE procedure on the basis of information-theoretic reasoning. We will show that the MLE is the estimator which causes the (parametric) model f to be as close as possible to the EPDF under the K-L divergence. From the data, we can form the EPDF estimator for the distribution:

$$\hat{f}\left(x\right) = \frac{1}{N} \sum_{n=1}^{N} \delta\left[x - x_n\right] \tag{4.4}$$

We will use the notation $\hat{X}$ for the random variable distributed according to the EPDF $\hat{f}$. The K-L divergence between the EPDF and the model $f\left(x;p\right)$ is:

$$D\left[\hat{X},X\right] = E_{\hat{X}}\left[-\ln f\left(\hat{X};p\right)\right] - E_{\hat{X}}\left[-\ln f\left(\hat{X}\right)\right] = H_{\hat{X}}\left[X\right] - H\left[\hat{X}\right] \tag{4.5}$$

Now, suppose we wish to estimate the parameter by minimizing this K-L divergence:

$$\hat{p} = \arg\min_{p} D\left[\hat{X}, X\right] = \arg\min_{p} H_{\hat{X}}\left[X\right] \qquad (4.6)$$

This holds because only the cross-entropy term $H_{\hat{X}}\left[X\right]$ depends upon the parameter p. Expanding out the above, we get:

$$
\begin{aligned}
E_{\hat{X}}\left[-\ln f\left(\hat{X}; p\right)\right] &= -\int \frac{1}{N} \sum_{n=1}^{N} \delta\left[x - x_n\right] \ln f\left(x; p\right) dx \\
&= -\frac{1}{N} \sum_{n=1}^{N} \int \delta\left[x - x_n\right] \ln f\left(x; p\right) dx \\
&= -\frac{1}{N} \sum_{n=1}^{N} \ln f\left(x_n; p\right) \qquad (4.7)
\end{aligned}
$$

which is the NLL. Therefore:

$$\hat{p} = \arg\min_{p} D\left[\hat{X}, X\right] = \arg\min_{p} \left(-\frac{1}{N} \sum_{n=1}^{N} \ln f\left(x_n; p\right)\right) \qquad (4.8)$$

which is the MLE for p. Hence, the minimum K-L divergence estimator with respect to the EPDF, is the same as the MLE.

Aside from parameter inferences, one of the most important applications of probabilistic modelling is making *predictions*, that is, using the model to infer quantitative aspects of random variables not already seen in the given data. Given the data points $x_1, \ldots, x_N$ we will have obtained optimal parameter values $\hat{p}$, which then allow us to make optimal predictions $\hat{x}$ about a new data point $\tilde{x}$:

$$\hat{x} = \arg\max_{\tilde{x}} f\left(\tilde{x}, x_1, \ldots, x_N; \hat{p}\right)$$

For two or more models which are not *nested*, that is, where none of the models is a special case of any other which can be obtained by e.g. fixing some of the parameter values, it is reasonable to compute the (maximum) *likelihood ratio*:

$$l = \frac{f_1\left(x_1, \ldots, x_N; \hat{p}_1\right)}{f_2\left(x_1, \ldots, x_N; \hat{p}_2\right)} \qquad (4.9)$$

where f_1, f_2 are the likelihoods of the two models. It follows that if $l > 1$ we would prefer model f_1 over f_2, and if $l < 1$ we would make the other choice. Note that this simple procedure *does not* work for nested models because in that case, the model with more parameters can always have a larger likelihood. More sophisticated methods which take into account this expected improvement in model fit for nested models, such as the *likelihood ratio test*, need to be used instead. Alternatively, we can use *regularization* methods, which have the advantage that they can be applied to nested or non-nested models.

Loss functions and empirical risk estimation

Given a statistical model, it is often the case that we can associate it with some kind of measure of discrepancy which is a function of the data and the parameters, known as a *loss function*, $L(x, p)$. Simple examples include the *square loss* $L(x, p) = \frac{1}{2}(x - p)^2$ and the *absolute loss* $L(x, p) = |x - p|$.

Given a loss function, the *expected loss*, or *risk* for a given random variable is defined as:

$$E_X[L(X, p)] = \int L(x, p) f(x) \, dx \qquad (4.10)$$

Empirical risk is the name given to an estimator for this quantity which arises when the distribution is estimated from the data using the ECDF, where the associated EPDF is:

$$f(x) \approx \frac{1}{N} \sum_{n=1}^{N} \delta[x - x_n] \qquad (4.11)$$

giving:

$$
\begin{aligned}
E_X[L(X, p)] &\approx \int L(x, p) \frac{1}{N} \sum_{n=1}^{N} \delta[x - x_n] \, dx \\
&= \frac{1}{N} \sum_{n=1}^{N} \int L(x, p) \delta[x - x_n] \, dx \qquad (4.12) \\
&= \frac{1}{N} \sum_{n=1}^{N} L(x_n, p)
\end{aligned}
$$

In other words, the empirical risk is an equally weighted sum of losses: the sample mean loss associated with the data. The value of p which minimizes this empirical risk is known as the *empirical risk estimator* for p.

Next, we will assume that we can construct the distribution for the data using the loss function so that:

$$f(x; p) = Z^{-1} \exp(-L(x, p)) \qquad (4.13)$$

with the *normalizer* $Z = \int \exp(-L(x, p)) \, dx$. Then, assuming that the data is i.i.d., the joint likelihood of the entire set of data is:

$$f(\boldsymbol{x}; p) = \prod_{n=1}^{N} f(x_n; p) \qquad (4.14)$$

Inserting the distribution of the data, the joint likelihood becomes:

$$f(\boldsymbol{x}; p) = Z^{-N} \prod_{n=1}^{N} \exp(-L(x_n, p)) \qquad (4.15)$$

so that the NLL is:

$$
\begin{aligned}
-\ln f\left(\boldsymbol{x};p\right) &= N\ln Z + \sum_{n=1}^{N} L\left(x_n,p\right)\\
&\approx N\ln Z + N\,E_X\left[L\left(X,p\right)\right] \quad\quad (4.16)
\end{aligned}
$$

Finally, if we assume that Z is not a function of p, then the minimum of the NLL is attained at approximately the same value of p as that value for the expected loss. The conclusion we can draw (in this special case where the distribution can be defined in terms of the negative exponent of a loss function) is that MLE and empirical risk estimators for p, effectively coincide. This turns out to be the case for quite a wide range of distributions and parameters, as we will see later in this chapter.

Maximum a-posteriori and regularization

In maximum likelihood estimation we have a non-Bayesian model $f\left(\boldsymbol{x};\boldsymbol{p}\right)$ (or a conditional model $f\left(\boldsymbol{x}|\boldsymbol{p}\right)$) where the distribution over the parameters $\boldsymbol{p}$ is not explicitly known. The MLE maximizes this distribution with respect to $\boldsymbol{p}$. By contrast, there are situations where both the likelihood $f\left(\boldsymbol{x}|\boldsymbol{p}\right)$ and the prior $f\left(\boldsymbol{p}\right)$ are known instead. In this case, using Bayes' rule, we know (at least in principle) the posterior $f\left(\boldsymbol{p}|\boldsymbol{x}\right)$. *Maximum a-posteriori* estimation (MAP) estimates $\boldsymbol{p}$ by maximizing this posterior distribution with respect to $\boldsymbol{p}$. In practice, MAP estimation is quite similar to MLE: it involves optimization. But, it has conceptual (and practical) implications which reach far beyond MLE, which will be discussed in detail here and in the next section.

The MAP estimate of the parameter $\boldsymbol{p}$ is:

$$
\begin{aligned}
\hat{\boldsymbol{p}} &= \arg\max_{\boldsymbol{p}} f\left(\boldsymbol{p}|\boldsymbol{x}\right)\\
&= \arg\max_{\boldsymbol{p}} f\left(\boldsymbol{x}|\boldsymbol{p}\right) f\left(\boldsymbol{p}\right) \quad\quad (4.17)
\end{aligned}
$$

The second line follows from Bayes' rule, $f\left(\boldsymbol{p}|\boldsymbol{x}\right) = f\left(\boldsymbol{x}|\boldsymbol{p}\right) f\left(\boldsymbol{p}\right)/f\left(\boldsymbol{x}\right)$, and the fact that the denominator is independent of the parameter. So, the MAP estimate is the value of the parameter which maximizes the product of the (conditional) likelihood $f\left(\boldsymbol{x}|\boldsymbol{p}\right)$ and the prior. By contrast, the MLE is the parameter value which maximizes the conditional likelihood alone.

As with the MLE situation, a common example occurs when the joint likelihood of a set of i.i.d. data is $f\left(\boldsymbol{x};p\right) = \prod_{n=1}^{N} f\left(x_n|p\right)$, the MAP estimate for the parameter p is:

$$\begin{aligned}
\hat{p} &= \underset{p}{\arg\max} \left(f(p) \prod_{n=1}^{N} f(x_n|p) \right) \\
&= \underset{p}{\arg\min} \left(-\ln \left[f(p) \prod_{n=1}^{N} f(x_n|p) \right] \right) \\
&= \underset{p}{\arg\min} \left(-\ln f(p) - \sum_{n=1}^{N} \ln f(x_n|p) \right)
\end{aligned} \tag{4.18}$$

It is instructive to compare this to (4.3): the only difference is that the quantity to be minimized includes the negative log prior term $-\ln f(p)$, which is known as a *regularizer* or *penalty*. Here, the prior term reduces the sensitivity of the solution upon the likelihood of data x_n.

We have already encountered examples of this in section 2.2. For illustration, consider the case of univariate Gaussian distributions for both prior and likelihood, with the single data item x. We want the MAP estimate for the mean parameter of X, call it $\hat{\mu}$. We will assume the prior mean is zero with variance σ_0^2, and (without loss of generality), unit likelihood variance. These distributions can be written as $f(\mu) = Z_0^{-1} \exp\left(-\frac{1}{2\sigma_0^2}\mu^2\right)$ and $f(x|\mu) = Z^{-1} \exp\left(-\frac{1}{2}(x-\mu)^2\right)$. The MAP estimator is:

$$\hat{\mu} = \underset{\mu}{\arg\min} \left(\frac{1}{2}(x-\mu)^2 + \frac{1}{2\sigma_0^2}\mu^2 \right) \tag{4.19}$$

where we are ignoring the normalizers Z, Z_0 which do not depend upon μ. This is a convex, least-squares optimization problem, so the solution is unique and is found analytically to be:

$$\hat{\mu} = \frac{\sigma_0^2}{1+\sigma_0^2} x \tag{4.20}$$

This is exactly the 'ridge' shrinkage solution obtained in (2.20) in section 2.2. Understanding the behaviour of the ratio $\sigma_0^2 \left(1+\sigma_0^2\right)^{-1}$ is critical, in the two extreme cases when $\sigma_0^2 \to \infty$ and when $\sigma_0^2 \to 0$. In the first case, the ratio tends to 1, so that $\hat{\mu} \to x$. This coincides exactly with the MLE. In the other case, the ratio tends to zero and so $\hat{\mu} \to 0$ also. In the first case, $\hat{\mu}$ depends entirely upon the data, x, whereas, in the second case, it has no dependence upon the data at all. Otherwise, $\hat{\mu}$ lies between 0 and x, smoothly interpolated by σ_0^2.

Although this is a rather artificial example, it illustrates that a single data point will always be a poor estimate of the mean of X, and if our confidence that our prior 'guess' for the mean (that is, zero) is good — represented by σ_0^2 being small — then the MAP estimate would be a considerable improvement over this highly variable MLE.

There are a vast number of other circumstances where regularization, whether it arises from a likelihood and prior or not, is powerful and useful. For example, consider the least-squares problem $\hat{p} =$

$\arg\min_{\boldsymbol{p}} \left(\frac{1}{2} \|\mathbf{X}\boldsymbol{p} - \boldsymbol{y}\|_2^2 \right)$. It often happens that a small change in the elements of $\mathbf{X}$ and $\boldsymbol{y}$ leads to a large change in the solution (known as a *ill-conditioned* problem in the numerical analysis literature). This can make the solution unreliable. One way to improve the reliability is to control the size of the variation in the solution by including the L_2-norm of the vector $\boldsymbol{p}$ as part of the problem. In which case, we arrive at Tikhonov regularization as introduced in section 2.2:

$$\hat{\boldsymbol{p}} = \arg\min_{\boldsymbol{p}} \left(\frac{1}{2} \|\mathbf{X}\boldsymbol{p} - \boldsymbol{y}\|_2^2 + \frac{\gamma}{2} \|\boldsymbol{p}\|_2^2 \right) \tag{4.21}$$

When γ is large, this penalized least-squares solution tends to $\mathbf{0}$ and the influence of changes in $\mathbf{X}$ and $\boldsymbol{y}$ is reduced. Another important example of regularization arising in DSP are *smoothness priors* for noise removal from a digital signal: if we assume that the underlying signal we wish to recover changes slowly in time, then penalizing the square magnitude of the first or higher-order (discrete) derivative of the solution leads to a tractable regularization problem which can be very effective in practice.

To summarize: the essential idea is that MAP estimators are examples of regularized solutions, where a prior on the parameters is used to control the sensitivity of the MLE to variability in the data. In the next section we will see that the concept has many other interpretations stemming from the central role that it plays in *all* the quantitative sciences.

Regularization, model complexity and data compression

Perhaps one of the oldest principles in science is known as *Occam's razor* . This states (informally) that a (mathematical) model for a phenomena should be as simple as possible, but no simpler. Another informal way of saying this is that among all models that fit the data about some phenomena equally well, we would prefer the one that embodies the smallest number of assumptions. Ideally, we want a model that summarizes the data about the phenomena in an error-free way, so, it is not just another way of 'storing' the data. A model which merely stores the data is not really a useful model at all, because the data is expected to be disturbed by noise, for example. So, each set of data is a realization of a random variable, and each set of data will be different. That means that a model which reproduces the unique noise in one set of data, will not actually represent the general structure underlying all possible sets of data.

This seems like common sense, but it turns out to be surprisingly hard to provide a concise mathematical formulation of the concept in complete generality, because there are a large variety of mathematical models and sets of assumptions one could use to build them. It will suffice for the purposes of this book, however, to explore the concept probabilistically. For illustrative purposes, we will invoke a conditional model with K parameters $\boldsymbol{p}$ and assume that we know the like-

Named after William of Ockham, an English philosopher from the Medieval period.

lihood and the prior. Using Bayes' rule, the posterior distribution is $f(\boldsymbol{p}|\boldsymbol{x},\boldsymbol{y}) = f(\boldsymbol{x},\boldsymbol{y}|\boldsymbol{p})\,f(\boldsymbol{p})/f(\boldsymbol{x},\boldsymbol{y})$. We will assume that the model depends smoothly upon these parameters $\boldsymbol{p}$, and we have fixed data vectors $\boldsymbol{x},\boldsymbol{y}$. We will discuss several different, but essentially equivalent, ways of expressing this trade-off between fit to the data $\boldsymbol{y}$, and number, or strength, of assumptions.

The first instance of Occam's razor hinges on the number of *degrees of freedom*, or free variables, in a model. This will be the same as the K parameters in the model. If we are allowed to increase K without bound, for many models, using e.g. MLE, make the likelihood $f(\boldsymbol{x},\boldsymbol{y}|\boldsymbol{p})$ as large as we desire for any data. This may seem to be a good thing, but eventually a model with sufficient degrees of freedom can *exactly* reproduce any data – it is said to be *overfitted*. Similarly, if K is too small, then it will have a low likelihood and represent all possible data sets equally badly; it is *underfitted*. We cannot therefore rely on the MLE to pick the optimal model with just the right K.

A simple example is illustrated in Figure 4.1 for the polynomial *regression* (curve-fitting) model $y = \sum_{k=0}^{K} p_k x^k + \epsilon$ with $\epsilon \sim \mathcal{N}(0,\sigma)$, and the $K+1$ vector of parameters $\boldsymbol{p}$. (Twice) the NLL for N data points x_n, y_n is:

$$2E = N \ln\left(2\pi\sigma^2\right) + \frac{1}{\sigma^2}\sum_{n=1}^{N}\left(y_n - \sum_{k=0}^{K}\hat{p}_k x_n^k\right)^2 \qquad (4.22)$$

which is in the form of a normalizer term depending only upon N, σ and a model fit error term which is the sum-of-squares. We will assume that we have the MLE parameters $\hat{\boldsymbol{p}}$. The MLE fit error term goes to zero when $K+1=N$. This situation is shown in the case where the true model has $K=1$, where overfitting and underfitting for the wrong choice of K can be seen (Figure 4.1).

Given that we prefer models with smaller degrees of freedom, one approach would be to penalize the NLL for large values of K. In a now classic result, the Japanese statistician Hirotoku Akaike, using information-theoretic arguments, showed that the *expectation* of the sample NLL over all realizations of N draws from the distribution over the data, is biased downwards by K (Akaike, 1974). One can correct for this bias by adding it back to the NLL, leading to the *Akaike information criteria* (AIC) as a direct substitute for the fit error:

$$AIC = 2E + 2K \qquad (4.23)$$

By minimizing this quantity with respect to K, one can therefore obtain an unbiased estimate of the true degrees of freedom. This is shown in Figure 4.1 where we can see that, unlike $2E$, the AIC is minimized at the correct value of $K=1$, selecting the linear model.

While the AIC attempts to find the K which provides the smallest, expected NLL, this is not a Bayesian procedure and so does not take into account the uncertainty in the parameters. By contrast, the *Bayesian information criteria* (BIC) is a MAP estimator (Schwarz, 1978):

$$BIC = 2E + K \ln N \tag{4.24}$$

Like AIC this criteria favours a small numbers of parameters, but this penalty also increases with the sample size N. This expression can be derived under very general assumptions on the likelihood, assuming that the prior over K is flat (*uniform, non-informative*) or almost flat (Cavanaugh and Neath, 1999). Although AIC and BIC look superficially similar, they address quite different problems. Whereas AIC selects the model with the optimal likelihood, BIC will select the correct K (provided that the data was actually generated by our chosen likelihood for a particular value of K). But, both AIC and BIC only hold asymptotically as $N \to \infty$, so this is really only meaningful for very large sample sizes. Nonetheless, both of these criteria have found extensive, practical use in DSP and machine learning applications.

There are many, commonly occurring situations where the model does not have a 'discrete' number of parameters, each of which is a separate degree of freedom. Regularization or penalization is another instance of Occam's razor, which, as discussed in the previous section, can often be derived as the MAP solution to some combination of likelihood and prior. However, in some cases we can find a continuous quantity which interpolates K, known as the *effective degrees of freedom* $\tilde{K}$. Examples of this include Tikhonov regularization (4.21):

$$\tilde{K} = \sum_{k=1}^{K} \frac{s_k^2}{s_k^2 + \gamma} \tag{4.25}$$

where s_d are the (non-negative) *singular values* of the *hat matrix* $\mathbf{H} = \mathbf{X}\left(\mathbf{X}^T\mathbf{X} + \gamma\mathbf{I}\right)^{-1}\mathbf{X}^T$ (used when making predictions, i.e. $\hat{\mathbf{y}} = \mathbf{H}\mathbf{y}$, see Hastie *et al.*, 2009, pages 66-68). When $\gamma = 0$ (i.e. the non-regularized solution), then $\tilde{K} = K$, the number of model parameters, and when $\gamma \to \infty$, $\tilde{K} \to 0$. Hence, in this case, the effective degrees of freedom captures the sensitivity of the solution to the data, which decreases as the regularization is increased.

Occam's razor also has a much more general interpretation in terms of data compression as introduced in Section 1.5. Let us revisit the polynomial regression model (4.22) above. We want to represent, without loss, the sequence $S = (y_n)_{n=1}^{N}$ given the parameters $\mathbf{p}$ and the data $\mathbf{x}$. Firstly, to do this we will need to extend the coding point of view from discrete to continuous random variables. One straightforward approach to this is to discretize them to a precision δ, then the compressed representation of Y is $-\log_2 f(y) - \log_2 \delta$. The last term is a constant, independent of y. This discretization can be, for practical purposes, at the precision of e.g. floating point values represented in particular computing hardware. If we pose the following conditional mean Gaussian probabilistic model:

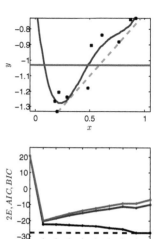

Fig. 4.1: **Exploring polynomial model fit while varying the number of degrees of freedom K. Top panel shows data points (black dots, $N = 10$) generated from a linear model, $y = -1.5 + 0.8\,x + \epsilon$ (true line, dashed cyan), x values drawn uniformly at random from the range $[0, 1]$ and zero-mean Gaussian errors ϵ with standard deviation $\sigma = 0.1$. Overfitted polynomial $K = 4$ (blue) and underfitted $K = 0$ constant (red). The bottom panel shows the negative log likelihood (NLL) $2E$ (black line), the AIC (blue line) and BIC (red line), along with the $K = 10$ "zero-error" fit (dashed black line). The NLL decreases with increasing K until it reaches the zero-error fit, while both AIC and BIC reach their minima at $K = 1$, the true number of degrees of freedom of the linear model.**

$$f(\boldsymbol{x}, \boldsymbol{y}|\boldsymbol{p}) = \prod_{n=1}^{N} \mathcal{N}\left(y_n - \sum_{k=0}^{K} p_k x_n^k, \sigma\right) \qquad (4.26)$$

then, this would have an expected compressed representation of length:

$$-\log_2 f(\boldsymbol{x}, \boldsymbol{y}|\boldsymbol{p}) = \frac{1}{\ln 2}\left[\frac{1}{2\sigma^2}\sum_{n=1}^{N}\left(y_n - \sum_{k=0}^{K} p_k x_n^k\right)^2\right] + C(N, \sigma, \delta)$$
$$(4.27)$$

This provides yet another interpretation of the NLL: up to scale ($1/\ln 2$, which only has to do with the choice of logarithmic base) and additive constant ($C(N, \sigma, \delta)$, depending upon the data length, standard deviation parameters, and discretization only) factors, it is the expected compressed length of the data $\boldsymbol{y}$, conditional on the parameters and $\boldsymbol{x}$. We can, therefore, think about the MLE as minimizing the compressed length of $\boldsymbol{y}$ by varying $\boldsymbol{p}$.

What if we cannot assume that we know the values of the parameter vector $\boldsymbol{p}$? These are, of course, essential for reconstructing $\boldsymbol{y}$ from $\boldsymbol{x}$. Assuming that the (typical) estimation error in the MLE for these parameters is $1/\sqrt{N}$, then we can encode them with a precision $\delta = 1/\sqrt{N}$. For K parameters, their code length will then be $-\log_2 f(p) + \frac{1}{2}\log_2 N$. Further, let us assume for simplicity that the prior is uniform over the maximum range of values of each parameter p. Then we arrive at the following code length for the posterior:

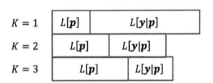

$K = 1$	$L[\boldsymbol{p}]$	$L[\boldsymbol{y}|\boldsymbol{p}]$
$K = 2$	$L[\boldsymbol{p}]$	$L[\boldsymbol{y}|\boldsymbol{p}]$
$K = 3$	$L[\boldsymbol{p}]$	$L[\boldsymbol{y}|\boldsymbol{p}]$

Fig. 4.2: The *minimum description length* (MDL) principle. K corresponds to the model complexity which in this case is the number of elements in the parameter vector p. The compressed code length for the parameter vector is $L[p]$, which increases with K as there are an increasing number of parameters to encode. The compressed code length for the fixed data y given the parameters is $L[y|p]$, which decreases with increasing K, as there are more degrees of freedom in the model to fit the data better. The total code length is the sum of these two, and is minimized for $K = 2$.

$$
\begin{aligned}
-\log_2 f(\boldsymbol{p}|\boldsymbol{x}, \boldsymbol{y}) =\ & \frac{1}{\ln 2}\left[\frac{1}{2\sigma^2}\sum_{n=1}^{N}\left(y_n - \sum_{k=0}^{K} p_k x_n^k\right)^2 + \frac{K}{2}\log_2 N\right] \\
& + \log_2 f(\boldsymbol{x}, \boldsymbol{y}) + C(N, K, \sigma, \delta) \qquad (4.28)
\end{aligned}
$$

The corresponding estimator $\hat{\boldsymbol{p}}$ is regularized by the priors, and it is known as the *minimum description length* (MDL) estimator (for this model). Interestingly, this estimator coincides with BIC (4.24) above (see Hansen and Yu, 2001 for an elaboration of this point). The general MDL principle is (Hansen and Yu, 2001):

$$MDL = -\log_2 f(\boldsymbol{y}|\boldsymbol{p}) - \log_2 f(\boldsymbol{p}) + C(\boldsymbol{y}) \qquad (4.29)$$

In practice, we can omit the constant $C(\boldsymbol{y})$ because we are usually interested in optimizing the MDL with respect to the parameters.

We can see that the MDL principle is just the information map applied to the posterior in Bayesian models. For this reason, it is sometimes argued that, since the Bayesian approach is more general, MDL provides no real benefit over probabilistic modelling (Mckay, 2003, page 352). However, the data compression interpretation – that the MDL estimator minimizes the compressed size of the posterior representation for $\boldsymbol{y}$ – deserves further explanation (Figure 4.2). Let us denote the first two compressed length terms in (4.29) as $L\,[\boldsymbol{y}|\boldsymbol{p}]$ and $L\,[\boldsymbol{p}]$ respectively. What we find, is that as the number of parameters K (or effective degrees of freedom) is increased, then $L\,[\boldsymbol{y}|\boldsymbol{p}]$ gets smaller (in fact, as discussed above, we can in some cases have $L\,[\boldsymbol{y}|\boldsymbol{p}] \to 0$ as $K \to N$). However, it must be that $L\,[\boldsymbol{p}]$ is increasing with K. Therefore, the sum of these two terms usually has a minimum. Thus, minimizing the MDL with respect to K addresses the problem of overfitting, much as regularization, penalization, BIC and AIC do.

Cross-validation and regularization

We have seen above that for non-Bayesian models, we want to find the MLE with respect to both the parameters and the model complexity, say, degrees of freedom K. But, without the application of Occam's razor in some form, this is prone to extreme overfitting, because that single dataset is only one realization of the underlying distribution. This is purely an artefact of only having a single dataset (a 'finite sample' effect), which biases the likelihood upwards. If, instead, we already had unbiased parameter MLEs $\hat{\boldsymbol{p}}$, then one could find the corresponding MLE for K:

$$\hat{K} = \arg\max_{K} f\left(\boldsymbol{x}; \hat{\boldsymbol{p}}, K\right) \qquad (4.30)$$

However, the vector of parameters $\boldsymbol{p}$ depends upon K, so for each K we need different parameter MLEs. But, due to the finite sample overfitting effect, we cannot simultaneously maximize the likelihood with respect to both $\boldsymbol{p}$ and K given $\boldsymbol{x}$. We can address this by the use of more data. For each fixed value of K in (4.30), we can maximize with respect to the parameters, over a *different* data set $\boldsymbol{x}'$, distinct from $\boldsymbol{x}$, drawn from the underlying distribution over the data:

$$\hat{\boldsymbol{p}} = \arg\max_{\boldsymbol{p}} f\left(\boldsymbol{x}'; \boldsymbol{p}, \hat{K}\right) \qquad (4.31)$$

This approach gets around the single sample overfitting effect by the use of different realizations of the data. Procedures like this which use different data realizations to ameliorate the sensitivity of complex models to single samples are known as *validation* or *out-of-sample* methods. In this context $\boldsymbol{x}'$ and $\boldsymbol{x}$ are called *training* and *validation* (or *test*) sets, respectively. Similarly, the likelihoods on the training and validation data are known as the *in-* and *out-of-sample* likelihoods.

In effect, out-of-sample methods enforce a kind of regularization, as introduced in the previous section. In fact, because the likelihood is a random variable, so are the parameter estimates, because they are functions of random variables. So, are these parameter estimates typical for the underlying (and inaccessible) distribution of the data, or not? In general, what is the distribution over these quantities? A single train-test dataset pair does not allow us to answer any of these critical

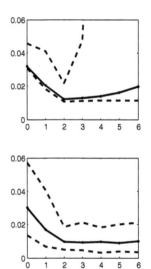

Fig. 4.3: Cross-validation for model complexity selection, for in-sample training set size of $N = 20$ (top panel) and $N = 980$ (bottom panel). K is the model complexity, here, the order of the fitted polynomial, the actual order of the underlying model is $K = 2$. 1000 samples of data are generated from the polynomial model with i.i.d. Gaussian errors, and the cross-validation is repeated by splitting the data randomly into test and training sets 100 times. The vertical axes represent out-of-sample empirical risk for the square loss function (which is proportional to the out-of-sample likelihood for this model). The dashed lines represent the maximum and minimum empirical risk, and the solid line the median risk.

questions. But, if we can repeat the above maximizations (4.30), (4.31) a large number of times over many different realizations x', x, then we can hope to estimate the posterior distributions $f(p|x)$ and $f(K|x)$.

Such a procedure is going to require a *lot* of data, more than we usually have in practice. The widely used *cross-validation* method attempts to implement this procedure, whilst also making efficient use of the given data. In *M-fold* cross validation, the original dataset is divided into M smaller subsets $x^{(m)}$. A subset $x^{(m)}$ is picked uniformly at random and used as test set, and the rest of the subsets, $x^{(l)}$ with $l \neq m$, are concatenated and used as the training set. If $M = N$, then the procedure is known as *leave-one-out*.

What is the effect on the likelihood, parameter and model complexity estimates, of the value M? Typically (this is the case for the MLE) the reliability of parameter estimates improves as the number of samples available for training increases. So, there is a strong case for leave-one-out, since it will lead to small variability in the parameters. However, then the test set will contain only one sample, so the out-of-sample variability will be large. On the other hand, if M is small, then the training set size is reduced, causing the parameters values to become highly variable. The choice of cross-validation scheme interacts with the model and the data leading to significant impact on estimated parameters such as model complexity and out-of-sample prediction errors. For example, using a simple polynomial model, we find that a small training set size leads to highly unreliable empirical risk estimates, but using the typical (median) risk would lead to a reliable estimate for the model order. On the other hand, we can obtain quite reliable empirical risk estimates using a large training set size, but this would lead to unreliable model order estimates (Figure 4.3). See Hastie *et al.* (2009, page 243) for further explorations of these effects.

One of the most serious limitations of out-of-sample techniques is that the training and test sets *must* be drawn from the same underlying distribution. If not, there are no guarantees that the parameter estimates maximize the expected out-of-sample likelihood, given any fixed model complexity. Typically, mismatch between train/test set distributions causes extreme downward bias in the out-of-sample likelihood. We have even found cases where carelessly splitting up the data according to some particular variable in the dataset causes a distributional mismatch between subsamples in cross-validation – in other words, that the *originally* i.i.d. dataset can be split up in such a way that causes unintended, internal distributional mismatch. It should be said that there are ways to adapt cross-validation for distributional mismatch (the so-called *covariate shift* problem), but the distributions must be known (Sugiyama *et al.*, 2007). One other requirement of validation methods is that the training and test sets are independent, otherwise information from the training set can 'leak' into the test set. There are simple adaptations which can mitigate this to a certain extent (Arlot and Celisse, 2010, page 66). Unfortunately, the uncritical use of cross-validation without checking whether the underlying assumptions hold is a common problem in the machine learning literature.

It may come as no surprise that since validation methods perform regularization, they are related to other MAP methods. In fact, leave-one-out and AIC have been shown to be equivalent (Stone, 1977). See Arlot and Celisse (2010) for other interesting relationships and further reading on cross-validation.

The bootstrap

Given a set of i.i.d. samples $x_1, \ldots, x_N$ drawn from X we often compute quantities such as summary statistics (mean, variance etc.) or estimate parameters of a probabilistic model fitted to this data. However, this is just one sample set, and provides only one value of the estimate, but we know that the estimates are random variables because they are computed from random variables. So, what is the distribution of these estimates? This is a tough problem to solve because we have to transform the distribution of the data through the function used to compute the estimate, which may be very complex. The *bootstrap* is an ingenious computational approach to approximately solving this problem: it is based around the idea that, assuming the ECDF is a good model for the distribution of X (Section 1.4), then new samples generated from this ECDF can be used to estimate the distribution of, for example, MLE parameters (4.1).

Sampling from the ECDF is very simple. We can consider the associated empirical distribution function to be a *mixture* of delta functions, where all the mass of the distribution lies under these delta functions, centred at each x_n. Then, issuing one of the samples x_n chosen uniformly at random amounts to drawing uniformly at random, i.i.d., from the ECDF. Using this procedure, we can generate new samples of any desired size.

While the bootstrap is remarkably simple and computationally straightforward, it has several drawbacks. The most critical is that the original samples $x_1, \ldots, x_N$ must be a representative draw from the underlying distribution. Any 'outliers', that is, samples which would be highly unlikely given the distribution, will distort the bootstrap samples and any estimates based on them. Similarly, while the ECDF is a consistent estimator of the CDF in the limit of an infinite amount of data, for any finite sample, the ECDF may be quite different from the underlying CDF, making the bootstraps unreliable.

Using the bootstrap to generate additional datasets for out-of-sample model selection, as in the section above, has a major limitation. Although bootstrap datasets are, in principle at least, draws from the underlying distribution for the data, an artefact of the method is that each generated bootstrap shares some data with the original dataset. This means that each bootstrap is not exactly a different realization of the underlying distribution, as is required to avoid the single sample overfitting effect. This would have the effect of biasing the likelihood upwards. Several techniques have been developed to mitigate this effect, but at the expense of substantially increased complexity (Hastie *et al.*, 2009, page 251).

4.3   Bayesian inference

The previous section was largely devoted to non-Bayesian and MAP inference. These frequentist inferences are known as *point estimates* because they estimate only a single, optimal set of parameters or predictions, ignoring uncertainty. By contrast, in the fully Bayesian approach, we wish to infer the full conditional distribution over all estimates.

Typically, in Bayes' rule, we will nearly always obtain the evidence probability through marginalization:

$$f(\boldsymbol{p}|\boldsymbol{x}_1,\ldots,\boldsymbol{x}_N) = \frac{f(\boldsymbol{x}_1,\ldots,\boldsymbol{x}_N|\boldsymbol{p})\,f(\boldsymbol{p})}{\int f(\boldsymbol{x}_1,\ldots,\boldsymbol{x}_N|\boldsymbol{p})\,f(\boldsymbol{p})\,d\boldsymbol{p}} \tag{4.32}$$

When the prior is some function of hyperparameters $\boldsymbol{q}$ then $\int f(\boldsymbol{x}_1,\ldots,\boldsymbol{x}_N|\boldsymbol{p})\,f(\boldsymbol{p};\boldsymbol{q})\,d\boldsymbol{p} = f(\boldsymbol{x}_1,\ldots,\boldsymbol{x}_N;\boldsymbol{q})$ is known as the *marginal likelihood*. So, given some data $\boldsymbol{x}_1,\ldots,\boldsymbol{x}_N$, the goal is to compute the above posterior. This distribution can then be summarized, by e.g. the mean and variance, or mode and some other measure of 'spread'. The evidence probability integral will usually be computationally intractable; fortunately, with multivariate sampling techniques such as MCMC we can often avoid the need to evaluate this quantity (see Section 3.4).

In the Bayesian framework, making predictions is not as simple as the frequentist case because the parameters are uncertain. Instead, taking into account the posterior of the parameters given the data, we can work out the *posterior predictive* distribution of the new data point:

$$f(\tilde{\boldsymbol{x}}|\boldsymbol{x}_1,\ldots,\boldsymbol{x}_N) = \int f(\tilde{\boldsymbol{x}}|\boldsymbol{p})\,f(\boldsymbol{p}|\boldsymbol{x}_1,\ldots,\boldsymbol{x}_N)\,d\boldsymbol{p} \tag{4.33}$$

This can be understood as marginalizing out the parameters from $f(\tilde{\boldsymbol{x}},\boldsymbol{p}|\boldsymbol{x}_1,\ldots,\boldsymbol{x}_N)$, which is the same as $f(\tilde{\boldsymbol{x}}|\boldsymbol{x}_1,\ldots,\boldsymbol{x}_N,\boldsymbol{p}) \times f(\boldsymbol{p}|\boldsymbol{x}_1,\ldots,\boldsymbol{x}_N)$, since the new data point $\tilde{\boldsymbol{x}}$ is conditionally independent of $\boldsymbol{x}_1,\ldots,\boldsymbol{x}_N$ given $\boldsymbol{p}$. Another way of explaining this is through the conditional dependencies $\boldsymbol{x}_1,\ldots,\boldsymbol{x}_N \to \boldsymbol{p} \to \tilde{\boldsymbol{x}}$, the posterior predictive distribution is obtained by integrating out (*collapsing*) $\boldsymbol{p}$ from this chain. Making predictions before receiving any data requires the *prior predictive distribution*, where the posterior $f(\boldsymbol{p}|\boldsymbol{x}_1,\ldots,\boldsymbol{x}_N)$ is replaced by the prior $f(\boldsymbol{p})$ in (4.33).

Computing the posterior can be viewed as 'updating' the prior given the data. What happens when new data, say, $\boldsymbol{x}_{N+1}$ becomes available? The new posterior, conditioned on $\boldsymbol{x}_1,\ldots,\boldsymbol{x}_N$ is:

$$f(\boldsymbol{p}|\boldsymbol{x}_{N+1},\boldsymbol{x}_1,\ldots,\boldsymbol{x}_N) = \frac{f(\boldsymbol{x}_{N+1}|\boldsymbol{p},\boldsymbol{x}_1,\ldots,\boldsymbol{x}_N)\,f(\boldsymbol{p}|\boldsymbol{x}_1,\ldots,\boldsymbol{x}_N)}{f(\boldsymbol{x}_{N+1}|\boldsymbol{x}_1,\ldots,\boldsymbol{x}_N)} \tag{4.34}$$

This identity can be proven by expanding out the conditional probabilities on the right hand side and cancelling, and of course it is exactly the posterior $f\left(\boldsymbol{p}|\boldsymbol{x}_1,\ldots,\boldsymbol{x}_{N+1}\right)$ that would be obtained if we had all the data available in the first place. However, notice that the prior $f\left(\boldsymbol{p}|\boldsymbol{x}_1,\ldots,\boldsymbol{x}_N\right)$ is the posterior from (4.32). We are using the posterior from the original dataset as the prior when the next data point becomes available. This procedure can of course be repeated on the arrival of each new data point. This crucial idea, known as *Bayesian updating*, is used to great effect in sequential DSP methods such the *Kalman filter* that we will encounter in later chapters.

Of course, writing down equations such as (4.32)-(4.34) is easy but actually computing the resulting distributions in closed form is usually extremely difficult, if possible at all. However, there turns out to be a remarkably broad (and closed) set of distributions for which all of these computations are tractable: the *exponential family* distributions. Within this class of distributions certain parametric forms of the prior are matched to the likelihood such that the posterior is in the same parametric family as the prior – the prior and likelihood are said to be *conjugate*. The parameters of the posterior are obtained through simple Bayesian updating formulae. Because these distributions have such mathematically convenient properties, they are a mainstay of Bayesian modelling and will be discussed in detail in Section 4.5.

In parametric Bayesian models there will typically be hyperparameters $\boldsymbol{q}$ which select different curves for the prior $f\left(\boldsymbol{p};\boldsymbol{q}\right)$. A 'pure' Bayesian would argue that this prior – its functional form and choice of hyperparameters – are 'subjective' in the sense that each modeller might choose differently. By contrast, *empirical Bayes* is an approach to estimating prior hyperparameters from the data itself, typically by maximum likelihood. This approach also goes by the name of *maximum likelihood type II*, *generalized maximum likelihood*, and the *evidence approximation* (see Bishop, 2006, pages 165-172). Examining the marginal likelihood of the data $f\left(\boldsymbol{x}_1,\ldots\boldsymbol{x}_N;\boldsymbol{q}\right)$ strongly suggests that we can estimate values for the hyperparameters using maximum likelihood based on the data alone:

$$\hat{\boldsymbol{q}} = \arg\max_{\boldsymbol{q}} f\left(\boldsymbol{x}_1,\ldots\boldsymbol{x}_N;\boldsymbol{q}\right) \qquad (4.35)$$

Empirical Bayes' estimates, under certain circumstances, have some intriguing properties. For illustration, consider the case of univariate Gaussian data with non-i.i.d. means, i.e. $X_n \sim \mathcal{N}\left(\theta_n,\sigma^2\right)$ with mean θ_n (we will assume that σ^2 is known). The distribution of these parameters is given by $\theta_n \sim \mathcal{N}\left(\mu,\tau^2\right)$. So, the hyperparameters in this case are $\boldsymbol{q} = \left(\mu,\tau^2\right)$. From known properties of the Gaussian we can compute that $X_n \sim \mathcal{N}\left(\mu,\sigma^2+\tau^2\right)$, so, it is clear that the X_n's are i.i.d. when the θ_n are integrated out. Now, substituting the MLE for μ (the sample mean of the X_n's) and an (unbiased) estimator for the ratio $\sigma^2/\left(\sigma^2+\tau^2\right)$, we get the following 'ridge' shrinkage estimator for the parameters (known as the *James-Stein estimator*):

$$\hat{\theta}_n = \left(\frac{(N-3)\,\sigma^2}{\sum_{n=1}^{N}(x_n-\mu)^2} \right)\mu + \left(1 - \frac{(N-3)\,\sigma^2}{\sum_{n=1}^{N}(x_n-\mu)^2} \right)x_n \qquad (4.36)$$

Remarkably, it can be shown that, when considering all parameters together, for $N \geq 4$ this estimator always has lower empirical risk under the squared loss, than the MLE, which is just x_n (Casella, 1985). This is counterintuitive because we know that the MLE has the lowest squared loss empirical risk when estimating any single parameter alone. The important point to bear in mind is that this improvement only applies to estimating parameters *jointly* by including more parameters to estimate, which is a different goal to the case of estimating any single parameter. We note that this phenomena is quite general and is not restricted to Gaussian variables, see for example Casella (1985) for a similar demonstration using binomially-distributed data with Beta priors. For further, more sophisticated applications of empirical Bayes in machine learning, see Bishop (2006, section 3.5).

4.4 Distributions associated with metrics and norms

We have seen in the previous section the fundamental importance of optimal probability calculations such as maximum likelihood and maximum a-posteriori to statistical inference. In the next few sections we will explore useful classes of distributions which find common usage in statistical machine learning and DSP, either explicitly or implicitly. Firstly, we will explore distributions which arise as a homomorphism between a metric and a density function, through the logarithmic information map $-\ln(x)$ (see Chapter 1). Consider that we have a density function $f(x;p)$ in the form:

$$f(x;p) = \frac{1}{Z}\exp\left(-d(x,p)\right) \qquad (4.37)$$

where the function d is a metric, and Z is a normalizing constant that does not depend on p. The metric d therefore takes the role of loss function $L(x,p) = d(x,p)$ (see Section 4.2). In this form, the logarithmic map creates a homomorphism between a metric space and a density function over the real line. Now, if we assume we have a set of N i.i.d. data points $\boldsymbol{x}$, the likelihood is $f(\boldsymbol{x};p) = \prod_{n=1}^{N} f(x_n;p)$ and the MLE for the parameter p is:

$$\hat{p} = \arg\max_{p} \prod_{n=1}^{N} f(x_n;p) = \arg\min_{p} \sum_{n=1}^{N} L(x_n,p) \qquad (4.38)$$

which is in the form of a minimizer for a *sum of losses* and indeed coincides with the empirical risk estimator (4.12). The choice of metric determines many of the properties of the MLE $\hat{p}$, such as how it is estimated and how it 'summarizes' the data.

Least squares

Perhaps the most widely encountered metric is the *square Euclidean distance*:

$$\frac{1}{2} \sum_{i=1}^{N} |x_i - p|^2 = \frac{1}{2} \| \boldsymbol{x} - p\mathbf{1} \|_2^2 \tag{4.39}$$

where $\mathbf{1}$ is the N-vector of ones. This is the L_2-norm loss induced by the Euclidean distance metric $d(x,p) = \frac{1}{2}|x - p|^2$, and in this case (4.38) becomes the *least-squares* estimator. What density function corresponds to this metric? The answer is the Gaussian distribution:

$$f(x;p) = \frac{1}{Z} \exp \left(-\frac{|x - p|^2}{2\sigma^2} \right) \tag{4.40}$$

(Note that the variance term in (4.39) is lost because the optimal value $\hat{p}$ is unaffected by scaling the sum of losses L). In this case, by taking the derivative of (4.39) with respect to p and setting this to zero, we obtain the analytically tractable solution $\hat{p} = \frac{1}{N} \sum_{n=1}^{N} x_n$, which is just the sample mean of the data.

Least squares has certain properties that make it statistically desirable. For example, this estimator can be shown to have the minimum variance amongst all unbiased estimators for the mean parameter of the Gaussian. However, least-squares estimators run into trouble when the data departs from the Gaussian assumption — there are more robust metrics as we will see below. This relationship between optimal estimators and metrics is such an important feature of many useful statistical models that we will explore several examples in depth next.

Least L_q-norms

The use of the generalized metric $d(x,p) = \frac{1}{q}|x - p|^q$, with $q \in \mathbb{R}$, gives us the following sums of losses:

$$\frac{1}{q} \sum_{n=1}^{N} |x_n - p|^q = \frac{1}{q} \| \boldsymbol{x} - p\mathbf{1} \|_q^q \tag{4.41}$$

Of course with $q = 2$ we recover sums of squares, but there exist some other very important special cases. In particular $q = 1$ corresponds to the *city-block* or *absolute distance* metric, and the corresponding sample estimator of the parameter q, is a sample *median* of the data (the value that lies 'in the middle' of all the data points). This can be seen from the equation that holds at the minimum of (4.38):

$$\frac{\partial}{\partial p} \sum_{n=1}^{N} |x_n - p| = \sum_{n=1}^{N} \text{sign}(x_n - p) = 0 \tag{4.42}$$

Each of the sign terms in the sum is $+1$ if $x_n > p$ and -1 if $x_n < p$, so the minimum is only obtained when there are as many terms larger than p as smaller: this works when N is odd (a different argument is needed in general, though). Interestingly, this creates an intimate connection between *rank-order statistics* and specific density functions, e.g. $q = 1$ is associated with the *Laplace distribution*:

$$f\left(x;p\right) = \frac{1}{Z}\exp\left(-\frac{|x-p|}{b}\right) \tag{4.43}$$

where b is the spread parameter analogous to the variance for the Gaussian. This distribution has *heavier tails* than the Gaussian, that is, it assigns larger probability to extreme value far from p. As a result, the median is much less likely to be adversely influenced by samples with extremely small or large values, whereas, the mean is heavily shifted towards these extreme values. This is one of the reasons why the median is known as a *robust estimator*, because it is largely insensitive to extreme-valued points that are uncharacteristic of the main body of the data.

Another very important special case is obtained at the limit when $q \to \infty$, the *max-norm* estimator:

$$\max_{n=1\ldots N} |x_n - p| = \|\boldsymbol{x} - \mathbf{1}p\|_\infty \tag{4.44}$$

In this case, the MLE is the *mid-range* $\hat{p} = \frac{1}{2}\left(\min_{n=1\ldots N} x_n + \max_{n=1\ldots N} x_n\right)$. This is an extremely efficient, but highly non-robust estimator of central tendency. It is efficient because it uses only two samples (the minimum and maximum), but non-robust because contamination of either of those affects the estimator.

L_q-norms for $q \geq 1$ are convex which means that we can apply all the techniques from Section 2.3 (and for $q = 1$, from Section 2.4). For $q < 1$, the norms are non-convex and non-differentiable – the case $q = 0$ has special importance in *sparse signal processing* which we will encounter in later chapters. There is no simple closed-form estimator for this norm and numerical methods will be severely challenged to find good quality approximate solutions. For this reason, the $q = 1$ estimator has a special place in machine learning and signal processing because it is the norm which is convex and yet is the 'closest' estimator to the L_0-norm.

Covariance, weighted norms and Mahalanobis distance

A sample estimate of the variance can be written in the form of an L_2-norm:

$$\mathrm{var}\left[X\right] = E\left[\left(X - \mu_X\right)^2\right] \approx \frac{1}{N}\|\boldsymbol{x} - \mathbf{1}\mu_X\|_2^2 \tag{4.45}$$

where $\mathbf{1}$ is the N-vector of ones. However, in matrix terms, this is also a (diagonal) *quadratic form* $\|\boldsymbol{x} - \mathbf{1}\mu_X\|_2^2 = \left(\boldsymbol{x} - \mathbf{1}\mu_X\right)^T \left(\boldsymbol{x} - \mathbf{1}\mu_X\right)$, which naturally raises the question about other diagonal quadratic forms. Indeed, we can write out the covariance between two random variables X, Y in this form:

$$\text{cov}[X, Y] = E[(X - \mu_X)(Y - \mu_Y)] \approx \frac{1}{N}(\boldsymbol{x} - \boldsymbol{1}\mu_X)^T(\boldsymbol{y} - \boldsymbol{1}\mu_Y)$$
(4.46)

An important normalized joint central moment is known as the *correlation coefficient*:

$$\rho_{X,Y} = \frac{\text{cov}[X, Y]}{\sqrt{\text{var}[X]\,\text{var}[Y]}}$$
(4.47)

So, a sample estimator of the correlation coefficient can be expressed in terms of quadratic forms:

$$\rho_{X,Y} \approx \frac{(\boldsymbol{x} - \boldsymbol{1}\mu_X)^T(\boldsymbol{y} - \boldsymbol{1}\mu_Y)}{\sqrt{(\boldsymbol{x} - \boldsymbol{1}\mu_X)^T(\boldsymbol{x} - \boldsymbol{1}\mu_X)(\boldsymbol{x} - \boldsymbol{1}\mu_Y)^T(\boldsymbol{x} - \boldsymbol{1}\mu_Y)}}$$
(4.48)

Sometimes we find that specific dimensions of the vector space for $\boldsymbol{x}$ have different importance in the sum of losses, in which case it can be useful to invoke the *weighted L_q-norm*:

$$\|\boldsymbol{u}\|_{q,\boldsymbol{w}}^q = \sum_{n=1}^{N} w_n |u_n|^q$$
(4.49)

for the N-element *weight vector* $\boldsymbol{w}$. If, for $q \geq 1$, the entries satisfy $w_n \geq 0$, then the norm is convex. This gives rise to the corresponding weighted L_q-norm empirical risk estimator for p:

$$\hat{p} = \arg\max_p \sum_{n=1}^{N}(w_n |x_n - p|^q)$$
(4.50)

In the L_2 case $q = 2$, we can associate this weighted sum of losses with a likelihood consisting of a product of univariate Gaussians having a different variance for each observation x_n, where each variance is $\sigma_n^2 = w_n^{-1}$. It is straightforward to show that the corresponding MLE is the *weighted mean* $\hat{p} = \frac{1}{N}\sum_{n=1}^{N}(x_n/\sigma_n^2)$. Thus, high-variance observations are given low weight, which influence the estimator less than low-variance observations. This makes intuitive sense as we would be less likely to trust an observation the higher the uncertainty in the observed value. For the Laplacian case where $q = 1$, we have the *weighted median* estimator, which is obtained by replicating the observations by the size of the weights before finding the middle value (Arce, 2005, page 13). Indeed, the weighted median associated with the Laplacian with varying spread b_n is the foundation of an entire class of nonlinear DSP methods we will discuss in a later chapter (Arce, 2005).

In the L_2-norm case, we can write out the weighted norm (4.49) as $\frac{1}{2}\|\boldsymbol{u}\|_{\mathbf{W}}^2 = \frac{1}{2}\boldsymbol{u}^T\mathbf{W}\boldsymbol{u}$, using the diagonal matrix $\mathbf{W}$ whose diagonal contains the weights. Now, in this representation, more flexibility is possible if $\mathbf{W}$ is a general (non-diagonal) matrix. In particular if $\mathbf{W}$ is positive semi-definite, then the associated distance is the Mahalanobis distance and the naturally associated density would be the multivariate Gaussian (Section 1.4). Then the weight matrix is the inverse of the covariance matrix for the random vector $\mathbf{X}$, i.e. $\mathbf{W} = \mathbf{\Sigma}^{-1}$.

The above representation of the multivariate normal in terms of the Mahalanobis distance in a vector space can be generalized considerably in the form of the *elliptically contoured* (or just *elliptical*) distributions. Consider the D-vector random variable $\mathbf{X}$, then, at a high level of generality, these are defined in terms of the CF of this vector with the special form:

$$\psi\left(\boldsymbol{s}\right) = \exp\left(i\boldsymbol{s}^T\boldsymbol{\mu}\right) G\left(\frac{1}{2}\boldsymbol{s}^T\boldsymbol{\Sigma}\boldsymbol{s}\right) \tag{4.51}$$

with parameters $\boldsymbol{\mu}$ (D entries) and $\boldsymbol{\Sigma}$ ($D \times D$ positive semi-definite). The function $G : [0, \infty) \to \mathbb{R}$ is called the *characteristic generator* of the distribution. When the associated density function does exist, it can be shown that it has the following form:

$$f\left(\boldsymbol{x}\right) = C\sqrt{|\boldsymbol{\Sigma}|}g\left(\frac{1}{2}\left(\boldsymbol{x} - \boldsymbol{\mu}\right)^T\boldsymbol{\Sigma}^{-1}\left(\boldsymbol{x} - \boldsymbol{\mu}\right)\right) \tag{4.52}$$

provided the function g, the *density generator*, satisfies $\int_0^\infty u^{D/2-1}g\left(u\right)du < \infty$. The constant C has a simple closed-form formula in terms of the latter integral and the gamma function (Landsman and Valdez, 2003). We will designate such an elliptical distribution with $\mathcal{EC}\left(\boldsymbol{\mu}, \boldsymbol{\Sigma}, g\right)$. The form (4.52) reduces to the multivariate normal for $g\left(u\right) = \exp\left(-u\right)$, and to several other families of distributions with different choices of g (Table 4.1) including a multivariate version of the Laplace distribution.

Distributional family	$g\left(u\right)$
Normal	$\exp\left(-u\right)$
Exponential-Power	$\exp\left(-r\,u^q\right),\; r, q > 0$
Student-t	$\exp\left(1 + \frac{2u}{m}\right)^{-(D+m)/2}$
Logistic	$\frac{\exp(-u)}{(1+\exp(-u))^2}$

Table 4.1: **Families of elliptically contoured distributions whose density function can be written in the form of (4.52), parameterized by the density generator** $g\left(u\right)$. **These families contain, for example, a multivariate Laplace distribution (exponential-power family with** $r = \sqrt{2}$ **and** $q = 1$**), and a multivariate Cauchy (Student-t family with** $m = 1$**).**

By construction, the elliptical distributions share many desirable properties of the multivariate normal. For example, they are invariant under linear transformations, which also implies that their marginals and conditionals are elliptical, with simple closed-form formulas for the resulting distributions. There is also a simple way to draw random variables from the distribution. For the random vector U uniformly distributed on the unit sphere in D dimensions, the univariate radius variable R has PDF $f(r) = 2C^{-1}r^{D-1}g(r^2)$, and $\Sigma = A^T A$ for a given full-rank matrix A, then $X = \mu + RA^T U$ has the $\mathcal{EC}(\mu, \Sigma, g)$ distribution. Furthermore, provided that $\int_0^\infty g(u)\,du < \infty$ and $\left|\frac{\partial G}{\partial u}(0)\right| < \infty$, then $E[X] = \mu$ and $\text{cov}[X] = \Sigma$ (Landsman and Valdez, 2003). Thus, the practical value of the elliptical distributions is that analytically tractable multivariate modelling is hard, and these distributions provide quite a broad set of statistical models with e.g. flexible behaviour for extreme values of the Mahalanobis distance.

4.5   The exponential family (EF)

A classical and recurring situation in DSP and machine learning is when we are given N i.i.d. data points x drawn from a random variable X with unknown distribution function $f(x)$. We are tasked to find the best distribution to model this data. This of course is not a solvable problem as stated, but if we ask: what is the distribution with the largest *entropy* that satisfies various *moment constraints*, then the problem is well-posed. We will see that this leads to the very large an important field of maximum entropy distributions and associated exponential families.

Maximum entropy distributions

For any distribution to be valid, we must have that $\int f(x)\,dx = 1$. Let us also assume we can compute sample moments $E[g_m(X)] = a_m$ for $m = 1, \dots M$. We can formalize this constrained maximum entropy problem as follows:

$$\text{maximize} \quad H[X] \tag{4.53}$$
$$\text{subject to} \quad \int f(x)\,dx = 1,$$
$$\int g_m(x)f(x)\,dx = a_m, \ , m = 1, \dots M$$

The Lagrangian for this problem is:

$$L(f, \nu) = H[X] - \nu_0 \int f(x)\,dx - \sum_{m=1}^{M} \nu_m \int g_m(x)f(x)\,dx \tag{4.54}$$

where ν is an $M + 1$ vector of multipliers. The (functional) derivative of this Lagrangian with respect to f is:

$$\frac{\partial L}{\partial f}(f, \boldsymbol{\nu}) = -1 - \ln f(x) - \nu_0 - \sum_{m=1}^{M} \nu_m g_m(x) \qquad (4.55)$$

Setting $\frac{\partial L}{\partial f}(f, \boldsymbol{\nu}) = 0$ gives the following solution at a turning point for L:

$$\hat{f}(x) = \exp\left(-1 - \nu_0 - \sum_{m=1}^{M} \nu_m g_m(x)\right) \qquad (4.56)$$

Now, since $H[X]$ is a concave function of f (Cover and Thomas, 2006, page 30), then (4.56) is the unique *maximum entropy* distribution that satisfies the sample moment constraints. The values of the multipliers $\boldsymbol{\nu}$ can then be chosen such that the constraints in (4.53) are satisfied. In the case of discrete random variables, $\hat{f}$ is the distribution which is 'closest' to uniform under the K-L divergence. We can also view $\hat{f}$ as the distribution with the largest Kolmogorov complexity under the constraints.

As a specific example, consider the single moment constraint on the mean, i.e. $g_1(x) = x$ with $M = 1$, when the sample space is the positive half of the real line $\mathbb{R}^+$. Then $\hat{f}(x) = \exp(-1 - \nu_0 - \nu_1 x)$ and applying the constraints (assuming $a_1 > 0$), we get that $\nu_1 = a_1^{-1}$ and $\nu_0 = \ln(a_1) - 1$. This leads to $\hat{f}(x) = \exp\left(\ln\left(a_1^{-1}\right) - a_1^{-1}x\right) = a_1^{-1}\exp\left(a_1^{-1}x\right)$. This is in fact the well-known *exponential distribution*.

Maximum entropy modelling is a rather compelling idea and has many proponents. However, the arguments put forward to justify the approach on first principles do not generally arise from, or have any specific, mathematical consequences. For example, there is no irrefutable reason why one might select the Shannon entropy here; different choices of entropies lead to entirely different distributions. In this book we will simply point out that there is a remarkably large family of distributions which have many highly convenient mathematical properties, known as the *exponential family* (EF), which are all maximum entropy distributions. These include the Gaussian, exponential, Gamma, Bernoulli, Dirichlet, multinomial and many others. Thus, EF distributions are widely used throughout statistical machine learning and DSP and we will explore their properties in detail in this section.

Sufficient statistics and canonical EFs

Sufficient statistics and exponential families are intimately related. A *statistic* is just a (vector) function $\boldsymbol{g}$ of some random data. A *sufficient* statistic for a parametric distribution is a function of the data drawn from that random variable which is all that is required to uniquely estimate the (vector) parameter $\boldsymbol{p}$ from the data. Formally: if the conditional distribution of the data given the statistic $\boldsymbol{g}$ is independent of $\boldsymbol{p}$, then $\boldsymbol{g}$ is a sufficient statistic for that parameter.

For appropriately regular distributions, the *Neyman factorization theorem* shows that the existence of a sufficient statistic implies that the distribution is in exponential family form $f(\boldsymbol{x}) f(\boldsymbol{p}, \boldsymbol{g}(\boldsymbol{x}))$, and the *Pitman-Koopman lemma* establishes the converse: any distribution in the exponential family form, has a sufficient statistic (Orbanz, 2009). In fact, all distributions in this class can be written as special cases of the general *canonical* form:

$$f(\boldsymbol{x}; \boldsymbol{p}) = \exp(\boldsymbol{p} \cdot \boldsymbol{g}(\boldsymbol{x}) - a(\boldsymbol{p}) + h(\boldsymbol{x})) \tag{4.57}$$

where $a(\boldsymbol{p}) = \ln \int \exp(\boldsymbol{p} \cdot \boldsymbol{g}(\boldsymbol{x}) + h(\boldsymbol{x})) \, d\boldsymbol{x}$ is the *log normalizer* or *partition function*, and $h(\boldsymbol{x})$ is the (natural log of the) *base measure*. The M entries of the parameter vector $\boldsymbol{p}$ are known as the *natural* or *canonical parameters*, and the individual functions g_m are the sufficient statistics. The loss function associated with this canonical form is just $L(\boldsymbol{p}) = -\boldsymbol{p} \cdot \boldsymbol{g}(\boldsymbol{x}) - h(\boldsymbol{x})$.

In this canonical form we will next show that the ML estimates for the natural parameters are particularly simple. Taking the derivative of the normalization function, we find:

$$
\begin{aligned}
\frac{\partial a}{\partial \boldsymbol{p}}(\boldsymbol{p}) &= \frac{\int \exp(\boldsymbol{p} \cdot \boldsymbol{g}(\boldsymbol{x}) + h(\boldsymbol{x})) \boldsymbol{g}(\boldsymbol{x}) \, d\boldsymbol{x}}{\int \exp(\boldsymbol{p} \cdot \boldsymbol{g}(\boldsymbol{x}) + h(\boldsymbol{x})) \, d\boldsymbol{x}} \\
&= \frac{\int \exp(\boldsymbol{p} \cdot \boldsymbol{g}(\boldsymbol{x}) - a(\boldsymbol{p}) + h(\boldsymbol{x})) \boldsymbol{g}(\boldsymbol{x}) \, d\boldsymbol{x}}{\int \exp(\boldsymbol{p} \cdot \boldsymbol{g}(\boldsymbol{x}) - a(\boldsymbol{p}) + h(\boldsymbol{x})) \, d\boldsymbol{x}} \\
&= E[\boldsymbol{g}(\boldsymbol{X})]
\end{aligned}
\tag{4.58}
$$

Using similar algebra, the second partial derivatives with respect to p_i, p_j can be shown to be $\operatorname{cov}[g_i(\boldsymbol{X}), g_j(\boldsymbol{X})]$ for all $i, j = 1, \ldots, M$. Now, for i.i.d. data, the NLL is:

$$
\begin{aligned}
L[\boldsymbol{p}] &= -\sum_{n=1}^{N} \ln f(\boldsymbol{x}_n; \boldsymbol{p}) \tag{4.59} \\
&= \sum_{n=1}^{N} [-\boldsymbol{p} \cdot \boldsymbol{g}(\boldsymbol{x}_n) - h(\boldsymbol{x}_n)] + N a(\boldsymbol{p})
\end{aligned}
$$

At a turning point of the above, the following holds:

$$\frac{\partial L}{\partial \boldsymbol{p}}[\boldsymbol{p}] = -\sum_{n=1}^{N} \boldsymbol{g}(\boldsymbol{x}_n) + N E[\boldsymbol{g}(\boldsymbol{X})] = 0 \tag{4.60}$$

from which we obtain:

$$\frac{1}{N} \sum_{n=1}^{N} \boldsymbol{g}(\boldsymbol{x}_n) = E[\boldsymbol{g}(\boldsymbol{X})] \tag{4.61}$$

Furthermore, since the second partial derivatives of the normalizer are a positive semi-definite covariance function, then the NLL (4.59) is convex, which shows that (4.61) *uniquely* determines the ML parameters. What this MLE equation says is that for EF distributions, the natural ML parameters are obtained by matching sample moments of the sufficient statistics with the corresponding analytical moments of the distribution. Now, typically, the natural parameters will be written in terms of some other more 'natural' parameterization and, therefore, finding the ML values of these nice parameters comes down to solving (4.61) in terms of them. This solution may not be obtainable analytically. However, the fact that the NLL (4.59) is convex tells us that we can always find the MLEs via some convex numerical optimization method from Chapter 2 such as gradient descent.

As an example, the univariate Gaussian is written in canonical form with sufficient statistics $g_1(x) = x$ and $g_2(x) = x^2$, and natural parameters $p_1 = \mu/\sigma^2$ and $p_2 = -1/(2\sigma^2)$ in terms of the more familiar mean and variance μ, σ^2. The base measure is $h(x) = 0$. This gives the log normalizer $a(\boldsymbol{p}) = -p_1^2/(4p_2) - \frac{1}{2}\ln(-p_2/\pi)$, which becomes $\mu^2/(2\sigma^2) + \frac{1}{2}\ln(2\pi\sigma^2)$:

$$
\begin{aligned}
f(x) &= \exp\left(\frac{\mu x}{\sigma^2} - \frac{x^2}{2\sigma^2} - \frac{\mu^2}{2\sigma^2} - \frac{1}{2}\ln\left(2\pi\sigma^2\right)\right) \\
&= \frac{1}{\sqrt{2\pi\sigma^2}}\exp\left(-\frac{(x-\mu)^2}{2\sigma^2}\right)
\end{aligned}
\tag{4.62}
$$

The ML equations (4.61) become:

$$
\frac{1}{N}\sum_{n=1}^{N} x_n = \mu
\tag{4.63}
$$

$$
\frac{1}{N}\sum_{n=1}^{N} x_n^2 = \mu^2 + \sigma^2
\tag{4.64}
$$

The first equation can be used to estimate μ, and given this, by rearranging the second equation, we get an explicit estimate for σ^2. From Table 4.2 we can see that the canonical form of the EF allows considerably generalized treatment of many widely used distributions in DSP and machine learning. The price paid for such uniform treatment is that many of the distributions do not appear in easily recognizable forms.

Conjugate priors

Perhaps the most important of the many mathematically convenient properties of EF distributions is that the EF is *closed under conjugation*, that is, each distribution is conjugate with some other distribution in the family. This follows from the relationship between sufficient statistics and the EF (Orbanz, 2009). To see this, consider the M-dimensional parameter EF likelihood function:

$$
f(\boldsymbol{x}|\boldsymbol{p}) = \exp\left(\boldsymbol{p}\cdot\boldsymbol{g}(\boldsymbol{x}) - a(\boldsymbol{p}) + h(\boldsymbol{x})\right)
\tag{4.65}
$$

Distribution	Sample space Ω	Sufficient statistics $\boldsymbol{g}(\boldsymbol{x})$	(Log) base measure $h(\boldsymbol{x})$	Natural parameters $\boldsymbol{p}$	Log partition $a(\boldsymbol{p})$
Exponential	$\mathbb{R}^+$	x	0	$-\lambda$	$-\ln(-p_1)$
Univariate Gaussian	$\mathbb{R}$	x, x^2	0	$\frac{\mu}{\sigma^2}, -\frac{1}{2\sigma^2}$	$-\frac{1}{4}p_1 p_2^{-1} p_1 - \frac{1}{2}\ln(-p_2 \pi^{-1})$
Multivariate Gaussian	$\mathbb{R}^D$	$\boldsymbol{x}, \boldsymbol{x}\boldsymbol{x}^T$	0	$\boldsymbol{\Sigma}^{-1}\boldsymbol{\mu}, -\frac{1}{2}\boldsymbol{\Sigma}^{-1}$	$-\frac{1}{4}\boldsymbol{p}_1^T \boldsymbol{p}_2^{-1} \boldsymbol{p}_1 - \frac{1}{2}\ln(-|\boldsymbol{p}_2|\pi^{-D})$
Gamma	$\mathbb{R}^+$	$\ln x, x$	0	$\alpha - 1, \beta$	$\ln\Gamma(p_1 + 1) - (p_1 + 1)\ln(p_2)$
Inverse Gamma	$\mathbb{R}^+$	$\ln x, x^{-1}$	0	$-(\alpha - 1), \beta$	$\ln\Gamma(-(p_1 + 1)) + (p_1 + 1)\ln(p_2)$
Log-normal	$\mathbb{R}^+$	$\ln x, (\ln x)^2$	$-\ln x$	$\frac{\mu}{\sigma^2}, -\frac{1}{2\sigma^2}$	$-\frac{1}{4}p_1 p_2^{-1} p_1 - \frac{1}{2}\ln(-p_2 \pi^{-1})$
Pareto	$[x_{\min}, \infty)$	$\ln x$	0	$-(\alpha + 1)$	$-\ln(-(p_1 + 1)) + (1 + p_1)\ln(x_{\min})$
Weibull	$\mathbb{R}^+$	x^k	$(k - 1)\ln x$	$-\lambda^{-k}$	$-\ln(-p_1 k)$
Bernoulli	$\{0, 1\}$	x	0	$\ln\left(\frac{\mu}{1-\mu}\right)$	$\ln(1 + \exp(p_1))$
Geometric	$\mathbb{Z}^+$	x	0	$\ln(1 - \mu)$	$-\ln(1 - \exp(p_1))$
Poisson	$\mathbb{Z}^+$	x	$-\ln(x!)$	$\ln(\lambda)$	$\exp(p_1)$
Categorical	$\{1, \ldots, K\}$	$\delta[x = 1], \ldots, \delta[x = K]$	0	$\ln\pi_1, \ldots, \ln\pi_K$	0
Dirichlet	$[0, 1]^K$	$\ln x_1, \ldots, \ln x_K$	0	$\alpha_1 - 1, \ldots, \alpha_K - 1$	$\sum_{k=1}^K \ln\Gamma(p_k + 1) - \ln\Gamma\left(\sum_{k=1}^K (p_k + 1)\right)$

Table 4.2: **An important selection of distributions from the exponential family, in canonical form** $f(x; p) = \exp(p \cdot g(x) - a(p) + h(x))$. **Note that the natural parameters are usually expressed as functions of the more 'recognizable' parameters such as** μ, σ^2 **(mean, variance) for the Gaussian or** λ **(rate) for the exponential.**

Then, the conjugate prior over the parameter vector $\boldsymbol{p}$ has the $M+1$-dimensional sufficient statistics p_m for $m = 1, \ldots, M$ and $-a(\boldsymbol{p})$ (Orbanz, 2009):

$$f(\boldsymbol{p}; \boldsymbol{q}) = \exp\left(\boldsymbol{q} \cdot (\boldsymbol{p}, -a(\boldsymbol{p})) - a_0(\boldsymbol{q})\right) \tag{4.66}$$

Given N i.i.d. observed data drawn from the likelihood, the joint distribution over the data and the parameters is:

$$
\begin{aligned}
f(\boldsymbol{x}_1, \ldots \boldsymbol{x}_N | \boldsymbol{p}) f(\boldsymbol{p}; \boldsymbol{q}) &= \prod_{n=1}^{N} \exp\left(\boldsymbol{p} \cdot \boldsymbol{g}(\boldsymbol{x}_n) - a(\boldsymbol{p}) + h(\boldsymbol{x}_n)\right) \times \\
&\quad \exp\left(\boldsymbol{q} \cdot (\boldsymbol{p}, -a(\boldsymbol{p})) - a_0(\boldsymbol{q})\right) \\
&= \exp\left(\boldsymbol{q}^N \cdot (\boldsymbol{p}, -a(\boldsymbol{p}))\right) \times \\
&\quad \exp\left(\sum_{n=1}^{N} h(\boldsymbol{x}_n) - a_0(\boldsymbol{q})\right) \tag{4.67}
\end{aligned}
$$

where $\boldsymbol{q}^N = \boldsymbol{q} + \left(\sum_{n=1}^{N} \boldsymbol{g}(\boldsymbol{x}_n), N\right)^T$. The term $\exp\left(\boldsymbol{q}^N \cdot (\boldsymbol{p}, -a(\boldsymbol{p}))\right)$ has the same form as the prior (4.66) but with 'updated' sufficient statistics $\boldsymbol{q} \leftarrow \boldsymbol{q}^N$. The evidence (marginal likelihood) is:

$$
\begin{aligned}
f(\boldsymbol{x}_1, \ldots, \boldsymbol{x}_N; \boldsymbol{q}) &= \int \prod_{n=1}^{N} \exp\left(\boldsymbol{p} \cdot \boldsymbol{g}(\boldsymbol{x}_n) - a(\boldsymbol{p}) + h(\boldsymbol{x}_n)\right) \times \\
&\quad \exp\left(\boldsymbol{q} \cdot (\boldsymbol{p}, -a(\boldsymbol{p})) - a_0(\boldsymbol{q})\right) d\boldsymbol{p} \\
&= \exp\left(\sum_{n=1}^{N} h(\boldsymbol{x}_n) - a_0(\boldsymbol{q})\right) \times \\
&\quad \int \exp\left(\boldsymbol{q}^N \cdot (\boldsymbol{p}, -a(\boldsymbol{p}))\right) d\boldsymbol{p} \\
&= \exp\left(\sum_{n=1}^{N} h(\boldsymbol{x}_n) + a_0(\boldsymbol{q}^N) - a_0(\boldsymbol{q})\right) \tag{4.68}
\end{aligned}
$$

so that the posterior is:

$$
\begin{aligned}
f(\boldsymbol{p} | \boldsymbol{x}_1, \ldots \boldsymbol{x}_N; \boldsymbol{q}) &= \exp\left(\boldsymbol{q}^N \cdot (\boldsymbol{p}, -a(\boldsymbol{p}))\right) \times \\
&\quad \exp\left(\sum_{n=1}^{N} h(\boldsymbol{x}_n) - a_0(\boldsymbol{q})\right) \times \\
&\quad \exp\left(-\sum_{n=1}^{N} h(\boldsymbol{x}_n) + a_0(\boldsymbol{q}) - a_0(\boldsymbol{q}^N)\right) \\
&= \exp\left(\boldsymbol{q}^N \cdot (\boldsymbol{p}, -a(\boldsymbol{p})) - a_0(\boldsymbol{q}^N)\right) \tag{4.69}
\end{aligned}
$$

As an example, let us look at the Bernoulli distribution for the likelihood. This has one sufficient statistic $g_1(x) = x$, with base measure $h(x) = 0$ and log normalizer $a(p) = \ln(1 + \exp(p))$, so we can write this out as $f(x|p) = \exp(px - \ln(1 + \exp(p)))$. This implies that the corresponding conjugate prior has 2-dimensional sufficient statistics $g_1(p) = p$ and $g_2(p) = -\ln(1 + \exp(p))$:

$$f(p; \boldsymbol{q}) \quad = \quad \exp(\boldsymbol{q} \cdot (p, -\ln(1 + \exp(p))) - a_0(\boldsymbol{q})) \quad (4.70)$$

Using the natural parameterization $p = \ln(\mu/(1 - \mu))$, we get that:

$$f(p; \boldsymbol{q}) = \exp(q_1 \ln(\mu) + (q_2 - q_1) \ln(1 - \mu) - a_0(\boldsymbol{q})) \quad (4.71)$$

which has sufficient statistics $\ln(\mu)$ and $\ln(1 - \mu)$ corresponding to the *Beta* distribution (this is the Dirichlet distribution with $K = 2$). The corresponding posterior is another Beta distribution, with updated sufficient statistics $q_1 \leftarrow q_1 + \sum_{n=1}^{N} x_n$ and $q_2 \leftarrow q_2 + N$. In the more recognizable parameterization, substituting $q_1 = \alpha - 1$ and $q_2 = \alpha_1 + \alpha_2 - 2$ gets us the posterior in the standard parameterized form of the Beta distribution:

$$f(\mu|x_1, \ldots x_N; \boldsymbol{\alpha}) \quad = \quad \mu^{\alpha_1 - 1 + N\bar{x}} (1 - \mu)^{\alpha_2 - 1 + N(1 - \bar{x})} \times$$
$$\frac{\Gamma(\alpha_1 + \alpha_2 + N)}{\Gamma(\alpha_1 + N\bar{x}) \Gamma(\alpha_2 + N(1 - \bar{x}))} \quad (4.72)$$

where $\bar{x}$ is the sample mean.

More complex EF conjugacy occurs with parameter vectors $\boldsymbol{p}$ with more than one element. We will look, in particular, at the univariate normal likelihood in the standard parameterization. Both the mean and variance parameters (μ, σ^2) can have priors. In this case, if instead of working with the variance we work with its reciprocal, the *precision* τ, then the conjugate prior of the joint (μ, τ) is the (*four-parameter*) *normal-Gamma* distribution $\mathcal{NG}$:

$$\begin{aligned} X &\sim \mathcal{N}(\mu, \tau) \\ (\mu, \tau) &\sim \mathcal{NG}(\mu_0, \tau_0, \alpha_0, \beta_0) \end{aligned} \quad (4.73)$$

This has the following density function:

$$(\mu, \tau) \sim \frac{1}{Z} \tau^{\alpha_0 - 1/2} \exp(-\beta_0 \tau) \exp\left(-\frac{\tau_0 \tau}{2}(\mu - \mu_0)^2\right) \quad (4.74)$$

where Z is a normalizer which is a function of $\alpha_0, \beta_0, \tau_0$. Then, it can be shown that the posterior after obtaining N samples of X is also normal-Gamma:

$$(\mu, \tau) \quad \sim \quad \mathcal{NG}\left(\frac{\tau_0 \mu_0 + N\bar{x}}{\tau_0 + N}, \tau_0 + N, \alpha_0 + \frac{N}{2}, \beta_0 + \right.$$
$$\left. \frac{1}{2}\left(N\bar{\sigma}^2 + \frac{\tau_0 N (\bar{x} - \mu_0)^2}{\tau_0 + N}\right)\right) \tag{4.75}$$

where $\bar{x}$ is the sample mean and $\bar{\sigma}^2$ the sample variance. As can be appreciated above, even in the EF, full Bayesian modelling is generally much more complex than frequentist modelling and the number of variables grows rapidly. Indeed, here we have four hyperparameters, two per parameter, and these degrees of freedom must be chosen somehow.

Finally, it is interesting to view EF conjugate updates from an information-theoretic view. The posterior Kolmogorov complexity is:

$$-\log_2 f(\boldsymbol{p}|\boldsymbol{x}_1, \ldots, \boldsymbol{x}_N; \boldsymbol{q}) = \frac{1}{\ln 2}\left[-\boldsymbol{q}^N \cdot (\boldsymbol{p}, -a(\boldsymbol{p})) + a_0(\boldsymbol{q}^N)\right] \tag{4.76}$$

Prior and posterior predictive EFs

Another very useful property of EF distributions is that, in Bayesian contexts, the posterior predictive density has, in principle at least, a straightforward form:

$$
\begin{aligned}
f(\tilde{\boldsymbol{x}}|\boldsymbol{x}_1, \ldots, \boldsymbol{x}_N; \boldsymbol{q}) &= \int f(\tilde{\boldsymbol{x}}|\boldsymbol{p}) f(\boldsymbol{p}|\boldsymbol{x}_1, \ldots, \boldsymbol{x}_N; \boldsymbol{q}) \, d\boldsymbol{p} \\
&= \exp\left(h(\tilde{\boldsymbol{x}}) - a_0(\boldsymbol{q}^N)\right) \times \\
&\quad \int \exp\left(\boldsymbol{p} \cdot \boldsymbol{g}(\tilde{\boldsymbol{x}}) - a(\boldsymbol{p}) + \boldsymbol{q}^N \cdot (\boldsymbol{p}, -a(\boldsymbol{p}))\right) d\boldsymbol{p} \\
&= \exp\left(h(\tilde{\boldsymbol{x}})\right) \exp\left(a_0\left(\boldsymbol{q}^N + (\boldsymbol{g}(\tilde{\boldsymbol{x}}), 1)^T\right) - \right. \\
&\quad \left. a_0(\boldsymbol{q}^N)\right) \tag{4.77}
\end{aligned}
$$

From this, we can obtain the prior predictive density very easily by setting $N = 0$:

$$
\begin{aligned}
f(\tilde{\boldsymbol{x}}; \boldsymbol{q}) &= \int f(\tilde{\boldsymbol{x}}|\boldsymbol{p}) f(\boldsymbol{p}; \boldsymbol{q}) \, d\boldsymbol{p} \tag{4.78} \\
&= \exp\left(h(\tilde{\boldsymbol{x}})\right) \exp\left(a_0\left(\boldsymbol{q} + (\boldsymbol{g}(\tilde{\boldsymbol{x}}), 1)^T\right) - a_0(\boldsymbol{q})\right)
\end{aligned}
$$

For example, taking the Bernoulli-Beta conjugate pair, we have $a_0(\boldsymbol{q}) = \ln \Gamma(q_1 + 1) + \ln \Gamma(q_2 - q_1 + 1) - \ln \Gamma(q_2 + 2)$ (note that this differs from the form in Table 4.2 because the table shows the 'standard' parameterization), and $h(\boldsymbol{x}) = 0$. Therefore, the posterior and prior predictive densities are, respectively (natural parameterization):

$$f\left(\tilde{x}|x_1,\ldots,x_N;\boldsymbol{\alpha}\right) = \frac{\Gamma\left(\alpha_1 + N\bar{x} + \tilde{x}\right)\Gamma\left(\alpha_2 + N\left(1-\bar{x}\right)+1-\tilde{x}\right)}{\left(\alpha_1+\alpha_2+N\right)\Gamma\left(\alpha_1+N\bar{x}\right)\Gamma\left(\alpha_2+N\left(1-\bar{x}\right)\right)}$$

$$= \frac{1}{\alpha_1+\alpha_2+N}\begin{cases}\alpha_2+N\left(1-\bar{x}\right), & \tilde{x}=0 \\ \alpha_1+N\bar{x}, & \tilde{x}=1\end{cases} \quad (4.79)$$

$$f\left(\tilde{x};\boldsymbol{\alpha}\right) = \frac{1}{\alpha_1+\alpha_2}\begin{cases}\alpha_2, & \tilde{x}=0 \\ \alpha_1, & \tilde{x}=1\end{cases} \quad (4.80)$$

These distributions are special cases of the so-called *Beta-Binomial* distribution with one trial (so that the binomial distribution simplifies to the Bernoulli). Similar calculus leads to the *Student-t* distribution as the predictive density for the *Normal-Gamma* conjugate pair, and the *Negative-Binomial* distribution for the *Poisson-Gamma* pair.

It is important to note that these predictive distributions are not necessarily EF densities. The primary advantage of this representation is that, once the conjugate prior log normalizer a_0 is known, the form of the predictive densities can be obtained easily without the need to compute difficult and often very complex integrals.

Conjugate EF prior mixtures

The EF family gives us a broad 'palette of likelihoods but to be analytically tractable, conjugacy is usually invoked, and so, the choice of likelihood determines the prior. This is obviously a limitation but we will see that *finite mixtures* of conjugate priors are also conjugate. Consider the weighted mixture of K priors $\sum_{k=1}^K w_k f\left(\boldsymbol{p};\boldsymbol{q}_k\right)$ with $w_k > 0$ and $\sum_{k=1}^K w_k = 1$, where all priors are conjugate to the likelihood $f\left(\boldsymbol{x}_1,\ldots,\boldsymbol{x}_N|\boldsymbol{p}\right)$. Writing the evidence distributions as $Z_k = f\left(\boldsymbol{x}_1,\ldots,\boldsymbol{x}_N;\boldsymbol{q}_k\right)$, the posterior is:

$$\begin{aligned}f\left(\boldsymbol{p}|\boldsymbol{x}_1,\ldots,\boldsymbol{x}_N;\boldsymbol{q}_1,\ldots,\boldsymbol{q}_K\right) &= \frac{f\left(\boldsymbol{x}_1,\ldots,\boldsymbol{x}_N|\boldsymbol{p}\right)\sum_{k=1}^K w_k f\left(\boldsymbol{p};\boldsymbol{q}_k\right)}{\int f\left(\boldsymbol{x}_1,\ldots,\boldsymbol{x}_N|\boldsymbol{p}\right)\sum_{i=1}^K w_i f\left(\boldsymbol{p};\boldsymbol{q}_i\right)d\boldsymbol{p}} \\ &= \frac{\sum_{k=1}^K w_k f\left(\boldsymbol{x}_1,\ldots,\boldsymbol{x}_N|\boldsymbol{p}\right)f\left(\boldsymbol{p};\boldsymbol{q}_k\right)}{\sum_{i=1}^K w_i Z_i} \\ &= \sum_{k=1}^K \frac{w_k Z_k}{\sum_{i=1}^K w_i Z_i} \times \quad (4.81) \\ &\quad f\left(\boldsymbol{p}|\boldsymbol{x}_1,\ldots,\boldsymbol{x}_N;\boldsymbol{q}_k\right)\end{aligned}$$

Thus, the posterior is a weighted mixture of the conjugate posteriors associated with each prior in the mixture. For EFs, this posterior mixture has the simple form:

$$
\begin{aligned}
f\left(\boldsymbol{p}|\boldsymbol{x}_1,\ldots,\boldsymbol{x}_N;\boldsymbol{q}_1,\ldots,\boldsymbol{q}_K\right) &= \sum_{k=1}^{K}\frac{w_k Z_k}{\sum_{i=1}^{K} w_i Z_i} \times \\
&\quad \exp\left(\boldsymbol{q}_k^N\cdot\left(\boldsymbol{p},-a\left(\boldsymbol{p}\right)\right)-a_0\left(\boldsymbol{q}_k^N\right)\right) \\
&= \left[\sum_{i=1}^{K} w_i \exp\left(a_0\left(\boldsymbol{q}_i^N\right)-a_0\left(\boldsymbol{q}_i\right)\right)\right]^{-1} \times \\
&\quad \sum_{k=1}^{K} w_k \exp\left(\boldsymbol{q}_k^N\cdot\left(\boldsymbol{p},-a\left(\boldsymbol{p}\right)\right)\right. \\
&\quad \left. -a_0\left(\boldsymbol{q}_k\right)\right)
\end{aligned}
\tag{4.82}
$$

where $\boldsymbol{q}_k^N = \boldsymbol{q}_k + \left(\sum_{n=1}^{N}\boldsymbol{g}\left(\boldsymbol{x}_n\right),N\right)^T$.

4.6   Distributions defined through quantiles

In Section 3.2 we encountered distributions whose quantile function is available in closed-form, in the context of random sampling. In this section we will explore this idea as an approach to statistical modelling by manipulating the quantile function (Gilchrist, 2000). As quantile functions must only be non-decreasing, they can be quite easily manipulated using several rules which generate new distributions (see Table 4.3). For example, given a random variable X, the *location-scale-shape(post) transform* $T\left(X\right) = \sigma X^\kappa + \mu$ "shapes" the random variable by κ, scales it by $\sigma > 0$ and shifts it by μ. This is a very widely used and simple approach to useful generalization of a "standard" distribution. Table 4.4 shows that, composing simple post-transformations of the standard uniform random variable on $[0,1]$ with this location-scale-shape transform, generates a host of important and useful distributions. An example is the Cauchy distribution with $\Phi\left(p\right) = \tan\left(\pi\left(p-\frac{1}{2}\right)\right)$ and $\kappa = 1$, and a somewhat more complex example is the Laplace distribution where $\Phi\left(p\right) = \mathrm{sgn}\left(\frac{1}{2}-p\right)\ln\left(1-2\left|\frac{1}{2}-p\right|\right)$ and $\kappa = 1$.

These rules expose very important relationships between distributions. Taking the exponential distribution with $Q_1\left(p\right) = -\ln\left(1-p\right)$, reflecting this to get $Q_2\left(p\right) = \ln\left(p\right)$, then applying the addition rule, we obtain:

$$
\begin{aligned}
Q\left(p\right) &= Q_1\left(p\right) + Q_2\left(p\right) \\
&= -\ln\left(1-p\right) + \ln\left(p\right) = -\ln\left(\frac{1-p}{p}\right)
\end{aligned}
\tag{4.83}
$$

which is the quantile function for the logistic distribution. Similarly, raising the exponential quantile function to the power κ gives the Weibull distribution. Gilchrist (2000, Chapter 6) explores many such relationships in detail.

Rule	Transformation	Conditions
Addition	$Q_1(p) + Q_2(p)$	
Multiplication	$Q_1(p) \times Q_2(p)$	$Q_1, Q_2 > 0$
Convex sum	$wQ_1(p) + (1-w)Q_2(p)$	$0 < w < 1$
Post-transform	$T(Q(p))$	T non-decreasing
Pre-transform	$Q(S(p))$	S non-decreasing, $S(0) = 0, S(1) = 1$
Reciprocal	$1/Q(1-p)$	$X \leftarrow 1/X$
Reflection	$-Q(1-p)$	$X \leftarrow -X$

Table 4.3: Transformation rules for quantile functions which produce new quantiles.

Expectations often have quite simple forms for these distributions, because the required definite integral is often easy to compute:

$$E[X] = \int_0^1 Q(p)\, dp \qquad (4.84)$$

and more generally, the k-th moment is:

$$E[X^k] = \int_0^1 (Q(p))^k\, dp \qquad (4.85)$$

So, for the location-scale-shape transform with $\kappa = 1$, $E[T(X)] = \mu + \sigma E[X]$. In particular, for base quantiles which are symmetric about $p = \frac{1}{2}$, then $E[T(X)] = \mu$. This is true of the Gaussian and Laplace distributions. The variance is:

$$
\begin{aligned}
\text{var}[X] &= \int_0^1 \left(Q(p) - \int_0^1 Q(p)\, dp \right)^2 dp \\
&= E[X^2] - E[X]^2 \qquad (4.86)
\end{aligned}
$$

Relationships among the non-central and central moments follow all the same rules as when the expectation is computed using the PDF (Gilchrist, 2000, page 71).

It is no surprise that there are simple analytical formulae for specific quantiles of distributions defined using the quantile function and transformation rules. The median is $Q\left(\frac{1}{2}\right)$, so that for location-scale-shape distributions, the median is $\mu + \sigma \left(T\left(\frac{1}{2}\right)\right)^\kappa$ and if, in particular $T\left(\frac{1}{2}\right) = 0$, then the median simplifies to μ (this is the case for the Gaussian, Cauchy, logistic and Laplace distributions, for example). If $T\left(\frac{1}{2}\right) = 1$ then the median is $\sigma + \mu$. Similarly, the *interquartile range* $Q\left(\frac{3}{4}\right) - Q\left(\frac{1}{4}\right)$, a useful measure of spread which plays a role similar to the standard deviation, often has very simple formulae for these distri-

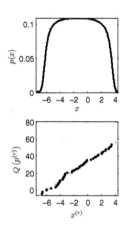

Fig. 4.4: Using quantile matching to estimate distributional parameters. A data set of $N = 50$ samples (bottom panel, black dots) was generated from the distribution with quantile function $Q(p) = \mu + \sigma\left(\ln\left(\frac{p}{1-p}\right) + \tau p\right)$ with $\sigma = 0.17$, $\mu = -5.6$ and $\tau = 50$. The PDF can be computed numerically (top panel), but this PDF is not available in a simple closed-form making maximum likelihood very difficult. Minimizing the squared error functional (4.87) with respect to σ, μ is a closed-form least-squares regression problem giving the reasonable estimates $\hat{\mu} = -5.96$ and $\hat{\sigma} = 0.18$ for the best fit line (bottom panel, dashed line) for the analytical quantiles $Q\left(p^{(r)}\right)$ (vertical axis) against the sample quantiles $x^{(r)}$ (horizontal axis).

Distribution	μ	κ	$\Phi(p)$
Pareto	0	$\kappa < 0$	$1 - p$
Exponential	0	1	$-\ln(1 - p)$
Weibull	0	$\kappa > 0$	$-\ln(1 - p)$
Logistic	μ	1	$-\ln((1 - p)/p)$
Log-Logistic	0	$\kappa > 0$	$p/(1 - p)$
Gumbel	μ	1	$-\ln(-\ln p)$

Table 4.4: Random variables whose quantile function is in the form $Q(p) = \mu + \sigma(\Phi(p))^{\kappa}$ where Φ is the "base" transformation, with the general restriction that the scale parameter $\sigma > 0$. Specific restrictions on the location μ and shape κ parameters for each distribution are also listed (for example, the exponential distribution must have $\mu = 0$).

butions: to illustrate the point, for the Laplace distribution it is easily obtained as $2\sigma \ln(2)$.

Section 4.2 indicates that finding the MLE requires knowledge of the PDF. With only the quantile function this is difficult, but we can instead turn to *quantile matching*. The approach makes use of the *sample quantiles*, also known as *order statistics* of a dataset. Consider an i.i.d. sample $x_1, \ldots, x_N$, then, sorting these samples in ascending order, the r-th value, $x^{(r)}$ for $r = 1, \ldots, N$ is known as the *r-th order statistic*. The special statistic corresponding to $Q\left(\frac{1}{2}\right)$, known as the *median*, has exactly half the data less than it, and half larger than it. So, for example, for N odd, this is $x^{((N+1)/2)}$. The order statistics occur at the corresponding probabilities $p^{(r)} = \frac{r}{N}$. We can therefore consider estimating the parameters of the distribution (e.g. σ, μ and κ for the location-scale-shape transformed quantile function) by minimizing some kind of *quantile matching error* with respect to the parameters. For example, the squared error is:

If N is even, the median is any value which lies in the middle of the ranked data.

$$E = \sum_{r=1}^{N} \left(Q\left(p^{(r)}\right) - x^{(r)} \right)^2 \qquad (4.87)$$

In the case of location-scale transformations, this is a standard linear regression problem with a closed-form solution (Figure 4.4).

4.7 Densities associated with piecewise linear loss functions

We have seen above that it is fruitful to consider that loss functions and distributions are intimately related. Many practical loss functions used in machine learning are *piecewise linear* (PWL), that is, they are functions constructed from a few linear segments joined together at *knots* where the lines intersect. PWL loss functions have considerable importance in *sparsity-promoting inference* situations, arising in many contexts in DSP and machine learning, which will be discussed in detail in later chapters.

An important example is the asymmetric *check* or *tick-loss* $L(x, \mu, q) = (x - \mu)(q - \mathbf{1}[x \leq \mu])$ which is a metric associated with an *asymmetric Laplace distribution*:

$$f(x; \mu, \sigma, q) = \frac{1}{\sigma} q(1-q) \exp\left(-\frac{1}{\sigma} L(x, \mu, q)\right) \tag{4.88}$$

where the *quantile* parameter $q \in [0, 1]$ and scale parameter $\sigma > 0$. Assuming we have N i.i.d, asymmetric Laplace distributed data points, the MLE for μ is the qth quantile. Quantile matching gives the same conclusion. The quantile function for the "standardized" form of this distribution (with $\mu = 0$, $\sigma = 1$) can be written as:

$$\Phi(p) = \begin{cases} \frac{1}{q-1} \ln\left(\frac{q}{p}\right) & p > q \\ \frac{1}{q} \ln\left(\frac{q-1}{p-1}\right) & p \leq q \end{cases} \tag{4.89}$$

so that $\Phi(p) = 0$, and application of the location-scale-shape transform $Q(p) = \mu + \sigma\Phi(p)$ from Section 4.6 gives us $Q(p) = \mu$.

The corresponding MLE for the scale parameter σ is:

$$\sigma = \frac{1}{N} \sum_{n=1}^{N} L(x_n, \mu, q) \tag{4.90}$$

which can be recognized as the sample expected *asymmetric* absolute error around the location parameter μ. This tick loss (and associated distribution) is often invoked where the data has asymmetric outliers and we want an estimate of the 'center' of the distribution of X. There are many applications of this in machine learning, in particular in *quantile regression*, and in *econometric time series analysis*, which overlaps substantially with DSP.

Another PWL loss function of particular importance is the three-segment ϵ-*insensitive error*:

$$L(x, \mu, \epsilon) = \begin{cases} 0 & |x - \mu| < \epsilon \\ |x - \mu| - \epsilon & \text{otherwise} \end{cases} \tag{4.91}$$

This function is similar to the absolute loss but positive or negative errors of at most $\epsilon > 0$ in magnitude, do not contribute to the loss. The insensitive region is known as the *margin*. This is a powerful idea which forms the basis of *support vector regression* (SVR), which gets a lot of use in statistical machine learning. The associated PDF is exponential on either side of the margin, and uniform inside this region.

If we make the general association with a PDF by applying the information map to a PWL loss function, $f(x; \mathbf{p}) = Z^{-1} \exp(-L(x, \mathbf{p}))$, it turns out to be particularly easy to sample from such distributions

Setting $q = \frac{1}{2}$ recovers the Laplace distribution, for which, as we have established earlier, the MLE for μ is just the median. As q approaches 1 and 0, the MLE for μ approaches the sample maximum and minimum, respectively.

(Görür and Teh, 2011). We can write all PWL loss functions in the form:

$$L(x, \boldsymbol{p}) = \max_{i=1,\ldots,K} [a_i x + b_i] \tag{4.92}$$

where $\boldsymbol{p} = (a_1, \ldots, a_K, b_1, \ldots, b_K)$ is a vector of $2K$ real-valued slope and intercept parameters. If the knots are located at $z_0 < z_1 < \cdots < z_K$ and the slope is a_i for $x \in [z_{i-1}, z_i]$, then the area Z_i under each exponential segment of the distribution is:

$$Z_i = \int_{z_{i-1}}^{z_i} \exp(a_i x + b_i)\, dx = \frac{1}{a_i} [\exp(a_i z_i + b_i) - \exp(a_{i-1} z_{i-1} + b_{i-1})] \tag{4.93}$$

Using these, we can draw a sample x from the distribution by first choosing a segment i with probability proportional to Z_i, then sampling a value from $[z_{i-1}, z_i]$ using the quantile function of that segment, e.g. choosing u from the uniform distribution on $[0, 1]$ and setting $x = \frac{1}{a_i} \ln(u \exp[a_i x_i] + (1 - u) \exp[a_i x_{i-1}])$.

In the degenerate case where $a_i = 0$ then that segment is just chosen uniformly on the interval $[z_{i-1}, z_i]$.

This information map association with a PDF is only really meaningful if the loss function is convex. Otherwise, Z is infinite and we cannot normalize the density function. Nonetheless (non-normalizable) PWL loss functions are useful on their own in machine learning and DSP. A particularly important example of a non-normalizable PWL loss is the *hinge loss* $L(x, \mu) = \max[1 - \mu x, 0]$ which is a crucial component of the *support vector machine* (SVM). This is an asymmetric loss that with the property that for $\mu = 1$ and $x > 1$, then $L(x, 1) = 0$ and for $x \leq 1$, then the loss increases linearly. The situation is reversed for $\mu = -1$. In the context of classification, this property assigns zero loss to predictions which are correctly classified, and increasing loss to misclassifications. Although ideally, we would like to assign the same loss value to all misclassified data points, the resulting *0-1 loss function* is non-convex (and, subgradients do not exist) which makes parameter inference very hard. By contrast, the hinge loss is convex with subgradients, which means we can use the relevant optimization methods from Section 2.4.

This leads to a more general observation. Inference problems involving PWL loss functions are quite tractable. In fact, they are solvable using convex non-differentiable optimization methods, for example, regression using the check loss function (known as *quantile regression* problems) can be solved using LP, and solutions to the SVM problem, which can be formulated as a hinge loss with L_2-norm regularizer, can be obtained using QP. An interesting fact shown by Rosset and Zhu (2007) is that all problems in the form:

$$\sum_{n=1}^{N} L(y_n, \boldsymbol{x}_n^T \boldsymbol{p}) + \gamma M(\boldsymbol{p}) \tag{4.94}$$

In fact, here, the loss function L, can, more generally, be piecewise quadratic.

where the loss L and regularization functions M are convex, non-negative and PWL, have PWL regularization paths with respect to $\gamma \geq 0$, exactly as for the TVD functional (2.47). This means that (in principle at least), we can readily compute the solutions for all values of γ.

4.8   Nonparametric density estimation

All the models given earlier are parametric. This is reasonable if the form of the density function is known or can be justified on the basis of other information. However, often it is not reasonable to make this assumption, and *nonparametric* density estimates are preferable.

Given some N i.i.d. data $\boldsymbol{x}_n$, the empirical density (EPDF) f_N (1.45) is always available and is perhaps the simplest nonparametric model. However, this model has no smoothness at all which makes it mostly unusable in practice. The next simplest density estimator is the well-known (one-dimensional, equal-bin sized) *histogram* which constructs a density estimate by normalizing the count of the number of data points which fall into a partition of the space of $\boldsymbol{x}$ into regions or *bins*:

$$f(x) = \frac{1}{N} \sum_{n=1}^{N} \sum_{k=1}^{K} \mathbf{1} \left[w(k-1) + x_{\min} \leq x < wk + x_{\min} \right] \qquad (4.95)$$

where K is the number of bins, the bin width is $w = \frac{x_{\max} - x_{\min}}{K}$ and $x_{\min}, x_{\max}$ are the minimum and maximum of the data, respectively. Extensions to multiple dimensions are straightforward, and it is also possible to have arbitrary bin sizes. The histogram estimator has the big advantage over the EPDF that is has some smoothness, i.e. it is constant in between the bin edges, and so, more useful in practice. Sampling from this estimator is simple: pick a bin with probability equal to the bin count over N, and sample uniformly from within that bin. Selecting the number of bins requires making some assumptions, in particular, assuming that the data is normal leads to the binomial *Sturge's rule* $K = 1 + \log_2 N$ (Scott, 1992, page 48).

The major problem with the histogram is that it is not continuous everywhere: it has jumps at the bin edges. This is often an unrealistic assumption; in reality most data does not exhibit these kinds of arbitrary discontinuities. One way around this is to place a smooth *kernel* PDF function $\kappa(\boldsymbol{x}; \boldsymbol{\mu})$ with mean $\boldsymbol{\mu}$ equal to the value of each sample (Silverman, 1998). This gives rise to the *kernel density estimate* (KDE):

$$f(\boldsymbol{x}) = \frac{1}{N} \sum_{n=1}^{N} \kappa(\boldsymbol{x}; \boldsymbol{x}_n) \qquad (4.96)$$

This is in fact an example of a *mixture distribution*. Sampling from this estimator involves picking one of the samples from the dataset with uniform probability, then sampling from the kernel density centred on that sample. A very widely-used choice of distributions is the (isotropic) multivariate Gaussian with covariance $\boldsymbol{\Sigma} = \sigma\mathbf{I}$ in which case σ is called the *bandwidth* of the estimator.

As with the histogram estimator, where κ has a (bandwidth) parameter, it must be chosen, and it determines the smoothness of the estimator.

4.9 Inference by sampling

There is a particular pattern to the previous sections: we choose a certain (parametric) model for the data, and, using some kind of procedure, whether by analytical formula or by numerical optimization, we infer the optimal parameter values by loss minimization, or the application of Bayes' rule. These are *deterministic* approaches, which are characterized by the fact that given any set of fixed data values (and specific initial guess value for an optimization procedure), the inferred parameter values will always be the same. There are advantages to deterministic inference (all computations are 'predictable'), but there are some serious limitations as well. In many cases, deterministic inference in the parametric model will be computationally infeasible or require unacceptable approximations.

An alternative is to sample from the model and use these samples to infer the values of the parameters. Let us explore how this works in practice. Consider a (generally fully Bayesian) model $f(\boldsymbol{x}, \boldsymbol{p}; \boldsymbol{q})$ for our data. We will typically have fixed observations $\boldsymbol{x}_1, \ldots, \boldsymbol{x}_N$. Similarly, we would have chosen the hyperparameters $\boldsymbol{q}$. Sampling methods can be used to draw values for the unknown parameters $\boldsymbol{p}$ which are consistent with the model evaluated at $\boldsymbol{x}_1, \ldots, \boldsymbol{x}_N, \boldsymbol{q}$ i.e. we will use a sampling method to draw samples from $f(\boldsymbol{x}_1, \ldots, \boldsymbol{x}_N, \boldsymbol{p}; \boldsymbol{q})$. We can then use summaries of the distribution over these parameters, for instance, we may be interested in the values of these parameters at the mean or mode.

MCMC inference

Section 3.4 explored some of the most popular MCMC methods such as Gibbs sampling and the MH algorithm. All of these methods can be adapted to perform parameter inference by the trick discussed above, i.e. by fixing the values of the known variables (for example, the data) and drawing samples from the rest. For example, for the K parameter values in $\boldsymbol{p}$, each of the $k = 1, 2, \ldots, K$ Gibbs steps in Algorithm 3.5 becomes:

$$p_k^{n+1} \sim f\left(p_k \,\middle|\, p_1^n, \ldots, p_{k-1}^n, p_{k+1}^n, \ldots, p_K^n, \boldsymbol{x}_1, \ldots, \boldsymbol{x}_N, \boldsymbol{q}\right) \qquad (4.97)$$

Assessing convergence in MCMC methods

The simplicity of sampling methods for inference described above gives us some indication as to why they are so popular in practice. However, we should not overlook the limitations. The underlying principle of MCMC methods is that, after an infinite number of samples, the sampling distribution of the draws from the random variables converges on the target distribution. The key problem with this is: how many samples suffice?

An important intuition is that starting values of the variables will be poor, and these will improve with subsequent iterations; these starting values (often called the "burn-in period", see Figure 4.5) should therefore be discarded. Immediately one can see that selecting too few samples for the burn-in period will put "bad" samples into the estimate of the target distribution, whereas, selecting too many burn-in samples will reduce the number of samples available to produce a good quality estimate. Drawing more samples than necessary leads to excessive computational burden. Also, because the samples are drawn from a Markov chain they are not independent, but the goal is to draw independent samples from the target distribution. There is no one solution for dealing with these difficulties; we will next discuss some popular approaches.

Firstly, we expect that at convergence the samples will be drawn from the target distribution. Therefore, after burn-in, if we take subsets of samples, all subsets should have approximately the same distribution. Comparing the distribution of these subsets (or the distribution of quantities such as moments estimated from these subsets) should give confidence that any particular subset is representative of the target distribution. This is the conceptual basis of the *Geweke test* (Geweke, 1991): excluding burn-in, test that the sample means of two subsets are (approximately) equal $\mu_1 \approx \mu_2$. Under the assumption that the target distribution is Gaussian, then the (*Student's*) *t-test* statistic:

$$t = \frac{\mu_1 - \mu_2}{\sqrt{\sigma_1^2/N_1 + \sigma_1^2/N_2}} \tag{4.98}$$

has the *Student-t* distribution with number of degrees of freedom which is a function of σ_1^2, σ_2^2 (the sample variances) and N_1, N_2 (the number of samples in the two subsets). Thus, the probability that the two means are equal can be computed and used to decide whether the subsets are likely to be equal in distribution or not. Of course, the assumption that the distribution is approximately normal may be unjustified and more general, "distribution-free" approaches may be required. For example, one can substitute the *Kolmogorov-Smirnov test* (Press, 1992, page 623) for the t-test in the subset approach: this test computes the maximum difference in ECDF for the two subsets instead, and is thus applicable to any univariate distribution.

Both tests above, and in fact many statistical estimators, assume that the samples are independent which is not true for MCMC samples. Reducing the dependency between samples in the sequence requires increasing the acceptance rate which depends in a complex way upon the relationship between the proposal and target distributions. A simple approach which reduces dependency at the expense of increased computational burden is to discard all but one out of every $M > 1$ samples, a process known as *thinning*.

Each run of the algorithm will produce different samples, but at con-

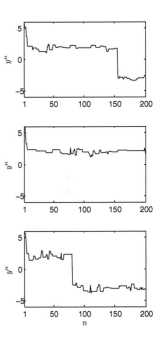

Fig. 4.5: Demonstrating the convergence properties of Metropolis-Hastings (MH) Markov-Chain Monte Carlo (MCMC). Each panel shows a separate run of **200 MH** iterations applied to the two-component Gaussian mixture model with Y variable parameters $\mu_1 = -3$, $\mu_2 = 2$, $\sigma = 0.3$, and X variable (Bernoulli) parameter $p = 0.6$. The proposal distribution is a single Gaussian $g(x|x') = \mathcal{N}(x; x', \tau)$ with standard deviation $\tau = 1.5$. Each run starts from the same initial condition $y^1 = 6$. The "burn-in" period of around 10 iterations can be seen at the start of each run. Notice that, from the second run alone, we would not be able to find the two modes in the distribution of Y.

vergence, samples across runs should come from the same distribution: the target. Of course, for any finite-length run this might be far from true – this effect is clearly shown in Figure 4.5 where although the target has two modes, one of the runs is restricted to one mode. This shows that it is not always possible to detect convergence from a single run.

This problem suggests a 'multiple-run' strategy, i.e. one where convergence is detected by analyzing several short runs from different starting conditions. This strategy has the advantage that the runs can be carried out entirely separately from each other, which is a natural fit for parallel computing architectures. Gelman (2004, page 296) starts from the observation that prior to convergence, the variance of estimators of quantities from samples in each run will be larger between each run than within them. Computing a function of the between- and within-run variances can be used to determine whether to continue to draw samples or to stick with the samples drawn so far (Gelman, 2004, page 297).

Cowles and Carlin (1996) is a comprehensive review of similar *convergence diagnostics* for MCMC inference. It concludes that no single diagnostic is able to detect all forms of non-convergence and therefore applying several diagnostics simultaneously over multiple parallel runs is necessary.

Probabilistic graphical models

5

Statistical machine learning and statistical DSP are built on the foundations of probability theory and random variables. Different techniques encode different dependency structure between these variables. This structure leads to specific algorithms for inference and estimation. Many common dependency structures emerge naturally in this way, as a result, there are many common patterns of inference and estimation that suggest general algorithms for this purpose. So, it becomes important to formalize these algorithms; this is the purpose of this chapter. These general algorithms can often lead to substantial computational savings over more brute-force approaches, another benefit that comes from studying the structure of these models in the abstract.

5.1 Statistical modelling with PGMs

Consider a set of N random vector variables organized as a matrix $\boldsymbol{X} \in \mathbb{R}^{D \times N}$, in the case where $D = 1$ this might represent a finite-length sample of a discrete-time stochastic process X_n, for example. In keeping with the rest of this book, the goal of machine learning is to estimate the parameters or values of unobserved variables in a (Bayesian) model for the full set of joint random variables $\boldsymbol{X}$. This will require us, in general, to take into consideration the complete set of probabilistic interactions between $N \times D$ variables, which is of order $O\left(N^2 D^2\right)$. Now, let us assume that the variables all have the same finite range, say $[0, 1]$ for simplicity. If it requires M random realizations from a single variable/parameter to estimate its value, then it will require M^2 values to estimate two variables simultaneously if they share some general probabilistic dependence, and in general M^L for L variables at once. In other words, the growth of the number of required observations is exponential in the number of mutually dependent variables we want to estimate. This phenomenon, known as the *curse of dimensionality*, means that data requirements for statistical model fitting are generally intractable for fully-dependent sets of random variables.

For this reason, nearly all practical machine learning and DSP assumes some kind of independence between the variables in the model, the simplest assumption being that the dimensions or the observations, or both, are i.i.d. This is an extreme assumption to make, and cannot

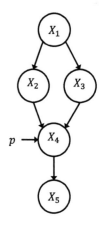

Fig. 5.1: A simple example of a probabilistic graphical model over 5 variables $X_1, X_2, \ldots, X_5$ with one hyperparameter p (not a random variable, and so not inside a circular node). The particular structure of the graph determines the conditional independence structure, e.g. that X_5 and X_4 do not depend upon X_1. From this structure, we can simplify the joint distribution to the factorization $f(x_1, x_2, \ldots, x_5) = f(x_1) f(x_2|x_1) f(x_3|x_1) f(x_4|x_2, x_3; p) \times f(x_5|x_4)$, where there are at most three variables involved in any kind of statistical inference on these variables.

Machine Learning for Signal Processing: Data Science, Algorithms, and Computational Statistics. Max A. Little.

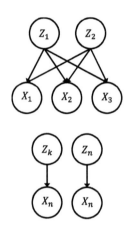

Fig. 5.2: Expressing repeated connectivity in probabilistic graphical models. The variables Z_1, Z_2 together influence all the variables X_1, X_2, X_3 (top graph). This bipartite graph is succinctly denoted by connecting a single Z_k node to a single X_n where $k \in 1, 2$ and $n \in 1, 2, 3$ (bottom left graph). An alternative structure has each X_n depend only upon one Z_n, this is denoted by a single link between one Z_n node to a single X_n node (bottom right graph).

generally be justified for DSP applications, for example. So, in practice, some forms of limited dependence are proposed; the art of model construction for statistical machine learning often comes down to judicious dependency choices.

These choices can be encoded as a *probabilistic graphical model* (PGM). A PGM is a graph (1.6) whose vertices V are a set of random variables in the model, and the (directed) edges E represent the dependency between these variables. So, the complete PGM represents model in which every variable is dependent upon every other. A Markov chain (1.4) is a tree in which there is a linear chain of directed edges going in one direction only.In this book we will only work with PGMs that are DAGs, because dependency loops are usually not needed in practice and introduce unnecessary complexities in inference. DAG PGMs allow factorization of the joint distribution over the random variables, as we describe next. To simplify the discussion (without loss of generality) consider the joint distribution over $D = 1$ dimensional random variables in the model $f(x_1, x_2, \ldots, x_N)$. Using the basic definition of conditional probability, we can factorize this into a product of conditional distributions:

$$
\begin{aligned}
f(x_1, x_2, \ldots, x_N) \;=\; & f(x_1 | x_2, \ldots, x_N) \times \\
& f(x_2 | x_3, \ldots, x_N) \times \\
& \vdots \\
& f(x_{N-1} | x_N) \times \\
& f(x_N)
\end{aligned}
\tag{5.1}
$$

since e.g. $f(x_{N-1} | x_N) f(x_N) = f(x_N) f(x_{N-1}, x_N) / f(x_N) = f(x_{N-1}, x_N)$ and so on up. This is known as the *chain rule* of probability. Note that there are other permutations by which this factorization can be done, for instance, for three variables, two alternative factorizations are $f(x_1, x_2, x_3) = f(x_1 | x_2, x_3) f(x_2 | x_3) f(x_3)$ and $f(x_3 | x_1, x_2) f(x_1 | x_2) f(x_2)$. Note that this holds *in general*, it is true for all sets of random variables, so this factorization is just another way of writing out the joint distribution. However, given a specific PGM, there will be certain *conditional independencies* implied by the graph (Figure 5.1). For example, in the three variable case, if X_1 does not depend upon X_3, the joint distribution simplifies to $f(x_1, x_2, x_3) = f(x_1 | x_2) f(x_2 | x_3) f(x_3)$. Whereas, in the three variable case, we required training data for all three variables simultaneously, we now find that we only need training data for at most two variables simultaneously, significantly reducing the demand for training data.

Generally, using the PGM, we can write down the product of N factorized distributions directly using knowledge of $\mathcal{P}(n)$, the set of indices of *parent variables* for random variable X_n:

$$
f(x_1, x_2, \ldots, x_N) = \prod_{n=1}^{N} f\left(x_n | (x_{n'})_{n' \in \mathcal{P}(n)}\right)
\tag{5.2}
$$

where a parent variable is simply a variable upon which a given variable depends. For N variables, this factorization reduces the training set data requirement from $O\left(M^N\right)$ to $O\left(M^L\right)$ where $L = \max_{n\in 1,2,\ldots,N} |\mathcal{P}(n)| + 1$ is the largest number of variables (variable and its parents) in any single conditional distribution in the factorization. Thus, for efficiency, L should generally be small which means reducing the maximum number of parents for any variable.

This information can be obtained directly from the adjacency matrix of the graph, it is just the maximum vertex in-degree.

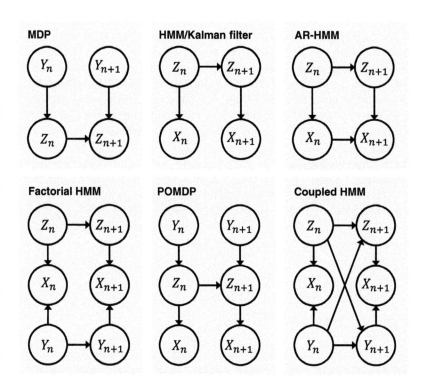

Fig. 5.3: Some more complex probabilistic graphical models. *Markov decision process* (MDP): a semi-Markov chain with hidden conditioning Y_n in which the state Z_n is observed. *Hidden Markov model* (HMM): Markov chain in which the state is usually hidden while the observations X_n depend upon the state; the same structure as the continuous-state Kalman filter. *HMM with autoregressive* (AR) *observations* (AR-HMM) in which the observations are temporally dependent upon previous observations. *Factorial HMM*: the observations X_n depend upon more than one hidden Markov chain. *Partially observable MDP* (POMDP): MDP where the semi-Markov chain states are hidden. *Coupled HMM*: An HMM with two or more semi-Markov chains, each acting upon each other.

A commonly encountered PGM in machine learning and DSP is a repetitive sub-graph, for example, a model with N variables that mutually depend upon K parameter variables but are otherwise independent. In those cases we use a succinct notational device that involves a single directed edge between the two types of variables to indicate this directional, bipartite relationship. For example, the graph $Z_k \to X_n$ indicates the bipartite model in which each of $n \in 1, 2, \ldots, N$ variables

Other textbooks and authors use so-called plate notation *but this does not convey more than the notation used here.*

X_n depend only upon each of $k \in 1, 2, \ldots, K$ variables Z_k, whereas the graph $Z_n \to X_n$ is a bipartite model in which each of n variables X_n depend only upon their corresponding Z_n (see Figure 5.2).

PGMs capture the dependency structure of a vast array of machine learning/DSP algorithms (Figure 5.4). For example, many basic models are exemplified by the pattern $Z_n \to X_n$. These are the simplest *hierarchical* Bayesian models. The case where Z_n is a discrete random variable describes methods such as *mixture models* and *classification* (see Section 6.3). The case where Z_n is continuous describes methods including various forms of *regression* and *probabilistic principal components analysis*.

Introducing data ordering into the model, most simply in the non-hierarchical, "first order" case $Z_n \to Z_{n+1}$ captures the structure of Markov chains (see Section 1.4), which include classical DSP methods such as *autoregressive* models. Where the dependency extends further back in time we obtain *higher-order* Markov chains; but we can always express these as first order chains to order D by *embedding* to obtain $\boldsymbol{X}_n \to \boldsymbol{X}_{n+1}$ where $\boldsymbol{X}_n = (X_n, X_{n-1}, \ldots, X_{n-D})$. Hierarchical models with time dependence such as $Z_n \to Z_{n+1} \to X_{n+1}$ include *hidden Markov models* (HMM) when the variables Z_n are discrete, and *Kalman filters* when the Z_n and X_n are continuous (and Gaussian in the simplest case). Hierarchical models with time dependence at both levels, $Z_n \to Z_{n+1}$ and $X_n \to X_{n+1}$ along with the hierarchical dependence $Z_n \to X_n$ represent *autoregressive* HMMs (Figure 5.3).

5.2   Exploring conditional independence in PGMs

The conditional independence structure of a PGM enables many practical insights which are of direct importance in statistical machine learning and probabilistic modelling in general.

Hidden versus observed variables

In machine learning and DSP applications it is nearly always the case that some, but not all of the PGM variables are observed, that is, we have realizations for only some of the nodes. The rest of the random variables are 'hidden' or 'latent' variables. For example, in an HMM, we do not usually observe the Markov state Z_n. Instead, we have data for the observed variables X_n, and a typical goal is to *infer* the distribution of the hidden states. By contrast, in (supervised) classification problems, there is data for both the hidden (class) variable Z_n and the feature data X_n to allow inference of the distribution of the model parameters. Then, the distribution over the class variable can be inferred using some subsequent data.

Whether a variable is hidden or observed thus depends upon the context, and it influences the nature of conditional independence between variables as we discuss next.

Directed connection and separation

Are two nodes in a PGM dependent, possibly conditional on some other node(s)? PGMs being acyclic graphs, influence between nodes is 'transmitted' through (undirected) paths between them. To answer the question, it suffices to consider the smallest 'unit' of transmission: any group of three nodes X, Y, Z connected by two edges. Then, every (undirected) path between nodes can be broken down into the constituent triplets of nodes that make up that path. In every such triplet, there are only three fundamentally different configurations of directed edges between them.

The first configuration is the simple *chain*, $X \to Y \to Z$. It is fairly obvious that X influences Z because X influences Y which then influences Z. The joint distribution is $f(x, y, z) = f(x) f(y|x) f(z|y)$. In this configuration, X, Z are not independent because:

$$f(x, z) = \int f(x) f(y|x) f(z|y) \, dy \qquad (5.3)$$
$$= f(x) \int f(y|x) f(z|y) \, dy$$

Another configuration has $X \leftarrow Y \leftarrow Z$; this is also a chain and all the arguments here apply with the role of X and Z reversed.

and since $\int f(y|x) f(z|y) \, dy$ is a function of X, this joint distribution cannot be written as a product of factors of X and Z alone. However, while X, Z are not independent, if we condition on Y they are, since:

$$f(x, z|y) = \frac{f(x) f(y|x) f(z|y)}{f(y)}$$
$$= \frac{f(y, x) f(z|y)}{f(y)} \qquad (5.4)$$
$$= f(x|y) f(z|y)$$

We can think about observation of Y "blocking" the influence of X on Z.

The second configuration is the *common cause*, $X \leftarrow Y \to Z$, that is, where X and Z depend mutually upon Y. It is less obvious in this situation that X, Z are dependent: here is an intuitive explanation. Consider Y to represent the status of your car's battery, X to represent whether your car's lights work, and Z whether the radio works. If you cannot get the lights to work, this influences whether the battery is working, and in turn this influences whether the radio will work. Formally, the joint distribution is $f(x, y, z) = f(y) f(x|y) f(z|y)$, so that $f(x, z) = \int f(y) f(x|y) f(z|y) \, dy$ which cannot be written as a product of factors in X and Z, so they are dependent. However, as with the chain, observation of Y makes X, Z independent:

$$f(x, z|y) = \frac{f(y) f(x|y) f(z|y)}{f(y)} \qquad (5.5)$$
$$= f(x|y) f(z|y)$$

Finally, consider the configuration $X \to Y \leftarrow Z$ where Y is a *common*

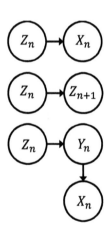

Fig. 5.4: Three basic probabilistic graphical model patterns. One-level hierarchical models (top graph) represent i.i.d. situations which have some kind of 'latent' structure on which the 'observed' data depends. Basic Markov models have (one time step) temporal dependence (middle graph). Two-level hierarchies (bottom graph) capture more complex i.i.d. models, where the latent structure is itself hierarchical.

effect of X, Z. In that situation the two causes of Y are independent: the joint distribution is $f(x, y, z) = f(x) f(z) f(y|x, z)$ so that:

$$
\begin{aligned}
f(x, z) &= \int f(x) f(z) f(y|x, z) \, dy \\
&= f(x) f(z) \int f(y|x, z) \, dy \qquad (5.6)\\
&= f(x) f(z)
\end{aligned}
$$

However, observation of Y makes the two causes mutually dependent, since:

$$
f(x, z|y) = \frac{f(x) f(z) f(y|x, z)}{f(y)} \qquad (5.7)
$$

This counterintuitive idea is known as explaining away *one cause by knowing something about another.*

which cannot be written as a product of factors in X, Z separately. Observation of a mutual effect means that the causes can mutually influence their distributions. If this seems strange, consider whether your car's engine works represented by Y. Two competing causes would be the working status of the battery Z and whether the fuel tank is full, X. If the engine does not start, then there are two potential reasons (flat battery, empty tank), but if the tank also turns out to be full, the battery is more likely to be flat than it would have been without observation of the tank. Thus, conditioning on the common effect forces dependence between previously independent causes.

Using these three cases, we can define the concept of *d-connection*; two nodes X, Z in a PGM are d-connected if the following rules hold:

(1) There is a path between them (ignoring edge directions),

(2) All nodes in chains on that path are unobserved,

(3) All nodes in common cause triplets have unobserved causes,

(4) Any common effect triplets have observed effects, or any common effect triplets have a descendent their effect node to an observed node.

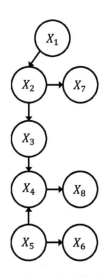

Any pairs of nodes for which one or more of the above conditions do not hold, are *d-separated*. Let us apply these ideas to the example in Figure 5.5. Nodes X_1, X_4 are d-connected, as are nodes X_4, X_6. However, X_1, X_6 are d-separated due to the common effect X_4. But, conditioned on X_4 or on X_8 the end nodes X_1, X_6 *are* d-connected. By contrast, conditioned on X_2, nodes X_1, X_4 are d-separated.

The Markov blanket of a node

For any given node, when conditioned on all the nodes in the network, are all nodes essential to determining this conditional distribution? The answer is generally no, and knowing which nodes are necessary to computing this conditional is important because in practice we can exploit the computational efficiencies entailed. We will first look at comput-

Fig. 5.5: A probabilistic graphical model illustrating d-connectivity. The long (undirected) path between X_1 and X_6 is d-separated by the common effect node X_4 so that X_1, X_6 are independent. But, conditioning on X_4 (or its descendent, X_8) makes X_1, X_6 dependent.

ing the conditional distribution of node X_i conditioned on all the other nodes:

$$f\left(x_i | (x_j)_{j \neq i}\right) = \frac{\prod_{j=1}^{N} f\left(x_j | (x_k)_{k \in \mathcal{P}(j)}\right)}{\int \prod_{j=1}^{N} f\left(x_j | (x_k)_{k \in \mathcal{P}(j)}\right) dx_i} \qquad (5.8)$$

where $\mathcal{P}(i)$ is the set of indices of variables that are parents of X_i. Now, all terms inside the integral in the denominator above which do not involve X_i at all can be taken outside the integral, and in doing so they cancel with the same conditionals in the numerator. We get:

$$f\left(x_i | (x_j)_{j \neq i}\right) = Z^{-1} f\left(x_i | (x_k)_{k \in \mathcal{P}(i)}\right) \times \qquad (5.9)$$
$$\prod_{j : i \in \mathcal{P}(j)} f\left(x_j | (x_k)_{k \in \mathcal{P}(j)}\right)$$

The normalization term Z is obtained by marginalizing out X_i:

$$Z = \int f\left(x_i | (x_k)_{k \in \mathcal{P}(i)}\right) \prod_{j : i \in \mathcal{P}(j)} f\left(x_j | (x_k)_{k \in \mathcal{P}(j)}\right) dx_i \qquad (5.10)$$

This can be stated as the conditional of X_i, given all the other nodes, is proportional to the product of: the node conditioned on its parents and all the children of X_i, in turn conditioned on their parents. This minimal set of nodes on which X_i depends is called the *Markov blanket* of X_i, or $\mathcal{M}(i)$. The Markov blanket effectively isolates the conditional distribution of this node from the rest of the network. We can write this as $f\left(x_i | (x_j)_{j \neq i}\right) = f\left(x_i | (x_j)_{j \in \mathcal{M}(i)}\right)$. The Markov blanket is extremely useful for practical inference algorithms such as Gibbs sampling, as we will discuss later.

5.3 Inference on PGMs

Solving inference problems for PGMs involves finding the distribution of variables (or summary value of those distributions) that cannot be observed in practice. There are several approaches to this, including computation of unknown distributions, maximizing distributions, and marginalization. With posterior computation for example, the variable of interest is the parent of some variables whose (conditional) distributions are known; then the aim is to compute the posterior distribution of the variable of interest. With maximum probability calculations, the goal is to maximize the distribution over the variable of interest; for a parent, this will correspond to MAP inference for example. Finally, in marginalization, we wish to integrate out a variable which generally simplifies inference on the PGM because it reduces the number of degrees of freedom in the PGM.

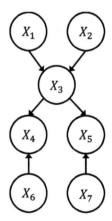

Fig. 5.6: An example Markov blanket. The distribution of X_3 conditioned on all the nodes in the network is independent of all these nodes apart from those depicted in this PGM: that is, the parents of X_3 (X_1, X_2), its children (X_4, X_5), and the children's other parents (X_6, X_7).

Exact inference

In general, exact inference in PGMs is analytically intractable. There are, however, important special cases where inference can be exact. The most commonly-encountered case is where the random variables of interest are discrete and take on finite ranges. In that case, we can usually use *variable elimination*, that we describe next. This is a very general idea and because it has very wide applicability across DSP and machine learning we will introduce it in the algebraically abstract form using semirings (see Section 1.1). Let us consider the situation in which we have a length-N vector of general variables $\boldsymbol{X}$, that each take values in the set X from the semiring $(\oplus, \otimes, X)$ where $\oplus, \otimes$ are generic operators. We also have M *local functions* $f_i : X^{N_i} \to X$ for $i \in 1, 2, \dots, M$ each defined on an N_i-length sub-vector of the variables, $\boldsymbol{X}_i$. We will denote by S_i the set of the indices of the variables making up that sub-vector and $S_1 \cup S_2 \cup \cdots \cup S_M = S$ where $S = \{1, 2, \dots, N\}$. These variable subsets can be overlapping, so $S_i \cap S_j \neq \emptyset$ in general.

From these local functions we form a *global function* as a (semiring) product, $f(\boldsymbol{X}) = \bigotimes_{i=1}^{M} f_i(\boldsymbol{X}_i)$. The problem that variable elimination solves is one of (semiring) summation to remove unwanted variables thereby obtaining a new, *semi-local* (or *semi-global*) *function*. For example, in the case where we want to retain only variables in some subset S_j of size N_j, we need to sum out the $N - N_j = L$ variables $S \backslash S_j$, the set of all variables excluding those in the subset S_j:

$$
g(\boldsymbol{X}_j) = \bigoplus_{\boldsymbol{X}_{S \backslash S_j} \in X^L} f(\boldsymbol{X}) \tag{5.11}
$$

$$
= \bigoplus_{\boldsymbol{X}_{S \backslash S_j} \in X^L} \bigotimes_{i=1}^{M} f_i(\boldsymbol{X}_i)
$$

We can obtain any semi-local function of any desired subset of the variables by the same procedure. This "marginalizing over a product" occurs frequently in machine learning. For example, we can recognize this as the same problem of marginalizing out variables in PGMs. In this form, this computation requires $(M+1) |X|^L$ (semiring) additions and multiplications, so it is an exponential $O\left(|X|^L\right)$ operation and only computationally tractable for small L. However, the fact that the operator $\otimes$ distributes over $\oplus$ means that we can often rearrange the expression and thereby reduce the computational effort required since:

$$
\bigoplus_{i=1}^{M} (a \otimes b_i) = a \otimes \bigoplus_{i=1}^{M} b_i \tag{5.12}
$$

for any arbitrary values $a, b_i \in X$. Here, the left hand side requires M multiplications and additions, but the right hand side only requires one multiplication and M additions, for a saving of $M - 1$ operations. In certain situations, these savings can mount up to make an intractable inference problem, tractable. For illustration consider the simple situation where each $S_i = \{i\}$ and $N = M$ and only one variable per local function, $f_i(X_i)$. Then, in the worst case, if we want to eliminate all the variables, we need to compute:

$$g = \bigoplus_{\boldsymbol{X} \in X^M} \bigotimes_{i=1}^{M} f_i(X_i) = \bigotimes_{i=1}^{M} \bigoplus_{X_i \in X} f_i(X_i) \qquad (5.13)$$

The first expression has exponential complexity, but the second only uses a total of $M|X|$ operations so we have an implementation which has linear complexity $O(M)$ (more pertinently, $O(|X|)$ if $|X| \gg M$). This is a dramatic reduction in computational effort and makes this computation tractable in practice.

In general, this method leads to useful computational savings when the variables appear approximately uniquely among the local functions, so that the products need only be evaluated over a small number of local functions and many variables can be factored out of the summations. For PGMs used in signal processing, where the local functions are conditional distribution functions, this situation arises where the signal, or some latent signal to be estimated, is in the form of a Markov chain (see Section 9.4).

We often need to compute several semi-local functions $g(\boldsymbol{X}_j)$ in one application. In doing so, we typically encounter the same intermediate, partial computation repeatedly; the smart thing to do is to retain these repeated partial computations and simply look up the results when they are required again. A useful method that formalizes the efficient computation of multiple semi-local functions for PGMs is known as the *junction tree* algorithm (Aji and McEliece, 2000). A junction tree (JT) is an undirected tree graph whose vertices v_i are the *maximal cliques* in the *triangulated, undirected moral graph* of the PGM. These cliques C_i are subsets of the PGM variable indices $\{1, 2, \ldots, N\}$. For each clique, there is an associated local function h_i. These local functions will be products of the PGM local (conditional distribution functions) f_i.

This general algorithmic strategy is known as dynamic programming.

A maximum weight, *minimum spanning tree* for that undirected graph with maximal cliques as vertices, is a JT, the edge weights of the tree being the number of shared variables between vertices. For details of the construction of junction trees, see Aji and McEliece (2000). Given a junction tree for a PGM, we can use a simple *message passing* algorithm to efficiently compute any semi-local function on the sets of variables C_i making up the junction tree vertices. These semi-local functions are (semi-ring) marginalized distributions over the variables in that clique.

Firstly, we pass *message functions* $\mu_{i \to j}$ between vertices v_i and v_j in the JT. Marking the desired variable clique vertex in the JT as the root of the tree creates directions for all edges in the JT such that messages can be passed from leaves to root through child vertices. Each JT vertex can only send out messages to its children once it has received all messages from its parents. Once the message passing is complete and the root vertex has received all messages, we can then use these messages to compute the semi-local functions for the desired JT vertex.

Algorithm 5.1 The junction tree algorithm for (semi-ring) marginalization inference of a single variable clique.

(1) *Initialization.* For the junction tree with K vertices v_i, $i = 1, 2, \ldots, K$, corresponding variable cliques $\boldsymbol{X}_{C_i}$ and clique functions h_i, set the *messages* $\mu_{i \to j} \left(\boldsymbol{X}_{C_i \cap C_j} \right)$ for all adjacent vertices $v_i \leftrightarrow v_j$ in the junction tree to the identity for the semi-ring operator $\otimes$,

(2) *Message passing.* Sequentially update all messages from junction tree leaves to root using $\mu_{i \to j} \left(\boldsymbol{X}_{C_i \cap C_j} \right) = \bigoplus_{\boldsymbol{X}_{C_i \setminus C_j} \in X^{|C_i \setminus C_j|}} h_i \left(\boldsymbol{X}_{C_i} \right) \bigotimes_{k : \{v_k \leftrightarrow v_i\} \wedge (k \neq j)} \mu_{k \to i} \left(\boldsymbol{X}_{C_k \cap C_i} \right)$, where the root v_l is the desired set of variables to retain, and $k : \{v_k \leftrightarrow v_i\} \wedge (k \neq j) :$ is the set of junction tree vertices v_k connected to v_i such that $k \neq j$,

(3) *Marginal function computation.* Use messages computed above to find $g_l \left(\boldsymbol{X}_{C_l} \right) = h_l \left(\boldsymbol{X}_{C_l} \right) \bigotimes_{k : \{v_k \leftrightarrow v_l\}} \mu_{k \to l} \left(\boldsymbol{X}_{C_k \cap C_l} \right)$, where $k : \{v_k \leftrightarrow v_l\}$ is the set of all junction tree vertices adjacent to v_l.

Here is a simple example (Figure 5.7): consider the random variables A, B, C, D in the PGM $A \leftarrow B \to C \to D$ with distribution functions $f(b)$, $f(a|b)$, $f(c|b)$ and $f(d|c)$. We are interested in marginalizing out A, B from the joint distribution to leave the distribution $f(c, d)$. A suitable JT is defined as $v_1 \leftrightarrow v_2 \leftrightarrow v_3$ with the corresponding cliques $C_1 = \{A, B\}$, $C_2 = \{B, C\}$ and $C_3 = \{D, C\}$. Thus the 'target' clique/vertex is C_3, obtaining the semi-local function $g_3(c, d)$. The associated clique functions are $h_1(a, b) = f(a|b) f(b)$, $h_2(c, b) = f(c|b)$ and $h_3(c, d) = f(d|c)$. The message passing sequence is $v_1 \to v_2$ and $v_2 \to v_3$.

The first message is $\mu_{1,2}(b) = \bigoplus_a h_1(a, b)$, followed by $\mu_{2,3}(c) = \bigoplus_b h_2(c|b) \otimes \mu_{1,2}(b)$. The desired semi-local function is then $g_3(c, d) = h_3(c, d) \otimes \mu_{2,3}(c)$. For the ordinary probability semiring $(+, \times, \mathbb{R}^+)$, this becomes:

$$
\begin{aligned}
g_3(c, d) &= f(d|c) \sum_b f(c|b) \left[\sum_a f(a|b) f(b) \right] \\
&= f(d|c) \sum_b f(c|b) f(b) \\
&= f(d|c) f(c) \\
&= f(c, d)
\end{aligned}
\tag{5.14}
$$

as required.

If the sample space of all these variables is X, this implementation requires $2\,|X|$ summations, whereas the brute-force implementation, $f\,(C,D) = \sum_{A,B} f\,(A,B,C,D)$ requires $|X|^2$ summations. More generally, message passing implementations such as this require at most $\sum_{i=1}^{K} d_i\,|X|^{|C_i|}$ operations, where d_i is the vertex degree of each of the K JT vertices (Aji and McEliece, 2000). Thus exact inference using message passing is exponential in the size of the largest JT clique. By contrast, the brute-force implementation is exponential in the number of local functions, $M\,|X|^M$. Thus, the complexity of the JT algorithm, and therefore how competitive it is with brute-force variable elimination, depends critically upon the largest *tree width* of the PGM, that is, how many variables (minus one) need to be kept together in the optimal JT. In practical situations where inference over more than a single variable clique are required, more computational savings can be had using dynamic programming to pre-compute messages between all adjacent JT vertices in both directions, where they will be subsequently used in computing the semi-local marginals. Aji and McEliece (2000) show that this inference over all JT vertices requires at most four times the number of operations for the single vertex.

Exact inference for discrete variables presented above is computationally tractable because the summations are always over a finite set X, resulting in a new set of discrete distribution functions defined by tables of probabilities. Can the JT algorithm work for continuous variables defined on infinite sets? For example, we might use the usual $(+, \times, \mathbb{R})$ semiring and replace the finite summations by integrals. Unfortunately, the answer in general is no, since the marginalization integrals involved are either not available in closed form, do not lead to simple closed-form formulae, or only lead to simple formulae which cannot be "replicated" in the sense that these formulae cannot themselves be subsequently integrated analytically. There is one special case, however, that is analytically tractable, that is *linear-Gaussian* models, where all the variables are Gaussian with mean values as a linear function of other variable(s). For instance, taking the model defined by the local conditionals $f\,(x_1|x_2) = \mathcal{N}\,(x_1; a\,x_2, \sigma^2)$ with $f\,(x_2) = \mathcal{N}\,(x_2; 0, \tau^2)$, the global function $f\,(x_1, x_2) = f\,(x_1|x_2)\,f\,(x_2)$ can be marginalized analytically to obtain another Gaussian. This can be extended to any number of Gaussian variables in linear functional combinations such as this. Thus, linear-Gaussian models such as the Kalman filter can exploit this fact for computational gains (see Section 7.5).

For an alternative description of the junction tree algorithm using *factor graphs* instead, see Bishop (2006, Section 8.4).

Approximate inference

Since exact inference is only really tractable in the special case of discrete or linear-Gaussian PGMs, for most PGMs we need to turn to approximate inference. Stochastic sampling methods such as Gibbs (Al-

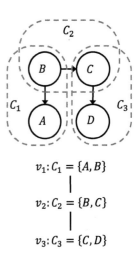

$v_1 \colon C_1 = \{A, B\}$

$|$

$v_2 \colon C_2 = \{B, C\}$

$|$

$v_3 \colon C_3 = \{C, D\}$

Fig. 5.7: A simple probabilistic graphical model (top) and an associated junction tree (bottom), over the four random variables A, B, C and D. Maximal cliques of the probabilistic model are shown (dashed grey boxes).

gorithm 3.5) are extremely useful for general PGMs (see Section 3.4). All we need is the conditional distribution for each variable conditioned on all the others, see (5.8) and simplifications. Sometimes, we may not have explicit methods for sampling from these product distributions, but generic methods such as Metropolis-Hastings (Algorithm 3.6) may well be suitable in practice to sample each Gibbs step. Indeed with MH, there is no need to expressly compute the denominator in the conditional distributions, which makes things much simpler. Any observed variables can simply be set to their realized values and Gibbs updates for these variables is skipped. Thus, "black box" approximate inference in PGMs can be very practical for a wide range of models and is a workhorse technique in machine learning and DSP as a result.

Nonetheless, the major drawback of Gibbs is that it is computationally intensive, and for very long signals or embedded applications it is usually intractable. In many applications it may suffice to have, not the full distribution joint distribution of all variables in the PGM, but the values of the variables that maximize the joint distribution, in other words the MLE. That computation itself is generally intractable, but a simple approximation, known as *iterative conditional modes* (ICM), that we describe next, is based on performing local maximum probability calculations instead:

Algorithm 5.2 Iterative conditional modes (ICM).

(1) *Initialization.* Choose a random starting point for each x_i^0, $i = 1, 2, \ldots, D$, set iteration number $n = 0$, and convergence tolerance $\epsilon > 0$,

(2) *Sequential maximization of all conditionals.* Compute all conditional distributions $f(x_i | x_1, x_2 \ldots x_{i-1}, x_{i+1}, \ldots, x_D)$ and obtain the sequence of values $x_i^{n+1} = \arg\max_x f\left(X_i = x \mid \boldsymbol{X}_J = \boldsymbol{x}_J^{n+1}, \boldsymbol{X}_K = \boldsymbol{x}_K^n\right)$ for $i = 1, 2, \ldots, D$, where $J = \{1, 2, \ldots, i-1\}$ and $K = \{i+1, i+2, \ldots, D\}$,

(3) *Compute (negative log) likelihood.* Find $E^{n+1} = -\sum_{i=1}^{D} \ln f\left(x_i^{n+1} \mid \left(x_j^{n+1}\right)_{j \in \mathcal{P}(i)}\right)$,

(4) *Iteration.* If $n > 0$ and $E^n - E^{n+1} < \epsilon$ then stop with solution $x_1^{n+1}, x_2^{n+1}, \ldots, x_D^{n+1}$, otherwise update $n \leftarrow n+1$ and go back to step 2.

ICM often converges very rapidly (in a dozen or so iterations) and can sometimes achieve the global MLE (Figure 5.8). Each maximization step implies that the NLL (of the joint distribution of the PGM) cannot increase with each iteration, so there is a fixed point towards which the iterations converge. There are, however, no general guarantees of convergence to the global MLE. Nonetheless, in many practical signal processing applications, rapidly obtaining a reasonably good solution is all that is needed in practice. To incorporate evidence, any variables that are observed can simply be fixed in the iteration.

Consider a general PGM defined on two distinct sets of variables: $\boldsymbol{X}$ and $\boldsymbol{Z}$, where the set $\boldsymbol{X}$ have known values and the $\boldsymbol{Z}$ variables are unknown for some reason. This may be because the values will never be observed or because they may be missing or the goal is to infer their distribution as part of the modelling process (often described as *latent* in the machine learning literature). We can write out the joint distribution as $f(\boldsymbol{x}, \boldsymbol{z}; \boldsymbol{p}) = f(\boldsymbol{x}; \boldsymbol{p}) f(\boldsymbol{z}|\boldsymbol{x}; \boldsymbol{p})$ with parameter vector $\boldsymbol{p}$. What we wish to do is to find the MLE $\boldsymbol{p}$ such that the complete likelihood $f(\boldsymbol{x}, \boldsymbol{z}; \boldsymbol{p})$ is maximized. Obviously, since the values of $\boldsymbol{Z}$ are not available we cannot compute this distribution to optimize it. Instead, we will want to optimize the marginal likelihood $f(\boldsymbol{x}; \boldsymbol{p}) = \int f(\boldsymbol{x}, \boldsymbol{z}; \boldsymbol{p}) \, d\boldsymbol{z}$.

This aim may be computationally difficult. As an alternative, let us assume that $\boldsymbol{Z}$ has the (arbitrary) marginal distribution $g(\boldsymbol{z})$ which is not a function of the parameters $\boldsymbol{p}$. Then the K-L divergence between $g(\boldsymbol{z})$ and the complete likelihood $f(\boldsymbol{x}, \boldsymbol{z}; \boldsymbol{p})$ is:

$$F(\boldsymbol{g}, \boldsymbol{p}) = D[\boldsymbol{Z}, (\boldsymbol{Z}, \boldsymbol{X})] = -\int g(\boldsymbol{z}) \ln\left(\frac{f(\boldsymbol{x}, \boldsymbol{z}; \boldsymbol{p})}{g(\boldsymbol{z})}\right) d\boldsymbol{z} \qquad (5.15)$$

This measures the deviation between $g(\boldsymbol{z})$ and the complete likelihood. The marginal NLL is $-\ln f(\boldsymbol{x}; \boldsymbol{p}) = \ln f(\boldsymbol{z}|\boldsymbol{x}; \boldsymbol{p}) - \ln f(\boldsymbol{x}, \boldsymbol{z}; \boldsymbol{p})$. Taking expectations $E_{\boldsymbol{Z}}[\cdot]$ on both sides, after some rearrangement, we get:

$$
\begin{aligned}
-\int g(\boldsymbol{z}) \ln f(\boldsymbol{x}; \boldsymbol{p}) \, dz &= -\int g(\boldsymbol{z}) \ln\left(\frac{f(\boldsymbol{x}, \boldsymbol{z}; \boldsymbol{p})}{g(\boldsymbol{z})}\right) dz \\
&\quad + \int g(\boldsymbol{z}) \ln\left(\frac{f(\boldsymbol{z}|\boldsymbol{x}; \boldsymbol{p})}{g(\boldsymbol{z})}\right) dz \qquad (5.16) \\
&= D[\boldsymbol{Z}, (\boldsymbol{Z}, \boldsymbol{X})] - D[\boldsymbol{Z}, \boldsymbol{Z}|\boldsymbol{X}]
\end{aligned}
$$

so that $-\ln f(\boldsymbol{x}; \boldsymbol{p}) = F(\boldsymbol{g}, \boldsymbol{p}) - D[\boldsymbol{Z}, \boldsymbol{Z}|\boldsymbol{X}]$ or $F(\boldsymbol{g}, \boldsymbol{p}) = -\ln f(\boldsymbol{x}; \boldsymbol{p}) + D[\boldsymbol{Z}, \boldsymbol{Z}|\boldsymbol{X}]$. Now, $D[\boldsymbol{Z}, \boldsymbol{Z}|\boldsymbol{X}] \geq 0$ so that $-\ln f(\boldsymbol{x}; \boldsymbol{p}) \leq F(\boldsymbol{g}, \boldsymbol{p})$ which is an upper bound for the marginal NLL $-\ln f(\boldsymbol{x}; \boldsymbol{p})$ that we can minimize as a proxy. However, this becomes equal to the marginal NLL when $g(\boldsymbol{z}) = f(\boldsymbol{z}|\boldsymbol{x}; \boldsymbol{p})$ since then $D[\boldsymbol{Z}, \boldsymbol{Z}|\boldsymbol{X}] = 0$. Similarly, we can also write $F(\boldsymbol{g}, \boldsymbol{p}) = E_{\boldsymbol{Z}|\boldsymbol{X}}[-\ln f(\boldsymbol{x}, \boldsymbol{Z}; \boldsymbol{p})] - H[\boldsymbol{Z}]$, so that with respect to the parameters p we only need to optimize the more tractable expected complete NLL $E_{\boldsymbol{Z}|\boldsymbol{X}}[-\ln f(\boldsymbol{x}, \boldsymbol{Z}; \boldsymbol{p})]$.

This suggests a simple two-step *coordinate descent* minimization pro-

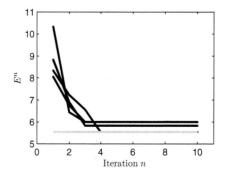

Fig. 5.8: Convergence of the iterative conditional modes (ICM) algorithm for approximate inference. Each curve is the value of the negative log likelihood (NLL) $E = -\ln f(x_1, x_2, x_3, x_4)$ of the joint variables for the simple PGM $X_1 \leftarrow X_2 \rightarrow X_3 \rightarrow X_4$, against algorithm iteration, for different random choices of starting values of the variables. The NLL converges quickly and in one case, achieves the global maximum likelihood value (grey line). Each variable in the model is discrete with sample space $|\Omega| = 8$. The conditional distributions are sampled uniformly at random.

cedure (where minimization over g is functional):

$$
\begin{aligned}
\boldsymbol{g}^{i+1} &= \underset{\boldsymbol{g}}{\arg\min}\, F\left(\boldsymbol{g}, \boldsymbol{p}^i\right) \\
\boldsymbol{p}^{i+1} &= \underset{\boldsymbol{p}}{\arg\min}\, F\left(\boldsymbol{g}^{i+1}, \boldsymbol{p}\right)
\end{aligned}
\tag{5.17}
$$

which simplifies to $\boldsymbol{p}^{i+1} = \arg\min_{\boldsymbol{p}} \left[-\int f\left(\boldsymbol{z}|\boldsymbol{x}; \boldsymbol{p}^i\right) \ln f\left(\boldsymbol{x}, \boldsymbol{z}; \boldsymbol{p}\right) d\boldsymbol{z} \right]$ since the first minimization step is just $\boldsymbol{g}^{i+1}\left(\boldsymbol{z}\right) = f\left(\boldsymbol{z}|\boldsymbol{x}; \boldsymbol{p}^i\right)$. This generic and widely used algorithm is known as *expectation-maximization* (E-M), the first minimization being known as the E-step and the second as the M-step. Since the coordinate descent procedure (5.17) can never increase the upper bound $F\left(\boldsymbol{g}, \boldsymbol{p}\right)$, the marginal NLL $-\ln f\left(\boldsymbol{x}; \boldsymbol{p}^{i+1}\right) \leq -\ln f\left(\boldsymbol{x}; \boldsymbol{p}^i\right)$ for all $i \geq 0$ so, provided the marginal NLL is bounded below, E-M is guaranteed to converge on some fixed point. The algorithm, detailed below, is of very general applicability for PGMs with latent or missing variables and is widely used in machine learning (for example, see the use of this for mixture modelling in Section 6.2):

Algorithm 5.3 *Expectation-maximization* (E-M).

---

(1) *Initialization.* Start with a randomized set of parameters $\boldsymbol{p}^0$, set iteration number $i = 0$, choose a small convergence tolerance $\epsilon > 0$,

(2) *E-step.* Compute the conditional probabilities $f\left(\boldsymbol{z}|\boldsymbol{x}; \boldsymbol{p}^i\right)$,

(3) *M-step.* Solve $\boldsymbol{p}^{i+1} = \arg\min_{\boldsymbol{p}} E_{\boldsymbol{Z}|\boldsymbol{X}}\left[-\ln f\left(\boldsymbol{x}, \boldsymbol{Z}; \boldsymbol{p}\right)\right]$,

(4) *Convergence check.* Calculate the value of the observed variable marginal NLL $E\left(\boldsymbol{p}^{i+1}\right) = -\ln\left[\int f\left(\boldsymbol{x}, \boldsymbol{z}; \boldsymbol{p}^{i+1}\right) d\boldsymbol{z}\right]$ and the improvement $\Delta E = E\left(\boldsymbol{p}^i\right) - E\left(\boldsymbol{p}^{i+1}\right)$ and if $\Delta E < \epsilon$, exit with solution $\boldsymbol{p}^{i+1}$,

(5) *Iteration.* Update $i \leftarrow i + 1$, go back to step 2.

For many models, solving the M-step may be carried out analytically (this is generally true for exponential families, for example), and if not analytically tractable then it can usually be solved numerically. It can be shown that the E-M algorithm has linear convergence rate (see Section 2.1), so, ΔE gets smaller with increasing iterations (Dempster *et al.*, 1977).

Of course, for general PGMs, the marginal likelihood is non-convex, so there can be multiple maxima and saddle points. E-M is essentially a greedy method and Wu (1983) proves that it converges on one of these stationary values of the marginal likelihood provided only that $E_{\boldsymbol{Z}}\left[-\ln f\left(\boldsymbol{x}, \boldsymbol{Z}; \boldsymbol{p}^i\right)\right] = -\int f\left(\boldsymbol{z}|\boldsymbol{x}; \boldsymbol{p}^i\right) \ln f\left(\boldsymbol{x}, \boldsymbol{z}; \boldsymbol{p}\right) d\boldsymbol{z}$ is continuous in both $\boldsymbol{p}$ and $\boldsymbol{p}^i$. This continuity condition usually holds in practice so while usually convergent, E-M cannot be expected to reach the *global*

MLE for p from any given initial guess p^0 (this is why we say it is an approximate PGM inference algorithm). It is not possible to predict in advance whether the algorithm will get trapped in some, sub-optimal, marginal likelihood stationary point from any given initial starting point p^0. The usual workaround for this problem is to use the random restarting heuristic (see Section 2.6) and keep the best solution found across restarts, that is, the one with the smallest E at convergence.

See Bishop (2006, Section 9.4) and Hastie *et al.* (2009, Section 8.5) for a different presentation of the same algorithm in terms of log likelihoods and maximization rather than minimization. In cases where the M-step is intractable, a variant of the algorithm known as *generalized* E-M allows for just a *decrease* rather than full minimization of $E_{\mathbf{Z}}\left[-\ln f\left(\mathbf{x}, \mathbf{Z}; p^i\right)\right]$ with respect to p on each iteration, see Bishop (2006, pages 454-455) for further discussion of *conditional* and *incremental* E-M variants.

It is possible to generalize the E-M algorithm very considerably. Most notably, E-M can be seen as an (important) special case of the *variational Bayes'* (VB) algorithm and this interesting relationship suggests a sibling of E-M which swaps the role of minimization and computation of posterior probabilities for the observed and unobserved variables (Kurihara and Welling, 2009).

Statistical machine learning

6

So far, the previous chapters in this book have contained mostly 'building blocks' which, although often valuable in their own right, become much more valuable when combined. This chapter describes in detail how the main techniques of statistical machine learning can be constructed from these components. It presents these concepts in a way which demonstrates how these techniques can be viewed as special cases of a more general probabilistic model which we fit to some data.

6.1 Feature and kernel functions

As a preliminary to this chapter, here we will briefly describe *feature* and *kernel functions*, two interrelated and extremely useful, mathematical tools which make such a regular appearance in machine learning and DSP that they deserve their own section. A feature function is just an arbitrary, nonlinear (vector) function which maps $\phi : \mathbb{R}^D \to \mathbb{R}^L$ where typically $L > D$. It takes as input an observed data value $x \in \mathbb{R}^D$ and thereby forms a general nonlinear map between the input and what is called the *feature space*. We will see that the primary value of this mapping is that it can allow simple linear machine learning algorithms to solve essentially nonlinear problems.

Kernel functions, in their most general form, are just functions of two input data values, $\kappa(x; x')$, but the most useful kernels for machine learning are *Mercer kernels*, which are symmetric functions that define a corresponding *positive definite Gram matrix* or *kernel matrix* $\mathbf{K}$ whose $N \times N$ entries are $k_{nn'} = \kappa(x_n; x_{n'})$, $n, n' \in 1, 2, \ldots, N$ for the data $x_n \in \mathbb{R}^D$. Such kernel functions can always be written in terms of an inner product of feature functions, $\kappa(x; x') = \phi(x)^T \phi(x')$. This is the content of *Mercer's theorem*, see Shawe-Taylor and Christianini (2004, Theorem 3.13). These features depend upon the eigenfunctions of the kernel function, so they can be infinite dimensional.

Mercer kernels satisfy several extremely useful properties (Shawe-Taylor and Christianini, 2004, page 75). Suppose we have kernels $\kappa_1, \kappa_2, \kappa_3$, real constant $a > 0$, and $\mathbf{A}$ is a positive semi-definite $D \times D$ matrix, and $\gamma : \mathbb{R}^D \to \mathbb{R}^D$. Then the following are all valid Mercer kernels:

(1) Additivity: $\kappa(x; x') = \kappa_1(x; x') + \kappa_2(x; x')$,

(2) Constant multiple: $\kappa(x; x') = a\,\kappa_1(x; x')$,

(3) Products: $\kappa(x; x') = \kappa_1(x; x')\,\kappa_2(x; x')$,

(4) Transformed parameters: $\kappa(x; x') = \kappa_3(\gamma(x); \gamma(x'))$, and,

Machine Learning for Signal Processing: Data Science, Algorithms, and Computational Statistics. Max A. Little.
© Max A. Little 2019. Published in 2019 by Oxford University Press. DOI: 10.1093/oso/9780198714934.001.0001

(5) Mahalanobis transform: $\kappa\left(\boldsymbol{x};\boldsymbol{x}'\right)=\boldsymbol{x}^T\mathbf{A}\boldsymbol{x}$.

Using these properties, we can take combinations of kernel functions and build entirely new ones from them. This is not an exhaustive list of properties.

6.2   Mixture modelling

All of the probabilistic models described in Chapter 4 are *unimodal*, that is they have one mode or "bump". This is often inadequate for real data which is very often *multimodal*, that is, having multiple peaks (see Figure 6.1). While it is clearly possible to create specific distributions which have multiple modes by design, a simple alternative is to use suitable *mixture distributions*. These are distributions which are a weighted linear combination of $K \geq 1$ component distributions $f\left(\boldsymbol{x};\boldsymbol{p}\right)$ with parameter vector $\boldsymbol{p}$ for the data $\boldsymbol{X} \in \mathbb{R}^D$. The mixture is written in the form:

$$f\left(\boldsymbol{x}\right)=\sum_{k=1}^{K}\pi_k f\left(\boldsymbol{x};\boldsymbol{p}_k\right) \tag{6.1}$$

where the *mixture weights* $0 \leq \pi_k \leq 1$ are normalized such that $\sum_{k=1}^{K}\pi_k = 1$. Typically, when the component distributions $f\left(\boldsymbol{x};\boldsymbol{p}\right)$ are unimodal, there will be (at most) K modes in $f\left(\boldsymbol{x}\right)$. There are K different parameter vectors, as each component has its own parameter vector.

How might one draw samples $\boldsymbol{x}$ from this mixture? It is convenient to view this as a probabilistic graphical model, with a latent or 'hidden' random variable Z. Since $\boldsymbol{x}$ can have each of the K possible parameters with probability π_k, then this is consistent with Z being categorically-distributed. Then ancestral sampling (Section 3.4) is applicable: simply draw a categorical variable $z = k$, called the *indicator variable*, $k \in 1,\ldots,K$ with probability π_k, then draw $\boldsymbol{x}$ from $f\left(\boldsymbol{x};\boldsymbol{p}_k\right)$.

The next question that arises: given N data points, $\boldsymbol{x}_1,\ldots\boldsymbol{x}_N$, how to estimate the parameter vectors? Each of these data points will be drawn from one of the mixture components, using indicators, $z_n = k$, representing that data point $\boldsymbol{x}_n$ was drawn from component k. Assuming that all these data points are i.i.d., the joint likelihood over all the data is:

$$f\left(\boldsymbol{x};\boldsymbol{p}\right)=\prod_{n=1}^{N}\sum_{k=1}^{K}\pi_k f\left(\boldsymbol{x}_n;\boldsymbol{p}_k\right) \tag{6.2}$$

where we include the categorical vector π as part of the parameter vector $\boldsymbol{p}$.

Gibbs sampling for the mixture model

The MLE for the parameters $\boldsymbol{p}_k$ is obtained by maximizing (6.2) with respect to these parameters. Unfortunately, the maximization cannot be performed in closed form. We therefore have to turn to approximate inference methods, and we will first augment this model with sufficient

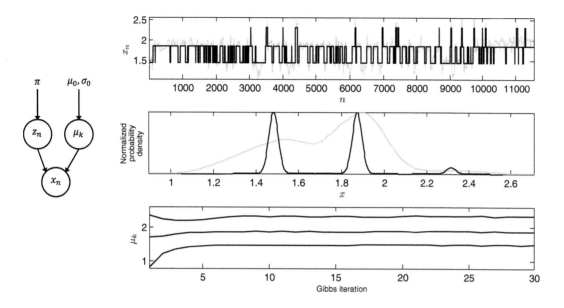

Fig. 6.1: A basic univariate Gaussian mixture model used to stratify a Gamma ray intensity signal measured from a drill hole (top right, grey line: source Kansas Geological Survey, University of Kansas) into $K = 3$ three different types of rock layer. Left: the probabilistic graphical model for the mixture. The MAP assignment over all Gibbs iterations is used to estimate the MAP stratification (top right, black lines). The (normalized density of the mixture model at convergence (middle right, black lines) is superimposed on the normalized density of the data (left middle, grey lines, kernel density estimate). Bottom right: the path to convergence of the mixture component means.

prior information to invoke the ubiquitous Gibbs sampler. To do this, we will place a (usually conjugate) prior on the mixture component parameters $\boldsymbol{p}_k$ with shared hyperparameters $\boldsymbol{q}$. Then, by considering the Markov blanket of each unknown variable $\boldsymbol{z}, \boldsymbol{p}$ in turn (Section 5.2), we have the following (block) Gibbs updates:

(1) Sample each component indicator z_n from $f\left(z_n | \boldsymbol{p}, \boldsymbol{x}; \boldsymbol{\pi}\right)$ for each $n = 1, 2, \ldots, N$,

(2) Sample each $\boldsymbol{p}_k$ from $f\left(\boldsymbol{p}_k | \boldsymbol{z}, \boldsymbol{x}; \boldsymbol{q}\right)$ for each $k = 1, 2, \ldots, K$.

For the purposes of illustration, it is instructive to look at the simplified special case of univariate $(D = 1)$ Gaussian data with component means $\boldsymbol{\mu}$ and shared, fixed variances σ^2. The extension to multivariate Gaussians is straightforward. The conjugate prior over the mean parameters is univariate Gaussian with mean and variance μ_0, σ_0^2, so the posterior for each component mean is also univariate Gaussian. This gives us the following conditional probabilities:

$$f\left(z_n = k | \boldsymbol{\mu}, \boldsymbol{x}; \boldsymbol{\pi}\right) = \frac{\pi_k \mathcal{N}\left(x_n; \mu_k, \sigma^2\right)}{\sum_{j=1}^{K} \pi_j \mathcal{N}\left(x_n; \mu_j, \sigma^2\right)} \qquad (6.3)$$

$$f\left(\mu_k | \boldsymbol{z}, \boldsymbol{x}; \mu_0, \sigma_0\right) = \mathcal{N}\left(\mu_k; \bar{\mu}_k, \bar{\sigma}_k\right) \qquad (6.4)$$

where:

$$\bar{\mu}_k = \frac{1}{\sigma^2 + \sigma_0^2 N_k} \left(\sigma^2 \mu_0 + \sigma_0^2 \sum_{n:z_n=k} x_n \right)$$

$$\bar{\sigma}_k = \frac{\sigma^2 \sigma_0^2}{\sigma^2 + \sigma_0^2 N_k} \tag{6.5}$$

Here, N_k is the number of z_n's assigned to the component k. As an example (nonlinear) DSP application, we can use this Gaussian mixture to perform (Bayesian) nonlinear noise removal from "spike-like" signals, something that is extremely difficult to do using classical linear DSP (Figure 6.1).

While Gibbs sampling is convenient, it suffers from the problem that it can be very slow, taking 100's if not 1000's of iterations to reach convergence (if convergence can be detected at all). This is generally not practical for embedded DSP applications where it is usually required to find a solution in a handful of iterations. A relatively simple solution is to avoid sampling altogether and use a deterministic algorithm instead.

E-M for mixture models

The goal of any deterministic inference algorithm is to minimize (or maximize) some objective function. In the case without priors for the parameters $\boldsymbol{p}$, the relevant objective function in this case is the *complete data* negative log-likelihood (NLL) for the mixture model:

$$-\ln f(\boldsymbol{x}, \boldsymbol{z}; \boldsymbol{p}, \boldsymbol{\pi}) = -\ln \prod_{n=1}^{N} \pi_{z_n} f(\boldsymbol{x}_n; \boldsymbol{p}_{z_n})$$

$$= -\sum_{n=1}^{N} \ln \pi_{z_n} f(\boldsymbol{x}_n; \boldsymbol{p}_{z_n}) \tag{6.6}$$

However, we do not know the values of z_n as these are unobserved, latent variables, so we cannot evaluate this objective directly. Instead, we can turn to the E-M algorithm to find an approximate solution (Section 5.3), for which we need the expected NLL with respect to (the posterior distribution of) $\boldsymbol{z}$:

$$E_{\boldsymbol{Z}|\boldsymbol{X}}\left[-\ln f(\boldsymbol{x}, \boldsymbol{Z}; \boldsymbol{p}, \boldsymbol{\pi})\right] = -\sum_{n=1}^{N} \sum_{k=1}^{K} \left[f(z_n = k | \boldsymbol{x}; \boldsymbol{p}, \boldsymbol{\pi}) \ln \pi_k f(\boldsymbol{x}_n; \boldsymbol{p}_k) \right]$$

$$\tag{6.7}$$

Notice that for this expression, we can differentiate with respect to each variable $\boldsymbol{p}_k$ separately since they are all independent, which means that we can minimize (6.7) with respect to each $\boldsymbol{p}_k$ in turn. In particular, for the Gaussian mixture model, this results in the following estimators

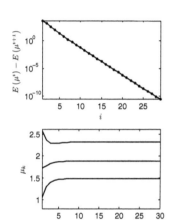

Fig. 6.2: Illustrating convergence of the expectation-maximization algorithm for Gaussian mixture modelling, applied to the problem in Figure 6.1. The difference in the observed data NLL (top panel) decreases exponentially since the convergence rate is linear; (bottom panel) the parameters converge on their (locally) optimal values fairly rapidly.

for the mean (Bishop, 2006, page 439):

$$\boldsymbol{\mu}_k = \frac{\sum_{n=1}^{N} f(z_n = k|\boldsymbol{x}; \boldsymbol{\mu}, \boldsymbol{\pi}) \boldsymbol{x}_n}{\sum_{m=1}^{N} f(z_m = k|\boldsymbol{x}; \boldsymbol{\mu}, \boldsymbol{\pi})} \tag{6.8}$$

Of course, this is not a closed-form expression since $f(z_n = k|\boldsymbol{x}; \boldsymbol{p}, \boldsymbol{\pi})$ depends upon all $\boldsymbol{p}_k$'s. E-M chooses (randomized) values for $\boldsymbol{p}_k$, then computes the $\boldsymbol{Z}$ probabilities, and then re-computes each $\boldsymbol{p}_k$ and so on. To assess convergence we compute the incomplete (observed data) NLL by marginalizing out the indicators:

$$E(\boldsymbol{p}) = -\sum_{n=1}^{N} \ln \sum_{k=1}^{K} \pi_k f(\boldsymbol{x}_n; \boldsymbol{p}_k) \tag{6.9}$$

We state the E-M algorithm for such (i.i.d.) mixture models here:

Algorithm 6.1 Expectation-maximization (E-M) for general i.i.d. mixture models.

(1) *Initialization.* Start with a randomized set of parameters $\boldsymbol{p}^0$, set iteration number $i = 0$,

(2) *E-step.* Compute the posterior probabilities $f(z_n = k|\boldsymbol{x}; \boldsymbol{p}^i)$ for all $k \in 1, 2, \ldots, K$ and $n \in 1, 2, \ldots N$,

(3) *M-step.* Solve $\boldsymbol{p}^{i+1} = \arg\min_{\boldsymbol{p}} E_{\boldsymbol{Z}|\boldsymbol{X}} \left[-\ln f(\boldsymbol{x}, \boldsymbol{Z}; \boldsymbol{p}^i) \right]$,

(4) *Convergence check.* Calculate the observed data NLL $E(\boldsymbol{p}^{i+1}) = -\sum_{n=1}^{N} \ln \sum_{k=1}^{K} \pi_k f(\boldsymbol{x}_n; \boldsymbol{p}_k^{i+1})$ and the improvement $\Delta E = E(\boldsymbol{p}^i) - E(\boldsymbol{p}^{i+1})$ and if $\Delta E < \epsilon$, exit with solution $\boldsymbol{p}^{i+1}$,

(5) *Iteration.* Update $i \leftarrow i + 1$, go back to step 2.

Note that in the above algorithm we have incorporated the mixture weights $\boldsymbol{\pi}$ into the parameters $\boldsymbol{p}$ as these are very often estimated at the same time using E-M, and this is straightforward (see Bishop, 2006, page 439 for details).

The linear convergence of E-M (see Section 2.1) can be seen clearly for the Gaussian mixture model (Figure 6.2). Setting a tolerance of $\epsilon = 10^{-6}$, for example, would achieve convergence in around 17 iterations, which will usually be a substantial improvement on Gibbs for embedded DSP applications. Nevertheless, E-M for mixture models can be very slow to converge when the modes in the density are not well separated.

For exponential family component distributions, $f(\boldsymbol{x}_n|z_n = k; \boldsymbol{p}_k) = \exp\left(\boldsymbol{p}_k^T \boldsymbol{g}(\boldsymbol{x}_n) - a(\boldsymbol{p}_k) + h(\boldsymbol{x}_n)\right)$ (see Section 4.5), the M-step for the component parameters has a simple form. We can minimize (6.7) with respect to each $\boldsymbol{p}_k$ using calculus:

$$\frac{\partial}{\partial \boldsymbol{p}_k} \left[-\sum_{n=1}^{N} \sum_{k=1}^{K} f\left(z_n = k|\boldsymbol{x}\right) \left[\boldsymbol{p}_k^T \boldsymbol{g}\left(\boldsymbol{x}_n\right) - a\left(\boldsymbol{p}_k\right) + h\left(\boldsymbol{x}_n\right)\right] \right] \tag{6.10}$$

$$= 0$$

$$\sum_{n=1}^{N} f\left(z_n = k|\boldsymbol{x}\right) \left(\boldsymbol{g}\left(\boldsymbol{x}_n\right) - a'\left(\boldsymbol{p}_k\right)\right) \tag{6.11}$$

$$= 0$$

giving $\sum_{n=1}^{N} f\left(z_n = k|\boldsymbol{x}\right) \boldsymbol{g}\left(\boldsymbol{x}_n\right) = \sum_{n=1}^{N} f\left(z_n = k|\boldsymbol{x}\right) a'\left(\boldsymbol{p}_k\right)$ and leading to:

$$a'\left(\hat{\boldsymbol{p}}_k\right) = \frac{\sum_{n=1}^{N} f\left(z_n = k|\boldsymbol{x}\right) \boldsymbol{g}\left(\boldsymbol{x}_n\right)}{\sum_{n=1}^{N} f\left(z_n = k|\boldsymbol{x}\right)} \tag{6.12}$$

Thus, for exponential family components, the M-step updates require computation of the posterior-weighted average of sufficient statistics. This can then be solved to find the values of the parameters $\hat{\boldsymbol{p}}_k$ for the subsequent E-step in terms of the weighted average sufficient statistics.

6.3   Classification

Given a fitted mixture model, we often need to solve the problem of predicting, given some new data $\hat{\boldsymbol{x}}$, which component would be the optimal representative for that data. In essence, this is the problem of *classification* which is one of the most intensely studied problems in machine learning. Note that in practical mixture modelling, we often have to solve the classification as part of an approximate parameter inference method. For example, we need to solve the MAP problem:

$$\begin{aligned}
\hat{z} &= \underset{k \in 1,2...,K}{\arg\max} \, f\left(z = k|\hat{\boldsymbol{x}}; \boldsymbol{p}, \boldsymbol{\pi}\right) \\
&= \underset{k \in 1,2...,K}{\arg\max} \, \left[\pi_k f\left(\hat{\boldsymbol{x}}; \boldsymbol{p}_k\right)\right] \\
&= \underset{k \in 1,2,...,K}{\arg\min} \, \left[-\ln \pi_k - \ln f\left(\hat{\boldsymbol{x}}; \boldsymbol{p}_k\right)\right]
\end{aligned} \tag{6.13}$$

Thus, classifying a new data point involves evaluating the posterior probability of the data point for all components, and picking the one with the highest posterior probability density. However, it is not essential to have a probabilistic model and we can instead express the classification problem as one of minimizing a loss function $L\left(\boldsymbol{x}, \boldsymbol{p}\right)$ biased by the negative log component (prior) probability π_k:

$$\hat{z} = \underset{k \in 1,2,...,K}{\arg\min} \, \left[-\ln \pi_k + L\left(\hat{\boldsymbol{x}}, \boldsymbol{p}_k\right)\right] \tag{6.14}$$

We would of course obtain the same thing if the component probability model is in the form $f\left(\boldsymbol{x}; \boldsymbol{p}\right) \propto \exp\left(-L\left(\boldsymbol{x}, \boldsymbol{p}\right)\right)$ where the normalizing constant does not depend upon $\boldsymbol{p}$. Both probabilistic (6.13) and non-probabilistic (6.14) *discriminants* are important in classification and we will investigate both in detail in the following sections.

The choice of the data distribution $f(x; p)$ or the loss function $L(x, p)$ largely determine the geometry of the *decision boundaries* in the space of the data, that, is, the set of points which separate pairs of classes. Consider, for example the two-class case $z \in 1, 2$ for the probabilistic model, the decision boundary B is the set of points where the joint probabilities are equal:

$$B = \left\{ x \in \mathbb{R}^D : \pi_1 f(x; p_1) = \pi_2 f(x; p_2) \right\} \qquad (6.15)$$

In the usual classification setting, we have *training data pairs* $\mathcal{D} = (x_n, z_n)_{n=1}^N$ using which we can use to estimate the parameters of the model p, by minimizing the complete data NLL:

$$
\begin{aligned}
E &= -\sum_{n=1}^{N} \ln \pi_{z_n} f\left(x_n; p_{z_n}\right) \\
&= -\sum_{k=1}^{K} \sum_{n: z_n = k} \left[\ln \pi_k + \ln f\left(x_n; p_k\right)\right] \qquad (6.16) \\
&= -\sum_{k=1}^{K} N_k \ln \pi_k + \sum_{k=1}^{K} \sum_{n: z_n = k} \ln f\left(x_n; p_k\right)
\end{aligned}
$$

Now, unlike mixture modelling, since the indicators are known, the above expression can (often) be minimized in closed-form. In particular, if the model for the observations is unimodal, then the following holds at the global minimum of the complete data NLL:

$$\frac{\partial E}{\partial p_k} = \sum_{n: z_n = k} \frac{\partial}{\partial p_k} \ln f(x_n; p_k) = 0 \qquad (6.17)$$

for each $k = 1, 2 \ldots K$. Similarly (using Lagrange multipliers to enforce the constraint that $\sum_{k=1}^{K} \pi_k = 1$), the MLE for the mixture weights can be obtained by minimizing E with respect to each π_k separately:

$$\hat{\pi}_k = \frac{N_k}{N} \qquad (6.18)$$

Quadratic and linear discriminant analysis (QDA and LDA)

One of the simplest classification methods arises when we consider multivariate Gaussian data x with parameters μ (mean) and Σ (covariance). Estimating the parameters is straightforward since (6.17) is solvable in closed form, in fact, it is just the MLE for the multivariate Gaussian:

$$\hat{\boldsymbol{\mu}}_k \quad = \quad \frac{1}{N_k} \sum_{n:z_n=k} \boldsymbol{x}_n \tag{6.19}$$

$$\hat{\boldsymbol{\Sigma}}_k \quad = \quad \frac{1}{N_k} \sum_{n:z_n=k} (\boldsymbol{x}_n - \hat{\boldsymbol{\mu}}_k)(\boldsymbol{x}_n - \hat{\boldsymbol{\mu}}_k)^T \tag{6.20}$$

Then, the negative log discriminant becomes:

$$\hat{z} \quad = \tag{6.21}$$

$$\underset{k\in 1,2,...,K}{\arg\min} \left[\frac{1}{2} (\hat{\boldsymbol{x}} - \hat{\boldsymbol{\mu}}_k)^T \hat{\boldsymbol{\Sigma}}_k^{-1} (\hat{\boldsymbol{x}} - \hat{\boldsymbol{\mu}}_k) + \frac{1}{2} \ln \left| \hat{\boldsymbol{\Sigma}}_k \right| - \ln \pi_k \right]$$

This method, obtained by using a multivariate Gaussian per mixture component, is known as *quadratic discriminant analysis* (QDA), because, up to a constant, the negative log discriminant is a so-called *quadratic form*. QDA is a reasonably complex method since it has $\frac{1}{2}KD(D+1) + KD = \frac{1}{2}KD(D+3)$ degrees of freedom arising from the covariance and mean vectors. The resulting decision boundaries B are (hyper) *conic sections* in D dimensions: this includes ellipsoids, spheres, hyperboloids and paraboloids, and degenerate examples such as hyperplanes.

Indeed, the latter degenerate hyperplane boundaries occur when the covariances across components are shared, i.e. when $\boldsymbol{\Sigma}_k = \boldsymbol{\Sigma}$ for all $k \in 1, 2, \ldots, K$. If this is constructed by design, so that:

$$\hat{\boldsymbol{\Sigma}} = \frac{1}{N} \sum_{k=1}^{K} \sum_{n:z_n=k} (\boldsymbol{x}_n - \hat{\boldsymbol{\mu}}_k)(\boldsymbol{x}_n - \hat{\boldsymbol{\mu}}_k)^T \tag{6.22}$$

This results in the classification method known as *linear discriminant analysis* (LDA) whose discriminant is:

$$\hat{z} \quad = \quad \underset{k\in 1,2,...,K}{\arg\min} \left[\frac{1}{2} (\hat{\boldsymbol{x}} - \hat{\boldsymbol{\mu}}_k)^T \hat{\boldsymbol{\Sigma}}^{-1} (\hat{\boldsymbol{x}} - \hat{\boldsymbol{\mu}}_k) - \ln \pi_k \right] \tag{6.23}$$

Typical LDA and QDA boundaries are depicted in Figure 6.3.

Logistic regression

In the two-class case, the posterior probability of the indicators can be written:

$$
\begin{aligned}
f\left(z=1|\boldsymbol{p},\boldsymbol{x}\right) &= \frac{\pi_1 f\left(\boldsymbol{x};\boldsymbol{p}_1\right)}{\pi_1 f\left(\boldsymbol{x};\boldsymbol{p}_1\right)+\pi_2 f\left(\boldsymbol{x};\boldsymbol{p}_2\right)} \\
&= \frac{1}{1+\frac{\pi_2 f\left(\boldsymbol{x};\boldsymbol{p}_2\right)}{\pi_1 f\left(\boldsymbol{x};\boldsymbol{p}_1\right)}} \\
&= \frac{1}{1+\exp\left(\ln\frac{\pi_2}{\pi_1}+\ln\frac{f\left(\boldsymbol{x};\boldsymbol{p}_2\right)}{f\left(\boldsymbol{x};\boldsymbol{p}_1\right)}\right)}
\end{aligned}
\tag{6.24}
$$

In the case of LDA the log ratio of the indicator posterior probabilities is:

$$
\begin{aligned}
\ln\frac{\pi_2}{\pi_1}+\ln\frac{f\left(\boldsymbol{x};\boldsymbol{p}_2\right)}{f\left(\boldsymbol{x};\boldsymbol{p}_1\right)} &= \ln\frac{\pi_2}{\pi_1}+\ln\left[\frac{\exp\left(-\frac{1}{2}\left(\boldsymbol{x}-\boldsymbol{\mu}_2\right)^T\boldsymbol{\Sigma}^{-1}\left(\boldsymbol{x}-\boldsymbol{\mu}_2\right)\right)}{\exp\left(-\frac{1}{2}\left(\boldsymbol{x}-\boldsymbol{\mu}_1\right)^T\boldsymbol{\Sigma}^{-1}\left(\boldsymbol{x}-\boldsymbol{\mu}_1\right)\right)}\right] \\
&= \ln\frac{\pi_2}{\pi_1}-\frac{1}{2}\left(\boldsymbol{\mu}_2+\boldsymbol{\mu}_1\right)^T\boldsymbol{\Sigma}^{-1}\left(\boldsymbol{\mu}_2-\boldsymbol{\mu}_1\right) \\
&\quad +\boldsymbol{x}^T\boldsymbol{\Sigma}^{-1}\left(\boldsymbol{\mu}_2-\boldsymbol{\mu}_1\right) \\
&= b_0+\boldsymbol{b}^T\boldsymbol{x}
\end{aligned}
\tag{6.25}
$$

if we define $b_0=\ln\frac{\pi_2}{\pi_1}-\frac{1}{2}\left(\boldsymbol{\mu}_2+\boldsymbol{\mu}_1\right)^T\boldsymbol{\Sigma}^{-1}\left(\boldsymbol{\mu}_2-\boldsymbol{\mu}_1\right)$ and $\boldsymbol{b}=\left(\boldsymbol{\Sigma}^{-1}\left(\boldsymbol{\mu}_2-\boldsymbol{\mu}_1\right)\right)^T$. Therefore:

$$
f\left(z=1|\boldsymbol{x};\boldsymbol{p}\right)=\phi\left(b_0+\boldsymbol{b}^T\boldsymbol{x}\right)
\tag{6.26}
$$

where:

$$
\phi\left(u\right)=\frac{1}{1+\exp\left(-u\right)}
\tag{6.27}
$$

which is known as the *logistic (sigmoid)* function for which $\phi:\mathbb{R}\to[0,1]$. Thus, in the two-class LDA case, the posterior indicator probability is a nonlinearly transformed, linear function of the data $\boldsymbol{x}$; the asymptotic behaviour of this function is $\phi\left(u\right)\to 0$ as $u\to-\infty$ and $\phi\left(u\right)\to 1$ as $u\to\infty$. This same logistic sigmoid concept can be extended to any number of classes.

Note that there are usually more degrees of freedom in the LDA model than the dimensionality of the parameters $b_0,\boldsymbol{b}$ alone. If the data $\boldsymbol{x}\in\mathbb{R}^D$ then LDA requires KD (mean vectors) along with $\frac{1}{2}D\left(D-1\right)$ (covariance matrix) and $K-1$ (indicator probability) parameters. By contrast, modelling the indicator probabilities directly as in (6.26) requires only $D+1$ parameters, usually a considerable simplification, which is the classification approach known as *logistic regression*. The parameters can be fitted using IRLS (see Section 2.3), directly minimizing the joint negative log posterior over all the data (Hastie *et al.*, 2009, page 121):

$$\left(\hat{b}_0, \hat{\boldsymbol{b}}\right) = \underset{(b_0, \boldsymbol{b})}{\arg\min} \left[-\sum_{n=1}^{N} \ln f\left(z_n | \boldsymbol{x}_n; b_0, \boldsymbol{b}\right) \right] \qquad (6.28)$$

For notational simplicity (and to motivate the next section), in the $K = 2$ two class case, if we use the labels $z \in \{-1, 1\}$, then we can write the categorical distribution as:

$$f\left(z | \boldsymbol{x}; b_0, \boldsymbol{b}\right) = p^{\frac{1}{2}(1+z)} \left(1 - p\right)^{\frac{1}{2}(1-z)} \qquad (6.29)$$

with parameter $p = \phi\left(b_0 + \boldsymbol{b}^T \boldsymbol{x}\right)$, which leads to:

$$E = \sum_{n=1}^{N} \left[\ln\left(\exp\left(-\left(b_0 + \boldsymbol{b}^T \boldsymbol{x}_n\right)\right) + 1\right) + \frac{1}{2}\left(b_0 + \boldsymbol{b}^T \boldsymbol{x}_n\right)(1 - z_n) \right] \qquad (6.30)$$

Since the decision boundary is the set where the log ratio of the posterior probabilities of the indicators is zero, we can see that logistic regression, like LDA, has linear decision boundaries, $B = \left\{ \boldsymbol{x} : b_0 + \boldsymbol{b}^T \boldsymbol{x} = 0 \right\}$. However, the reduced number of parameters in logistic regression is helpful to avoid overfitting and usually leads to better classification results because rather than modelling the joint distribution of all the random variables in the problem, we are finding the parameters that directly optimize some measure of the classification performance. However, since we do not know the distribution of the data given each class or the indicator probabilities (because the parameters of the Gaussian and categorical class model parameters are inextricably coupled together in b_0 and $\boldsymbol{b}$), this *discriminative modelling* approach is not *generative* in that we cannot evaluate or draw samples directly from the joint distribution over the full, underlying probabilistic model. This is a limitation in many settings. Later in this chapter we will discuss *generalized linear models* of which logistic regression is a special case, and where a different probabilistic interpretation for this method arises.

Support vector machines (SVM)

In this section we will explore one of the most widely-used and successful classification methods, which is essentially non-probabilistic. The MAP classification rule (6.13) chooses the class which maximizes the posterior probability of the indicator for a particular data point $\boldsymbol{x}$. For example, in the two class case with $z \in \{-1, 1\}$ we would assign $z = -1$ if $f\left(z = -1 | \boldsymbol{p}, \boldsymbol{x}\right) > f\left(z = 1 | \boldsymbol{p}, \boldsymbol{x}\right)$ and $z = 1$ otherwise. We can also consider the log ratio of posterior probabilities instead, which for logistic regression is:

$$\ln\left(\frac{f\left(-z | \boldsymbol{x}; \boldsymbol{p}\right)}{f\left(z | \boldsymbol{x}; \boldsymbol{p}\right)}\right) = z \boldsymbol{p}^T \boldsymbol{x} \qquad (6.31)$$

with the augmented vectors $\boldsymbol{x} \mapsto \begin{pmatrix} 1 & \boldsymbol{x} \end{pmatrix}^T$ and $\boldsymbol{p} = \begin{pmatrix} b_0 & \boldsymbol{b} \end{pmatrix}^T$. To correctly classify $\boldsymbol{x}$, we require $z\boldsymbol{p}^T\boldsymbol{x} \geq 0$, and we would want to find $\boldsymbol{p}$ that gives correct classifications for all the data, i.e. so that $z_n\boldsymbol{p}^T\boldsymbol{x}_n \geq 0$ *for all* $n = 1, 2\ldots, N$. If the data are actually separable by a linear hyperplane $\boldsymbol{p}^T\boldsymbol{x} = 0$, then there will be an infinite number of feasible solutions. However, the larger the log posterior decision ratio $z\boldsymbol{p}^T\boldsymbol{x}$, the more confident we can be that the decision is correct and will generalize beyond this data. This motivates finding $\boldsymbol{p}$ for the largest non-zero *decision margin* $m > 0$ that applies to all the data, e.g. $z_n\boldsymbol{p}^T\boldsymbol{x}_n \geq m$ for $n = 1, 2\ldots, N$. Here, we can always re-scale $\boldsymbol{p}$ to make m as large as we like, so we need to remove one degree of freedom from the problem. This can be done by redefining $m = 1/\|\boldsymbol{b}\|$, so now, maximizing m corresponds to minimizing the magnitude of $\boldsymbol{b}$, leading to the *maximum margin* classification QP (see 2.4):

$$\begin{aligned} \text{minimize} \quad & \frac{1}{2}\|\boldsymbol{b}\|_2^2 \\ \text{subject to} \quad & z_n\boldsymbol{p}^T\boldsymbol{x}_n \geq 1, \; n = 1, 2\ldots N \end{aligned} \tag{6.32}$$

Of course, in reality we will usually never encounter data that is exactly linearly separable. One solution is to tolerate some misclassified data points lying on the wrong side of the boundary. We can do this by changing the classification inequalities to $z_n\boldsymbol{p}^T\boldsymbol{x}_n \geq 1 - u_n$, where $u_n \geq 0$ for all $n = 1, 2\ldots N$, and penalizing the sum $\sum_{n=1}^{N} u_n$. This gives us the usual formulation of the (*soft-margin*) *support vector machine* (SVM) QP:

$$\begin{aligned} \text{minimize} \quad & \mathbf{1}^T\boldsymbol{u} + \frac{\lambda}{2}\|\boldsymbol{b}\|_2^2 \\ \text{subject to} \quad & z_n\boldsymbol{p}^T\boldsymbol{x}_n \geq 1 - u_n, \quad n = 1, 2\ldots N \\ & u_n \geq 0, \quad\quad\quad\quad\quad n = 1, 2\ldots N \end{aligned} \tag{6.33}$$

where the parameter $\lambda > 0$ controls the trade-off between minimizing the misclassification rate (for λ small) and maximizing the classification margin (λ large).

This regularization aspect of the SVM is more clearly represented if we combine the constraints in (6.33) as $u_n \geq \max\left[0, 1 - z_n\boldsymbol{p}^T\boldsymbol{x}_n\right]$ to get the objective function:

$$E = \sum_{n=1}^{N} \max\left[0, 1 - z_n\boldsymbol{p}^T\boldsymbol{x}_n\right] + \frac{\lambda}{2}\|\boldsymbol{b}\|_2^2 \tag{6.34}$$

The function $\max\left[0, 1 - z\boldsymbol{p}^T\boldsymbol{x}\right]$ is known as the *hinge loss*, and in the next section we will discuss its importance to classification problems in general. Arguably, it is the properties of this function which are the source of the very substantial success of the SVM classifier. SVM classifiers are popular and perform well in practice (see for example, Figure 6.3).

Because of the vast evidence for the value of SVMs in practical machine learning, they have been the subject of intense study since their inception in the mid 1990's (Cortes and Vapnik, 1995). Much effort has

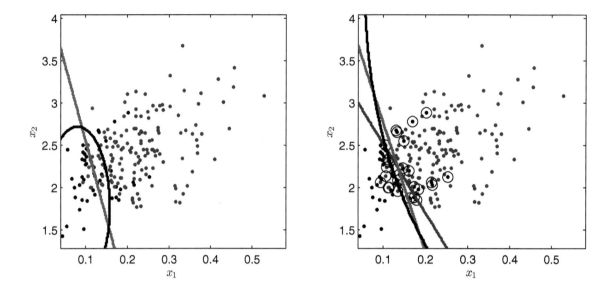

Fig. 6.3: Illustrating the decision boundaries found using various widely-used classifiers. The training x data contains two features from voice recordings of Parkinson's disease patients (blue dots), and healthy individuals (black dots). Linear discriminant analysis (LDA, left panel red line) achieves a classification accuracy (mean 0-1 loss) on this data of **86.7%**, whereas quadratic discriminant analysis (QDA, left panel black curve) finds an ellipse with less somewhat less accuracy (**83.6%**). By contrast, maximum margin classification (right panel, red line) finds a similar line to LDA but this line is better with **87.7%** accuracy. A linear support vector machine (SVM, right panel, blue line) finds yet another linear boundary with **86.2%** accuracy, and finally, a version of the classic SVM but with quadratic boundaries (right panel, black curve, c.f. Boyd and Vandenberghe, 2004, page 430) finds a hyperbolic boundary obtaining **87.2%** accuracy. There are **26** support vectors for this hyperbolic boundary (black circles) out of $N = 195$ data points.

been invested into finding ways to speed up the algorithms to solve the specific SVM QP (6.33), including the very popular *sequential minimal optimization* (SMO) method (Platt, 1999). There are many variants, for example, extensions to the multiple class case and to dependent output vectors and other discrete objects, known as *structural* SVMs (Tsochan-taridis *et al.*, 2005). One of the most important variants is the use of the so-called *kernel trick*, whereby dot products arising in the dual formulation of (6.33) are replaced by generalized products of functions which have the effect of mapping a nonlinear classification boundary in the input space into a higher-dimensional space where the data may be linearly separable, and so solvable using the support vector principle (Cortes and Vapnik, 1995).

Perhaps one of the only major drawbacks of the SVM is that it is essentially non-probabilistic; the hinge loss does not derive from any distribution. Because of this, SVMs are not particularly useful when estimates of classification uncertainty are required, for example, as they will be in most practical applications, and the only way to obtain approximate confidence intervals is by resampling e.g. cross-validation.

Classification loss functions and misclassification count

When addressing classification we need to ask: what is the essential problem we want to solve? Approaches such as LDA and QDA fit a probabilistic model (a Gaussian mixture model with known component assignments), whereas logistic regression and SVMs fit the parameters of the decision boundary directly. Implicit in all these boundary fitting methods is a specific loss function to minimize, optionally with some kind of regularization on these parameters. Therefore, we can examine these methods in terms of a general regularization equation:

$$\hat{\boldsymbol{p}} = \arg\min_{\boldsymbol{p}} \left[\sum_{n=1}^{N} L\left(z_n, \boldsymbol{x}_n; \boldsymbol{p}\right) + \frac{\lambda}{2} \|\boldsymbol{p}\|_2^2 \right] \qquad (6.35)$$

where $\boldsymbol{p}$ is a vector of boundary parameters, L is the loss function and $\lambda = 0$. For non-regularized methods we set $\lambda = 0$.

Specifically, in the two class case with $z \in \{-1, 1\}$, logistic regression uses the *logistic (log) loss*, $L\left(z, h\right) = \ln\left(\exp\left(-h\left(\boldsymbol{p}, \boldsymbol{x}\right)\right) + 1\right) + \frac{1}{2} h\left(\boldsymbol{p}, \boldsymbol{x}\right)\left(1 - z\right)$, where $h\left(\boldsymbol{p}, \boldsymbol{x}\right)$ is some scalar function (often affine $h\left(\boldsymbol{p}, \boldsymbol{x}\right) = \boldsymbol{b}^T \boldsymbol{x} + b_0$). For the (linear) support vector machine it is the hinge loss $L\left(z, h\right) = \max\left[0, 1 - z\, h\left(\boldsymbol{p}, \boldsymbol{x}\right)\right]$.

However, ideally we want to minimize the *classification error* which is quantified using the *0-1 loss* $L\left(z, h\right) = \delta\left[z - sign\left(-h\left(\boldsymbol{p}, \boldsymbol{x}\right)\right)\right]$ (where δ is the Kronecker delta); using this loss in (6.35) (with $\lambda = 0$) minimizes the number of misclassified data points. While this might be considered to be the ideal loss function for a classifier, it is not continuous and not convex either. Therefore, minimizing (6.35) with the 0-1 loss is a very difficult optimization problem which can, for all practical problems, only ever be solved approximately. However, a remarkable fact is that the SVM hinge loss is a *convex upper bound* on the 0-1 loss, that is, for all $h \in \mathbb{R}$, $\delta\left[z - \text{sign}\left(-h\right)\right] \leq \max\left[0, 1 - z\, h\right]$ (Figure 6.4). Therefore, the SVM solves a *convex relaxation* of the 0-1 loss, so it is a good thing to use if restricted to the world of convex functions.

With the SVM therefore, points that are correctly classified and more than a normalized unit distance from the boundary are effectively ignored in finding the boundary parameters. By contrast, the log loss underestimates the 0-1 loss around the decision boundary. Furthermore, points that are far from the boundary and yet correctly classified will also contribute to the total loss. At the same time, the hinge loss penalizes points that are misclassified more heavily than the log loss. These facts about the hinge loss give some intuition for why the SVM usually outperforms methods such as LDA, QDA and logistic regression in practice.

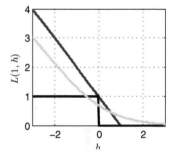

Fig. 6.4: **(Two-class) loss functions for classification: the 'ideal' 0-1 loss or misclassification count (black), the logistic (log) loss arising from logistic regression (grey), and the hinge loss occurring in the maximum margin and support vector machine (SVM) classifiers (blue). The horizontal axis is the (often affine) function of the parameters p defining the classification boundary and the data x, $h\left(x, p\right)$. Here we are showing the $z = 1$ case, for the other class $z = -1$ these functions are exactly mirrored about 0. It can be seen that the hinge loss is a convex upper bound on the 0-1 loss.**

Which classifier to choose?

Empirical studies generally support the conclusion that highly nonlinear, nonparametric classifiers and those that come closest to solving the 0-1 loss problem, perform better on average than simpler models such as LDA and logistic regression (Hastie *et al.*, 2009). For example: in every application of LDA, the linear SVM usually performs better. This would seem to indicate that there is little point in using a simple classifier.

However, these conclusions are generally obtained using sampling validation methods (typically, a single round of train-test splitting, or multiple cross-validation runs, see Section 4.2). The problem with this approach is that very often, the distribution of the data on which the method is validated, differs importantly from the distribution of the data in deployment (Hand, 2006). When this happens, we can no longer rely on the validation performance results: essentially, all bets are off regarding the eventual performance of the classifier in deployment. Because of this, Hand (2006) advocates the use of simpler classifiers because they tend to hold up better to this kind systematic variation in the final data than more complex classifiers.

This is a practical heuristic, but note that the goal of modelling should always be one of finding the *correct* level of complexity for the model. A simpler classifier may indeed be less sensitive to unknown variation in the data distribution, but we risk creating a model that essentially makes equally poor classifications for all data. A better solution might be to continuously re-train a classifier on all data available to-date, as it becomes available. We should always reserve the possibility of selecting between a simpler and a more complex classifier based upon the performance estimates using *all* the available evidence.

6.4   Regression

In basic classification problems, it is (implicitly or explicitly) assumed that there are hidden categorical variables Z_n which indicate the class for each data item x_n. By contrast, in *regression* problems, it is assumed that these hidden variables Z are *continuous*, and can indeed be multivariate. Regression is a well-known technique in classical statistics, where the hidden variables are called *covariates*. There are a very large number of different methods which can arise from different choices of the distributions for X (and the hidden variables Z upon which they may be assumed to depend).

Linear regression

Perhaps the simplest and most ubiquitous regression method is known as (*multiple*) *linear regression* which can be derived from the assumption that X depends upon Z through the *linear expectation* model $E[X|Z] = WZ$, where $\mathbf{W} \in \mathbb{R}^{D \times M}$ is the parameter matrix, for $X \in \mathbb{R}^D$ and $Z \in \mathbb{R}^M$. Assuming that X is multivariate Gaussian with covariance matrix $\mathbf{\Sigma} \in \mathbb{R}^{D \times D}$, for i.i.d. data the likelihood is:

$$f(x; \mathbf{W}, \mathbf{\Sigma}) = \prod_{n=1}^{N} \mathcal{N}(x_n; \mathbf{W}z_n, \mathbf{\Sigma}) \tag{6.36}$$

from which we get the NLL:

$$
\begin{aligned}
-\ln f\left(\boldsymbol{x}; \mathbf{W}, \boldsymbol{\Sigma}\right) &= -\ln \prod_{n=1}^{N} \mathcal{N}\left(\boldsymbol{x}_n; \mathbf{W}\boldsymbol{z}_n, \boldsymbol{\Sigma}\right) \\
&= \frac{1}{2}\sum_{n=1}^{N}\left(\boldsymbol{x}_n - \mathbf{W}\boldsymbol{z}_n\right)^T \boldsymbol{\Sigma}^{-1}\left(\boldsymbol{x}_n - \mathbf{W}\boldsymbol{z}_n\right) \\
&\quad + \frac{N}{2}\ln|\boldsymbol{\Sigma}| + \frac{ND}{2}\ln\left(2\pi\right)
\end{aligned}
\tag{6.37}
$$

We can write this in the form $E = \frac{1}{2}\|\mathbf{X} - \mathbf{W}\mathbf{Z}\|_{\mathrm{F}}^2 + \frac{N}{2}\ln|\boldsymbol{\Sigma}| + C(N, D)$, where $\|\mathbf{A}\|_{\mathrm{F}} = \sqrt{\operatorname{tr}\left(\mathbf{A}^T\mathbf{A}\right)}$ (which is known as the *Frobenius norm* of the matrix $\mathbf{A}$), $\mathbf{X}$ is the $D \times N$ matrix where each column is one of $\boldsymbol{x}_n$, $\mathbf{Z}$ is the $M \times N$ matrix with columns containing $\boldsymbol{z}_n$ and where C is a normalizing function. Using the properties of the trace (Section 1.3) and rules of scalar-by-matrix derivatives (Petersen and Pedersen, 2008), we can show that:

$$
\hat{\mathbf{W}} = \mathbf{X}\mathbf{Z}^T\left(\mathbf{Z}\mathbf{Z}^T\right)^{-1}
\tag{6.38}
$$

Perhaps because of the simplicity of this estimate which involves only linear algebra, linear regression is one of the mainstays of statistics, DSP and machine learning. It is nearly always the case, however, that such a model is unrealistic: that there is a purely linear model underlying the data and that the random deviations around this model are Gaussian is a rather specific set of assumptions and later we will see that these assumptions can be relaxed.

Bayesian and regularized linear regression

The most complex part of computing the linear regression estimate (6.38) is inverting the matrix $\mathbf{Z}\mathbf{Z}^T$. Often, this will not be possible because it does not have a unique inverse: this can happen if $\mathbf{Z}$ has any linearly dependent rows, for example. In general, the simple linear regression problem is often only solvable under some kind of constraint or regularization, and this motivates a Bayesian approach.

For simplicity let us consider the univariate case $D = 1$ so that $x_n \in \mathbb{R}$ is the data, and assume that the parameter vector $\boldsymbol{w} \in \mathbb{R}^M$ is multivariate Gaussian:

$$
f\left(\boldsymbol{w}\right) = \mathcal{N}\left(\boldsymbol{w}; \boldsymbol{w}_0, \boldsymbol{\Sigma}_0\right)
\tag{6.39}
$$

In this case, by conjugacy (see Section 4.5), assuming $\boldsymbol{\Sigma} = \mathbf{I}$, the posterior is also multivariate Gaussian:

$$
\begin{aligned}
f\left(\boldsymbol{w}|\mathbf{X}\right) &= \mathcal{N}\left(\boldsymbol{w}; \hat{\boldsymbol{w}}_N, \hat{\boldsymbol{\Sigma}}_N\right) \\
\hat{\boldsymbol{\Sigma}}_N &= \left(\mathbf{Z}\mathbf{Z}^T + \boldsymbol{\Sigma}_0^{-1}\right)^{-1} \\
\hat{\boldsymbol{w}}_N^T &= \left(\boldsymbol{w}_0^T\boldsymbol{\Sigma}_0^{-1} + \mathbf{X}\mathbf{Z}^T\right)\hat{\boldsymbol{\Sigma}}_N
\end{aligned}
\tag{6.40}
$$

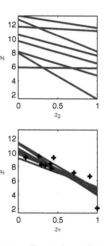

Fig. 6.5: Bayesian linear regression. The two-dimensional parameter vector w is given a multivariate Gaussian prior with mean $w_0 = \begin{bmatrix} 8 & -3 \end{bmatrix}$ and covariance $\Sigma_0 = 6\mathbf{I}$. The (affine) regression line is $x = w_0 + w_1 z_2$. The top panel shows several regression lines drawn from the prior. The bottom panel shows $N = 10$ data points x_n drawn from the likelihood $\mathcal{N}\left(x|w_0 + w_1 z_2, 1.0\right)$ (black crosses). Also shown are several regression lines drawn from the posterior, whose parameters $\hat{w}_N$ and $\hat{\Sigma}_N$ are computed using conjugate updates (6.40).

Here, the posterior mode is just $\hat{\boldsymbol{w}}_N$, which is therefore the MAP solution. The special case $\boldsymbol{w}_0 = \mathbf{0}$ we have encountered before: it is the ridge estimator (2.17). The posterior covariance $\hat{\boldsymbol{\Sigma}}_N$ depends critically upon the choice of prior covariance $\boldsymbol{\Sigma}_0$, and this prior covariance plays the role of regularization parameter. For illustration, if we set $\boldsymbol{\Sigma}_0 = \gamma \mathbf{I}$, then as $\gamma \to 0$, then the prior covariance dominates over the estimate obtained from the data, $\left(\mathbf{Z}\mathbf{Z}^T\right)^{-1}$, and vice versa as $\gamma \to \infty$.

Even simple Bayesian linear regression such as this allows us to properly account for the balance between the uncertainty in the data, and the uncertainty in our prior estimates for the parameters, and therefore it does not suffer from the non-uniqueness of the frequentist estimate (Figure 6.5). This approach can be made much more sophisticated, for example, in the above estimates we do not model the likelihood covariance $\boldsymbol{\Sigma}$ as a random variable, but we can do this by putting a conjugate *normal-Wishart* prior on the pair $(\boldsymbol{w}, \boldsymbol{\Sigma})$ (Bishop, 2006, page 102).

While the linear-Gaussian models above are computationally simple, they are obviously limited to situations where it is plausible to argue that the likelihood is Gaussian and the prior is conjugate to this likelihood. This is quite restrictive: there are many situations where the prior might be chosen according to different criteria. Here is an example: consider a linear model with univariate data that is Gaussian, but with (independent, zero median) Laplacian prior, the posterior is:

$$f\left(\boldsymbol{w}|\boldsymbol{x}\right) \propto \prod_{n=1}^{N} \mathcal{N}\left(x_n; \boldsymbol{w}^T\boldsymbol{z}_n, \sigma^2\right) \prod_{i=1}^{M} \exp\left(-\frac{1}{b}|w_i|\right) \qquad (6.41)$$

whose negative log posterior is:

$$-\ln f\left(\boldsymbol{w}|\boldsymbol{x}\right) = \frac{1}{2\sigma^2} \sum_{n=1}^{N} \left(x_n - \boldsymbol{w}^T\boldsymbol{z}_n\right)^2 + \lambda \sum_{i=1}^{M} |w_i| + C\left(\sigma, \lambda, N, \boldsymbol{x}\right)$$
$$(6.42)$$

where $C\left(\sigma, \lambda, N, \boldsymbol{x}\right)$ is a normalizing function and $b^{-1} = \lambda > 0$ is the reciprocal of the Laplace spread parameter. Minimizing this with respect to $\boldsymbol{w}$ cannot be performed analytically, but this is an L_2-L_1-mixed norm Lasso problem which can be solved numerically using QP (see Section 2.4).

Linear-in parameters regression

While linear (or affine) models are easily treated within the analytical frameworks described above, if the relationship between $\boldsymbol{Z}$ and $\boldsymbol{X}$ is nonlinear, then our job is much more complicated. This is because there are an infinite number of possible kinds of nonlinearities. Ideally, we want a regression method to discover the nonlinear relationship itself, but this is generally difficult, and we will discuss approaches to this later. There is, however a simple modification to extend the linear models above to arbitrary, but chosen kinds of nonlinearity if we replace or augment the input variables with transformed versions. Consider the case of one-dimensional $D = 1$ output variables X, the linear regression model is just:

$$E[X|\boldsymbol{Z}] = x(\boldsymbol{z}) = \sum_{i=1}^{M} w_i z_i \qquad (6.43)$$

but using the arbitrary nonlinear *basis functions* $h_i(z)$ we can create a nonlinear relationship:

$$x(\boldsymbol{z}) = \sum_{i=1}^{M} w_i h_i(z_i) \qquad (6.44)$$

Now, the ML and MAP parameter estimates are just those obtained by replacing the data for $\boldsymbol{Z}$ with the data $\boldsymbol{h}(\boldsymbol{z}) = (h_1(z_1), h_2(z_2), \ldots, h_M(z_M))^T$. It is entirely possible to have a different number Q of basis functions than the dimension of the data, for example for one-dimensional data we can have $\boldsymbol{h}(z) = (h_1(z), h_2(z), \ldots, h_Q(z))^T$. Widely used choices of bases include polynomials $h_i(z) = z^{i-1}$, and *Gaussian basis functions* $h_i(z) = \exp\left(-s(z - \mu_i)^2\right)$ where $\mu_i \in \mathbb{R}$ is a location parameter and $s > 0$ the bandwidth parameter which controls the sharpness of the peak of the basis function (Figure 6.6).

Although polynomials are simple and have no additional parameters to set, they extrapolate very badly outside of the range of the input data because the bases go to $\pm\infty$ at the extremes as $z \to \pm\infty$. By contrast, Gaussian radial bases go to zero at the extremes and are thus often a better controlled choice in practice. Of course, we then have to choose the values of the parameters μ and s, and there is no way to do this within the simple framework of linear-in-parameters regression.

Generalized linear models (GLMs)

The basic linear-Gaussian model works well if we can assume that the distribution of $\boldsymbol{X}$ is (multivariate) Gaussian. This is not always the case. Therefore, having a set of tools which work for non-Gaussian distributions is helpful. In general, the regression problem for arbitrary distributions is not tractable. However, if the distribution comes from the exponential family – which is a broad class of distributions (see Section 4.5) – then the non-Gaussian regression problem can be readily solved using tools we have already developed.

Consider the univariate $D = 1$ case, then the prediction equation for the linear-Gaussian regression model is just $\mu = E[X|\boldsymbol{Z}] = \boldsymbol{w}^T \boldsymbol{Z}$. Since for the Gaussian, $\mu \in \mathbb{R}$, this makes sense. But, for general distributions, the mean can occupy a different range of $\mathbb{R}$. Let us look at the case of one-dimensional parameter vector exponential families (known as the *one-parameter exponential families*), for which the likelihood is:

$$f(x; p) = \exp(pg(x) - a(p) + h(x)) \qquad (6.45)$$

We will make this into a linear prediction model by setting $p = \boldsymbol{w}^T \boldsymbol{z}$:

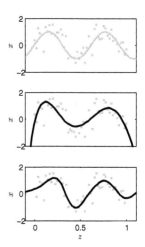

Fig. 6.6: Linear-in-parameters regression for the sine function $x(z) = \sin(10.23z)$ **(top, grey line). The likelihood variance** $\sigma = 0.5$, **and is used to generate the** $N = 50$ **data points** x_n **(top, grey dots) on the range** $z \in [0, 1]$. **Polynomial basis regression** $h(z) = (z^0, z^1, \ldots, z^5)^T$ **(middle, black line) fits the data well, but extrapolates away from the data with extreme fluctuations going rapidly to** $-\infty$. **Similarly, Gaussian basis function regression (bottom, black line) with** 10 **equispaced location parameter values** μ **on the range** $[0, 1]$ **and bandwidth** $s = 20.0$ **fits well for the given data, but safely extrapolates to zero outside the range of the data.**

$$f(x; \boldsymbol{w}, \boldsymbol{z}) = \exp\left(\boldsymbol{w}^T \boldsymbol{z} g(x) - a\left(\boldsymbol{w}^T \boldsymbol{z}\right) + h(x)\right) \qquad (6.46)$$

As shown in (4.58), the mean is the first derivative of the log normalizer $a(p)$, e.g. $\mu = \frac{\partial a}{\partial p}(p) = E[X|\boldsymbol{Z}]$. If we write the mean function as $\frac{\partial a}{\partial p}(p) = \phi(p) = \mu$, then we can solve to find μ in terms of $\boldsymbol{w}^T \boldsymbol{Z}$, e.g. $\mu = \phi^{-1}\left(\boldsymbol{w}^T \boldsymbol{Z}\right)$. The function ϕ^{-1}, which maps the linear predictor onto the mean, is called the (*canonical*) *link* function for the GLM. Thus, the choice of exponential family determines the canonical link.

This construction generalizes the linear model, in the following way. For the case of known variance $\sigma^2 = 1$, the one-parameter Gaussian has log normalizer $a(p) = \frac{1}{2}p^2$ so that simply $\phi(p) = p$ giving $\mu = \boldsymbol{w}^T \boldsymbol{Z}$, which is the indeed the one-dimensional Gaussian-linear model.

In the Bernoulli case, we have $a(p) = \ln(1 + \exp(p))$ giving the predictor:

$$\mu = \frac{1}{1 + \exp\left(-\boldsymbol{w}^T \boldsymbol{Z}\right)} \qquad (6.47)$$

This is the same relationship that we find in logistic regression (6.26). Thus, logistic regression is in fact a special kind of GLM with canonical link:

$$\phi^{-1}(u) = \ln\left(\frac{u}{1-u}\right) \qquad (6.48)$$

which is also known as the *logit* function.

The next question which arises is ML parameter estimation: in the case of the Gaussian we have simple linear-Gaussian regression which can be solved analytically as explained earlier. For the non-Gaussian case, things are more complex. Given i.i.d. data the NLL $E = -\ln f(\boldsymbol{x}; \boldsymbol{w})$ is:

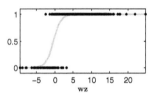

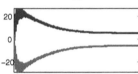

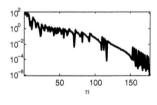

$$
\begin{aligned}
E &= -\sum_{n=1}^{N} \ln f\left(x_n; \boldsymbol{z}_n, \boldsymbol{w}\right) \qquad (6.49) \\
&= -\sum_{n=1}^{N} \boldsymbol{w}^T \boldsymbol{z}_n g(x_n) + \sum_{n=1}^{N} a\left(\boldsymbol{w}^T \boldsymbol{z}_n\right) - \sum_{n=1}^{N} h(x_n)
\end{aligned}
$$

This is a convex objective function (see Section 4.5), and the partial derivatives with respect to the parameters are:

$$\frac{\partial E}{\partial \boldsymbol{w}} = -\sum_{n=1}^{N} \boldsymbol{z}_n^T g(x_n) + \sum_{n=1}^{N} \boldsymbol{z}_n^T \phi\left(\boldsymbol{w}^T \boldsymbol{z}_n\right) \qquad (6.50)$$

Fig. 6.7: Logistic regression as a generalized linear model (GLM). The two-dimensional parameter vector $w = \begin{bmatrix} 5.1 & -5.7 \end{bmatrix}^T$. Plotted against $w^T z_n$ are the $N = 300$ randomly generated training data points (z_n, x_n) (top, black points), and the inverse link function (top, grey line). Using gradient descent with backtracking line search (see Section 2.3), the parameter estimates reach a convergence tolerance $\epsilon = 10^{-6}$ in around 160 iterations (middle, horizontal axis is number of gradient descent iterations), as is evident from the evolution of the value of the negative log likelihood (bottom).

This NLL and its gradient are all that is required to construct a useful gradient descent method for solving the general GLM ML problem (Figure 6.7), although iteratively reweighted least squares is widely used in this application too (see Bishop, 2006, page 207 and also Hastie *et al.*, 2009, page 120).

GLMs have very substantial flexibility and are therefore widely used in practice. Consider *count data*, e.g. $X \in 0, 1, 2, \ldots \infty$, then the Poisson distribution is a natural choice, whose canonical link is $\phi^{-1}(u) = \ln(u)$. Similarly real, positive data, $X \in \mathbb{R}$ with $X > 0$ suggests an exponential distribution (if there is no mode) or the gamma distribution (with one mode) both with reciprocal link $\phi^{-1}(u) = u^{-1}$.

Bayesian versions of GLMs can also been proposed, most notably by L_1-regularization (Hastie *et al.*, 2009, page 125):

$$E = \sum_{n=1}^{N} \left[-\boldsymbol{w}^T \boldsymbol{z}_n g\left(x_n\right) + a\left(\boldsymbol{w}^T \boldsymbol{z}_n\right) - h\left(x_n\right) \right] + \gamma \|\boldsymbol{w}\|_1 \quad (6.51)$$

with regularization parameter $\gamma > 0$. The objective function is convex but nonlinear in general, see for example a specialized interior-point algorithm in the case of logistic regression (Koh *et al.*, 2007).

Nonparametric, nonlinear regression

In all the regression models we have encountered so far, the structure of the model places strong constraints on the relationships between $\boldsymbol{Z}$ and $\boldsymbol{X}$ that it can represent. It may be that these constraints are too strong in some applications. Here we will consider nonparametric models that place only very mild restrictions on the form of the regression relationship.

Consider the following (univariate) kernel density estimate for the joint variables (Z, X), which assumes that the variables are independent:

$$f(z, x) = \frac{1}{N} \sum_{n=1}^{N} \kappa\left(z; z_n\right) \kappa\left(x; x_n\right) \quad (6.52)$$

The marginal distribution for Z is:

$$
\begin{aligned}
f(z) &= \frac{1}{N} \sum_{n=1}^{N} \kappa\left(z; z_n\right) \int \kappa\left(x; x_n\right) dx \\
&= \frac{1}{N} \sum_{n=1}^{N} \kappa\left(z; z_n\right) \quad (6.53)
\end{aligned}
$$

from which we can compute the conditional of X given Z:

$$f(x|z) = \frac{\sum_{n=1}^{N} \kappa\left(z; z_n\right) \kappa\left(x; x_n\right)}{\sum_{n'=1}^{N} \kappa\left(z; z_{n'}\right)} \quad (6.54)$$

The regression model is the expected value of the conditional:

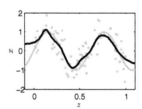

Fig. 6.8: The nonparametric kernel regression model is formed from a locally weighted average of the data **X**. The weighting functions κ are kernel PDFs, here we use univariate Gaussians with variance $\sigma^2 = 2.5 \times 10^{-3}$. The generated data is from the same function as Figure 6.6

$$E[X|Z] = \int x \frac{\sum_{n=1}^{N} \kappa(z; z_n) \kappa(x; x_n)}{\sum_{n'=1}^{N} \kappa(z; z_{n'})} dx$$

$$= \frac{1}{\sum_{n'=1}^{N} \kappa(z; z_{n'})} \sum_{n=1}^{N} \kappa(z; z_n) \int x \kappa(x; x_n) dx$$

$$= \frac{\sum_{n=1}^{N} \kappa(z; z_n) x_n}{\sum_{n'=1}^{N} \kappa(z; z_{n'})} \tag{6.55}$$

which holds provided the kernel PDF $\kappa(z, \mu)$ is chosen to have expected value μ. This gives rise to (nonparametric) *kernel regression*:

$$x(z) = \frac{\sum_{n=1}^{N} \kappa(z; z_n) x_n}{\sum_{n'=1}^{N} \kappa(z; z_{n'})} \tag{6.56}$$

which is formed from a normalized, weighted sum over all the data points. Typically, the kernel PDF is chosen to have large density near the mean and decreasing density with increasing distance to the mean. Then, kernel regression produces a regression curve which is a smooth, locally weighted average of the data x_n for each z near z_n (Figure 6.8).

Kernel regression is a particularly broad framework because of the large variety of algorithms it encompasses. For example, we can obtain the *local average* with the *uniform* or *box* kernel:

$$\kappa(z; \mu) = \delta[\|z - \mu\| \leq \sigma] \tag{6.57}$$

Similarly, we can define the *K-nearest neighbour* kernel:

$$\kappa(z; \mu) = \delta\left[\|z - \mu\| \leq \left\|z^{(K)} - \mu\right\|\right] \tag{6.58}$$

where $z^{(K)}$ is the data item which is the K most distant from z; all data further away than this are given zero kernel weight.

Note that the extension of kernel regression to multiple dimensions is straightforward, but we have to take into account the curse of dimensionality: we are unlikely to have sufficient data to populate the input space sufficiently to make this kind of interpolation practical for anything more than a handful of dimensions.

Kernel regression as defined above produces a locally *constant* regression curve, which is the kernel weighted local mean. This is simple but problematic in regions where the data z is sparse because general smooth functions are badly underfitted by the mean. A more sophisticated alternative is to use *locally linear* kernel regression, where we perform a kernel weighted local linear fit to the data. We can derive this as the minimizer for the local kernel weighted sum-of-squares error:

$$E(z) = \sum_{n=1}^{N} \kappa(z; z_n) [x_n - w_0(z) - w_1(z) z_n]^2 \tag{6.59}$$

This holds because the expected value is the minimizer for the squared error. The locally constant predictor has $w_1(z) = 0$. We need to minimize this squared error for each value z for which we want to make predictions about x, e.g. we must solve $\hat{\boldsymbol{w}}(z) = \arg\min_{\boldsymbol{w}(z)} E(z)$. This is just a weighted least squares problem (Section 2.2) and the predictor is $x(z) = \hat{w}_0(z) + \hat{w}_1(z)z$. Of course, this much better predictor is not without cost – normally $O(N^3)$ per prediction – whereas locally constant kernel regression requires only $O(N)$ per prediction.

The smoothness of the kernel, often controlled by a single bandwidth parameter (e.g. the standard deviation σ of the Gaussian) plays a substantial role in determining the smoothness of the regression curve. For most practical kernel functions, as the bandwidth goes to zero, the kernel approaches a Dirac delta. Thus, the predictor goes exactly through each data point x_n and so has zero prediction error. This is of course overfitted, so we need some way to regularize the solution. For this purpose, cross-validation is widely used (Section 4.2).

At the other extreme, when the kernel bandwidth is sufficiently large that it causes the kernel to be effectively constant for all z within the range of the data, then local constant kernel regression just fits the (global) mean, and local linear kernel regression becomes (global) linear regression. Therefore, we can view kernel regression as parametric regression models where the parameters can vary with Z. Just how sensitive they are to Z is determined by the bandwidth.

Kernel regression methods, as with most nonparametric methods, often make much more accurate predictors in practice than their parametric counterparts. However, they are essentially frequentist, and so we have very limited control over what we expect about the form of the regression curve in advance. Interestingly, these ideas, put into a suitable form, can be given a fully Bayesian treatment which is discussed in detail in Chapter 10.

There are deep connections between the nonparametric regression methods discussed in this section and *linear digital filters* from DSP theory (see Chapter 7). In the one-dimensional covariate case, when Z is the "time-like" variable for a signal which is discretized at uniform intervals $z_n = n\Delta t$, then many of these regression algorithms have direct counterparts as digital filters. The filtering operation is a matter of making predictions using the regression algorithm for every value of the signal z_n.

Variable selection

Often, in regression problems, we are faced with a large number of potential variables z_i, $i = 1, \ldots, M$ which could be included in the model. However, although we might be tempted to include all the variables, even for data where there is no relationship at all between $\boldsymbol{X}$ and $\boldsymbol{Z}$, the prediction error is biased downwards – ultimately to exactly zero when $M \geq N$ (because the problem is no longer overdetermined, see Figure 6.9). Thus, Occam's razor suggests we should trade off the num-

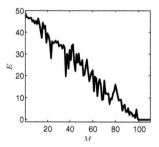

Fig. 6.9: Even for randomly generated data where there is no relationship at all between input Z and output variables X it is possible to lower the prediction error $E = \frac{1}{2}\|\mathbf{X} - \mathbf{WZ}\|_{\mathrm{F}}^2$ in a linear regression problem to zero. This is obviously misleading and is a strong justification for variable selection. Here there are $N = 100$ examples, and the number of input variables $M = 1, \ldots, N+10$. All entries for the data $\mathbf{X}$ and $\mathbf{Z}$ are zero mean, unit variance i.i.d. Gaussian random numbers.

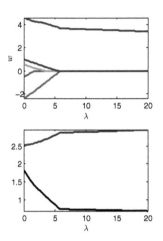

Fig. 6.10: The variable selection properties of Lasso regression. This is a univariate, $M = 5$ variable linear regression problem with parameter vector $w = \begin{bmatrix} 5.0 & -2.2 & 0 & 0 & 0 \end{bmatrix}^T$ and likelihood standard deviation $\sigma = 1$. Thus, the output X only depends upon the first two variables, the rest are superfluous. As the regularization parameter λ increases, so the individual values of the parameters $w_i, i = 1, \ldots, M$ shrink towards zero (top panel, where each differently coloured line represents one of the parameter vector values), meaning that they are selected out of the model. Bottom panel: the value of the prediction error $E = \|\mathbf{X} - \mathbf{WZ}\|_F^2$ (blue line) increases with the regularization parameter λ while the the average absolute parameter value $J = \frac{1}{M}\|w\|_1$ decreases (bottom panel). A value of $\lambda \approx 2.5$ correctly selects the two important variables with one incorrect selection, and at $\lambda \approx 5.0$ Lasso correctly selects the first variable but incorrectly removes the other.

ber of input variables against the prediction error. This motivates the topic of *variable selection*. Variable selection is important to all machine learning problems, not just regression, but the principles are the same and we will explore them in the context of regression here.

Variable selection is essentially a combinatorial optimization problem (Section 2.6), so we should not expect it to be easily solved. Indeed, finding the optimal subset of variables requires an exhaustive exploration of all 2^M possible subsets which is only feasible when M is very small, so that various heuristics are typically used instead. *Forward stepwise* variable selection starts with no variables (i.e. the empty subset $A_0 = \varnothing$), and selects the one which produces the smallest regularized prediction error, estimated by e.g. cross-validation or AIC (4.23) where $K = M$ (see Section 4.2). In the terminology of combinatorial optimization, this is just greedy search with best improvement, where the neighbourhood on each iteration $\mathcal{N}(A_{n+1})$ is the set of variables which are not already in A_n (Hastie *et al.*, 2009, page 58). *Backward stepwise* is a similar greedy procedure where the selection begins instead with the full set of variables, and removes the one on each iteration which produces the smallest prediction error (Hastie *et al.*, 2009, page 58). One can readily apply more sophistication here using e.g. simulated annealing or tabu search. If, instead of using the prediction error as a measure of the quality of a variable subset we use the magnitude of the correlation of the candidate variable with the *residual error* of the current subset $e = \hat{x} - x$, forward stepwise becomes *forward stagewise selection* (Hastie *et al.*, 2009, page 60).

Let us consider an explicit regularization approach to variable selection in the context of linear regression:

$$E = \frac{1}{2\sigma^2}\sum_{n=1}^{N}\left(x_n - \boldsymbol{w}^T\boldsymbol{z}_n\right)^2 + \gamma\sum_{i=1}^{M}|w_i|^q \qquad (6.60)$$

with the regularization parameter $\gamma > 0$ and $q \geq 0$. With $q = 0$ we obtain the L_0-loss which is $|u|^0 = 1$ if $u = 0$ and zero otherwise. In this case, the regularization term simply counts the number of non-zero coefficients e.g. variables included in the model. This is non-convex and computationally intractable, but if instead we choose $q = 1$ we obtain the Lasso problem earlier which is desirable since the L_1-loss is the closest convex alternative to the L_0-loss. Lasso regression therefore combines elements of variable selection with regression, in a computationally tractable way (Figure 6.10). The optimal value of λ can be chosen using cross-validation.

Many other variable selection methods can be viewed as "variations on a theme" of the above techniques. For example, *least-angle regression* (LAR), with a simple modification, can be seen as an approximate computation for the entire Lasso regression path (Hastie *et al.*, 2009, page 76). Although ridge regression is recovered for the case $q = 2$ in (6.60), in general, all the parameters will be non-zero except at the theoretical limit $\gamma \to \infty$, so it is not reasonable to view this as a variable selection method.

6.5   Clustering

Classification is a supervised technique where we have labelled training
data which allows us to learn a statistical model for each class. As
with mixture modelling, in *clustering* we do not know the label for each
data point $\boldsymbol{x}_1, \ldots, \boldsymbol{x}_N$, so it is *unsupervised* and we have to estimate
both a deterministic assignment of each point to a unique class, and
we also have to learn the parameters of each class model. There are a
very large number of clustering techniques in statistical machine learning
and signal processing, but we will begin our discussion with one of the
simplest which illustrates the main ideas.

K-means and variants

We have seen earlier that E-M usually provides a considerable reduction
in computational effort over Gibbs sampling for the Gaussian mixture
model, because the number of iterations required to reach convergence
is often at least an order of magnitude smaller. Nonetheless, E-M still
requires, for instance, the computation of a weighted mean over all observations in the M-step, which alone is order $O(NK)$. Here we will
show that an even simpler algorithm, one of the most ubiquitous algorithms for clustering known as *K-means*, is in fact a restricted case of
E-M and Gibbs sampling for the mixture model. Considering the Gaussian mixture model, if we take the covariance to be $\sigma^2 \mathbf{I}$ and then take
the limit $\sigma^2 \to 0$, the posterior indicator probability becomes (Bishop,
2006, page 443):

$$f(z_n = k | \boldsymbol{\mu}, \boldsymbol{x}; \boldsymbol{\pi}) \to \begin{cases} 1 & \text{if } k = \arg\min_{j \in 1,2,\ldots K} \left(\boldsymbol{x}_n - \boldsymbol{\mu}_j \right)^2 \\ 0 & \text{otherwise} \end{cases} \quad (6.61)$$

In words: if $\boldsymbol{x}_n$ is closer to component mean $\boldsymbol{\mu}_k$ than any other component mean, then the probability for $z_n = k$ is 1, otherwise it is 0. Thus,
Gibbs samples for z_n will amount to direct assignments of observations
$\boldsymbol{x}_n$ to their current closest components. Similarly, the component mean
posterior probabilities are:

$$f(\mu_k | \boldsymbol{z}, \boldsymbol{x}; \boldsymbol{\mu}_0, \sigma_0) \to \delta \left[\boldsymbol{\mu}_k - \frac{1}{N_k} \sum_{n:z_n=k} \boldsymbol{x}_n \right] \quad (6.62)$$

so that the Gibbs step simply becomes $\boldsymbol{\mu}_k = \frac{1}{N_k} \sum_{n:z_n=k} \boldsymbol{x}_n$. What this
implies is that Gibbs samples for the component means are always just
the mean of all the observations currently assigned to that component.
Together, these two steps comprise the K-means algorithm:

Algorithm 6.2 The K-means algorithm.

---

(1) *Initialization.* Start with a randomized set of parameters $\boldsymbol{\mu}_k^0$, $k \in 1, 2, \ldots, K$, set iteration number $i = 0$ and convergence tolerance $\epsilon > 0$,

(2) *Update indicators.* Compute $z_n^{i+1} = \arg\min_{j \in 1, 2, \ldots, K} \left\| \boldsymbol{x}_n - \boldsymbol{\mu}_j^i \right\|_2^2$,

(3) *Update component means.* Calculate $\boldsymbol{\mu}_k^{i+1} = \frac{1}{N_k} \sum_{n:z_n^{i+1}=k} \boldsymbol{x}_n$ for $k \in 1, 2, \ldots, K$,

(4) *Convergence check.* Calculate the clustering fit error $E\left(\boldsymbol{\mu}^{i+1}\right) = \sum_{k=1}^{K} \sum_{n:z_n^{i+1}=k} \left\| \boldsymbol{x}_n - \boldsymbol{\mu}_k^{i+1} \right\|_2^2$ and the improvement $\Delta E = E\left(\boldsymbol{\mu}^i\right) - E\left(\boldsymbol{\mu}^{i+1}\right)$ and if $\Delta E < \epsilon$, exit with solution $\boldsymbol{\mu}^{i+1}, \boldsymbol{z}^{i+1}$,

(5) *Iteration.* Update $i \leftarrow i + 1$, go back to step 2.

K-means offers noticeable computational saving over E-M in that the K-means equivalent to the M-step is $O(N)$. Together with the component mean updates, K-means requires $O(NK)$ per iteration, whereas E-M takes around twice that effort per iteration. Also, K-means typically converges in the same or fewer iterations than E-M. Despite these computational savings, K-means has many practical drawbacks, in particular, that we often arrive at the situation where one or more of the components has no data points assigned to it, in which case, the algorithm does not produce a sensible result. Similarly, there is no sense in which the K-means objective function E is an NLL of an underlying probabilistic model, it is just a sum of square losses:

$$E = \sum_{k=1}^{K} \sum_{n:z_n=k} \left\| \boldsymbol{x}_n - \boldsymbol{\mu}_k \right\|_2^2 \tag{6.63}$$

Comparisons of this quantity across data sets, for example, would be meaningless. K-means (at least the "canonical" version of the algorithm presented here) requires that the clusters are essentially spherical and all share the same radius and density of observations, these assumptions are very rarely satisfied in practice (Raykov *et al.*, 2016*a*).

It is not difficult to show that K-means eventually reaches a *fixed point*, that is, a configuration of the z_n (and therefore values of $\boldsymbol{\mu}_k$) which is unchanged under the iteration. One proof is that the objective function (6.63) cannot increase under the iteration since, with the indicators held fixed, the cluster means minimize the objective function, and conversely with cluster means fixed, the choice of indicators also minimizes the objective. Combining this with the fact that the number of clustering configurations is finite (but extremely large!), implies that there must be a configuration which is locally optimal given any starting condition, that is, from which the iteration cannot lower the objective function any further. So, in practice, the convergence check can just compare the current with the previous indicators: if they have not changed then the iteration can be stopped (but of course, this does not allow us to terminate the algorithm if we are satisfied with the given accuracy).

This relationship between K-means and E-M or Gibbs sampling obtained by shrinking the variance of the observation density to zero has more recently been called the *small-variance asymptotic* assumption and the concept has been applied to many models beyond simple mixture densities, including sophisticated infinite-dimensional variants (Jiang and Jordan, 2012) which we will discuss in Chapter 10.

K-means has a long historical pedigree, its conceptual origins go back at least as far as the 1950's (Lloyd, 1982, reprint of 1957 Bell Telephone labs paper). In *quantization* theory for signal processing, the *Lloyd-Max algorithm* is K-means but where the distribution of the data points $f(x)$ is known, a fact discussed in detail later (Section 8.2). So, we can derive K-means by replacing the known distribution in Lloyd-Max with its empirical density estimate.

K-means is simple but the use of the Euclidean metric can be problematic. We have shown earlier that it can be derived under the assumption that the components are spherical multivariate Gaussian, and shrinking the variance in each direction to zero. This entails that, everything else being equal, the clusters must be spherical. Similarly, because of this variance shrinking, we lose the effect of the indicator priors so that the clusters must have equal radius and equal density. These are often unrealistic assumptions in practice. We can therefore make some improvements in certain circumstances by changing the cluster component models. If instead of starting with the multivariate Gaussian density model for the cluster components we assume that each dimension is independent of the others and Laplace distributed (Section 4.4) with equal spread, then the indicator update step in Algorithm 6.2 becomes:

$$z_n^{i+1} = \operatorname*{arg\,min}_{j \in 1,2,...,K} \left\| \boldsymbol{x}_n - \boldsymbol{\mu}_j^i \right\|_1 \tag{6.64}$$

and the component update step changes to:

$$\mu_{k,d}^{i+1} = \operatorname{median}_{n:z_n^{i+1}=k} x_{n,d} \tag{6.65}$$

and the clustering fit error is:

$$E = \sum_{k=1}^{K} \sum_{n:z_n^{i+1}=k} \left\| \boldsymbol{x}_n - \boldsymbol{\mu}_k^{i+1} \right\|_1 \tag{6.66}$$

The resulting algorithm is known as *K-medians* for the obvious reason. Similar adaptations can be effected by the use of other parametric distributional models for the cluster component densities (Chapter 4),

Most practical software implementations of K-means use this check, since it avoids the need to choose a convergence tolerance ϵ.

which is particularly useful where there are closed-form maximum likelihood expressions then the component/cluster parameter updates (for example, exponential families, Section 4.5).

Geometrically, for the component parameter updates, K-medians selects a representative data point from the data assigned to every cluster in each dimension separately. However, it is usually more reasonable to assume that the dimensions are not independent. The problem then becomes one of finding the optimal cluster representative that minimizes the observed data NLL given the current cluster assignments. The resulting algorithm is known as *K-medoids* or *partitioning around medoids* in the literature. Unfortunately, this local minimization becomes non-convex and combinatorial, whereby the only guaranteed way of finding the representative is to search exhaustively for the representative in each cluster which minimizes the observed data NLL, and this will usually have prohibitively large computational complexity. An alternative is some discrete heuristic, such as greedy search (Section 2.6) initialized with the K-medians solution, which is often used in K-medoids implementations. Of course, we gain computational speed but we lose the certainty that the component parameter updates are true minimizers of the observed data NLL, given the current cluster assignments, that we have in K-means and E-M for mixture models.

*Note that strictly speaking, the K-medoids observed data NLL should **not** coincide with the K-medians NLL (6.66), because K-medoids does not posit a particular parametric component density model. In fact, any norm (e.g. L_2, L_∞) could be used here, but this will produce a different clustering solution.*

Soft K-means, mean shift and variants

Another perspective on the K-means clustering algorithm 6.2 views it as an iteration which maintains a set of centroids $\boldsymbol{\mu}_k$, $k = 1, 2, \ldots, K$, such that on each iteration, each of these centroids is replaced by the means of every data point which is closest in the Euclidean metric to that centroid. If we define an assignment function κ (Little and Jones, 2011*b*):

$$\kappa\left(\boldsymbol{x}; \boldsymbol{\mu}\right) = \delta\left[\boldsymbol{\mu} - \underset{m \in 1,2,\ldots,K}{\arg\min} \left\|\boldsymbol{x} - \boldsymbol{\mu}_m\right\|_2^2\right] \qquad (6.67)$$

then this function takes the value 1 where $\boldsymbol{\mu}$ is the closest centroid to $\boldsymbol{x}$, and 0 otherwise. In terms of this function, K-means can be written simply as (Little and Jones, 2011*b*):

$$\boldsymbol{\mu}_k^{i+1} = \frac{\sum_{n=1}^{N} \kappa\left(\boldsymbol{x}_n; \boldsymbol{\mu}_k^i\right) \boldsymbol{x}_n}{\sum_{n=1}^{N} \kappa\left(\boldsymbol{x}_n; \boldsymbol{\mu}_k^i\right)}, \text{ for } k \in 1, 2, \ldots, K \qquad (6.68)$$

Note that technically there is actually no need to check for convergence with K-means since it will converge on a fixed point, provided one is happy to continue iterating to reach that fixed point.

This very compact formulation invites us to consider other assignment functions, in particular, the *soft* Gaussian assignment:

$$\kappa\left(\boldsymbol{x}; \boldsymbol{\mu}\right) = \frac{\exp\left(-\beta \left\|\boldsymbol{x} - \boldsymbol{\mu}\right\|_2^2\right)}{\sum_{j=1}^{K} \exp\left(-\beta \left\|\boldsymbol{x} - \boldsymbol{\mu}_j\right\|_2^2\right)} \qquad (6.69)$$

with the parameter $\beta > 0$. Now, instead of replacing the centroids with the mean of only those data points which are closest given the current value of the centroid, the centroids are updated with a *weighted mean* of *all* data points, where the weighting decreases as data points get further away from the cluster centroid. The rate at which the weight decreases with Euclidean distance is controlled by β; as β increases so the rate of decrease is faster. In fact, as $\beta \to \infty$ then the soft function above coincides with the K-means function (6.67). The resulting iteration is known as *soft K-means* clustering (Little and Jones, 2011*b*). One advantage of this algorithm is that, provided β is sufficiently small, we avoid situations where a centroid ends up with no data points assigned to it, a problem that can arise with K-means.

This is very similar to a technique known as fuzzy C-means *clustering* (Dunn, 1973) *based on the interesting and well-developed theory of fuzzy sets.*

Soft K-means is intimately related to the E-M algorithm for mixture modelling (Section 6.2), as we will now show. Assume that we have a multivariate Gaussian mixture model with equal and diagonal covariance matrices, where the variance in each dimension is the same, $\sigma^2 = \frac{1}{2\beta}$, and with uniform priors over the indicators. In which case, we can see that (6.69) is in fact the posterior probability of the indicator z given $\boldsymbol{x}$:

$$f\left(z = k | \boldsymbol{x}; \boldsymbol{\mu}, \beta\right) = \frac{f\left(\boldsymbol{x}; \boldsymbol{\mu}_k, \beta\right)}{\sum_{j=1}^{K} f\left(\boldsymbol{x}; \boldsymbol{\mu}_j, \beta\right)} \tag{6.70}$$

and (6.68) is the corresponding M-step in the E-M algorithm. In other words, this formulation of K-means (or soft K-means) wraps up the E- and M-steps of E-M into one. Clearly soft K-means converges for the same reason that E-M does.

If, instead of choosing $K < N$ we set $K = N$ and use the following assignment kernel in the iteration (6.68):

$$\kappa\left(\boldsymbol{x}; \boldsymbol{\mu}\right) = \delta\left[\|\boldsymbol{x} - \boldsymbol{\mu}\|_2^2 \leq \sigma\right] \tag{6.71}$$

then we obtain an algorithm known as *mean shift* which is widely used in image signal processing. The iteration is usually initialized with $\boldsymbol{\mu}_n^0 = \boldsymbol{x}_n$ for $n \in 1, 2, \ldots, N$. The parameter σ controls the extent to which data points which are close to each of the N mean parameters $\boldsymbol{\mu}$, are included in the update of that parameter value on each iteration. As with soft K-means, we can define a soft version of mean shift by using e.g. the simple Gaussian soft assignment function:

$$\kappa\left(\boldsymbol{x}; \boldsymbol{\mu}\right) = \exp\left(-\beta \|\boldsymbol{x} - \boldsymbol{\mu}\|_2^2\right) \tag{6.72}$$

Cheng (1995) lays out a set of sufficient conditions for convergence of mean shift to a fixed point of the iteration, so that eventually $\boldsymbol{\mu}_k^{i+1} = \boldsymbol{\mu}_k^i$ for all $k \in 1, 2, \ldots N$ for some finite $i \geq 1$. By defining κ in terms of a *profile* function h, $\kappa\left(\boldsymbol{x}; \boldsymbol{\mu}\right) = h\left(\|\boldsymbol{x} - \boldsymbol{\mu}\|_2^2\right)$, then convergence occurs provided h is non-negative, non-increasing, piecewise continuous and has a finite integral, $\int_0^\infty h\left(r\right) dr < \infty$. This is true for many practical assignment functions. The effective "coverage" of the kernel h of the data points, that is, the diameter of the region in the space of the data for which the value of the kernel is large, determines the clustering behaviour. This is controlled by σ in (6.71) and β in (6.72).

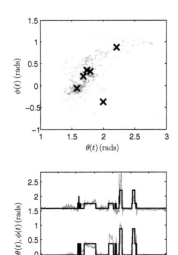

Fig. 6.11: Mean shift clustering applied to 3-axis accelerometry data of human movement. The raw sensor data is converted to spherical coordinates (top panel, grey dots, azimuth $\theta(t)$ horizontal, elevation $\phi(t)$ vertical). Applying soft mean shift with $\beta = 0.2$ and Gaussian kernel (6.72) gives the cluster centroids above (black crosses). These centroids are useful as estimates of the accelerometer device orientation with respect to gravity. Azimuth and elevation is plotted against time in the bottom panel, raw sensor data (grey lines) and soft mean shift output (black lines). Convergence was achieved in 15 iterations.

If the kernel coverage is small so that only a single point is included in the iteration for each cluster mean value, then a trivial fixed point – where there are N clusters centred on each data point – is reached in one iteration. Alternatively, if the kernel is non-zero for all data points for all mean values, then they all converge to the same, single cluster centroid. So, mean shift is most useful for intermediate values of the coverage, when it identifies local modes in the density of the data (Cheng, 1995). This is why it is justifiable to say that mean shift is a clustering method. Mean shift has similar "mode-seeking" behaviour to mixture modelling, but does so without the need to identify the number of modes (see Figure 6.11). This is the primary advantage of mean shift: it is a truly nonparametric clustering method, but the limitation is that each iteration requires $O\left(N^2\right)$ operations, which is prohibitive for large N. This contrasts with the complexity of K-means which can be considerably less expensive at $O\left(K\,N\right)$. We will discuss similar, but more statistically-motivated nonparametric clustering methods in Chapter 10.

Much as there are many variants of K-means, so there are many variants of mean shift. The mode-seeking behaviour of mean shift can also be derived starting with a nonparametric kernel density estimate (Section 4.8), where each iteration moves each parameter $\boldsymbol{\mu}_n$ along the direction of steepest ascent of the local gradient of the estimated density. By replacing the mean with the data point that maximizes this ascent direction instead, we obtain the *medoid shift* algorithm (Vedaldi and Soatto, 2008). Alternatively, *quick shift* moves each parameter to the nearest of the neighbouring data points that increase the kernel density estimate (Vedaldi and Soatto, 2008).

Semi-supervised clustering and classification

All the clustering methods described earlier share the common feature that they are unsupervised, expecting the data to be *unlabelled*. That is, when estimating the parameters, the labels are determined from the data. This contrasts with supervised classification where the expectation is that all the data is labelled. There are many situations in which the data is *partially labelled*, that is, some of the data has labels and the rest do not. This is known as a *semi-supervised* problem, and we can address this using some simple modifications to the clustering algorithms. For example, with K-means, we can fix some of the labels to their known values, and allow others to be updated during the usual K-means loop. In which case, K-means acts like a (degenerate likelihood, spherical) LDA classifier for the labelled data. Where this differs from LDA is that the estimates for the parameters for each class will include some unlabelled data as well. Provided that the classes for the unlabelled data have been correctly identified, this makes the estimate of the cluster parameters more statistically reliable. This can lead to substantial improvements to the performance of a clustering algorithm (Figure 6.12).

Choosing the number of clusters

Unlike classification, parametric clustering methods such as K-means and variants present us with the problem of how to choose the number of clusters K. Unfortunately, we cannot simply choose the K that minimizes the clustering fit error (e.g. (6.63) or (6.66)) because E always decreases for increasing K (until, at the extreme when $K = N$ then $E = 0$). Therefore, this is a model selection problem for which we need some kind of complexity control. The obvious solutions are AIC and/or BIC (see Section 4.2). The difficulty here is that K-means is not a probabilistic model, and we need to invoke a specific model for the cluster components in order to define a likelihood. For example, Pelleg and Moore (2000) suggest the spherical, isotropic multivariate Gaussian with variance σ^2 (which is not unreasonable since, as discussed earlier, we can derive K-means as a degenerate form of the Gaussian mixture model), whose complete data NLL is:

Only in principle: we would of course have to find the globally optimal solution for each K to see that increasing K always decreases E.

$$E = -\ln \prod_{n=1}^{N} \pi_{z_n} \mathcal{N}\left(\boldsymbol{x}_n; \boldsymbol{\mu}_{z_n}, \sigma^2 \mathbf{I}\right) \qquad (6.73)$$

leading to the BIC:

$$
\begin{aligned}
BIC \;=\; & \frac{1}{\hat{\sigma}^2} \sum_{n=1}^{N} \left\| \boldsymbol{x}_n - \boldsymbol{\mu}_{z_n} \right\|_2^2 - 2 \sum_{k=1}^{K} N_k \ln \hat{\pi}_k \qquad (6.74) \\
& - N D \ln(2\pi) + 2N D \ln \hat{\sigma} + K \ln N
\end{aligned}
$$

where the variance and indicator prior MLEs can be used:

$$\hat{\sigma}^2 \;=\; \frac{1}{ND} \sum_{n=1}^{N} \left\| \boldsymbol{x}_n - \boldsymbol{\mu}_{z_n} \right\|_2^2 \qquad (6.75)$$

$$\hat{\pi}_k \;=\; \frac{N_k}{N} \qquad (6.76)$$

Running K-means is now a matter of searching for the value of K that minimizes 6.74, simultaneously with the component means $\boldsymbol{\mu}_k$. This is an even more difficult optimization problem than K-means; one approach is to start with a minimum value of, e.g. $K = 2$, run K-means to convergence, and then increase K by one. The problem with this is that we have to run K-means "from scratch" each time; a more efficient heuristic involves increasing K by splitting existing clusters after running to convergence until some maximum number of clusters $K_{\max}$ is exceeded (Pelleg and Moore, 2000). Of course, this heuristic does not guarantee finding the optimal parameters, and indeed any other combinatorial optimization with K-means "in the loop" to find the optimal parameters given some candidate K would be suitable here (Section 2.6).

For an overview of different regularization approaches to selecting K, see Jain (2010).

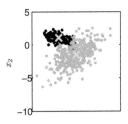

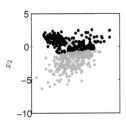

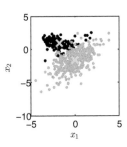

**Fig. 6.12: Semi-supervised cluster-
ing can improve on purely unsupervised clustering. A Gaussian
mixture model generates data from
two overlapping clusters (top panel,
black and grey dots) with prior probabilities $\pi_1 = 0.8$ (mixture component
centroids in yellow). This substantial cluster imbalance leads the unsupervised K-means to produce a poor
clustering solution (middle panel),
with expected 0-1 loss of 75% for
all the data. By contrast, with only
20% of the cluster indicators fixed
to their known values, the clustering
solution achieved by semi-supervised
K-means is substantially improved
(bottom panel), achieving expected
0-1 loss of 94% for the unlabelled
data.**

Other clustering methods

Given the importance and ubiquity of clustering in applications, a truly
vast number of algorithms have been invented over the years. It is not
possible to summarize all of them here, but they basically break down
into several different types. In the previous section we have explored
methods which are based on parametric cluster models and iterative assignment and cluster parameter improvement (K-means and variants).
However, there are many methods which are based on nonparametric
density estimation, which include mean shift and variants, and algorithms such as *DBSCAN* attempt to find high-density regions based on
specifying a neighbourhood and cluster membership size. *Graph clustering* and *topological* methods construct a weighted graph from the
pairwise distances between data points, where the edge weights are the
distances. These algorithms partition the data points into two or more
clusters by minimizing the *cut size* between clusters, that is, the sum of
the distances along all the edges connecting the clusters. Finding this
minimum cut is undertaken using several techniques including random
walks on the graph, or spectral decomposition on the connectivity matrix of the graph, or a transformation of that matrix. A readable and
wide-ranging review of clustering algorithms is contained in Jain (2010).

6.6 Dimensionality reduction

As described above, mixture models, clustering and classification posit
the existence of a discrete, hidden categorical indicator variable Z, and
one primary goal is to infer the value of this variable given the data
x. Regression, by contrast, supposes that these hidden variables are
multidimensional and continuous, and the primary goal is to predict the
value of X. However, regression does not answer the question of what to
do when we have X but want to infer Z, a kind of "reverse regression"
problem. This is the topic of this section. We assume that $X \in \mathbb{R}^D$ and
$Z \in \mathbb{R}^M$.

While we will start off by considering the important situation when
$D = M$ but this will quickly lead us on to the case where $M < D$ which
is known as *dimensionality reduction* and has applications across count-
less areas of machine learning and signal processing. In dimensionality
reduction, we are considering that the *dimensionality* of the observed
data x is somehow much smaller than D (Figure 6.13). This notion of
dimensionality is synonymous with the number of coordinates (degrees of
freedom) needed to fully and unambiguously locate each observed data
point. Many dimensionality reduction algorithms have the goal of finding a geometrical transformation and/or projection of the observed data
space such that the smallest amount of information is lost in performing
this transformation.

In practice, for most data D is too large that the observed data cannot
be plotted on a graph, for instance. What dimensionality reduction
promises is a substantial compression of the observed signal down onto

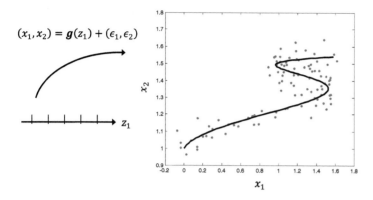

$(x_1, x_2) = \boldsymbol{g}(z_1) + (\epsilon_1, \epsilon_2)$

Fig. 6.13: *Dimensionality reduction* assumes that the data $\boldsymbol{X} \in \mathbb{R}^D$ (grey dots) actually exists on a set with smaller dimensionality $M < D$. This figure shows the case where $D = 2$ but $M = 1$, and the function $g : \mathbb{R} \to \mathbb{R}^2$ maps the one-dimensional latent space of the variable Z_1 into the two-dimensional space of the observed data (black curve, right panel), after the addition of the two-dimensional noise process ϵ. The goal of dimensionality reduction is to recover the values of the coordinates of the latent space z_1 from the observed data points (x_1, x_2).

a space with much smaller dimensionality, often small enough that it can be viewed directly. The implied model of the signal is that the intrinsic dimensionality of the observed data is indeed much smaller than $D = N$ (Figure 6.13).

Principal components analysis (PCA)

Perhaps the simplest and most ubiquitous dimensionality reduction technique is *principal components analysis* (PCA). This method proposes the assumption that the data lies on a *hyperplane* of dimension $M < D$. For example $M = 1$ implies that the data lies on a line, even if $D \geq 2$. The standard form of PCA which we present first does not necessarily entail a fully Bayesian probabilistic model (much like K-means clustering above), but we will see one extension of the idea which is fully Bayesian.

Let us start off by assuming the $\boldsymbol{X}$ is a multivariate Gaussian with $D \times D$ covariance matrix $\boldsymbol{\Sigma}$ (for now, we will assume that $\boldsymbol{Z}$ is not a random variable). We will also assume (without loss of generality) that the mean of $\boldsymbol{X}$ is zero, e.g. $\boldsymbol{\mu} = \boldsymbol{0}$. Now, we can find a rotation of the data which makes all the transformed coordinates independent and univariate Gaussian, and orders the coordinates by descending variance in each of the D dimensions. First, consider the case where the latent variable is scalar, so that $Z = \boldsymbol{X}^T \boldsymbol{w}$, where $\boldsymbol{w} \in \mathbb{R}^D$. Now, we can

We can always shift the data to the origin by subtracting off the sample mean.

expand out the variance of Z:

$$
\begin{aligned}
\operatorname{var}[Z] &= E\left[Z^2\right] - E[Z]^2 \\
&= E\left[\boldsymbol{w}^T \boldsymbol{X}\boldsymbol{X}^T\boldsymbol{w}\right] - E\left[\boldsymbol{w}^T\boldsymbol{X}\right] E\left[\boldsymbol{X}^T\boldsymbol{w}\right] \qquad (6.77)\\
&= \boldsymbol{w}^T E\left[\boldsymbol{X}\boldsymbol{X}^T\right]\boldsymbol{w} - \boldsymbol{w}^T E\left[\boldsymbol{X}\right] E\left[\boldsymbol{X}^T\right]\boldsymbol{w} \\
&= \boldsymbol{w}^T \boldsymbol{\Sigma} \boldsymbol{w}
\end{aligned}
$$

We wish to maximize $\operatorname{var}[Z]$ by varying $\boldsymbol{w}$, but to make this a well-posed problem, we need to put some constraints on $\boldsymbol{w}$, the simplest being that it is of unit length. This gives us the constrained optimization problem:

$$
\begin{aligned}
\text{maximize} \quad & \boldsymbol{w}^T \boldsymbol{\Sigma}\boldsymbol{w} && (6.78)\\
\text{subject to} \quad & \|\boldsymbol{w}\| = 1
\end{aligned}
$$

with the Lagrangian $L(\boldsymbol{w}, \lambda) = \boldsymbol{w}^T \boldsymbol{\Sigma}\boldsymbol{w} + \lambda\left(\boldsymbol{w}^T\boldsymbol{w} - 1\right)$. The gradient is $\nabla_{\boldsymbol{w}} L(\boldsymbol{w}, \lambda) = 2\boldsymbol{\Sigma}\boldsymbol{w} - 2\lambda\boldsymbol{w}$ such that the solution to (6.78) is:

$$
\boldsymbol{\Sigma}\boldsymbol{w} = \lambda\boldsymbol{w} \qquad (6.79)
$$

But this is just the eigenvector equation for the covariance of $\boldsymbol{X}$, so it follows that $\operatorname{var}[Z] = \lambda\boldsymbol{w}^T\boldsymbol{w} = \lambda$. Hence, the maximum-variance solution is simply obtained by finding the largest eigenvalue and corresponding eigenvector of $\boldsymbol{\Sigma}$. The vector $\boldsymbol{w}$ is known as the *first principal component* (PC) of $\boldsymbol{X}$, and λ is the largest eigenvalue. We will label these $\boldsymbol{w}_1$ and λ_1 and the projection down onto the first PC $Z_1 = \boldsymbol{X}^T\boldsymbol{w}_1$ accordingly. Next, we consider finding the second-largest variance direction by finding the vector $\boldsymbol{w}_2$ giving $Z_2 = \boldsymbol{X}^T\boldsymbol{w}_2$. As with the first PC we will require it to have unit length, but we also want it to be orthogonal to the first PC. Therefore we want to maximize $\boldsymbol{w}_2^T \boldsymbol{\Sigma}\boldsymbol{w}_2$ subject to the constraints $\boldsymbol{w}_2^T\boldsymbol{w}_2 = 1$ and $\boldsymbol{w}_1^T\boldsymbol{w}_2 = 0$. This has the Lagrangian $L(\boldsymbol{w}_2, \lambda_2, \mu) = \boldsymbol{w}_2^T \boldsymbol{\Sigma}\boldsymbol{w}_2 + \lambda_2\left(\boldsymbol{w}_2^T\boldsymbol{w}_2 - 1\right) + \mu\boldsymbol{w}_1^T\boldsymbol{w}_2$. Now, we must have that $\mu = 0$ since otherwise the constraint that $\|\boldsymbol{w}_1\| = 1$ would be violated, which means that we have another eigenvector equation $\boldsymbol{\Sigma}\boldsymbol{w}_2 = \lambda_2\boldsymbol{w}_2$ and so $\operatorname{var}[Z_2] = \lambda_2\boldsymbol{w}_2^T\boldsymbol{w}_2 = \lambda_2$.

We can continue on with this procedure to find a set of D eigenvalues which we can place in the $D \times D$ diagonal matrix $\boldsymbol{\Lambda}$ with the diagonals in decreasing size of λ. Similarly we have the set of corresponding orthonormal eigenvectors which we can organize to form the columns of the matrix $\mathbf{W} \in \mathbb{R}^{D \times D}$ which will be orthonormal by design. These two matrices together represent the full diagonalization of the covariance matrix:

$$
\boldsymbol{\Sigma} = \mathbf{W}^T \boldsymbol{\Lambda} \mathbf{W} \qquad (6.80)
$$

Since we are assuming that the data is multivariate Gaussian and we are imposing the constraints that each of principal component directions are orthogonal, it follows that each of the projections $Z_i = \boldsymbol{X}^T \boldsymbol{w}_i$, $i = 1, 2, \ldots D$ are independent, univariate Gaussian random variables with corresponding variances λ_i. Given some data $\boldsymbol{x}_1, \boldsymbol{x}_2, \ldots, \boldsymbol{x}_N$, the MLE for $\boldsymbol{\Sigma}$ would be the sample covariance matrix. Thus, we can view PCA as a way to transform this data into $\mathbf{Z} = \mathbf{W}^T \mathbf{X}$ where the matrix $\mathbf{X} \in \mathbb{R}^{D \times N}$ contains each data point $\boldsymbol{x}_n$ as a separate column. Then the matrix $\mathbf{Z} \in \mathbb{R}^{D \times N}$ contains columns with each of the N data points in the new, independent coordinate system.

Geometrically speaking, for multivariate Gaussian RVs, PCA finds a rotation in the space of $\boldsymbol{X}$ such that each axis of every equiprobable hyper-ellipsoid lies along one of the coordinate axes (see Section 1.4). This transformation on its own does not represent a simplification or dimensionality reduction, but this can be achieved simply by keeping the largest M eigenvalues and corresponding eigenvectors, which means keeping the first M dimensions of the transformed (latent) variables $\boldsymbol{z}_n$. In that case, the data is reduced down to a set of N vectors with M independent dimensions.

By doing dimensionality reduction with PCA in this way, we are now throwing away some information, so how good is this dimensionally-reduced representation? An alternative, "error-based" view of PCA can address this question. Suppose we have an orthogonal basis $\boldsymbol{v}_1, \boldsymbol{v}_2, \ldots, \boldsymbol{v}_D \in \mathbb{R}^D$ for the space of $\boldsymbol{X}$ and we wish to find an $M < D$ dimensional approximation which has minimum square loss empirical risk. Written in this basis, each data point is:

$$\boldsymbol{x}_n = \sum_{i=1}^{D} a_{ni} \boldsymbol{v}_i \tag{6.81}$$

where $a_{ni} = \boldsymbol{v}_i^T \boldsymbol{x}_n$ is the coefficient of basis vector i. Approximating $\boldsymbol{x}_n$ using only M of the basis vectors, the square loss empirical risk is:

$$
\begin{aligned}
E &= \frac{1}{N} \sum_{n=1}^{N} \left\| \boldsymbol{x}_n - \sum_{i=1}^{M} a_{ni} \boldsymbol{v}_i \right\|_2^2 \\
&= \frac{1}{N} \sum_{n=1}^{N} \left\| \sum_{i=M+1}^{D} a_{ni} \boldsymbol{v}_i \right\|_2^2 \\
&= \frac{1}{N} \sum_{n=1}^{N} \sum_{i=M+1}^{D} a_{ni}^2 \\
&= \frac{1}{N} \sum_{n=1}^{N} \sum_{i=M+1}^{D} \boldsymbol{v}_i^T \boldsymbol{x}_n \boldsymbol{x}_n^T \boldsymbol{v}_i = \sum_{i=M+1}^{D} \boldsymbol{v}_i^T \mathbf{S} \boldsymbol{v}_i
\end{aligned}
\tag{6.82}
$$

where $\mathbf{S} = \frac{1}{N} \mathbf{X} \mathbf{X}^T$ is the sample covariance matrix.

We now want to minimize E with respect to the basis vectors, subject to the constraint that they each have unit length. This leads to the Lagrangian $L(\mathbf{V}, \boldsymbol{\lambda}) = \sum_{i=M+1}^{D} \boldsymbol{v}_i^T \mathbf{S} \boldsymbol{v}_i + \sum_{i=M+1}^{D} \lambda_i \left(\boldsymbol{v}_i^T \boldsymbol{v}_i - 1 \right)$ which

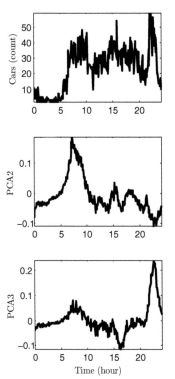

Fig. 6.14: **Principal components analysis (PCA) applied to traffic count data. The number of cars through a particular highway junction is captured every 5 minutes (top panel). This captures travelling between a local sport stadium, but also local residential and other traffic. PCA is applied to these counts over 175 days. The second principal component picks out the rush-hour morning work traffic (middle panel), and the picks out late-night sporting events at the stadium. The linear model is well-justified here since car counts are clearly additive, so, the daily measured traffic can be understood in terms of these component traffic flows.**

is minimized at:

$$\mathbf{S}\boldsymbol{v}_i = \lambda_i \boldsymbol{v}_i, \quad i = M+1, M+2, \dots D \tag{6.83}$$

Clearly this is the same set of eigenvalue/eigenvector equations as above but using the sample covariance matrix rather than the model covariance $\boldsymbol{\Sigma}$. The empirical risk is simply $E = \sum_{i=M+1}^{D} \lambda_i$. Since in practice we substitute the sample covariance estimate for $\boldsymbol{\Sigma}$, the empirical risk minimization and minimum latent variance diagonalization formulations of PCA coincide. The basis vectors $\boldsymbol{v}_i$ can be identified with the weight vectors $\boldsymbol{w}_i$ from the minimum variance projection. PCA is well-justified for DSP in linear phenomena, see Figure 6.14 for an example usage in physical traffic modelling.

Probabilistic PCA (PPCA)

Both formulations of PCA: observed data covariance diagonalization and empirical risk minimization, are non-probabilistic. This is a problem in that there is no principled likelihood with which to estimate the quality of the fit of the lower-dimensional model, we cannot draw samples from the model, and we cannot readily use the algorithm inside other probabilistic procedures, among other limitations. We will justify a probabilistic approach to PCA (PPCA for short) by following the linear regression model in Section 6.4, e.g. $\boldsymbol{X} = \mathbf{W}\boldsymbol{Z} + \boldsymbol{\epsilon}$ where $\mathbf{W} \in \mathbb{R}^{D \times M}$. This differs from the usual interpretation of regression in that the observed data are considered to be "explained" using their lower-dimensional hidden counterparts, and M is typically much smaller than D. We will assume that the observation noise $\boldsymbol{\epsilon}$ is spherical multivariate Gaussian. This gives us the following model for the observed variables:

$$f(\boldsymbol{x}|\boldsymbol{z}) = \mathcal{N}\left(\boldsymbol{x}; \mathbf{W}\boldsymbol{z}, \sigma^2 \mathbf{I}\right) \tag{6.84}$$

We next need some model for the hidden variables. Because we want these variables to be easy to interpret, we will assume that they are standard multivariate Gaussian, e.g. $f(\boldsymbol{z}) = \mathcal{N}(\boldsymbol{z}; \mathbf{0}, \mathbf{I})$. Therefore:

$$f\left(\boldsymbol{x}, \boldsymbol{z}; \mathbf{W}, \sigma^2\right) = \mathcal{N}\left(\boldsymbol{x}; \mathbf{W}\boldsymbol{z}, \sigma^2 \mathbf{I}\right) \mathcal{N}\left(\boldsymbol{z}; \mathbf{0}, \mathbf{I}\right) \tag{6.85}$$

and, using the marginalization properties of multivariate Gaussians (Section 1.4):

$$f\left(\boldsymbol{x}; \mathbf{W}, \sigma^2\right) = \mathcal{N}\left(\boldsymbol{x}; \mathbf{0}, \mathbf{Q}\right) \tag{6.86}$$

with the $D \times D$ covariance matrix $\mathbf{Q} = \mathbf{W}\mathbf{W}^T + \sigma^2 \mathbf{I}$.

Since we are interested in estimating the parameters $\mathbf{W}$ and σ^2, and the distribution over $\boldsymbol{Z}$ does not depend upon these parameters, we can estimate them by minimizing the NLL of $\boldsymbol{X}$ independently of $\boldsymbol{Z}$ instead.

This is:

$$
\begin{aligned}
E &= -\ln \prod_{n=1}^{N} f\left(\boldsymbol{x}_n; \mathbf{W}, \sigma^2\right) \\
&= \frac{1}{2} \sum_{n=1}^{N} \boldsymbol{x}_n^T \mathbf{Q}^{-1} \boldsymbol{x}_n \\
&+ \frac{ND}{2} \ln(2\pi) + \frac{N}{2} \ln|\mathbf{Q}|
\end{aligned}
\tag{6.87}
$$

It can be shown that the MLEs for $\mathbf{W}$ and σ^2 using (6.87) are(Tipping and Bishop, 1999):

$$
\hat{\mathbf{W}} = \mathbf{U}\left(\boldsymbol{\Lambda} - \hat{\sigma}^2\mathbf{I}\right)^{\frac{1}{2}}
\tag{6.88}
$$

$$
\hat{\sigma}^2 = \frac{1}{D-M} \sum_{i=M+1}^{D} \lambda_i
\tag{6.89}
$$

where $\boldsymbol{\Lambda}$ is the $M \times M$ diagonal matrix of the largest M eigenvalues λ_i of the $D \times D$ sample covariance matrix $\mathbf{S} = \frac{1}{N}\mathbf{X}\mathbf{X}^T$, and $\mathbf{U}$ is the $D \times M$ matrix whose columns are the corresponding M eigenvectors. Thus, as with non-probabilistic PCA, the estimates for $\mathbf{W}$ and σ^2 require only the diagonalization of the covariance matrix (but see Tipping and Bishop, 1999 for an approach to parameter estimation which uses E-M instead). We can also see that $\hat{\sigma}^2$ is just the squared error entailed by the projection from D down to M dimensions.

When using this dimensionality reduction model, we obviously want to visualize the values of the latent variables. The posterior distribution for the latent variable is:

$$
f\left(\boldsymbol{z}|\boldsymbol{x}; \mathbf{W}, \sigma^2\right) = \mathcal{N}\left(\boldsymbol{z}; \mathbf{P}^{-1}\mathbf{W}^T\boldsymbol{x}, \sigma^2\mathbf{P}^{-1}\right)
\tag{6.90}
$$

with the $M \times M$ matrix $\mathbf{P} = \mathbf{W}^T\mathbf{W} + \sigma^2\mathbf{I}$. It is reasonable to summarize this distribution over the latent variables using their mean, knowing that this coincides with their ML values since for the multivariate Gaussian its only mode coincides with its mean:

$$
E[\boldsymbol{Z}|\boldsymbol{X}] = \left(\mathbf{W}^T\mathbf{W} + \sigma^2\mathbf{I}\right)^{-1}\mathbf{W}^T\boldsymbol{X}
\tag{6.91}
$$

It is also illuminating to see what happens when the variable $\sigma^2 \to 0$. In that case, the observation model $f(\boldsymbol{x}|\boldsymbol{z})$ collapses down onto its mean and there is no longer any well-defined likelihood, as with PCA. Also as with PCA, the projection of the data down onto the latent variables (6.91) becomes the linear regression $\boldsymbol{Z} = \left(\mathbf{W}^T\mathbf{W}\right)^{-1}\mathbf{W}^T\boldsymbol{X} = \mathbf{W}^T\boldsymbol{X}$ which coincides with the PCA projection (up to a scale factor in each direction). Thus, PCA can be recovered as a special case of probabilistic PCA, much like K-means is a special "degenerate" case of spherical Gaussian mixture modelling.

There are some peculiarities which need to be understood with PPCA.

In particular, when σ^2 is large, the projection (6.91) is biased towards the origin $\mathbf{0}$. This is the regularization effect of the standard, zero-mean multivariate Gaussian prior on the latent variables. Thus, if σ^2 is large, we can interpret the low-dimensional projection as unreliable, as we would in any Bayesian analysis.

Nonlinear dimensionality reduction

So far, the methods above have all entailed a *linear* model relating the hidden to the observed variables. In practice, there are many real signal processing and machine learning problems for which this assumption is too restrictive. In this section we will extend the model to cover various nonlinear relationships.

One of the simplest methods involves replacing the entries in the covariance matrix $\boldsymbol{\Sigma}$ in PCA with kernel functions applied to the data, much as in kernel regression (Section 6.4). The expectation is that, exploiting Mercer's theorem (Section 6.1), in this new, possibly higher dimensional *feature space* $\boldsymbol{\phi}(\boldsymbol{x})$, there is a rotation from which projection can be found that separates out the important directions, thereby making it meaningful to apply PCA in this new space. In this feature space, the kernel matrix $\mathbf{K}$ can be diagonalized and has positive real eigenvalues which can be ordered by magnitude, as with PCA.

Often, we only have access to the kernel function $\kappa(\boldsymbol{x}; \boldsymbol{x}')$ from which we can compute the kernel matrix, not the feature functions, but we need to find the projection of an observation down onto the unknown feature function to find the associated latent value $\boldsymbol{z}$. Denote $\mathbf{U}$ by the $N \times N$ matrix of eigenvectors, and $\boldsymbol{\Lambda}$ the $N \times N$ diagonal matrix of eigenvalues of $\mathbf{K}$. Let us assume that there is a *kernel design matrix* $\boldsymbol{\Phi} \in \mathbb{R}^{L \times N}$ with columns containing the feature values $\boldsymbol{\phi}(\boldsymbol{x}_n)$. The eigenvectors of this matrix can be written as $\mathbf{V} = \boldsymbol{\Phi} \mathbf{U} \boldsymbol{\Lambda}^{-\frac{1}{2}}$. Although we may not be able to find these eigenvectors, Murphy (2012, page 494) shows that the size $1 \times N$ projection of any $\boldsymbol{x}$ onto these N eigenvectors is given by:

$$Z(\mathbf{x}) = \boldsymbol{\phi}(\boldsymbol{x})^T \mathbf{V} = \boldsymbol{k}(\boldsymbol{x})^T \mathbf{U} \boldsymbol{\Lambda}^{-\frac{1}{2}} \qquad (6.92)$$

where the length N vector $\boldsymbol{k}(\boldsymbol{x})$ has elements $k_n(\boldsymbol{x}) = \kappa(\boldsymbol{x}; \boldsymbol{x}_n)$ for $n = 1, 2, \ldots, N$. We can compute the value of all the latent variables in the observed data, $\mathbf{Z} = \mathbf{K}^T \mathbf{U} \boldsymbol{\Lambda}^{-\frac{1}{2}}$ which contains each of the N data points mapped into the latent feature space. Furthermore, by retaining only $M < D$ of the largest eigenvalues of $\mathbf{K}$ in $\boldsymbol{\Lambda}' \in \mathbb{R}^{M \times M}$ and the corresponding eigenvectors in $\mathbf{U}' \in \mathbb{R}^{N \times M}$, then $\mathbf{Z}' = \mathbf{K}^T \mathbf{U}' \boldsymbol{\Lambda}'^{-\frac{1}{2}}$ projects down onto only the most important directions, as with PCA but in feature space, to obtain the dimensionally-reduced representation $\mathbf{Z}' \in \mathbb{R}^{N \times M}$. Finally, removing the mean of the vectors in feature space is necessary as a pre-processing step, Murphy (2012, page 494) shows that this can be accomplished by computing $\bar{\mathbf{K}} = \mathbf{H} \mathbf{K} \mathbf{H}$ using the *centering matrix* $\mathbf{H} = \mathbf{I} - (1/N) \mathbf{1}_N \mathbf{1}_N^T$ with $\mathbf{1}_N$ being a vector of length N containing all ones. Using this centred kernel matrix $\bar{\mathbf{K}}$ in place of the original matrix $\mathbf{K}$, this nonlinear dimension reduction algorithm is

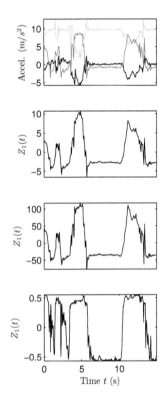

Fig. 6.15: Kernel principal components analysis (KPCA) applied to human 3-axis accelerometry movement data (raw data, top panel). Understanding device orientation under rotational motion relative to the Earth's gravitational field direction is most parsimoniously represented in spherical, rather than the Cartesian, coordinate system of the raw device data. This is an essentially nonlinear transformation, which justifies the use of KPCA (third and bottom panels, quadratic polynomial and Gaussian kernels with $\sigma = 1$, respectively). The linear, PCA projection is shown for comparison (second panel).

known as *kernel principal components analysis* (KPCA), see Figure 6.15 for an example sensor signal processing application.

For PCA, the model is linear and *global* that is, applies to the entire, infinite observed and lower-dimensional latent variable spaces. For certain problems, this is not necessarily a good model. A much more general model (more general than KPCA) involves the concept of a *manifold*, that is, a space which (informally) locally behaves like Euclidean space. So, globally the model is nonlinear (and may also have different global topological structure to Euclidean space), but near every point, like PCA, it is linear. For example, the surface of a torus (donut shape) which is contained in three or more dimensions has only two coordinates which can be mapped to a finite subset of the two-dimensional plane with boundaries that are connected together (top with bottom, left with right). Thus, if the observed, high-dimensional data actually comes from the surface of a torus, we should be able to find a representation of that data which is two-dimensional and yet loses no information about the original data.

More formally, a manifold is a topological space for which there is a continuous, invertible function which maps it to Euclidean space around every point.

This manifold embedding idea is central to *locally linear embedding* (LLE) described by Roweis and Saul (2000) which we discuss next. First, for each observed data item $\boldsymbol{x}_n$, find the K nearest neighbours,

$$\mathcal{N}_K(\boldsymbol{x}_n) = \{n' \in 1, 2, \ldots, N : \|\boldsymbol{x}_n - \boldsymbol{x}_{n'}\| \leq \|\boldsymbol{x}_n - \boldsymbol{x}_{(K)}\| \wedge n' \neq n\},$$

where $\boldsymbol{x}_{(K)}$ is the data item which is the K most distant from $\boldsymbol{x}_n$. Then, approximate each data item as a linear sum of its nearest neighbours $\mathbf{X} \approx \mathbf{AX}$ where the matrix $\mathbf{A} \in \mathbb{R}^{N \times N}$ contains the local linear model parameters $a_{nn'}$ for $n, n' \in 1, 2, \ldots, N$. If a data item $\boldsymbol{x}_{n'}$ is not in the neighbourhood of $\boldsymbol{x}_n$ i.e. $n' \notin \mathcal{N}_K(\boldsymbol{x}_n)$, then $a_{nn'} = 0$. So, these parameters relate each observed data item only to their nearest neighbours. The justification for this local linear modelling assumption is reasonable if the data comes from a manifold. The parameters $\mathbf{A}$ of that approximation are obtained by minimizing the global *approximation error* $E(\mathbf{A}) = \|\mathbf{X} - \mathbf{XA}\|_{\mathrm{F}}^2$ using the data matrix $\mathbf{X} \in \mathbb{R}^{D \times N}$ with each column containing one $\boldsymbol{z}_n$. We can think about this approximation as fitting a locally linear regression model for each data item. The solution can be obtained using the normal equations for each data observation and its neighbours $i \in \mathcal{N}_K(\boldsymbol{x}_n)$. By enforcing the constraint $\sum_{n' \in \mathcal{N}_K(\boldsymbol{x}_n)} a_{nn'} = 1$ for all $n \in 1, 2, \ldots, N$ Roweis and Saul (2000) obtain a practical solution in terms of the local $K \times K$ covariance matrices $\mathbf{X}_n^T \mathbf{X}_n$ with the locally centred data matrix $\mathbf{X}_n$ having columns $\boldsymbol{x}_n - \boldsymbol{x}_{n'}$ for $n' \in \mathcal{N}_K(\boldsymbol{x}_n)$.

By fitting these local regression models, for each data item we have a representation of its local space $\{\boldsymbol{x}_{n'}\}$, $n' \in \mathcal{N}_K(\boldsymbol{x}_n)$, captured in the model $\mathbf{A}$. Now, we can use this to obtain a reduced-dimensional *embedded* latent representation of each data item that respects its local model, $\mathbf{Z} \approx \mathbf{AZ}$ for $\boldsymbol{z}_n \in \mathbb{R}^M$. The matrix of embedded representations $\mathbf{Z} \in \mathbb{R}^{M \times N}$ can be obtained by minimizing the global *embedding error* $E(\mathbf{Z}) = \|\mathbf{Z} - \mathbf{ZA}\|_{\mathrm{F}}^2$. Solving this requires applying some constraints. As with PCA, we assume that the embedded data items have unit (sample) covariance $\frac{1}{N}\mathbf{ZZ}^T = \mathbf{I}$ and the latent representations are centred at

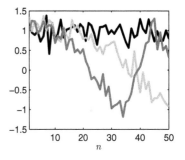

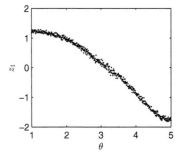

Fig. 6.16: Locally linear embedding (LLE) applied to *chirp* signals (that is, nonstationary sinusoids with linearly increasing frequency over time) with additive i.i.d. Gaussian noise (top panel, black, light grey, dark grey correspond to increasing values of chirp frequencies θ). Given $N = 500$ signals with $D = 50$ samples each with chirp frequencies of $\theta \in [1,5]$, LLE with $K = 100$ neighbours with embedding dimension $M = 1$ can relatively easily identify the chirp frequency from the signal (bottom panel), although the recovery is not entirely linear.

the origin, $\sum_{n=1}^{N} \boldsymbol{z}_n = \mathbf{0}$. Roweis and Saul (2000) show that the solution is found by computing the smallest $M + 1$ eigenvalues and associated eigenvectors of the matrix $\mathbf{M} = (\mathbf{I} - \mathbf{A})^T (\mathbf{I} - \mathbf{A})$. The eigenvector corresponding to the smallest eigenvalue contain all 1's by construction and is discarded, the rest of the eigenvectors form the latent embedded variables.

LLE can tease out the underlying organization in some quite complex signals (see Figure 6.16) and has demonstrated effectiveness in digital image processing, for example (Roweis and Saul, 2000). However, unlike PCA/PPCA it lacks a simple way to map arbitrary points in the latent space $\boldsymbol{Z}$ back to the original space $\boldsymbol{X}$, and it is not probabilistic like PPCA. It is also computational challenging: the embedding parameter eigendecomposition is in general $O(N^3)$ which dominates over finding nearest neighbours $O(N^2)$ and computing the approximation parameters, $O(NK^3)$. So, for large N, LLE is not very practical, however, we generally require a lot of very close neighbours to it effective.

Since the description of the original LLE method, manifold-based dimensionality reduction has grown into a very mature subfield, for a comprehensive overview read van der Maaten *et al.* (2009).

Linear-Gaussian systems and signal processing

7

Linear systems theory, based on the mathematics of vector spaces, is the backbone of all "classical" DSP and a large part of statistical machine learning. The basic idea – that linear algebra applied to a signal can of substantial practical value – has counterparts in many areas of science and technology. In other areas of science and engineering, linear algebra is often justified by the fact that it is often an excellent model for real-world systems. For example, in acoustics the theory of (linear) wave propagation emerges from the concept of *linearization* of small pressure disturbances about the equilibrium pressure in classical fluid dynamics. Similarly, the theory of electromagnetic waves is also linear. Except when a signal emerges from a justifiably linear system, in DSP and machine learning we do not have any particular correspondence to reality to back up the choice of linearity. However, the mathematics of vector spaces, particularly when applied to systems which are time-invariant and jointly Gaussian, is highly tractable, elegant and immensely useful.

7.1 Preliminaries

There are a few key fundamental ideas that will be used in this section. These are mathematical preliminaries to this section and they will form important building blocks for later sections as well. There are certain special functions and signals that occur in DSP applications, and we also need the basics of complex numbers and the marginal and conditional properties of jointly multivariate normal distributions.

Delta signals and related functions

The first special signals are the so-called *delta functions* written as $\delta[\cdot]$. The operator brackets indicate that for these objects it is only really consistent to think of them as operators (in fact, functionals which map signals onto the real line) rather than as functions, for reasons that we will discuss shortly. The easiest to define is the discrete-time *Kronecker* delta, which produces the discrete *infinite impulse signal* $\delta = (\ldots, 0, 0, 1, 0, 0, \ldots)$. We can write the Kronecker delta as $\delta[n] = 1$ if $n = 0$ and zero for $n \neq 0$. Then the infinite impulse signal is $\delta_n = \delta[n]$. Despite this definition, the primary value of such an object is what it

Machine Learning for Signal Processing: Data Science, Algorithms, and Computational Statistics. Max A. Little.
© Max A. Little 2019. Published in 2019 by Oxford University Press. DOI: 10.1093/oso/9780198714934.001.0001

does to a *test signal* $\boldsymbol{x}$ under infinite summation, in particular:

$$\sum_{n \in \mathbb{Z}} \delta\left[n - m\right] x_n = x_m \qquad (7.1)$$

for $m \in \mathbb{Z}$. In other words, the Kronecker delta "evaluates" the signal at the index m and the infinite summation collapses away. This is known as the *sift property* of the Kronecker delta. This might seem like a needlessly complex way just to pick out x_m, but in later sections we will encounter the need to handle many such infinite summations and this simplification plays a critical role.

We can integrate the Kronecker delta to obtain the so-called *unit step function*:

$$\sum_{n=-\infty}^{m} \delta\left[n\right] = u_m = \begin{cases} 0 & m < 0 \\ 1 & m \geq 0 \end{cases} \qquad (7.2)$$

This function has substantial value as a way to limit the non-zero range of a test function of interest, for example.

In continuous time, we can define very similar objects, in particular the analogue of the Kronecker delta, called the *Dirac* delta. To define this object, we will want it to behave as the Kronecker delta under integration:

$$\int_{\mathbb{R}} \delta\left[t - s\right] f\left(t\right) dt = f\left(s\right) \qquad (7.3)$$

Now, in order for this to be true, this "function" must be infinite at zero, zero everywhere else, and yet integrate up to 1. Unfortunately, it cannot then be a function in the classical sense. A larger class of objects which includes the Dirac delta are *tempered distributions*, which are actually functionals: for a classic introduction to such objects, see **Lighthill (1958)**. For the practical DSP purposes in this book, treating (7.3) as the defining property of the Dirac delta does what we need.

As with the Kronecker delta, we can integrate the Dirac to obtain the so-called *Heaviside step function*:

$$\Phi\left(t\right) = \int_{-\infty}^{t} \delta\left[u\right] du = \begin{cases} 0 & t < 0 \\ 1 & t \geq 0 \end{cases} \qquad (7.4)$$

which is of course the continuous-time analogue of the unit step. We will use the same notation δ for both Kronecker and Dirac delta, distinguishing them by the type of the argument (e.g. discrete argument for Kronecker, and continuous argument for Dirac).

Another signal which is of significant importance in DSP is the *sinc function* is defined as (see Figure 7.1):

$$\text{sinc}\left(x\right) = \frac{\sin\left(\pi x\right)}{\pi x} \qquad (7.5)$$

which, we will see, is in fact intimately related to the (unit) *rectangular pulse function*:

$$\begin{aligned} \text{rect}\,(x) \;&=\; \Phi\,(x+\pi) - \Phi\,(x-\pi) \\[4pt] &=\; \begin{cases} 1 & |x| < \pi \\ 0 & \text{otherwise} \end{cases} \end{aligned} \tag{7.6}$$

Complex numbers, the unit root and complex exponentials

While probability, statistics and machine learning rarely deal with *complex numbers*, they are fundamental to the theory of linear systems. Complex numbers allow all algebraic equations with arbitrary coefficients to be solved, for example, one cannot solve the equation $x^2+1=0$ without using the complex unit $i = \sqrt{-1}$ such that $i^2 = -1$. Then, the solution to our equation is $x = \pm i$, such that $x^2 = i^2 = -1$. All complex numbers consist of a *real* part a and an *imaginary* part b, so that the complex number $z = a + ib$ where $a, b \in \mathbb{R}$. This is the so-called Cartesian or *rectangular* form, and each number occupies the point (a, b) on the *complex plane*. The real and imaginary parts are $\text{Re}\,(z) = a$ and $\text{Im}\,(z) = b$ respectively. The *complex conjugate* $\bar{z} = a - ib$ simply changes the sign of the imaginary part.

In this Cartesian form, addition is simple: $z_1 + z_2 = (a + ib) + (c + id) = (a + c) + i\,(b + d)$. Multiplication is more complex: $z_1 \times z_2 = (a + ib)\,(c + id)$ which, after expanding brackets gives $(ac - bd) + i\,(bc + ad)$. A similar Cartesian formula exists for division, but multiplication and division are much simpler in the *polar form*, which just involves a coordinate transformation from Cartesian planar coordinates to polar coordinates. In this form, each number is located by the *argument-magnitude* (or *phase angle-absolute value*) coordinate pair $(\phi, r) \in \mathbb{R}^2$, where $r = \sqrt{a^2 + b^2} = z\bar{z}$ is the magnitude of the number z, and $\phi = \arg\,(z)$ is the angle (in radians) that the vector from the origin $(0, 0)$ in the complex plane to the point (a, b) makes. The angle is usually taken to be the principal value of $\phi = \arctan\,(b/a)$ in the range $[-\pi, \pi]$. In the polar form, $z = r\,(\cos\phi + i\sin\phi)$ and using trigonometric sum formulae, we get that $z_1 \times z_2 = r_1 r_2\,(\cos\,(\phi_1 + \phi_2) + i\sin\,(\phi_1 + \phi_2))$. *Euler's formula* creates a link between the natural logarithm of a complex number and the polar form, $r\exp\,(i\phi) = r\,(\cos\phi + i\sin\phi)$. Thus, $z^n = r^n\,(\cos\,(n\phi) + i\sin\,(n\phi))$ for $n \in \mathbb{Z}$, a relationship known as *de Moivre's formula*.

The above formula allows us to solve the polynomial $z^N = 1$ for the complex number z and any $N \in 1, 2, 3, \ldots$. This equation has N solutions called the N *roots of unity* which can be found by setting $\phi = \frac{2\pi}{N}$ and $r = 1$ then:

$$\left(r\exp\left(\frac{2\pi i}{N}\right)\right)^n \;=\; \exp\left(2\pi i \frac{n}{N}\right) \tag{7.7}$$

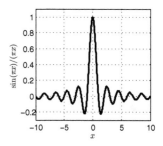

Fig. 7.1: The *sinc* function $\text{sinc}\,(x) = \sin\,(\pi x)/\pi x$ which plays a critical role in bandlimited sampling and reconstruction. It has the *interpolation property* $\text{sinc}\,(n) = \delta\,[n]$ (Kronecker delta) for $n \in \mathbb{Z}$ (of fundamental importance in sampling applications, section 8.1) and also $\text{sinc}\,(x) \to 0$ for $x \to \pm\infty$. The continuous-time Fourier transform of sinc is the *rectangular (pulse) function* $\text{rect}\,(x) = 1$ for $|x| < \pi$, $\text{rect}\,(x) = 0$ for $|x| \geq \pi$. Another way to appreciate the sinc function is that it is the impulse response of the *ideal low-pass filter*.

for each $n \in 0, 1, \ldots N - 1$. Now, setting $z = \exp\left(2\pi i n / N\right)$ we get that $z^N = \exp\left(2\pi i n\right) = 1$ since $\cos\left(2\pi n\right) = 1$ and $\sin\left(2\pi n\right) = 0$. We will see that these complex exponentials, evenly distributed around the unit circle in the complex plane, are fundamental to the analysis of linear time-invariant systems, which we investigate next.

Marginals and conditionals of linear-Gaussian models

Let us consider the "generative" model $\boldsymbol{X} = \mathbf{A}\boldsymbol{Z} + \boldsymbol{V}$, with $\boldsymbol{Z} \sim \mathcal{N}\left(\boldsymbol{\mu}_{\boldsymbol{Z}}, \boldsymbol{\Sigma}_{\boldsymbol{Z}}\right)$ and $\boldsymbol{V} \sim \mathcal{N}\left(\mathbf{0}, \boldsymbol{\Sigma}\right)$, which defines a linear relationship between the generally multivariate random vector variables $\boldsymbol{X}$ and $\boldsymbol{Z}$ represented by the matrix $\mathbf{A}$ with appropriate number of rows and columns to so that the model is consistent with the sizes of the two variables. It will be useful to know the corresponding joint, marginal and conditional distributions:

(1) *Marginals*: We have $f\left(\boldsymbol{z}\right) = N\left(\boldsymbol{z}; \boldsymbol{\mu}_{\boldsymbol{Z}}, \boldsymbol{\Sigma}_{\boldsymbol{Z}}\right)$ by definition and we can show (using linearity properties of expectations) that $f\left(\boldsymbol{x}\right) = \mathcal{N}\left(\boldsymbol{x}; \mathbf{A}\boldsymbol{\mu}_{\boldsymbol{Z}}, \mathbf{A}\boldsymbol{\Sigma}_{\boldsymbol{Z}}\mathbf{A}^T + \boldsymbol{\Sigma}\right)$,

(2) *Joint distribution*: "Stacking" the variables into one large vector (using the marginals above and linearity properties of covariance), we can write:

$$f\left(\boldsymbol{x}, \boldsymbol{z}\right) = \mathcal{N}\left(\begin{pmatrix} \boldsymbol{z} \\ \boldsymbol{x} \end{pmatrix}; \begin{pmatrix} \boldsymbol{\mu}_{\boldsymbol{Z}} \\ \mathbf{A}\boldsymbol{\mu}_{\boldsymbol{Z}} \end{pmatrix}, \begin{pmatrix} \boldsymbol{\Sigma}_{\boldsymbol{Z}} & \boldsymbol{\Sigma}_{\boldsymbol{Z}}\mathbf{A}^T \\ \mathbf{A}\boldsymbol{\Sigma}_{\boldsymbol{Z}} & \mathbf{A}\boldsymbol{\Sigma}_{\boldsymbol{Z}}\mathbf{A}^T + \boldsymbol{\Sigma} \end{pmatrix}\right) \tag{7.8}$$

(3) *Conditionals*: From Gaussian affine invariance, $f\left(\boldsymbol{x}|\boldsymbol{z}\right) = \mathcal{N}\left(\boldsymbol{x}; \mathbf{A}\boldsymbol{z}, \boldsymbol{\Sigma}\right)$, and the corresponding posterior distribution is:

$$f\left(\boldsymbol{z}|\boldsymbol{x}\right) = N\left(\boldsymbol{z}; \boldsymbol{\Sigma}_{\boldsymbol{Z}|\boldsymbol{X}}\left(\mathbf{A}^T\boldsymbol{\Sigma}^{-1}\boldsymbol{x} + \boldsymbol{\Sigma}_{\boldsymbol{Z}}^{-1}\boldsymbol{\mu}_{\boldsymbol{Z}}\right), \boldsymbol{\Sigma}_{\boldsymbol{Z}|\boldsymbol{X}}\right) \tag{7.9}$$

with $\boldsymbol{\Sigma}_{\boldsymbol{Z}|\boldsymbol{X}} = \left(\boldsymbol{\Sigma}_{\boldsymbol{Z}}^{-1} + \mathbf{A}^T\boldsymbol{\Sigma}^{-1}\mathbf{A}\right)^{-1}$.

These conditional properties can be derived from the following general conditioning rule for multivariate Gaussians. For a jointly normally distributed pair of vectors with distribution:

$$f\left(\boldsymbol{u}_1, \boldsymbol{u}_2\right) = \mathcal{N}\left(\begin{pmatrix} \boldsymbol{u}_1 \\ \boldsymbol{u}_2 \end{pmatrix}; \begin{pmatrix} \boldsymbol{\mu}_1 \\ \boldsymbol{\mu}_2 \end{pmatrix}, \begin{pmatrix} \boldsymbol{\Sigma}_{11} & \boldsymbol{\Sigma}_{12} \\ \boldsymbol{\Sigma}_{21} & \boldsymbol{\Sigma}_{22} \end{pmatrix}\right) \tag{7.10}$$

then $f\left(\boldsymbol{u}_1|\boldsymbol{u}_2\right) = \mathcal{N}\left(\boldsymbol{u}_1; \boldsymbol{\mu}_{1|2}\left(\boldsymbol{u}_2\right), \boldsymbol{\Sigma}_{1|2}\right)$ with:

$$\begin{aligned} \boldsymbol{\mu}_{1|2}\left(\boldsymbol{u}_2\right) &= \boldsymbol{\mu}_1 + \boldsymbol{\Sigma}_{12}\boldsymbol{\Sigma}_{22}^{-1}\left(\boldsymbol{u}_2 - \boldsymbol{\mu}_2\right) \\ \boldsymbol{\Sigma}_{1|2} &= \boldsymbol{\Sigma}_{11} - \boldsymbol{\Sigma}_{12}\boldsymbol{\Sigma}_{22}^{-1}\boldsymbol{\Sigma}_{21} \end{aligned} \tag{7.11}$$

where we have made the explicit functional dependence of the mean $\boldsymbol{\mu}_{1|2}$ on the second variable.

7.2   Linear, time-invariant (LTI) systems

To begin, we will start with the definition of a *system*. This is a mathematical operation H which takes a discrete-time (generally infinite length) signal $\boldsymbol{x} = (\dots, x_{-2}, x_{-1}, x_0, x_1, x_2, \dots)$ as input and outputs another signal $\boldsymbol{y} = (\dots, y_{-2}, y_{-1}, y_0, y_1, y_2, \dots)$; we will write $\boldsymbol{y} = H[\boldsymbol{x}]$. Linear systems are indeed just examples of linear operators as defined by (1.9). A *linear, time-invariant* (LTI) system has the additional property that time shifting the input *commutes* with the operation. By time shifting, we mean $T_d[\boldsymbol{x}] = (\dots, x_{-2-d}, x_{-1-d}, x_{-d}, x_{1-d}, x_{2-d}, \dots)$, i.e. an operator that shifts the indices of the time series backwards by d so that $x_n \mapsto x_{n-d}$. Thus, for an LTI operator, time-invariance means that:

Of course, all real-world signals are finite in length. This is the subject of the next section.

$$H[T_d[\boldsymbol{x}]] = T_d[H[\boldsymbol{x}]] \qquad (7.12)$$

for all $d \in \mathbb{Z}$.

Convolution and impulse response

Such LTI operators H have an important special property: they can be represented as a *convolution* with a particular signal $\boldsymbol{h}$:

$$H[\boldsymbol{x}] = \sum_{m \in \mathbb{Z}} h_m x_{n-m} = \boldsymbol{h} \star \boldsymbol{x} \qquad (7.13)$$

where the *convolution operator* of $\boldsymbol{x} \star \boldsymbol{y}$ is defined as:

$$C_{\boldsymbol{y}}[\boldsymbol{x}] = \sum_{m \in \mathbb{Z}} x_m y_{n-m} \qquad (7.14)$$

How does this arise, and what is $\boldsymbol{h}$? To answer, we will use the sift property of the (Kronecker) delta function:

$$
\begin{aligned}
x_n &= \sum_{m \in \mathbb{Z}} x_m \delta[n - m] \\
&= \sum_{m \in \mathbb{Z}} x_m T_m[\boldsymbol{\delta}] \qquad (7.15)
\end{aligned}
$$

Now, applying the system operator H:

$$
\begin{aligned}
H[\boldsymbol{x}] &= H\left[\sum_{m \in \mathbb{Z}} x_m T_m[\boldsymbol{\delta}]\right] \\
&= \sum_{m \in \mathbb{Z}} x_m H[T_m[\boldsymbol{\delta}]] \\
&= \sum_{m \in \mathbb{Z}} x_m T_m[H[\boldsymbol{\delta}]] \qquad (7.16) \\
&= \sum_{m \in \mathbb{Z}} x_m T_m[\boldsymbol{h}]
\end{aligned}
$$

so that:

$$H\left[\boldsymbol{x}\right] \;=\; \sum_{m\in\mathbb{Z}} x_m h_{n-m}$$

This same function arises in a vast number of other disciplines where it goes by the name of the Green's function (physics) or point spread function (image analysis).

The signal $\boldsymbol{h} = H\left[\boldsymbol{\delta}\right]$ is known as the *impulse response* of the operator. Convolution is commutative and associative:

$$
\begin{aligned}
\boldsymbol{x} \star \boldsymbol{y} &= \sum_{m\in\mathbb{Z}} x_m y_{n-m} = \sum_{m\in\mathbb{Z}} x_{n-m} y_m \\
&= \boldsymbol{y} \star \boldsymbol{x} \\
\boldsymbol{z} \star \left(\boldsymbol{x} \star \boldsymbol{y}\right) &= \left(\boldsymbol{z} \star \boldsymbol{x}\right) \star \boldsymbol{y}
\end{aligned}
\tag{7.17}
$$

The convolution identity is the impulse signal:

$$\boldsymbol{x} \star \boldsymbol{\delta} = \boldsymbol{x} \tag{7.18}$$

so that time-shifting is equivalent to convolution with the shifted impulse signal:

$$T_d\left[\boldsymbol{x}\right] = T_d\left[\boldsymbol{x} \star \boldsymbol{\delta}\right] = \boldsymbol{x} \star T_d\left[\boldsymbol{\delta}\right] \tag{7.19}$$

The discrete-time Fourier transform (DTFT)

The convolution operator is one of the most fundamental ideas in classical LTI DSP, and it has far-reaching consequences, particularly because wherever it is used, we can invoke the equally fundamental *Fourier transform* (FT). If we apply the convolution $C_{\boldsymbol{y}}\left[\boldsymbol{x}\right]$ to the complex exponential signal $x_n = \exp\left(i\omega n\right)$ with $\omega \in \mathbb{R}$, we get:

$$
\begin{aligned}
C_{\boldsymbol{y}}\left[\boldsymbol{x}\right] &= \sum_{m\in\mathbb{Z}} x_{n-m} y_m \\
&= \sum_{m\in\mathbb{Z}} \exp\left(i\omega\left(n-m\right)\right) y_m \\
&= \sum_{m\in\mathbb{Z}} \exp\left(-i\omega m\right) y_m \, \exp\left(i\omega n\right) \\
&= Y\left(\omega\right) \boldsymbol{x}
\end{aligned}
\tag{7.20}
$$

where:

$$Y\left(\omega\right) = F_{\omega}\left[\boldsymbol{y}\right] = \sum_{n\in\mathbb{Z}} y_n \exp\left(-i\omega n\right) \tag{7.21}$$

is called the *discrete time Fourier transform* (DTFT) of $\boldsymbol{y}$, itself a linear operator mapping a signal into the complex plane for each ω. The infinite summation (7.21) converges uniformly if the signal $\boldsymbol{y}$ is absolutely summable, $\sum_{n\in\mathbb{Z}} |y_n| < \infty$ but is convergent in the mean-square sense if the signal is square summable, $\sum_{n\in\mathbb{Z}} |y_n|^2 < \infty$. So, in practice we have to find ways to prevent infinite discrete-time signals from growing in an unbounded way at the extremes.

The relationship (7.20) is an eigenvalue equation with the convolution operator playing the role of matrix, the $Y(\omega)$ being eigenvalues and the complex exponentials the eigenfunctions. Thus in fact, the discrete FT *diagonalizes* the discrete convolution operator, and the complex exponentials are the basis for this diagonalizing transformation. Since Fourier transforming a signal involves a linear transformation onto a new basis of complex exponentials, we can say that the "Fourier transform diagonalizes the convolution operator". Indeed because of this, the convolution of two signals is the inverse Fourier transform of the product of their Fourier transforms, as we will see later.

This convolution property is key to the power of the FT. The output of any linear system can be represented as a convolution with the impulse response of the system, and that convolution is a simple multiplication of the FT of the input signal with the FT of the impulse response, for each value of ω. This variable ω in the DTFT is known as the *frequency variable* (in units of radians per second), and it is "dual" to the time index n. In physics and engineering circles, the signal written in terms of the (complex) frequency coefficients indexed by ω is known as the *frequency domain* representation, and in terms of n it is the *time domain* representation.

For discrete-time signals, the DTFT is *periodic* (that is, repeating) with period 2π. To see this, let $j \in \mathbb{Z}$:

$$
\begin{aligned}
Y(\omega + 2\pi j) &= \sum_{n\in\mathbb{Z}} y_n \exp\left(-i\left(\omega + 2\pi j\right)n\right) \\
&= \sum_{n\in\mathbb{Z}} y_n \exp\left(-i\omega n\right)\exp\left(-i2\pi j n\right) \qquad (7.22) \\
&= \sum_{n\in\mathbb{Z}} y_n \exp\left(-i\omega n\right) \\
&= Y(\omega)
\end{aligned}
$$

since both $n, j \in \mathbb{Z}$ so that $\exp\left(-i2\pi j n\right) = 1$. Therefore, due to the discreteness of the index n, the unique frequency range for discrete-time signals is $[-\pi, \pi]$, outside this range frequencies are *folded* onto a frequency within the range.

Just as we can transform a signal $\boldsymbol{x}$ into the frequency domain $X(\omega)$ using the FT, so we can undo the transform and return to the time domain. To see how, note that for each value of ω, $X(\omega)$ is just the inner product of the signal with the complex exponential of frequency ω. Thus, the signal can be reconstructed from a superposition of a continuum of complex exponentials. To derive the *inverse discrete-time Fourier transform* (IDTFT) which performs this reconstruction, we first note that on the relevant frequency interval $[-\pi, \pi]$, the complex exponentials form an orthogonal system:

$$
\int_{-\pi}^{\pi} \exp\left(i\omega n\right) d\omega = 2\pi\delta\left[n\right] \qquad (7.23)
$$

expressed using the Kronecker delta. Now, we want to form a superposition of complex exponentials with amplitudes $X(\omega)$:

$$
\begin{aligned}
\int_{-\pi}^{\pi} X(\omega)\exp(i\omega n)\,d\omega &= \int_{-\pi}^{\pi}\left[\sum_{m\in\mathbb{Z}} x_m\exp(-i\omega m)\right]\exp(i\omega n)\,d\omega \\
&= \sum_{m\in\mathbb{Z}} x_m\int_{-\pi}^{\pi}\exp(i\omega(n-m))\,d\omega \qquad (7.24)\\
&= 2\pi\sum_{m\in\mathbb{Z}} x_m\delta[n-m] \\
&= 2\pi x_n
\end{aligned}
$$

where we have assumed the FT is convergent so that we can swap the order of summation and integration. The above says that the IDTFT is:

$$
\boldsymbol{x} = F_n^{-1}[X(\omega)] = \frac{1}{2\pi}\int_{-\pi}^{\pi} X(\omega)\exp(i\omega n)\,d\omega \qquad (7.25)
$$

and $F_n^{-1}[F_\omega[\boldsymbol{x}]] = \boldsymbol{x}$ so that the IDTFT and DTFT are inverses of one another.

LTI operators induce a whole host of time-frequency symmetries which manifest themselves in the FT representation. Firstly, since the transform is linear, then the FT of the (weighted) sum of any number of signals is the (weighted) FTs of the individual signals. As discussed above, one of the most important properties is that convolution in time corresponds to multiplication in frequency:

$$
\begin{aligned}
F_\omega[\boldsymbol{x}\star\boldsymbol{y}] &= \sum_{n\in\mathbb{Z}}\left[\sum_{m\in\mathbb{Z}} x_m y_{n-m}\right]\exp(-i\omega n) \\
&= \sum_{m\in\mathbb{Z}} x_m\left[\sum_{n\in\mathbb{Z}} y_{n-m}\exp(-i\omega n)\right] \qquad (7.26)\\
&= \sum_{m\in\mathbb{Z}} x_m\exp(-i\omega m)\sum_{n\in\mathbb{Z}} y_{n-m}\exp(-i\omega(n-m)) \\
&= X(\omega)Y(\omega)
\end{aligned}
$$

The corresponding time-frequency symmetry is that multiplication in time corresponds to convolution in frequency:

$$
\begin{aligned}
F_\omega \left[\boldsymbol{x} \circ \boldsymbol{y} \right] &= \sum_{n \in \mathbb{Z}} x_n y_n \exp\left(-i\omega n\right) \\
&= \sum_{n \in \mathbb{Z}} \left[\frac{1}{2\pi} \int_{-\pi}^{\pi} X\left(\omega'\right) \exp\left(i\omega' n\right) d\omega' \right] y_n \exp\left(-i\omega n\right) \\
&= \frac{1}{2\pi} \int_{-\pi}^{\pi} X\left(\omega'\right) \left[\sum_{n \in \mathbb{Z}} y_n \exp\left(-i\left(\omega - \omega'\right) n\right) \right] d\omega' \\
&= \frac{1}{2\pi} \int_{-\pi}^{\pi} X\left(\omega\right) Y\left(\omega - \omega'\right) d\omega' \qquad (7.27) \\
&= X\left(\omega\right) \star Y\left(\omega\right)
\end{aligned}
$$

This convolution property is commonly used to analyze LTI systems. Given the impulse response $\boldsymbol{h}$ of the system, the effect of the system on an input signal $\boldsymbol{x}$ is $Y\left(\omega\right) = H\left(\omega\right) X(\omega)$. In other words, the FT of the system output $\boldsymbol{y}$ is the product of the FT of the input with that of the impulse response $H\left(\omega\right)$. The latter is known in engineering circles as the *transfer function*, the effect of which is to modify only the magnitude and phase angle of the FT of the input signal at each frequency component ω. Thus, in principle, an LTI system can independently enhance, attenuate or delay any frequency component in time, but it cannot create frequency components at the output that have zero magnitude in the input. Such manipulations of the FT of the input are known as *filtering* operations, and the design of LTI systems with specific transfer functions is one of the central topics in classical DSP.

Shifting in time corresponds to multiplication by a complex exponential:

$$
\begin{aligned}
F_\omega \left[T_m \left[\boldsymbol{x} \right] \right] &= \sum_{n \in \mathbb{Z}} x_{n-m} \exp\left(-i\omega n\right) \\
&= \sum_{l \in \mathbb{Z}} x_l \exp\left(-i\omega \left(l + m\right)\right) \qquad (7.28) \\
&= \exp\left(-i\omega m\right) \sum_{l \in \mathbb{Z}} x_l \exp\left(-i\omega l\right) \\
&= \exp\left(-i\omega m\right) X\left(\omega\right)
\end{aligned}
$$

and by defining $w_n = \exp\left(i\omega' n\right)$ then it is not hard to show the corresponding time-frequency symmetry: multiplication by a complex exponential corresponds to shifting in frequency:

$$
\begin{aligned}
F_\omega \left[\boldsymbol{w} \circ \boldsymbol{x} \right] &= \sum_{n \in \mathbb{Z}} x_n \exp\left(-i\left(\omega - \omega'\right) n\right) \\
&= X\left(\omega - \omega'\right) \qquad (7.29)
\end{aligned}
$$

Reversing a signal (using the *reversal operator* $R\left[\boldsymbol{x}\right] = \boldsymbol{y}$ so that $y_n =$

x_{-n}) corresponds to flipping the sign of the frequency variable:

$$
\begin{aligned}
F_\omega \left[R \left[\boldsymbol{x} \right] \right] &= \sum_{n \in \mathbb{Z}} x_{-n} \exp \left(-i\omega n \right) \\
&= \sum_{m \in \mathbb{Z}} x_m \exp \left(-i \left(-\omega \right) m \right) \qquad (7.30) \\
&= X \left(-\omega \right)
\end{aligned}
$$

and by undoing the above, flipping the sign of the frequency variable corresponds to time reversal:

$$
\begin{aligned}
F_n^{-1} \left[X \left(-\omega \right) \right] &= \frac{1}{2\pi} \int_{-\pi}^{\pi} X \left(-\omega \right) \exp \left(i\omega n \right) d\omega \\
&= \frac{1}{2\pi} \int_{-\pi}^{\pi} X \left(\omega' \right) \exp \left(i\omega' \left(-n \right) \right) d\omega' \qquad (7.31) \\
&= R \left[\boldsymbol{x} \right]
\end{aligned}
$$

Integrating the sample-by-sample (complex conjugate) product of two signals is equivalent to a similar integral in the frequency domain, known as *Plancherel's theorem*:

$$
\begin{aligned}
\langle \boldsymbol{x}, \boldsymbol{y} \rangle &= \sum_{n \in \mathbb{Z}} x_n \left[\frac{1}{2\pi} \int_{-\pi}^{\pi} \bar{Y} \left(\omega \right) \exp \left(-i\omega n \right) d\omega \right] \\
&= \frac{1}{2\pi} \int_{-\pi}^{\pi} \bar{Y} \left(\omega \right) \left[\sum_{n \in \mathbb{Z}} x_n \exp \left(-i\omega n \right) \right] d\omega \qquad (7.32) \\
&= \frac{1}{2\pi} \int_{-\pi}^{\pi} X \left(\omega \right) \bar{Y} \left(\omega \right) d\omega
\end{aligned}
$$

From this, we can get *Parseval's relation*:

$$
\begin{aligned}
\langle \boldsymbol{x}, \boldsymbol{x} \rangle &= \| \boldsymbol{x} \|_2^2 = \sum_{n \in \mathbb{Z}} |x_n|^2 \\
&= \frac{1}{2\pi} \int_{-\pi}^{\pi} |X \left(\omega \right)|^2 \, d\omega
\end{aligned}
$$

This shows the interesting observation that the energy (sum of squares) of a signal can be obtained by integration in the frequency domain.

A very similar notion to convolution is *cross-correlation*:

$$
r_{\boldsymbol{xy}} \left(m \right) = \langle \boldsymbol{x}, T_m \left[\bar{\boldsymbol{y}} \right] \rangle = \sum_{n \in \mathbb{Z}} x_n \bar{y}_{n-m} \qquad (7.33)
$$

for $d \in \mathbb{Z}$. Now, by the time-shift property, $F_\omega \left[T_d \left[\boldsymbol{y} \right] \right] = \exp \left(-i\omega m \right) Y \left(\omega \right)$, and by Plancherel's theorem:

$$
\begin{aligned}
r_{\boldsymbol{xy}}\left(m\right) &= \frac{1}{2\pi}\int_{-\pi}^{\pi} X\left(\omega\right)\bar{Y}\left(\omega\right)\exp\left(i\omega m\right)d\omega \\
&= F_n^{-1}\left[X\left(\omega\right)\bar{Y}\left(\omega\right)\right] \tag{7.34}
\end{aligned}
$$

This also implies that $F_\omega\left[r_{\boldsymbol{xy}}\left(m\right)\right] = X\left(\omega\right)\bar{Y}\left(\omega\right)$, which is known as the *cross-correlation theorem*. Specializing to the case of *autocorrelation* $r_{\boldsymbol{xx}}\left(m\right)$, we obtain the well-known *Wiener-Khintchine* theorem:

$$
\begin{aligned}
r_{\boldsymbol{xx}}\left(m\right) &= \frac{1}{2\pi}\int_{-\pi}^{\pi} X\left(\omega\right)\bar{X}\left(\omega\right)\exp\left(i\omega m\right)d\omega \\
&= \frac{1}{2\pi}\int_{-\pi}^{\pi}\left|X\left(\omega\right)\right|^2\exp\left(i\omega m\right)d\omega \tag{7.35} \\
&= F_m^{-1}\left[\left|X\left(\omega\right)\right|^2\right]
\end{aligned}
$$

or similarly, $F_\omega\left[r_{\boldsymbol{xx}}\left(m\right)\right] = \left|X\left(\omega\right)\right|^2$. In other words, the FT of the auto-correlation of a signal is the magnitude of its FT, which is an extremely useful property in practice.

Differentiation in frequency has the following FT:

$$
\begin{aligned}
\frac{dX}{d\omega}\left(\omega\right) &= \frac{d}{d\omega}\sum_{n\in\mathbb{Z}} x_n\exp\left(-i\omega n\right) \\
&= \sum_{n\in\mathbb{Z}} x_n\exp\left(-i\omega n\right)\frac{d}{d\omega}\left(-i\omega n\right) \\
&= -i\sum_{n\in\mathbb{Z}} n\,x_n\exp\left(-i\omega n\right) \tag{7.36} \\
&= -i\,F_\omega\left[\left(n\,x_n\right)_{n\in\mathbb{Z}}\right]
\end{aligned}
$$

alternatively:

$$
F_\omega\left[\left(n\,x_n\right)_{n\in\mathbb{Z}}\right] = i\frac{dX}{d\omega}\left(\omega\right) \tag{7.37}
$$

and while there is no direct discrete counterpart to differentiation in time, differencing in time is straightforward using the linearity and time shift properties of the FT:

$$
\begin{aligned}
F\left[\boldsymbol{x} - T_1\left[\boldsymbol{x}\right]\right] &= X\left(\omega\right) - \exp\left(-i\omega\right)X\left(\omega\right) \\
&= \left(1 - \exp\left(-i\omega\right)\right)X\left(\omega\right) \tag{7.38}
\end{aligned}
$$

We will see this idea exploited to convert difficult *difference equations* into easily-solved algebraic ones later.

Using these properties, we can derive expressions for various important classes of signals. The first is the infinite impulse signal $\delta\left[n\right]$:

$$
F_\omega\left[\boldsymbol{\delta}\right] = \sum_{n\in\mathbb{Z}}\delta\left[n\right]\exp\left(-i\omega n\right) = 1
$$

so it follows that $F_n^{-1}[1] = \boldsymbol{\delta}$. Similarly:

$$
\begin{aligned}
F_n^{-1}\left[\delta\left[\omega\right]\right] &= \int_{-\pi}^{\pi} \delta\left[\omega\right] \exp\left(i\omega n\right) d\omega \\
&= \mathbf{1}
\end{aligned}
\tag{7.39}
$$

Here, δ is the Dirac delta and $\mathbf{1}$ is the constant signal, $x_n = 1$. This means that for the constant spectrum, $F_\omega[\mathbf{1}] = \delta[\omega]$, i.e. the constant signal has Dirac delta FT. Another important signal is the unit step function $x_n = 0$ for $n < 0$ and $x_n = 1$ for $n \geq 0$, which we will write as u_n. With this function we can find the FT of functions which would otherwise not be absolutely summable such as the *geometric sequence* $x_n = a^n u_n$ for $a \in \mathbb{C}$ and $|a| < 1$:

$$
\begin{aligned}
F_\omega\left[\boldsymbol{x}\right] &= \sum_{n \in \mathbb{Z}} a^n u_n \exp\left(-i\omega n\right) \\
&= \sum_{n=0}^{\infty} \left(a\, e^{-i\omega}\right)^n \\
&= \frac{1}{1 - a\, \exp\left(-i\omega\right)}
\end{aligned}
\tag{7.40}
$$

Finite-length, periodic signals: the discrete Fourier transform (DFT)

The above exposition lays out most of the important properties of discrete-time convolution and FT for infinite-length signals. But, in practical DSP applications, the signal is not infinite but finite with length N. But, if one assumes that the signal is *periodic* with period N, then we can write $x_n = x_{n+N}$ for all $n \in \mathbb{Z}$ and $N > 1$, and the finite length subset over the index range, say, $n = 0, 1, \ldots N - 1$ contains all the information in this infinite periodic signal. This is the *periodic extension* assumption and it has many highly convenient mathematical properties, discussed in this section.

We wish to compute the FT for this finite-length signal x_n for $n \in 0, 1, \ldots, N - 1$, but the periodicity restricts the range of frequencies in the following way. The complex exponential must satisfy $\exp\left(i\omega n\right) = \exp\left(i\omega\left(n + N\right)\right) = \exp\left(i\omega n\right) \exp\left(i\omega N\right)$ from which we require that $\exp\left(i\omega N\right) = 1$. Now, the only way this can be true is if $\omega N = 2\pi k$ where $k \in \mathbb{Z}$, i.e. an integer multiple of 2π. In which case:

$$
\omega = \frac{2\pi k}{N}
\tag{7.41}
$$

In fact, this set of exponential vectors, $w_n^{N,k} = \exp\left(2\pi i k n / N\right)$, form an orthogonal basis for the vector space $\mathbb{R}^N$ (and also more generally $\mathbb{C}^N$) of discrete-time signal of length N. To show this, take any $j, k \in$

$0, 1, \ldots, N-1$:

$$
\begin{aligned}
\left\langle \boldsymbol{w}^{N,j}, \boldsymbol{w}^{N,k} \right\rangle &= \sum_{n=0}^{N-1} w_n^{N,j} \bar{w}_n^{N,k} \\
&= \sum_{n=0}^{N-1} \exp\left(2\pi i \frac{j-k}{N} \right)^n \qquad (7.42) \\
&= \begin{cases} N & a = 1 \\ \frac{1-a^N}{1-a} & a \neq 1 \end{cases}
\end{aligned}
$$

where $a = \exp\left(2\pi i\,(j-k)/N \right)$. Now, $a = 1$ if $j = k$. Otherwise, $a^N = \exp\left(2\pi i\,(j-k) \right) = 1$ since $j - k$ is integer, which means that $1 - a^N = 0$. But in this case also $a \neq 1$ because $(j-k)/N$ cannot be an integer since $N \geq 2$ and $1 \leq |j-k| \leq N - 1$, so that $1 - a \neq 0$. Therefore we have:

$$
\left\langle \boldsymbol{w}^{N,k}, \boldsymbol{w}^{N,j} \right\rangle = N\delta\,[j-k] \qquad (7.43)
$$

which demonstrates orthogonality. Furthermore, all vectors are non-zero, $\boldsymbol{w}^{N,k} \neq \boldsymbol{0}$. This means that together these N vectors form an orthogonal basis.

As with the DTFT, there is a finite-length, periodic version of the convolution theorem (7.26). Since all signals are periodic with indices in the range $0, 1, \ldots, N - 1$, convolution becomes *circular*:

$$
\boldsymbol{x} \star_N \boldsymbol{y} = \sum_{m=0}^{N-1} x_m y_{(n-m) \mod N} \qquad (7.44)
$$

where $a \mod b$ for $a, b \in \mathbb{Z}$ is the remainder after dividing a by b. The range of the sum matches that of the indices of the length N signal $\boldsymbol{x}$, and the modular difference $(n-m) \mod N$ ensures that the indices for $\boldsymbol{y}$ are in range and satisfy the periodicity condition. We can write the circular convolution operator as a finite matrix:

$$
\mathbf{C}_{\boldsymbol{y}}^N = \begin{pmatrix}
y_0 & y_{N-1} & y_{N-2} & \cdots & y_1 \\
y_1 & y_0 & y_{N-1} & \cdots & y_2 \\
y_2 & y_1 & y_0 & \cdots & y_3 \\
\vdots & \vdots & \vdots & \ddots & \vdots \\
y_{N-1} & y_{N-2} & y_{N-3} & \cdots & y_0
\end{pmatrix} \qquad (7.45)
$$

i.e. it has entries $y_{(j-k) \mod N}$ for row j and column k. As with discrete convolution, it is straightforward to show that the complex discrete exponentials are eigenvectors of the discrete circular convolution operator:

$$
\begin{aligned}
\mathbf{C}_y^N \boldsymbol{w}^{N,k} &= \sum_{m=0}^{N-1} w_{(n-m) \mod N}^{N,k} y_m \\
&= \sum_{m=0}^{N-1} \exp\left(2\pi i k\,(n-m) \mod N/N\right) y_m \qquad (7.46) \\
&= \sum_{m=0}^{N-1} \exp\left(-2\pi i k m/N\right) y_m \exp\left(2\pi i k n/N\right) \\
&= Y_k \boldsymbol{w}^{N,k}
\end{aligned}
$$

where $Y_k = \sum_{m=0}^{N-1} \exp\left(-2\pi i k m/N\right) y_m = \langle \boldsymbol{w}^{N,k}, \boldsymbol{y}\rangle$ are the eigenvalues. The set of N eigenvalues, one for each $k \in 0, 1, \ldots, N-1$, are the *discrete Fourier coefficients* for $\boldsymbol{y}$. Together, they form "the" *discrete Fourier transform* (DFT), a linear operator mapping $F : \mathbb{C}^N \to \mathbb{C}^N$:

$$
F_k\left[\boldsymbol{x}\right] = \sum_{n=0}^{N-1} x_n \exp\left(-2\pi i k \frac{n}{N}\right) = \boldsymbol{X} \qquad (7.47)
$$

where $\boldsymbol{X} = (X_0, X_1, \ldots, X_{N-1})$, which we can also write simply as $F_k\left[\boldsymbol{x}\right] = \langle \boldsymbol{x}, \boldsymbol{w}^{N,k}\rangle$. So, discrete-time Fourier analysis (7.21) for finite signals with periodic extension is just a coordinate transformation into the discrete complex exponential basis.

We can also write the DFT operator as an explicit, $N \times N$ matrix $\mathbf{F}$ with entries:

The specific polynomial structure of this matrix makes it a Vandermonde matrix.

$$
f_{kn} = w_n^{N,k} \qquad (7.48)
$$

This matrix is obviously orthogonal because every row is one member of the complex orthogonal basis set $\boldsymbol{w}^N$. Its inverse $\mathbf{F}^{-1}$ is straightforward: we require $\mathbf{F} \times \mathbf{F}^{-1} = \mathbf{I}$ and for this to be true, $\mathbf{F}^{-1}$ is just the matrix formed by putting the complex conjugates $\bar{\boldsymbol{w}}^{N,k}$ into each row and scaling by N e.g. $f_{nk}^{-1} = \frac{1}{N} \bar{w}_n^{N,k}$ since, multiplying each row j of $\mathbf{F}$ by each column k of $\mathbf{F}^{-1}$ gives us the inner product $\frac{1}{N} \langle \boldsymbol{w}^{N,j}, \boldsymbol{w}^{N,k}\rangle = \delta\left[j - k\right]$ by (7.43), and indeed these are just the 1's on the diagonal of the identity matrix. Thus in fact $\mathbf{F}^{-1} = \frac{1}{N} \bar{\mathbf{F}}^T$, the conjugate transpose of $\mathbf{F}$. So, we can define the *inverse discrete Fourier transform* (IDFT):

$$
F_n^{-1}\left[\boldsymbol{X}\right] = \frac{1}{N} \sum_{k=0}^{N-1} X_k \exp\left(2\pi i n \frac{k}{N}\right) = \boldsymbol{x} \qquad (7.49)
$$

such that $F_n^{-1}\left[F_k\left[\boldsymbol{x}\right]\right] = \boldsymbol{x}$. As with the DTFT, the DFT is periodic in

Property	Time domain (infinite)	Frequency domain (infinite DTFT)	Time domain (finite)	Frequency domain (finite, DFT)
	$x_n, y_n,$ $n \in \mathbb{Z}$	$X(\omega), Y(\omega),$ $\omega \in [-\pi, \pi]$	$x_n, y_n,$ $n \in 0, 1, \ldots N-1$	$X_k, Y_k,$ $k \in 0, 1, \ldots N-1$
Periodicity		$X(\omega) = X(\omega + 2\pi)$	$x_n = x_{n+N}$	$X_k = X_{k+N}$
Time shift	$x_n \mapsto x_{n-m},$	$\exp(-i\omega m) X(\omega)$	$x_n \mapsto$ $x_{(n-m) \bmod N}$	$\exp(-2\pi i km/N) X_k$
Frequency shift	$x_n \mapsto \exp(i\omega' n) x_n$	$X(\omega) \mapsto$ $X(\omega - \omega')$	$x_n \mapsto$ $\exp(2\pi i mn/N) x_n$	$X_k \mapsto$ $X_{(k-m) \bmod N}$
Convolution	$\boldsymbol{x} \star \boldsymbol{y}$	$X(\omega) Y(\omega)$	$\boldsymbol{x} \star_N \boldsymbol{y}$	$\boldsymbol{X} \circ \boldsymbol{Y}$
Multiplication	$\boldsymbol{x} \circ \boldsymbol{y}$	$X(\omega) \star Y(\omega)$	$\boldsymbol{x} \circ \boldsymbol{y}$	$\boldsymbol{X} \star_N \boldsymbol{Y}$
Wiener-Khintchine	$r_{\boldsymbol{xx}}(m)$	$|X(\omega)|^2$	$r_{\boldsymbol{xx}}^N(m)$	$|X_k|^2$
Plancherel theorem	$\sum_{n \in \mathbb{Z}} x_n \bar{y}_n$	$\frac{1}{2\pi} \int_{-\pi}^{\pi} X(\omega) \bar{Y}(\omega) \, d\omega \; \langle \boldsymbol{x}, \boldsymbol{y} \rangle$		$\frac{1}{N} \langle \boldsymbol{X}, \boldsymbol{Y} \rangle$

Table 7.1: Relationships between time and frequency domain properties of discrete-time infinite (DTFT) and finite (DFT) Fourier transforms.

frequency, but with period N:

$$\begin{aligned} X_{k+N} &= \sum_{n=0}^{N-1} x_n \exp\left(-2\pi i (k+N) \frac{n}{N}\right) \\ &= \sum_{n=0}^{N-1} x_n \exp\left(-2\pi i k \frac{n}{N}\right) \exp(-2\pi i n) \quad (7.50) \\ &= X_k \end{aligned}$$

Most of the properties of the DTFT described earlier apply to the DFT, after substituting finite summations for infinite summations, and circular convolutions for convolutions (Table 7.1). For example, the convolution theorem for the DFT is just $F_k[\boldsymbol{x} \star_N \boldsymbol{y}] = \boldsymbol{X} \circ \boldsymbol{Y}$ with the symmetric frequency counterpart $\boldsymbol{X} \star_N \boldsymbol{Y} = F_k[\boldsymbol{x} \circ \boldsymbol{y}]$, and Parseval's relation is $\sum_{n=0}^{N-1} x_n \bar{y}_n = \frac{1}{N} \sum_{k=0}^{N-1} X_k \bar{Y}_k$. For the DTFT of special signals we have introduced earlier and others, see Table 7.2.

Continuous-time LTI systems

Fourier transforms and convolutions are equally as applicable to continuous-time as discrete-time systems. The only major technical difficulty with the continuous-time Fourier transform is the need to work with e.g. tempered distributions such the Dirac delta "function". The key value of using tempered distributions is that they are a closed set under action of the continuous-time Fourier transform which we introduce in this section, and yet continuous-time signals as we usually encounter them in practice are also included within this set. The presentation below will implicitly assume that all signals are such objects.

In continuous time, entirely analogous arguments can be put forward as in discrete time, thus we can use the sift property of the Dirac delta function to represent a signal $f(t)$ using an integral:

$$f(t) = \int_{\mathbb{R}} f(t)\, \delta[t - \tau]\, d\tau \tag{7.51}$$

and apply the continuous-time system operator H to get:

$$
\begin{aligned}
H\left[f(t)\right] &= H\left[\int_{\mathbb{R}} f(\tau)\, T_{\tau}\left[\delta[t]\right] d\tau\right] \\
&= \int_{\mathbb{R}} f(\tau)\, T_{\tau}\left[h(t)\right] d\tau \tag{7.52} \\
&= \int_{\mathbb{R}} f(\tau)\, h(t - \tau)\, d\tau
\end{aligned}
$$

where $H\left[\delta[t]\right] = h(t)$ is the continuous-time impulse response function, which gives us the general, continuous-time convolution operator:

$$C_g\left[f(t)\right] = \int_{\mathbb{R}} f(\tau)\, g(t - \tau)\, d\tau = f(t) \star g(t) \tag{7.53}$$

From this we can show diagonalization in the Fourier basis by convolving against the signal $f(t) = \exp(i\omega t)$:

$$
\begin{aligned}
C_g\left[f(t)\right] &= \int_{\mathbb{R}} \exp\left(i\omega(t - \tau)\right) g(\tau)\, d\tau \\
&= \int_{\mathbb{R}} \exp\left(-i\omega\tau\right) g(\tau)\, d\tau \exp(i\omega t) \tag{7.54} \\
&= Y(\omega)\, f(t)
\end{aligned}
$$

so that the corresponding Fourier transform is:

$$F_{\omega}\left[g(t)\right] = G(\omega) = \int_{\mathbb{R}} g(t) \exp(-i\omega t)\, dt \tag{7.55}$$

with inverse transform:

$$g(t) = F_t^{-1}\left[G(\omega)\right] = \frac{1}{2\pi}\int_{\mathbb{R}} G(\omega) \exp(i\omega t)\, d\omega \tag{7.56}$$

The continuous-time FT integal (7.55) does not converge for all signals. Sufficient, but not necessary conditions for convergence are ab-

Function	Time domain (infinite discrete)	Frequency domain (infinite DTFT)	Time domain (infinite continuous)	Frequency domain (continuous)
	$n \in \mathbb{Z}$	$\omega \in [-\pi, \pi]$	$t \in \mathbb{R}$	$\omega \in \mathbb{R}$
Impulse signal	δ	1	$\delta[t]$	1
Constant	1	$\delta[\omega]$	1	$\delta[\omega]$
Unit step	u	$\frac{1}{1-\exp(-i\omega)} + \pi\delta[\omega]$	$\Phi(t)$	$\frac{1}{i\omega} + \pi\delta[\omega]$
Exponential	$u_n a^n, |a| < 1$	$\frac{1}{1-a\exp(-i\omega)}$	$\Phi(t)\exp(-at)$	$\frac{1}{a+i\omega}$
Rectangle pulse	$\text{rect}\left(\frac{\pi}{W}n\right)$	$\frac{\sin\left(\omega\left(W+\frac{1}{2}\right)\right)}{\sin\left(\frac{1}{2}\omega\right)}$	$\text{rect}\left(\frac{\pi}{W}t\right)$	$2W\,\text{sinc}\left(\frac{W}{\pi}\omega\right)$
Sinc	$2W\,\text{sinc}\left(\frac{W}{\pi}n\right)$	$\text{rect}\left(\frac{\pi}{W}\omega\right)$	$\frac{W}{\pi}\text{sinc}\left(\frac{Wt}{\pi}\right)$	$\text{rect}\left(\frac{\pi}{W}\omega\right)$
Gaussian			$C(\sigma)\exp\left(-\frac{1}{2}\left(\frac{t}{\sigma}\right)^2\right)$	$\exp\left(-\frac{1}{2}(\sigma\omega)^2\right)$

Table 7.2: Fourier transform pairs for the discrete-time infinite (DTFT) and continuous time Fourier transforms. Here the Gaussian normalization factor is $C(\sigma) = \frac{1}{\sqrt{2\pi\sigma^2}}$.

solute integrability $\int_{\mathbb{R}} |f(t)|\, dt < \infty$, a finite number of minima and maxima on any bounded interval of $\mathbb{R}$, and a finite number of finite discontinuities in any bounded interval (Proakis and Manolakis, 1996).

This FT shares most of the properties of the DTFT, with suitable modifications. Most important are the convolution and multiplication properties e.g. $F_\omega[x(t) \star y(t)] = X(\omega)Y(\omega)$ and $F_\omega[x(t)y(t)] = X(\omega)\star Y(\omega)$. There is also a version of Plancherel's theorem:

$$\int_{\mathbb{R}} f(t)\bar{g}(t)\, dt = \frac{1}{2\pi} \int_{\mathbb{R}} F(\omega)\bar{G}(\omega)\, d\omega \qquad (7.57)$$

Useful pairs of continuous-time FT functions are shown in Table 7.2.

Heisenberg uncertainty

A fundamental fact about Fourier analysis is that the longer the extent of some event in a signal in time, the more concentrated it is in the frequency domain. This means that if we want accurate information about the frequency content of an "event", then we must expect to compromise our ability to localize that event in time. The converse is

It turns out that this effect has a physical counterpart that is central to the physics of quantum mechanics, of which Werner Heisenberg was one of the leading proponents in the early 20th century.

true: if we wish to get accurate information about the point in time where an event occurred, we must expect to compromise on our ability to describe the frequency content in detail. This time-frequency trade-off is known as *Heisenberg uncertainty*, which we explore in this section in the context of the continuous-time FT.

To quantify this effect, we need a generic way of measuring the spread of a signal in time or frequency. One way to measure the extent of a signal is using the concept of the second raw moment $E\left[X^2\right]$ from probability theory, which, if the signal has zero mean, is the same as the variance. However, to use this measure, we will need to ensure that our signals are probability densities, that is, they are non-negative everywhere and normalized. Non-negativity can be achieved by squaring, e.g. $|f(t)|^2$, and for such a distribution, normalization is equivalent to having unit L_2-norm, $\int_{\mathbb{R}} |f(t)|^2\, dt = 1$. If this function is normalized, it follows that $\int_{\mathbb{R}} |F(\omega)|^2\, d\omega = 2\pi$ by Plancherel's theorem. Assuming then that the squared function has zero mean, the variance in time and frequency are:

$$\sigma_t^2 = \int_{\mathbb{R}} t^2 |f(t)|^2\, dt \tag{7.58}$$

$$\sigma_\omega^2 = \frac{1}{2\pi} \int_{\mathbb{R}} \omega^2 |F(\omega)|^2\, d\omega \tag{7.59}$$

Using these moments, Mallat (2009, Theorem 2.6, page 44) shows that they must satisfy the *Heisenberg uncertainty relation*:

$$\sigma_t^2 \sigma_\omega^2 \geq \frac{1}{4} \tag{7.60}$$

This shows that the time-frequency trade-off measured in terms of spread are reciprocally related. What about the lower bound $\frac{1}{4}$? If this is true, then the signal $f(t)$ must satisfy the following differential equation for some $\alpha \in \mathbb{C}$:

$$f'(t) = -2\alpha t\, f(t) \tag{7.61}$$

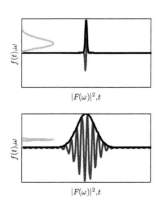

Fig. 7.2: Heisenberg uncertainty for the Fourier transform. For signals $f(t)$ (blue curve, black line is the Gaussian function bounding the signal) which have limited extent in the time domain (top panel), their corresponding Fourier transform has large extent in the frequency domain (grey curve, not to scale). Conversely, signals which have large extent in the time domain (bottom panel) have small extent in frequency.

which has the general solution $f(t) = C \exp\left(-\alpha t^2\right)$ for some $C \in \mathbb{C}$. This is indeed the Gaussian distribution (Table 7.2), which is the unique function which achieves the lower Heisenberg bound (Mallat, 2009, page 45). Up to scale factors in both time and frequency axis, this is invariant under the Fourier transform. These scaling factors, both depending upon the same positive real variable σ act reciprocally to each other in time and frequency: in time the scale factor is $\alpha = \sigma^2$, whereas in frequency it is $\alpha = \sigma^{-2}$ (Figure 7.2).

Because of time invariance (which implies frequency invariance) we can always translate the signal to centre it at some time t_0 and frequency ω_0 and the principle (7.60) still holds. So, the FT always trades off time against frequency resolution in a way which is independent of time-frequency location. We will see later that by dropping time-invariance we can create linear transforms other than Fourier that can extract quite different time-frequency trade-offs (Section 7.6).

While Heisenberg uncertainty applies to the continuous-time FT, there is no direct discrete-time counterpart for the DTFT and DFT. Nonetheless, it is intuitively obvious that something similar to this principle should apply in the discrete-time case. For example, if we consider the discrete-time sampling interval to very small, then the discrete signal will be a close approximation to some continuous-time analogue under appropriate smoothness constraints, and so the uncertainty principle should hold at least approximately under most situations.

To address Heisenberg uncertainty in discrete time, it is worth noting that the variance of the signal and its FT is not the only way in which we can quantify the uncertainty effect. For example, we can measure the support of a discrete signal as the maximum number of consecutive zeros, call this $N(\boldsymbol{x})$, and using this measure, a discrete signal with a small number of zeros would have a correspondingly large $N(\boldsymbol{X})$ under the DFT. So, an appropriate discrete uncertainty principle is that $N(\boldsymbol{x}) + N(\boldsymbol{X}) \leq N - 1$. Pei and Chang (2016) show that a kind of "discrete Gaussian" function can be considered as a discrete counterpart to the Gaussian since it is bell-shaped, non-negative, invariant under the DFT and satisfies $N(\boldsymbol{x}) + N(\boldsymbol{X}) = N - 1$. This means that these discrete functions are indeed optimal under this measure of spread. Generalized uncertainty principles such as this are an active area of current research.

Gibb's phenomena

For many classes of signals that are sufficiently smooth (e.g. those in l_2), Fourier analysis is meaningful and the FT of a function can always be inverted to recover the original function. However, to make this possible we require a large (and potentially infinite) number of non-zero Fourier coefficients to represent the signal. In practice, all empirical Fourier representations of any signal will be finite and thus we can never fully capture all the coefficients and we must truncate the representation somehow. There are also very practical examples of signals not in l_2 or l_1 for which the Fourier series representation has coefficients which slowly decrease in magnitude with increasing frequency, for which truncation leads to a poor approximation, for example the Heaviside step function.

With such truncation the reconstructed signal will suffer from spurious oscillations called *Gibb's phenomena*. For the particular example of the Heaviside step function $f(t) = \Phi(t)$, using the truncation $F(\omega) = 0$ for $|\omega| > \omega_0$ where $\omega_0 > 0$ and applying the inverse FT, the reconstructed signal is (Mallat, 2009, Theorem 2.8, page 48):

$$\hat{f}(t) = \int_{-\infty}^{\omega_0 t} \frac{\sin(t')}{\pi t'} dt' \qquad (7.62)$$

See Figure 7.3 for an illustration. This function has oscillations which increase in frequency with ω_0, but with maximum amplitude that does not depend upon ω_0. Gibb's phenomena and Heisenberg uncertainty are unavoidable facts of life with classical LTI DSP, it is only by dropping the assumptions of either time-invariance or linearity or both that we can circumvent these limitations. For example, we can construct nonlinear DSP algorithms that can work efficiently for other classes of signals, such as those with bounded *total variation* $\int_{\mathbb{R}} |f'(t)| \, dt$ which do not have an efficient representation in terms of the FT (Little and Jones, 2011*a*).

Transfer function analysis of discrete-time LTI systems

Because all information about discrete-time LTI systems is contained in the DTFT, we can predict the response of the system to any input signal using *transfer function* (TF) *analysis*. The DTFT of the output of the system is:

$$Y(\omega) = H(\omega) X(\omega) \tag{7.63}$$

where $X(\omega)$ is the DTFT of the input and the DTFT of the impulse response of the system is $H(\omega)$, known as the *transfer function* (TF). Assume the input is a unit-magnitude complex exponential signal $x_n = \exp(i\phi n)$ which has DTFT $X(\omega) = 2\pi\delta[\omega - \phi]$, and write out the complex value of the TF at frequency ω in polar form $H(\omega) = R(\omega) \exp(i\Theta(\omega))$. Then the output of the system is:

$$
\begin{aligned}
y &= \int_{-\pi}^{\pi} \delta[\omega - \phi] R(\omega) \exp(i\Theta(\omega)) \exp(i\omega n) \, d\omega \\
&= \int_{-\pi}^{\pi} \delta[\omega - \phi] R(\omega) \exp(i[\omega n + \Theta(\omega)]) \, d\omega \tag{7.64} \\
&= R(\phi) [\exp(i[\phi n + \Theta(\phi)])]_{n \in \mathbb{Z}} \tag{7.65}
\end{aligned}
$$

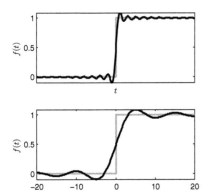

Fig. 7.3: The time-domain effect of truncating the Fourier transform of a Heaviside step function $f(t) = \Phi(t)$ (grey curve) is to introduce spurious oscillations known as *Gibb's phenomena* (black curve). These oscillations lead to "under-" and "overshoot", and cause the location of the step to be smeared out.

which is another complex exponential of magnitude $R(\phi)$ and argument $\phi n + \Theta(\phi)$. So, the effect on any complex exponential of frequency ϕ is to multiply its magnitude by the magnitude $R(\phi)$ of the TF at that frequency, and to shift the complex exponential radian phase by the phase of the TF $\Theta(\phi)$ at that frequency. No new frequency components are created at the output. Since we can write out any signal as a superposition of such complex exponentials and the system is linear (such that it affects each frequency component separately), we can understand the full behaviour of the system across the whole frequency range by "probing" it at each frequency separately. This is the central idea TF analysis.

There are a vast array of physical systems that can be modelled using (*linear, constant-coefficient*) *ordinary differential equations* (ODEs). Examples include (passive) electronic circuits involving resistors, capacitors and inductors, or laminar flow of non-viscous liquid in pipes of varying widths along with storage tanks and reservoirs etc. These models can all be written in terms of an LTI ODE system which is a differential equation such as $f'(t) + a\,f(t) = g(t)$ where $g(t)$ is the input to the system. Such systems can perform useful actions such as reducing the magnitude of high-frequency or low-frequency components (known as *low-pass* or *high-pass filters* respectively). TF analysis using the continuous-time FT is straightforward using the differential property of the Fourier transform:

$$F\left[\frac{d^j f}{dt^j}(t)\right] = (i\omega)^j F(\omega) \tag{7.66}$$

which allows us to turn all LTI ODEs into algebraic TFs. We can construct discrete-time counterparts which are amenable to DSP on computer, which can be expressed as discrete-time *difference equations*:

$$\sum_{j=0}^{P} a_j y_{n-j} = \sum_{k=0}^{Q} b_k x_{n-k} \tag{7.67}$$

supplemented with the P initial conditions $y_{-j} = C_j$ for $j \in 0, 1, \dots P - 1$. We will take the $P + Q + 2$ coefficients a_j, b_k to be real variables. The signal x is considered as the input to the system. Analysis of the behaviour of such systems is simplified considerably if we take the DTFT of both sides and apply the time shift property:

This restriction to real-valued coefficients is not strictly necessary, but it simplifies the presentation as we generally assume the signals to be real-valued.

$$\left[\sum_{j=0}^{P} a_j \exp(-i\omega j)\right] Y(\omega) = \left[\sum_{k=0}^{Q} b_k \exp(-i\omega k)\right] X(\omega) \tag{7.68}$$

which implies that:

$$H(\omega) = \frac{Y(\omega)}{X(\omega)}$$
$$= \frac{\sum_{k=0}^{Q} b_k z^{-k}}{\sum_{j=0}^{P} a_j z^{-j}} \tag{7.69}$$

where $z = \exp(i\omega)$. The function $H(\omega)$ is the transfer function of the system and for difference equations it is a ratio of polynomials in z. By the fundamental theorem of algebra, we can rewrite these polynomials in terms of the roots $z_k, p_j \in \mathbb{C}$ of the numerator and denominator (*zeros* and *poles*):

Use of powers of the complex variable z is central to the so-called z-transform, which is the discrete-time analogue of the Laplace transform.

$$H(z) = \frac{\prod_{k=1}^{Q}(z - z_k)}{\prod_{j=1}^{P}(z - p_j)} \tag{7.70}$$

From this we can see that the placement of the zeros and poles in the complex plane determine the properties of the TF of the difference equation system. For example, the simple difference equation:

$$y_n - a\, y_{n-1} = x_n \tag{7.71}$$

with $a \in (-1, 1)$ has the TF:

$$H(z) = \frac{1}{1 - a z^{-1}} \tag{7.72}$$

which has the single pole $z_1 = a$. The magnitude of the TF at any given frequency is:

$$|H(\omega)| = \frac{1}{\sqrt{(1 - a\exp(-i\omega))(1 - \exp(i\omega))}}$$
$$= \frac{1}{\sqrt{1 - 2a\cos\omega + a^2}} \qquad (7.73)$$

and the argument (phase angle) is:

$$\arg H(\omega) = \arctan\left(\frac{a\sin\omega}{a\cos\omega - 1}\right) \qquad (7.74)$$

From a plot of the magnitude of the TF (Figure 7.4), we can see that with $a > 0$, lower frequency components (those closer to zero) are amplified whereas higher frequency components (those closer to π) are attenuated. This is the defining characteristic of a *low-pass filter* (LPF). The converse is true when $a < 0$, i.e. high frequency components are amplified and lower frequency components are attenuated, making the this system act like a high-pass filter. The design of discrete difference systems with specific filtering properties is an important topic in classical LTI DSP addressed in Section 7.3.

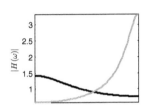
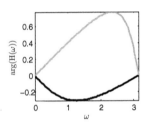

Fig. 7.4: Transfer function analysis of the discrete-time difference equation system $y_n - a\,y_{n-1} = x_n$ with $a = 0.3$ (black curves) and $a = -0.7$ (grey curves). A complete picture of the behaviour of the system can be obtained from both the amplitude (top) and phase/complex angle (bottom) response of the system, computed from the associated transfer function $H(z) = \left(1 - a\,z^{-1}\right)^{-1}$.

Fast Fourier transforms (FFT)

Because there are N frequency values which require N multiplications and N additions, computing the DFT for a signal of length N requires N^2 multiply-add operations, an $O(N^2)$ transform. While this complexity class is not as bad as exponential, it turns out that there are much more efficient ways to compute the DFT which exploit the inherent symmetries (group structure) in the algorithm. These algorithms all go by the name of *fast Fourier transforms* (FFTs). We will explore one approach which illustrates the general idea that these symmetries allow the DFT to be factorized into a product of smaller-sized DFTs, such that the total number of multiply-add operations is substantially reduced.

Recall that the DFT can be expressed as a matrix multiplication using the size $N \times N$ DFT matrix $\mathbf{F}_N$ (7.48). This matrix is *dense* in that all entries are non-zero so that none of the multiply-add operations required to carry out the DFT can be avoided. Let us assume, however, that $N = K \times M$ with $K > 1$ and $K \in \mathbb{N}$ so that N has a whole-number factorization. Then, using the facts that the DFT is cyclic and that the terms in the Fourier transform summation can be reorganized at will, we can rewrite the DFT matrix using the following matrix product(Puschel, 2003):

$$\mathbf{F}_N = (\mathbf{F}_K \otimes \mathbf{I}_M)\,\mathbf{T}_M^N\,(\mathbf{I}_K \otimes \mathbf{F}_M)\,\mathbf{L}_K^N \qquad (7.75)$$

We will explain each term in this product separately. The first and third terms are large matrices created using the *Kronecker* (matrix) product $\mathbf{C} = \mathbf{A} \otimes \mathbf{B}$ with $\mathbf{C}_{ij} = a_{ij}\mathbf{B}$. This is the matrix created by multiplying each entry a_{ij} by $\mathbf{B}$ and replacing each a_{ij} with the resulting matrix.

For example, since:

$$\mathbf{F}_2 = \begin{bmatrix} 1 & 1 \\ 1 & -1 \end{bmatrix} \tag{7.76}$$

then:

$$\mathbf{I}_2 \otimes \mathbf{F}_2 = \begin{bmatrix} 1 & 1 & 0 & 0 \\ 1 & -1 & 0 & 0 \\ 0 & 0 & 1 & 1 \\ 0 & 0 & 1 & -1 \end{bmatrix} \tag{7.77}$$

which is a 4×4 block-diagonal matrix. The second term is known as the $N \times N$ diagonal *twiddle* (factor) matrix $\mathbf{T}_M^N$ which has diagonal entries $w_N^{i \times j}$ for $i = 0, 1, \ldots, K - 1$ and $j = 0, 1, \ldots, M - 1$ using the complex exponentials $w_N = \exp(-2\pi i / N)$. The first term $\mathbf{L}_K^N$ is a special (stride) *permutation matrix* which re-organizes the input vector according to the scheme $iK + j \mapsto jM + i$ for $i = 0, 1, \ldots K - 1$ and $j = 0, 1, \ldots M - 1$, for example:

$$\mathbf{L}_2^4 = \begin{bmatrix} 1 & 0 & 0 & 0 \\ 0 & 0 & 1 & 0 \\ 0 & 1 & 0 & 0 \\ 0 & 0 & 0 & 1 \end{bmatrix} \tag{7.78}$$

At first glance, this factorization does not appear to have achieved anything: we now have four matrix multiplications where we originally had only one! But, matrices such as (7.77), (7.78) and $\mathbf{T}_M^N$ are highly *sparse* in that a great deal of the entries are zero. This means that the number of multiply-add operations is much smaller than a full, dense, matrix multiplication. In fact, vector multiplication by both $\mathbf{T}_M^N$ and $\mathbf{L}_K^N$ can be computed in $O(N)$ operations. Furthermore, if N is highly composite (that is, it has many factors) then (7.75) gives us an efficient, recursive way of breaking the size N DFT down into a set of K smaller DFTs $\mathbf{F}_M$ which require only $O(M^2)$ operations each, much less than the $O(N^2)$ for the full size N DFT.

Even better, the permutation matrix, as its name suggests, requires no multiplications at all as it simply re-orders the elements of the vector it multiplies.

How many times do we have to do this recursion? It requires $O(\ln N)$ recursive steps. To see why, it is helpful to look at a highly composite case, i.e. when $N = 2^r$ for some natural number $r > 0$. The first application of (7.75) gives us $\mathbf{F}_{2^r}$ in terms of $\mathbf{F}_{2^{r-1}}$ and $\mathbf{F}_2$ (by taking $K = 2$ so that $M = N/2$). Then the next application gives $\mathbf{F}_{2^{r-1}}$ in terms of DFTs of half the size, $\mathbf{F}_{2^{r-2}}$ and so on, down to $\mathbf{F}_2$, which is the trivial $O(1)$ base case (7.76) (which takes one addition and one subtraction operation). Thus we need to do r, which is $\log_2 N$, recursions to reach the terminal case. At each stage, we require $O(N)$ operations (twiddle factor and permutations) so that the total complexity is $O(N \log_2 N)$. Thus, the $K = 2$ factorized, recursive FFT is a very substantial computational improvement over the naive DFT implementation.

Algorithm 7.1 Recursive, Cooley-Tukey, radix-2, decimation in time, fast Fourier transform (FFT) algorithm.

---

(1) *Definition.* Function $fftrecurse2$ with input vector $\boldsymbol{x} \in \mathbb{C}^N$ of length $N = 2^r$ and providing DFT output $\boldsymbol{X} \in \mathbb{C}^N$,

(2) *Recurse.* Split $\boldsymbol{x}$ into even and odd element vectors of size $N/2$, $\boldsymbol{x}^{\mathrm{e}} = (x_1, x_3, \ldots, x_{N-1})$ and $\boldsymbol{x}^{\mathrm{o}} = (x_2, x_4, \ldots x_N)$, call $\boldsymbol{X}^{\mathrm{e}} = fftrecurse2(\boldsymbol{x}_{\mathrm{e}})$ and then $\boldsymbol{X}^{\mathrm{o}} = fftrecurse2(\boldsymbol{x}_{\mathrm{o}})$,

(3) *Combine.* For $k = 1, 2, \ldots, N/2$, set $w = \exp(-2\pi i\,(k-1)/N)$ and $X_k = X_k^e + wX_k^o$, $X_{k+N/2} = X_k^e - wX_k^o$,

(4) *Finish.* Return DFT vector $\boldsymbol{X}$.

Not the only people to spot the factorization as the mathematician Gauss used the essential idea as far back as 1805. However, he did not apparently notice its full importance.

This factorization approach is known as the *Cooley-Tukey* FFT, after the first two investigators to properly explore and popularize this idea. It has been enormously influential: in fact, the invention of this FFT in the 1960's is often credited with launching the entire field of DSP as it rendered the DFT, and with it digital spectral analysis, practical for the very limited computational hardware of the time. It can be shown that in general, for any *radix* (the size of the base case in the recursion), that the resulting Cooley-Tukey algorithm is $O(N \ln N)$. As a result, even today, the Cooley-Tukey FFT is fundamental to practically all large-scale DSP efforts. The implications are much broader. Wherever finite, discrete LTI systems are applicable, there will be the need for the DFT so that the FFT is valuable, and this also arises in machine learning.

As emphasized above, the special structure of the permutation and twiddle matrices and the simplicity of the DFT base case lend themselves to very efficient and simplified FFT implementations (see Algorithm 7.1 for an example).

The FFT is an enormously useful discovery but a word of caution is required. Although the factorization is always available where N is non-prime, if N does not have many, small factors, the FFT may not be much of a computational gain, as pointed out by Press (1992, page 509). In these situations, it is more efficient to stick with the radix-2 implementation and *zero pad* the data. Let us assume the data is length N which is not a power of two; zero padding appends the data with zeros up to length 2^r for $r = \lceil \log_2 N \rceil$. This now allows us to use the radix-2 FFT algorithm 7.1, for example. Because the entries N to 2^r of the signal $\boldsymbol{x}$ are zero, the DFT formula (7.47) is unchanged except that now the scale of the frequency variable k has changed, in fact the frequency resolution *increases* because the signal is longer (due to Heisenberg uncertainty).

Note that of course the zero-padded radix-2 FFT and the smaller-length DFT do not lead to exactly the same spectrum (as the input is different), and zero-padding uses "fictional" data which changes the interpretation of the spectrum due to changes in the periodicity of the data, for example. However, this may not matter much in practice, and may be offset by dramatic computational savings. Consider the signal which has (*Mersenne prime*) length $N = 131071$ with no factors, for which there is no FFT and the DFT requires around $N^2 \approx 1.7 \times 10^{10}$ operations. The next largest radix-2 FFT is of length $N = 131072 = 2^{17}$ and only requires around $N \log_2 N \approx 2.2 \times 10^6$ operations. In other words, simply zero-padding with one additional input value leads to four orders of magnitude improvement in computational effort, with negligible deviation from the $N = 131071$ length spectrum.

Because there are multiple underlying symmetries in the DFT, there are myriad possibilities for how to implement computational efficiencies, so it is wrong to talk about "the" FFT. For example, while the factorization 7.75 involves first permuting the input data before applying the smaller DFTs and twiddle factors, it is possible using appropriate matrix algebra identities, to re-organize the computations such that the permutation occurs at the end (Puschel, 2003):

$$\mathbf{F}_N = \mathbf{L}_M^N \left(\mathbf{I}_K \otimes \mathbf{F}_M\right) \mathbf{T}_M^N \left(\mathbf{F}_K \otimes \mathbf{I}_M\right) \tag{7.79}$$

Because it is the frequency components which are permuted, this is known as a *decimation in frequency* (DIF) Cooley-Tukey FFT as opposed to (7.79) which is called *decimation in time* (DIT). Similarly, while the above computations are recursive, one can expand out the recursion and flatten the brackets, to give an iterative algorithm for which the radix-2 case with $N = 2^r$ for $r \in \mathbb{N}$ with $r \geq 0$ is:

$$\mathbf{F}_N = \left(\prod_{j=1}^r \left(\mathbf{I}_{2^{j-1}} \otimes \mathbf{F}_2 \otimes \mathbf{I}_{2^{r-j}}\right) \cdot \left(\mathbf{I}_{2^{j-1}} \otimes \mathbf{T}_{2^{r-j}}^{2^{r-j+1}}\right) \right) \mathbf{R}_2^N \tag{7.80}$$

The matrix $\mathbf{R}_2^N$ turns out to be the *bit-reversal* permutation, that is, the permutation that reorders a vector by reversing the r-bit length binary representation of each element's position. So, for example, for $r = 4$, the element in position $13 = 1101_2$ (counting elements from zero), is swapped with the element in position $1011_2 = 11$; position 7 is swapped with position 14 etc. There is a corresponding DIF version of this iterative FFT, similarly, one can extend this in a straightforward way to any radix in which case the permutation becomes an analogous digit-reversal in the number base of the radix (Puschel, 2003).

All the above algorithms can be derived using basic, direct algebraic manipulations of the DFT formula (7.47), but these manipulations all require apparently special tricks and are not "automatic". Instead, the above sparse matrix representations flow from a unified treatment for all common FFTs, including the *prime-factor, Rader, Bluestein* transforms, and transforms not based on the complex exponentials (such as the discrete cosine and sine transforms – DCT and DST). This unified treatment is based on an abstract representation of the discrete time LTI filtering problem as a particular *polynomial algebra* with associated transformations of that algebra; the Chinese remainder theorem splits the algebra into a product of simpler, one-dimensional algebras, of which the DFT Vandermonde matrix is a special case (Puschel, 2003). Fast algorithms arise where a decomposition of the polynomial is possible (which, for the DFT occurs when N is composite, for example).

FFTs are a vast area of signal processing research and practice; for a theoretical perspective see Puschel and Moura (2008) and for its practical use in DSP, see Proakis and Manolakis (1996, chapter 6) and Press (1992, chapters 12 and 13).

7.3   LTI signal processing

Practical LTI DSP techniques that can perform functions such as filtering out unwanted signals or detection of specific signals, are based on the theoretical tools described above. Filtering involves the manipulation of a signal to change the frequency or time domain properties of the signal. For example, we can amplify (increase) or attenuate (decrease) the amplitude of specific frequencies using combinations of low- and high-pass digital filters. Designing and implementing precision filters with specific characteristics is the topic of this section.

Rational filter design: FIR, IIR filtering

We have seen above that any discrete-time filter can be represented using the following linear difference system (7.67) where the signal y is the output and x is the input. The goal of *rational filter design* is to find "optimal" values of the filter parameter vectors a and b. Of course, what counts as optimal depends upon the criteria applied. There are many such design criteria, depending upon the desired application. Typically, DSP engineers are interested in providing constraints on the frequency or time-domain response of the filter to specific frequency ranges. Alternatively, a "template" frequency or time-domain signal is given and the goal is to find parameters that lead to a filter with response that matches this signal as closely as possible.

However, we need to identify a few properties of such filters which are critical for design purposes. Firstly, if $P = 0$, then there is no *recursive* or *feedback* element to the filter so that the impulse response is only as long as Q: it is a *finite impulse response* (FIR) or *moving average* (MA) filter. Otherwise, there is some feedback element to the filter so that the impulse response is not finite, creating an *infinite impulse response* (IIR) or *autoregressive* (AR) filter. Secondly, any IIR filter that has poles which are outside the unit circle in the complex plane will be *unstable* in the following sense: any non-zero input causes the output signal y to grow without bound. This occurs because, assuming that the poles p_j

are all distinct, the time-domain solution to (7.67) is in the form:

$$y_n = \sum_{j=1}^{P} B_j p_j^n + v_n \qquad (7.81)$$

where v_n is a term which is determined solely by the form of the input signal x. The values B_j are constants which are determined by the initial conditions of the filter. Assume that the input signal is bounded e.g. $\sum_{n=0}^{\infty} |x_n| < \infty$. Any one pole can be written in polar form $p_j = r_j \exp(i\phi_j)$ with $r_j > 0$, so that $p_j^n = r_j^n \exp(i n\phi_j)$. It follows that if $r > 1$, then $r^n \to \infty$ as $n \to \infty$. Therefore, if any of the poles are outside the unit circle where $|z| = 1$, the output of the IIR filter will grow in magnitude indefinitely. Thus, for practical purposes, ensuring the stability of an IIR filter by paying close attention to the pole locations is critical and is also a significant practical limitation.

In practice, these initial conditions are usually chosen to be zero. But there are circumstances where they must be non-zero, for example, when processing a signal in windows such that the boundary conditions have to match at the end of one window and the beginning of the next.

Let us explore a simple design example. We wish to find an LPF which has the shortest possible finite impulse response such that the filter responds quickly to changes in the input. At the same time, we want the response at zero frequency to have unit amplitude, that is, the value of constant signals is unchanged. The simplest (useful) FIR filter has the following TF with $Q = 1$, $P = 0$ and (for convenience, without loss of generality, $a_0 = 1$):

$$H(z) = b_0 + b_1 z^{-1} \qquad (7.82)$$

whose TF square magnitude is:

$$|H(\omega)|^2 = b_0^2 + 2b_0 b_1 \cos\omega + b_1^2 \qquad (7.83)$$

Now, for this to be a low-pass filter, it must be that:

$$\frac{\partial}{\partial\omega} |H(\omega)|^2 < 0 \qquad (7.84)$$

for $\omega \in [0, \pi]$ which means that the amplitude is decreasing with increasing frequency over the Nyquist range. Now, the required gradient is $\frac{\partial}{\partial\omega} |H(\omega)|^2 = -2b_0 b_1 \sin\omega$ and since $\sin\omega \geq 0$ for $\omega \in [0, \pi]$, we deduce that b_0 and b_1 must be of the same sign in order to satisfy (7.84). Choosing $b_0 > 0$ requires that $b_1 > 0$, for example. Next, the unit amplitude criteria requires that:

$$|H(1)| = 1 \qquad (7.85)$$

(since $\exp(i\,0) = 1$) from which we get that:

$$|b_0 + b_1| = 1 \qquad (7.86)$$

Let us choose $b_0 = c$ for $c \in \left[\frac{1}{2}, 1\right]$, then we get that $b_1 = 1 - c$. The resulting difference system is:

$$y_n = c\, x_n + (1 - c)\, x_{n-1} \tag{7.87}$$

and the low-pass gradient is $-2c\,(1 - c) \sin \omega$. This has two zeros at $\omega = 0$ and $\omega = \pi$; and since we have constrained $\omega = 0$ to be a maxima, it follows that the other frequency is the minima of the amplitude response. At the minima, the amplitude is:

$$|H\,(\pi)| = 2c - 1 \tag{7.88}$$

so that the LPF has zero magnitude at the Nyquist frequency π if $c = \frac{1}{2}$. Thus, c controls the "sharpness" of the filter, that is, how rapidly it decreases in amplitude with increasing frequency and also how much it attenuates the signal at the Nyquist frequency. This can be seen from the magnitude response (see Figure 7.5). We have not placed any criteria on the time-domain response except for limiting the impulse response, the phase response is (Figure 7.5):

$$\arg H\,(\omega) = - \arctan \left(\frac{(c - 1) \sin \omega}{(c - 1) \cos \omega - c} \right) \tag{7.89}$$

An important observation is that when $c = \frac{1}{2}$, the phase response is:

$$
\begin{aligned}
\arg H\,(\omega) &= - \arctan \left(\frac{\sin\,(\omega)}{\cos\,(\omega) + 1} \right) \\
&= - \frac{\omega}{2}
\end{aligned}
\tag{7.90}
$$

How do we interpret this? While the phase response gives the phase shift applied to each complex sinusoidal input, by contrast the *phase delay*:

$$\tau\,(\omega) = - \frac{\arg H\,(\omega)}{\omega} \tag{7.91}$$

gives the time delay in sampling intervals at any frequency. To see why, as discussed above, the response of the filter to the complex sinusoidal input $x_n = \exp\,(i\phi n)$ is:

$$
\begin{aligned}
y_n &= |H\,(\phi)| \exp\,(i\,[\phi n + \arg H\,(\phi)]) \\
&= |H\,(\phi)| \exp\,(i\phi\,[n - \tau\,(\phi)])
\end{aligned}
\tag{7.92}
$$

which shows that the complex sinusoid is delayed by $\tau\,(\phi)$ at frequency ϕ, in samples.

For the simple FIR above, the phase delay is the constant $\frac{1}{2}$, which tells us that there is a half-sampling interval time delay for all frequencies. This is a very valuable property, particularly for signals with strong time localization where all frequency components need to be kept in sync. Only FIR filters with symmetric or antisymmetric coefficient sequences have this *linear-phase* property (Proakis and Manolakis, 1996, chapter 8), all IIR filters only satisfy this property approximately for

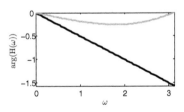

Fig. 7.5: **TF analysis of the simple FIR low-pass filter** $y_n = c\,x_n + (1 - c)\,x_{n-1}$ **for** $c = \frac{1}{2}$ **(black curves) and** $c = 0.8$ **(grey curves); amplitude (top) and phase response (bottom) computed from the associated TF** $H\,(z) = c + (1 - c)\,z^{-1}$. **At** $\omega = 0$, **the filter has unit amplitude. For** $c = \frac{1}{2}$, **the filter has zero amplitude at the Nyquist frequency** $\omega = \pi$. **Also for** $c = \frac{1}{2}$, **the filter has** *linear-phase* **whereby all frequencies are delayed by the same number of sampling intervals.**

certain ranges of frequencies. Thus, linear-phase FIR filters are very often preferred in many precision DSP applications where computational
effort is of less concern. FIR filter design thus tends to be dominated by
linear-phase techniques.

The example above shows what can be done with carefully "hand-
crafted" solutions, but this does not lend itself to most applications. We
will next explore a simple, but systematic approach to FIR filter design.
Assume we have a desired TF, $H(\omega)$. As an example, let us look at an
ideal LPF:

$$H(\omega) = \begin{cases} \exp(-i\omega) & 0 \leq |\omega| \leq \omega_c \\ 0 & \text{otherwise} \end{cases} \tag{7.93}$$

for the *cutoff* frequency $0 < \omega_c < \pi$. Using the Fourier inversion integral
(7.25), the impulse response of the filter is:

$$
\begin{aligned}
h_n &= \frac{1}{2\pi} \int_{-\omega_c}^{\omega_c} \exp(i\omega[n-1])\, d\omega \\
&= \text{sinc}\left(\frac{\omega_c}{\pi}[n-1]\right)\frac{\omega_c}{\pi} \tag{7.94}
\end{aligned}
$$

This impulse response is infinitely long, but we can truncate it to finite
length to obtain the coefficient sequence of the corresponding FIR filter,
e.g. $b_k = h_{k-\tau}$ for $k \in 0,1,\ldots Q$. For Q even we choose $\tau = \frac{Q-2}{2}$, to
ensure that the FIR filter is centred. The result is an FIR filter with
an approximation to the desired TF $H(\omega)$. It is linear-phase because
(7.94) is symmetric about $n = 1$.

How good is the approximation? Clearly, if Q were infinite then the
FIR filter would be exact, so the error in the approximation comes entirely from the truncation. The truncation can be written as the multiplication of the infinite impulse response with the rectangular *window
function* $w_n = \text{rect}\left(\frac{2\pi n}{Q}\right)$. Thus, the DTFT of the truncated impulse
response is:

$$
\begin{aligned}
\hat{H}(\omega) &= \sum_{n \in \mathbb{Z}} h_n w_n \exp(-i\omega n) \\
&= \sum_{n=-Q/2}^{Q/2} h_n \exp(-i\omega n) \tag{7.95}
\end{aligned}
$$

*For Q even we have an odd-length
FIR. We can also have even-
length FIRs and would need to
adjust the centering and windowing appropriately.*

Therefore, the approximate FIR TF is a reconstruction of the desired
TF $H(\omega)$ but with the higher frequency Fourier components removed.
The effect of this truncation is to introduce unavoidable Gibb's phenomena (see Section 7.2) in to the FIR amplitude response $\hat{H}(\omega)$, the
influence of these distortions is reduced by increasing the FIR length
$Q+1$. Implementing an FIR filter for a signal of length N is an $O(NQ)$
convolution operation, so, there is a trade-off between increasing the
computational effort and increasing the accuracy of the filtering approximation (see Figure 7.6). Of particular interest in practice is the slope of

the "transition band", that is, the rate of change of the filter amplitude with respect to frequency around the cutoff frequency ω_c. A long FIR will have a thinner transition band which more closely approximates the infinitely fast transition of the ideal low-pass filter TF (7.93).

While the FIR filter described above has linear phase, it has a phase delay equal to approximately $Q/2$ samples (because it is centred by τ). Thus, while a very long FIR can have a rapid transition band, it will necessarily have a long impulse response and thereby introduces a long phase delay into the filtering. This long phase delay can be critical limitation in many applications. Recursive (IIR) filters, on the other hand, can be designed to have much shorter phase delay along with very low computational requirements. Design of such filters tends to be dominated by modifications of techniques found in continuous time models, particularly in electronic engineering (Proakis and Manolakis, 1996, chapter 8). We will take a closer look at some continuous-time filters.

Firstly, much as with discrete time, we can transform an ordinary, constant-coefficient differential equation into a ratio of polynomials by using the derivative property of the Fourier transform. This gives us the following continuous-time TF:

$$H(s) = \frac{\sum_{k=0}^{Q} b_k s^k}{\sum_{j=0}^{P} a_j s^j} \tag{7.96}$$

where $s = i\omega$ and $\omega \in \mathbb{R}$. Both polynomials can be factorized to give a product representation in terms of Q poles and P zeros $p_j, z_k \in \mathbb{C}$. As a simple example, the ODE $y'(t) - a\,y(t) = x(t)$ with $a \in \mathbb{R}$ has TF:

$$H(s) = \frac{1}{s-a} \tag{7.97}$$

with one pole at $p_1 = a$. The amplitude response is:

$$|H(\omega)| = \frac{1}{\sqrt{a^2 + \omega^2}} \tag{7.98}$$

and phase response $\arg H(\omega) = \arctan(\omega/a)$. This model produces a very simple LPF for any a since $\frac{d}{d\omega}|H(\omega)| = -\omega\left(a^2 + \omega^2\right)^{-\frac{3}{2}} < 0$ for all $\omega \geq 0$ which means that the amplitude continuously decreases with increasing frequency.

Historically, filters have often been designed by finding various simple functions with desired shapes, for which the roots are easily obtained. For example, let us look at the simple polynomial ω^{2M} for $M \in \mathbb{N}$ which goes to positive infinity as ω gets large in magnitude. This has the property that as $M \to \infty$, so the function becomes more and more "flat" near zero, and the transition to infinity occurs more and more rapidly. Using this polynomial, the filter magnitude function:

$$A(\omega) = \frac{1}{\sqrt{1 + \omega^{2M}}} \tag{7.99}$$

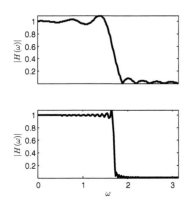

Fig. 7.6: TF (amplitude response) analysis of a linear-phase, FIR low-pass filter with cutoff frequency $\omega_c = 1.7$ obtained by truncating the impulse response of an ideal low-pass filter to length $Q+1$. For a short FIR (top panel, $Q = 20$) the Gibb's phenomena introduced by the truncation are very noticeable, and the transition between unit amplitude and zero amplitude at the cutoff frequency is slow. This is much improved for the longer FIR (bottom panel, $Q = 120$) which has faster Gibb's ripples and much more abrupt cutoff transition.

The use of the continuous-time complex variable s is entirely analogous to the use of z in discrete-tme, in fact s is the Laplace transform variable.

has a useful flat region near zero and a transition to zero as ω gets large in magnitude. This magnitude function is easily factorized into the poles $p_k = \exp\left(\pi i \frac{2k+M-1}{2M}\right)$ for $k = 1, 2 \ldots M$ lying on the unit circle in the negative real half complex plane. Since the magnitude function is real, these poles occur in complex conjugate pairs (M even), or complex conjugate pairs with the root $k = \frac{M+1}{2}$ equal to -1 (for M odd). The resulting design is known as the *Butterworth* LPF of order M. The corresponding transfer function is obtained by inserting these poles into (7.96). The LPF can be shifted to have cutoff frequency ω_c simply substituting $\omega \mapsto \omega/\omega_c$ into (7.99) with the poles now all being scaled by ω_c as a result.

To implement such filters digitally, we have to somehow discretize the continuous-time transfer function to find a corresponding digital version. In other words, we have to convert the continuous-time ordinary differential model into a discrete-time difference model. This takes us into the extremely broad area of *numerical methods for differential equations*. The set of methods we will concentrate on are based on finding ways to directly replace the derivatives (or integrals) in the original system with discrete counterparts. Because this process will not, in general, produce a filter with exactly the same behaviour, there will be certain trade-offs to be negotiated. Let us start with perhaps the simplest approach, the so-called *backward Euler* method which replaces the derivative with a finite difference:

$$\frac{df}{dt}(t) \approx \frac{f(t) - f(t - \Delta t)}{\Delta t} \tag{7.100}$$

where Δt is the sampling interval. The TF of this differentiator is therefore $H(z) = \frac{1}{\Delta t}\left(1 - z^{-1}\right)$. This converts the simple LPF $y'(t) - a\,y(t) = x(t)$ to the discrete difference system:

$$\frac{y_n - y_{n-1}}{\Delta t} - a\,y_n = x_n \tag{7.101}$$

which can be rearranged to:

$$y_n - \frac{1}{1 - a\Delta t}y_{n-1} = \frac{\Delta t}{1 - a\Delta t}x_n \tag{7.102}$$

which is an IIR with parameters $a_0 = 1$, $a_1 = -(1 - a\Delta t)^{-1}$ and $b_0 = \Delta t(1 - a\Delta t)^{-1}$. This new TF has the single real pole $p_1 = (1 - a\Delta t)^{-1}$, whereas the original continuous-time TF has the pole $p_1 = a$. The discrete-time condition $|p_1| < 1$ implies that $a < 0$ is sufficient for stability, which coincides with the stability condition for the continuous-time filter. The new TF is that of a single pole LPF (since there is only one stable, positive pole).

The Euler mapping above is simple but it is not a very accurate discretization – the (single sample interval) discrepancy (local error) between $y(t)$ and y_n is of the order $o\left(\Delta t^2\right)$ – which places a significant requirement on Δt to be very small to get a digital filter which behaves like the continuous-time version. To improve accuracy, we can turn to other discretizations, for example, we can use the *trapezoidal method*

Note that for practical purposes, here the poles must be chosen to have negative real parts. Otherwise, the output of the filter is growing with increasing time, not decreasing. This is the continuous-time analogue of the condition that the poles must lie inside the unit circle for a discrete-time IIR filter to be stable.

This simply follows from the usual definition of the derivative encountered in elementary calculus.

for numerical integration which has better single sample interval discrepancy of order $o\left(\Delta t^3\right)$. Starting with the fundamental theorem of calculus:

$$\int_{t-\Delta t}^{t} \frac{df}{dt}\left(\tau\right) d\tau = f\left(t\right) - f\left(t - \Delta t\right) \tag{7.103}$$

this method uses the geometrical formula for a trapezoidal area under the curve as approximation to the above integral:

$$\frac{\Delta t}{2}\left[\frac{df}{dt}\left(t\right) + \frac{df}{dt}\left(t - \Delta t\right)\right] \approx f\left(t\right) - f\left(t - \Delta t\right) \tag{7.104}$$

with differentiator TF:

$$H\left(z\right) = \frac{2}{\Delta t}\left(\frac{1 - z^{-1}}{1 + z^{-1}}\right) \tag{7.105}$$

Using this mapping, the simple LPF $y'\left(t\right) - a\,y\left(t\right) = x\left(t\right)$ above, becomes the discrete difference system:

$$y_n - y_{n-1} = \frac{\Delta t}{2}\left[a\,y_n + x_n + a\,y_{n-1} + x_{n-1}\right] \tag{7.106}$$

which can be rearranged to:

$$\left[1 - \frac{a\Delta t}{2}\right] y_n - \left[1 + \frac{a\Delta t}{2}\right] y_{n-1} = \frac{\Delta t}{2}\left[x_n + x_{n-1}\right] \tag{7.107}$$

which is another IIR. Structurally, this differs from the Euler IIR above in that the input is formed from the average of two neighbouring values, but the single real pole has the value $p_1 = \frac{2 + a\Delta t}{2 - a\Delta t}$. As above, provided $a < 0$ then this pole $|p_1| < 1$ ensuring stability, mirroring the stability condition for the continuous-time filter. This mapping technique is also known as a *bilinear transform*.

These stability and other properties of Euler and bilinear transforms can be understood in general to inform the filter design choices. The continuous-time differentiator has the transfer function $H\left(s\right) = s$, whereas for the backward Euler differentiator it is $H\left(z\right) = 1 - z^{-1}$ and for the bilinear differentiator it is $H\left(z\right) = \left(1 - z^{-1}\right)\left(1 + z^{-1}\right)^{-1}$. Thus, given the continuous-time filter TF, the discretization is obtained by substituting e.g. $s \mapsto 1 - z^{-1}$ for the Euler method.

We have chosen the sampling interval $\Delta t = 1$ for the Euler method and $\Delta t = 2$, which can be done without loss of generality because we can always rescale the frequency variable ω appropriately.

What is the effect of this transformation on the stability and frequency properties of the filter? To understand this, we need to know the effect of the transformation on the imaginary (frequency) axis line $s = i\omega$ for $\omega \in \mathbb{R}$. This can be expressed as the equation $|s - 1| = |s + 1|$, which has the *inverse points* $s = 1$ and $s = -1$, so the resulting equation under

Property	Backward Euler	Trapezoidal (bilinear)
Discretization map	$\frac{1}{\Delta t}\left(1 - z^{-1}\right)$	$\frac{2}{\Delta t}\left(\frac{1-z^{-1}}{1+z^{-1}}\right)$
Imaginary axis image	Circle centre $\frac{1}{2}$, radius $\frac{1}{2}$	Unit circle
Frequency map	$\arctan\left(\omega\Delta t\right)$	$2\arctan\left(\frac{1}{2}\omega\Delta t\right)$
Discrete frequency range	$\left[-\frac{\pi}{2}, \frac{\pi}{2}\right]$	$\left[-\pi, \pi\right]$
Local discretization error	$o\left(\Delta t^2\right)$	$o\left(\Delta t^3\right)$

Table 7.3: Properties of the backward Euler and trapezoidal (bilinear) discretizations widely used to convert continuous-time differential filter models into digital counterparts with sampling rate Δt. Both methods preserve stability so that if the continuous-time filter is stable, so is the corresponding digital filter. Both methods map the continuous-time imaginary frequency axis uniquely to some circle in the complex plane without aliasing. Both methods are exact at zero frequency but shrink the frequency range for higher frequencies.

the Euler transformation is (Priestley, 2003, page 24):

$$
\begin{aligned}
\left|1 - z^{-1} + 1\right| &= \left|1 - z^{-1} - 1\right| \\
\left|\frac{2z - 1}{z}\right| &= \left|\frac{1}{z}\right| \\
\left|z - \frac{1}{2}\right| &= \frac{1}{2}
\end{aligned}
\tag{7.108}
$$

which is the equation of a circle in the complex plane with centre $z = \frac{1}{2}$ and radius $\frac{1}{2}$. Applying similar reasoning gives the unit circle $|z| = 1$ for the bilinear method. What this means is that poles and zeros in the continuous-time TF which have negative real parts are mapped inside the respective circle for the discretization method, as above. Roots in the positive half are mapped outside this circle; in particular for the bilinear mapping this means that a continuous-time filter which has stable poles is guaranteed to have a stable digital counterpart.

While we can guarantee stability, frequencies $\omega \in \mathbb{R}$ must be mapped under the discretization to the digital frequency range $[-\pi, \pi]$. The mapping should be one-one otherwise some frequencies in the continuous-time model $\mathbb{R}$ will be *aliased* (i.e. more than one continuous-time frequency will be mapped onto one particular digital frequency), so this frequency mapping cannot be linear. For the Euler method, from the differentiator relationship $s = 1 - z^{-1}$ we get that $z = (1 - i\omega\Delta t)^{-1}$ for $s = i\omega$, from which the digital frequency is $\arg z = \arctan\left(\omega\Delta t\right)$. While this mapping is one-one, we have that the digital frequency range $[-\pi/2, \pi/2]$, in other words, it is not possible to map to the whole of the digital frequency range. This is a major limitation of the Euler method for filter design, indeed it is not even possible to discretize continuous-time HPF filters using this method as a consequence of this fundamentally limited frequency range.

By contrast, the bilinear method can be shown to have the frequency

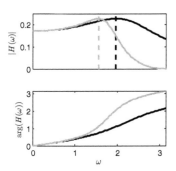

Fig. 7.7: TF illustration of the application of the bilinear transform discretization with $\Delta t = 1$. The continuous-time model is a bandpass filter (top: amplitude response, black curve), a simple conjugate-pair pole resonator, with a single peak frequency (at the dashed black line). The corresponding digital filter obtained by application of the bilinear transform has "warped" frequency and phase response (grey curves and dashed line at new peak frequency) since the bilinear transform maps the whole of the infinite imaginary frequency axes (continuous domain) onto the unit circle (discrete domain).

Actually, it is straightforward to demonstrate, using calculus, that the (positive) peak frequency of this filter is located at $\phi = \sqrt{\beta^2 - \alpha^2}$, at which the peak amplitude is $\left[(4\beta^2)(\beta - \phi)(\beta + \phi) \right]^{-2}$, so frequency and amplitude are coupled. But this information can be used to create a filter with truly independently controllable peak frequency and gain.

mapping $2 \arctan(\omega \Delta t/2)$ which maps $\mathbb{R} \to [-\pi, \pi]$ (Proakis and Manolakis, 1996, page 678). Thus, this method preserves the full range of zero and pole frequencies possible. The resulting arctangent mapping is close to linear around 0 frequency, and becomes increasingly nonlinear as $\omega \to \infty$. Comprehensive properties of both Euler and trapezoidal discretizations are summarized in Table 7.3. For both Euler and bilinear methods, higher order derivatives in the continuous-time model s^k where k is the derivative order, are mapped simply by making the substitution inside the power. For example, s^k is mapped to $\frac{1}{\Delta t^k} \left(1 - z^{-1} \right)^k$ using the Euler method.

As a practical example of discretization, let us look at using the bilinear transform to map a simple *resonator* or *bandpass filter* – that is, a filter which can selectively enhance a specific band of frequencies – onto a corresponding digital version. In principle, a single complex pole is all that is required to create such a resonator, but to ensure that the input/output is real-valued, we have to use a complex-conjugate pair of poles. The TF for this simple continuous-time resonator is:

$$H(s) = \frac{1}{(s-p)(s-\bar{p})} \tag{7.109}$$

Writing the pole as $p = \alpha + i\beta$ gives roughly independent control over the frequency ($\beta \in \mathbb{R}$) and amplitude near that frequency ($\alpha \in \mathbb{R}$), which can be chosen positive without loss of generality. Application of the bilinear mapping with $\Delta t = 1$ produces a corresponding digital resonator (Figure 7.7) with a single peak. As can be seen, at low frequencies the continuous-time and discrete-time filter TFs agree, and as the frequency increases, they begin to differ in both amplitude and phase response substantially. For this particular filter, since there is a single peak, it is possible to choose Δt to map the peak frequency ϕ exactly (provided $\phi < \pi$) by setting $\Delta t = (2/\phi) \tan(\phi/2)$, but we cannot avoid the inevitable distortion to the TF which is a consequence of the discretization error.

Digital filter recipes

There are a vast and fascinating array of digital filters for all manner of applications, it is not possible to detail them all here, so we describe the main categories and some criteria for choosing between them. Many filter designs start out life as LPFs, and there are straightforward transformations in the discrete variable z for converting these to HPFs or bandpass filters (Proakis and Manolakis, 1996, section 8.4). Thus, we can organize the "zoo" somewhat by picking an LPF design which satisfies certain TF characteristics, such as whether the amplitude is flat over a desired frequency region and the slope of the amplitude decrease (*transition*) around the LPF cutoff frequency.

For example, the Butterworth filter described earlier has *maximally flat* amplitude response nearly up to the cutoff frequency, which is a desirable property, above the cutoff frequency the amplitude response does not transition very rapidly with increasing frequency. By contrast

(*Type I*) *Chebyshev* filters, for the same number of poles, have amplitude "ripple" or fluctuations at frequencies below the cutoff frequency, but the amplitude decreases much more quickly than Butterworth filters above the cutoff frequency, which is also desirable (Proakis and Manolakis, 1996, page 683). Alternatively, *elliptic* filters have fluctuating amplitude response either side of the cutoff frequency, but for a given number of poles have optimally rapid transition around the cutoff.

IIR filters, while in general far more computationally tractable than long FIR filters, always involve a trade-off between rate of TF amplitude transition around the cutoff frequency (which is, typically, increased in rate with increasing number of filter poles) and the extent of nonlinearity in the phase response. A simple trick (*forward-backward filtering*) which can be used in situations where access to the entirety of a finite-length signal is available, is to run the IIR filter in the forwards direction of increasing time index to the end of the signal, and then run the output of this filtering operation backwards through the filter again to undo the nonlinear time delay. Care must be taken to deal with the fact that the amplitude response will be squared as a result. Of course, this might end up leading to more intensive computational effort than simply using an appropriate, linear phase FIR filter.

It suffices to say that nonlinear phase distortion is often an issue that effectively rules out the use of IIR filtering in many practical applications (for instance, in re-sampling or sample rate conversion applications, see Section 8.1, or in scientific signal processing applications where structures in a signal of different frequency must remain synchronized), but in some cases this property can actually be usefully exploited. So-called *all-pass* filters are deliberately designed such that the amplitude response is flat, with a specific delay curve in the phase response. These can be designed through judicious choice of mutually "cancelling" poles and zeros (Proakis and Manolakis, 1996, page 350).

Another extremely useful, but simple filter is the *comb filter*, so-called because the amplitude response resembles a comb with regularly spaced "teeth". There are two types of basic digital comb filter: FIR and IIR. The FIR form has TF $H(z) = 1 - a z^{-D}$ where $D \geq 1$ (and usually, $0 < a < 1$). This polynomial has D regularly spaced zeros around the unit circle, $z_k = a^{1/D} \exp(2\pi i \, k/D)$ for $k = 0, 1, \ldots, D-1$ which give rise to the teeth. Similarly, the corresponding IIR version has TF $H(z) = (1 - a z^{-D})^{-1}$ with D equally spaced poles. The FIR form can be used to effectively cancel out *harmonics* of unwanted interference, for example, AC power-line noise bleeding into an electrocardiogram signal in a hospital setting. Combining the two can produce combs with increased frequency selectivity:

$$H(z) = \frac{1 - z^{-D}}{1 - a z^{-D}} \tag{7.110}$$

where as $a \to 1$, the combs become arbitrarily "sharp". Note that since D is integer, for a fixed sampling interval Δt, comb filtering cannot be ideally tuned to an infinity of possible frequencies. One simple solution to this problem is to create effective *fractional* delay by linearly interpolating between z^{-D} and $z^{-(D+1)}$, for instance with the transfer function $H(z) = 1 - a\,z^{-D} - (1-a)\,z^{-(D+1)}$. This departs from the ideal comb response and the location of the zeros cannot be computed analytically in general. An alternative approach is to solve for a least-squares FIR filter design, with the desired fractional delay properties (Pei and Tseng, 1998), but this gives up the simplicity and $O(1)$ computational efficiency of the above, simple comb design.

Finally, comb filtering has a very interesting application in so-called *waveguide physical synthesis*, used to simulate various physical wave phenomena. Basically, for any physical system which can be represented using wave physics (such as linear acoustics), the delay property of the comb filter is an accurate, discrete representation of wave transport. So, for example, a plucked string instruments sets up pairs of counter-travelling waves which carry the pluck disturbance along the string and reflects them back at the string mounts. These waves interfere with each other and create the harmonic oscillation which is at a frequency determined by the wave speed in the string and the distance between the mounts. Therefore, a surprisingly accurate, computationally cheap simulation of most string instruments (guitar, piano etc.) can be based on the cascaded use of simple comb filters such as (7.110). The book by Smith (2010) contains an excellent, in-depth exposition of this concept and related ideas, while Proakis and Manolakis (1996) is a comprehensive source of techniques for general digital filter design.

Fourier filtering of very long signals

In the above sections, it was shown that while IIR filters can have small transition bands for low computational complexity, the inevitably non-linear phase characteristics make linear-phase FIR filters desirable in practice. However, this comes at a computational cost, requiring order $O(N\,Q)$ operations where N is the signal length and Q the FIR filter length. In order to achieve small transition bands, FIR filters need to have long impulse responses to avoid the Gibb's phenomenon, which means that Q must be large.

To gain the advantages of both linear-phase and small transition bands at low computational cost, filtering in the Fourier domain is a viable alternative, since one can leverage the computational advantages of an FFT algorithm. Since the FIR filter is a convolution operation $\boldsymbol{y} = \boldsymbol{b} \star \boldsymbol{x}$, it follows from the properties of the DTFT, that $\boldsymbol{y} = F_n^{-1}[\boldsymbol{B} \circ \boldsymbol{X}]$, using the DTFT of the impulse response of the FIR filter $\boldsymbol{B}$ and the signal $\boldsymbol{X}$. However, in practice the signal $\boldsymbol{x}$ is finite length, which means that we are restricted to using the DFT, computed efficiently with an FFT algorithm. This in turn means that the frequency domain product $\boldsymbol{B} \circ \boldsymbol{X}$ corresponds to the DFT of the *circular* convolution $\boldsymbol{b} \star_N \boldsymbol{x}$, rather than the required *non-circular* FIR filter convolution.

This problem, of using circular convolutions to implement non-circular convolution, can be solved by *zero-padding*, that is, by appending an appropriate number of zeros to the two sequences. Specifically, if the signal x has length N and the signal y has length M, and we treat this situation as if $x_n = 0$ for $n < 1$ and $n > N$ and $y_m = 0$ for $m < 1$ and $m > M$, then $x \star y = z$ is a signal such that $z_l = 0$ for $l < 1$ and $l > N+M-1$. We can then say that z has length $L = N+M-1$. Now, extend x to length L by appending $M - 1$ zeros, and similarly extend y to length L by appending $N - 1$ zeros, then $x \star_N y = x \star y = z$ (Proakis and Manolakis, 1996, section 5.3). So, $z = F_n^{-1}[X \circ Y]$ as required.

This leads to a simple algorithm for fast FIR filtering. Zero-pad vector b in (7.69) by appending $N - 1$ zeros to create the length $N + Q$ signal $h' = (b_0, b_1, \ldots, b_Q, 0, \ldots, 0)$. Next, zero pad the input signal x with Q zeros to create a new signal x' of length $N + Q$. Compute the DFTs H' of h' and X' of x' using an appropriate FFT. Compute $Y' = X' \circ H'$ and invert this result using an appropriate inverse FFT, to obtain $y' = F_n^{-1}[Y'] = b \star x$ as required. The computational cost (for radix 2 FFTs) is $3(N + Q)\log_2(N + Q) + N + Q$, which behaves as $O(N \log_2 N)$ provided $Q < N$. So, if Q is substantially less than $\log_2 N$, there is a computational saving to be obtained over naive FIR filtering.

A signal of the same length as the input can be obtained by truncating this, $y = (y_1, y_2, \ldots, y_N)$.

The *overlap-add* algorithm offers further reductions in computational cost by breaking the signal up into a number of smaller blocks and circularly convolving each block separately in the frequency domain, before combining them to reconstruct the output. Assume, for the sake of clarity, that the length of the input signal satisfies $N = KL$ so that we can define K new signals of length N but with non-zero segments of length L, defined as:

$$x_n^k = \begin{cases} x_n & (k-1)L < n \le kL \\ 0 & \text{otherwise} \end{cases} \tag{7.111}$$

for $k = 1, 2, \ldots K$. Then x can be written as a superposition of these new signals:

$$x_n = \sum_{k=1}^{K} x_n^k \tag{7.112}$$

Now, applying the FIR convolution to this superposition, we get:

$$y = b \star x = b \star \sum_{k=1}^{K} x^k$$
$$= \sum_{k=1}^{K} b \star x^k = \sum_{k=1}^{K} y^k \tag{7.113}$$

Explained another way, since convolution is a linear operation, the convolution of the entire signal with b is equivalent to the sum of the convolution of each of the K separate signals with b. This is known as the *overlap-add method* for convolution of long signals, for the following reason: the output signals y^k, like the x^k, have the property that $y_n^k = 0$ for $n \le (k-1)L$. However, the non-zero segment of each is extended by

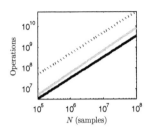

Fig. 7.8: Number of operations required to compute precision FIR filter convolution of length $Q = 500$, for long input signals of different lengths N. Brute-force FIR implementation (blue dotted line) takes the longest, and the computational effort is reduced by about half by using FFT-based convolution instead (grey line) although the improvement decreases for longer length signals. However, the overlap-add method (Algorithm 7.2) with block length $L = 1000$ achieves more than an order of magnitude improvement in computational effort (black line).

Q samples due to the convolution. So, while the non-zero segments in (7.112) do not overlap, the non-zero segments of y^k do. In practice, this means that we must add together the last Q samples in each non-zero output segment with the start of the following one.

This construction may not seem advantageous, nonetheless, we gain substantial computational efficiency because the non-zero segment of each y^k are now easily computed using the circular convolution implemented with an appropriate FFT, as above, see Algorithm 7.2.

Of course, we only need to compute H', the $L+Q$ zero-padded DFT of b, once at the start, thereafter we need K FFT-IFFT operations and K, $L+Q$ multiplications. For radix 2 FFT size, the computational complexity is therefore approximately $K((L+Q)\log_2(L+Q) + L + Q)$, and if Q is much smaller than L, this behaves as $O(N \log_2 L)$. By contrast to the simple FFT-based method above which computes only one set of long FFTs, provided L is much less than N, the savings can be significant in practice (Figure 7.8).

Note that of course, we do not actually need an FIR filter parameter vector, we can specify the frequency response of the filter directly in the Fourier domain to determine the spectrum H. Thus, the overlap-add method forms the basis of many so-called *direct Fourier filtering* DSP applications.

Algorithm 7.2 Overlap-add FIR convolution.

(1) *Initialization.* Set block size L, input signal x of size N samples, zero-pad $Q+1$-length FIR parameter vector b up to length $L + Q$ to obtain $h' = (b_0, b_1, \ldots, b_Q, 0, \ldots, 0)$, compute FFT H', set zero output signal y of length $N + Q$, set sample index $n = 1$,

(2) *Select input block.* Extract input block and zero pad with Q samples, to obtain $x' = (x_n, x_{n+1}, \ldots, x_{n+L-1}, 0, \ldots, 0)$, compute FFT X',

(3) *Block convolution.* Compute the circular convolution $y' = F_n^{-1}[H' \circ X']$. Overlap-add to output signal $y_{n+i-1} \leftarrow y_{n+i-1} + y'_i$ for $i = 1, \ldots L + Q$,

(4) *Iteration.* Update $n \leftarrow n + L$, go back to step 2 for $n < N$.

Kernel regression as discrete convolution

In Section 6.4, it was demonstrated that nonparametric kernel regression involves forming predictions as a weighted sum of the input covariates. If the covariate Z and the output X are both one-dimensional "time-like" variables and discretized on a uniform sampling grid, i.e. $z_n = n\Delta z$ where Δz is the discretization interval, and the regression kernels $\kappa\left(z;z_n\right)$ depend only upon the differences between z and z_n, i.e. $\kappa\left(z;z_n\right)=\kappa\left(z-z_n\right)$, then kernel regression is just a discrete convolution obtained at each sampling point n:

$$
\begin{aligned}
\hat{x}_n &= \frac{\sum_{m\in\mathbb{Z}}\kappa\left(\left[n-m\right]\Delta z\right)x_m}{\sum_{m'\in\mathbb{Z}}\kappa\left(\left[n-m'\right]\Delta z\right)} \\
&= \sum_{m\in\mathbb{Z}}h_{n-m}x_m = \boldsymbol{h}\star\boldsymbol{x}
\end{aligned}
\tag{7.114}
$$

where $\hat{x}_n$ is the regression prediction at sampled point z_n, using the discrete filter with impulse response:

$$
h_n = \frac{\kappa\left(n\Delta z\right)}{\sum_{n'\in\mathbb{Z}}\kappa\left(n'\Delta z\right)}
\tag{7.115}
$$

We can now use TF analysis to understand (uniformly-sampled) kernel regression from a spectral (frequency) point of view. In particular, the simple box kernel $\kappa\left(z\right)=\delta\left[\left|z\right|\le\sigma\right]$ with bandwidth $\sigma>0$ which performs local averaging (or *moving averaging* in the time series analysis context) has the TF:

$$
H\left(\omega\right) = \frac{\sin\left(\omega\left(\frac{\sigma}{\Delta z}+\frac{1}{2}\right)\right)}{\left(2\frac{\sigma}{\Delta z}+1\right)\sin\left(\frac{1}{2}\omega\right)}
\tag{7.116}
$$

This filter acts as an LPF, smoothing away details and leaving the larger-scale (lower-frequency) fluctuations in the input (Figure 7.9). For a fixed discretization interval Δz, the bandwidth σ controls the extent of the smoothing; the effective cutoff frequency of the LPF is inversely proportional to σ, therefore as σ increases, smaller length scale fluctuations are more attenuated. Similarly, the widely-used Gaussian kernel $\kappa\left(z\right)=\exp\left(-\sigma^{-2}z^2\right)$ has an approximately Gaussian amplitude response (Figure 7.9).

This TF perspective on kernel regression tells us that unlike local average regression, Gaussian kernel regression does not exhibit ripples in the amplitude response; as a form of averaging therefore, the Gaussian kernel would seem to be superior to the box kernel. But in practice (as for all infinite impulse response kernels) one has to truncate the kernel. For the above kernels which are symmetric, this truncation creates a linear-phase FIR filter which keeps features at different length scales in the input from being shifted in location, which is clearly an important consideration to produce a meaningful regression. Nonetheless, the truncation induces unavoidable Gibbs phenomena in the amplitude response which are worsened as the kernel is truncated more severely. This suggests that we make the FIR filter as long as possible, but this will increase the computational effort required.

Similarly, we cannot avoid Heisenberg uncertainty. While the LPF regression kernels above can smooth away irrelevant, random fluctuations if we make σ large, this will also smear out the location of any

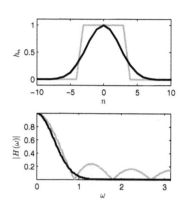

Fig. 7.9: Impulse response (top) and transfer function (amplitude response, bottom) of discretized kernel regression with Gaussian (black curves) and box (local average, grey curves) kernels. Both use bandwidth parameter $\sigma=0.7$ and discretization interval $\Delta z=0.2$. Here, the frequency variable ω corresponds to the length scale of the regression variables X,Z.

important features which are represented in the smaller scale (higher frequency) detail of the input data. Thus we can see that (equispaced) kernel regression, as an example of an LTI method, suffers from all the inherent trade-offs of LTI signal processing. Finally, we can construct HPF regression kernels which attenuate larger-scale features and amplify small-scale features.

We will explore further connections between DSP and Bayesian regression in the next section, but on this topic see also Candès (2006).

7.4 Exploiting statistical stability for linear-Gaussian DSP

The rather special properties of the multivariate Gaussian: invariance under affine transformations, and Gaussian marginals and conditionals with closed-form mean and covariance expressions, can be combined with LTI systems to make a very mathematically complete and useful set of results on which to build a theory of *stochastic linear DSP*, which is the topic of this section.

Discrete-time Gaussian processes (GPs) and DSP

In this section we will only be concerned with discrete-time processes; we will make use of continuous-time processes in later chapters which use the GP model for purposes such as regression.

Consider a countably infinite set of RVs X_n, $n \in \mathbb{Z}$ which we will denote as the infinite random vector $\boldsymbol{X}$, having the unique property that all the f.d.d.s are multivariate Gaussian. The marginalization property of the multivariate Gaussian, explored above, implies that these f.d.d.s are sufficient to define the stochastic process, known as a (discrete-time) *Gaussian process* (GP). As it based upon the multivariate Gaussian, mean and autocovariances suffice to fully characterize the process. From the statistical stability property of the Gaussian, we can predict that if a GP is input to an LTI digital filter, the output will be another GP. Furthermore, we can exactly predict the effect on the mean and autocovariance functions of the process.

We need some useful definitions for discrete-time processes. For stochastic signals, the cross-correlation is defined as a second statistical moment and is a fundamental quantity in stochastic DSP, $r_{\boldsymbol{XY}}(\tau) = E[X_n Y_{n-\tau}]$, and the autocorrelation is a special case, $r_{\boldsymbol{XX}}(\tau) = E[X_n X_{n-\tau}]$. A process is *weakly stationary* if the mean and autocorrelation is time translation-invariant, so that only the difference in time τ matters. Along with the mean function, the autocorrelation fully characterizes a GP. Autocorrelation is symmetric, $r_{\boldsymbol{XX}}(\tau) = r_{\boldsymbol{XX}}(-\tau)$, and bounded, $r_{\boldsymbol{XX}}(\tau) \in [-r_{\boldsymbol{XX}}(0), r_{\boldsymbol{XX}}(0)]$. By the Wiener-Khintchine theorem, the DTFT of the autocorrelation is the square of the magnitude of the FT, $S_{\boldsymbol{XX}}(\omega) = |X(\omega)|^2 = F_\omega[r_{\boldsymbol{XX}}(\tau)]$, which is known as the *power spectral density* (PSD, a.k.a. *power spectrum*) of the process.

The simplest GP is the zero mean, i.i.d. process in which each element of the process has distribution $X_n \sim \mathcal{N}(0, \sigma^2)$. Because any two time instances of the process are uncorrelated, this has the trivial autocorrelation function $r_{\boldsymbol{XX}}(\tau) = \sigma^2 \delta[\tau]$. The corresponding PSD is just $S_{\boldsymbol{XX}}(\omega) = \sigma^2$. Any signal which has a *flat* PSD like this is called 'white noise', and it is of such importance that we will give this signal the special notation $\boldsymbol{W}$ with elements W_n.

Given a signal with a particular PSD $S_{\boldsymbol{XX}}(\omega)$, it is very useful to know what happens to this signal when it is input to a general (stable) IIR filter with TF (7.69). The answer turns out to be quite simple (Wornell and Willsky, 2004, Chapter 4):

$$
\begin{aligned}
S_{\boldsymbol{YY}}(\omega) &= H(\omega) H(-\omega) S_{\boldsymbol{XX}}(\omega) \\
&= |H(\omega)|^2 S_{\boldsymbol{XX}}(\omega) \qquad (7.117)
\end{aligned}
$$

where the second line simply follows from the fact that $\exp(-i\omega) = \overline{\exp(i\omega)}$ and $a\bar{a} = |a|^2$ for $a \in \mathbb{C}$. Intuitively: the output is obtained by convolving the input signal with the impulse response of the filter, and in the frequency domain this convolution is multiplication. The fact that this only constrains the *magnitude* of the TF follows from the fact that the autocorrelation is time-invariant, containing only time-delayed covariance information which is a statistical expectation averaging over all time. It follows that for a white noise input $\boldsymbol{W}$, the output PSD is just $S_{\boldsymbol{YY}}(\omega) = \sigma^2 |H(\omega)|^2$. Thus, the stable digital LTI filter applied to discrete-time white noise acts to 'shape' the spectrum to the square magnitude of the TF of the filter.

This raises other possibilities. If the input signal $\boldsymbol{X}$ has the PSD $|H(\omega)|^2$, then we can run this through the *inverse filter* with TF $H(z)^{-1}$ which, for rational TFs, swaps the role of poles and zeros, and will *decorrelate* the input signal, thereby 'flattening' the spectrum. This inverse filter is known as a *whitening* filter, whereas the corresponding forward filter is known as the *synthesis* filter.

Since we have a fully probabilistic model of the signal, we can use this to perform optimal estimates of the GP parameters. For example, in the trivial case of the non-zero mean i.i.d. white noise of finite length N with mean μ, the MLE $\hat{\mu}$ for the parameter is just the sample mean $\hat{\mu} = \frac{1}{N} \sum_{n=1}^{N} x_n$. A similar argument applies for the variance. Much more interesting is the case of the non-i.i.d. GP for which $X_n \sim \mathcal{N}\left(\mu_n, \sigma^2\right)$. Now, the MLE for the mean signal is:

$$
\begin{aligned}
\hat{\mu}_n &= \operatorname*{arg\,min}_{\mu_n} \left(-\sum_{n=1}^{N} \ln \mathcal{N}\left(x_n; \mu_n, \sigma^2\right) \right) \\
&= \operatorname*{arg\,min}_{\mu_n} \left(\sum_{n=1}^{N} (x_n - \mu_n)^2 \right) \qquad (7.118) \\
&= \operatorname*{arg\,min}_{\mu_n} \left((x_n - \mu_n)^2 \right) = x_n
\end{aligned}
$$

Of course, such an estimator is very poor since it is based on a single realization and has variance σ^2 which cannot be improved.

We can do much better for other GPs. Consider the situation where the mean is "slowly varying" such that $\mu_{n-\tau/2} \approx \mu_n \approx \mu_{n+\tau/2}$ for all n, for some time delay $\tau > 0$ and even. In which case, we can assume that the signal is i.i.d. in the length $\tau + 1$ time windows $\left[n - \frac{\tau}{2}, n + \frac{\tau}{2}\right]$. Under these assumptions, the (approximately) optimal parameter estimator is the mean in each window, e.g. $\hat{\mu}_n = \frac{1}{\tau+1} \sum_{m=n-\tau/2}^{n+\tau/2} x_n$. This is known as the *moving* (or *running*) *average* (MA) FIR filter with $Q = \tau$ coefficients and $b_k = Q^{-1}$. This collapses down to the non i.i.d. case above when $\tau = 0$. The variance of this estimator is $\sigma^2/(\tau+1)$, from which we can see that as τ increases, the estimator becomes more certain (reliable). However, there is clearly an inherent trade-off between responsiveness and variance. If τ is small, then the moving average can respond quickly to changes in the mean, but because the variance will be large, the estimate will be unreliable. In practice, typically we do not know the ideal τ, so it becomes a matter of choice of trade-off in any one application.

A more complex task is that of estimating the parameters of a digital IIR or FIR TF which we will look at next. Before doing this however, it will help to explain an important simplification in linear, minimum mean-square estimation. This simplification is the fact that the difference between the MLE for a linear-Gaussian model and the inputs to that linear model are orthogonal. To make this statement precise, note, as discussed in Section 4.4, the square error is (up a constant) the NLL for linear-Gaussian models. Now, consider the sum of square errors for the model $Y = \boldsymbol{a}^T \boldsymbol{X}$ for $\boldsymbol{a}, \boldsymbol{X} \in \mathbb{R}^P$, which is $E = E_{Y,\boldsymbol{X}}\left[\left(Y - \boldsymbol{a}^T \boldsymbol{X}\right)^2\right]$, then the gradient of the expected square error with respect to each a_k is:

$$
\begin{aligned}
\frac{\partial E}{\partial a_k} &= \frac{\partial}{\partial a_k} E_{Y,\boldsymbol{X}}\left[\left(Y - \boldsymbol{a}^T \boldsymbol{X}\right)^2\right] \\
&= E_{Y,\boldsymbol{X}}\left[\frac{\partial}{\partial a_k}\left(Y - \boldsymbol{a}^T \boldsymbol{X}\right)^2\right] \qquad (7.119) \\
&= -2 E_{Y,\boldsymbol{X}}\left[\left(Y - \boldsymbol{a}^T \boldsymbol{X}\right) X_k\right]
\end{aligned}
$$

so that at the global minimum of E we have $E_{Y,\boldsymbol{X}}[\epsilon X_k] = 0$ for $k = 1, 2, \ldots, P$ for the *error process* $\epsilon_n = Y_n - \boldsymbol{a}^T \boldsymbol{X}_n$. In other words, the error process is orthogonal to the input signals X_k for each k. This is known as the *orthogonality principle* which is a powerful shortcut for linear-Gaussian MLEs, as we shall see. We can also show that the implication works the other way: if the orthogonality condition above holds, then $Y = \boldsymbol{a}^T \boldsymbol{X}$ minimizes the sum of square errors E (Vaidyanathan, 2007, Appendix A.1).

If $\boldsymbol{X}$ is zero-mean, then in addition, the input variables are uncorrelated with the error.

Let us now try to estimate the parameters of a digital FIR TF with white noise input $\boldsymbol{W}$ of variance σ^2. The model is:

$$Y_n = \sum_{k=0}^{Q} b_k W_{n-k} \qquad (7.120)$$

where we choose $a_0 = 1$ without loss of generality. Now, using the orthogonality principle, the MLE satisfies:

$$E\left[\left(Y_n - \sum_{j=0}^{Q} \hat{b}_j W_{n-j}\right) W_{n-k}\right] = 0 \qquad (7.121)$$

for $k = 0, 1, \ldots, Q$ from which it follows that:

$$E\left[Y_n W_{n-k}\right] = \sum_{j=0}^{Q} \hat{b}_j E\left[W_{n-j} W_{n-k}\right] \qquad (7.122)$$

and the solution is $\hat{b}_k = r_{\boldsymbol{YW}}(k)$. The simplification of the right hand side above is due to $E\left[W_{n-j} W_{n-k}\right] = r_{\boldsymbol{WW}}(j-k) = \delta[j-k]$. Thus, the FIR MLE parameters are obtained in practice by computing $Q+1$ empirical cross-correlation estimates.

We can similarly apply this principle to the IIR TF (7.67) with white noise input of variance σ^2 and no FIR component (so, $Q = 0$ and $b_0 = 1$), and $P \geq 1$ with $a_0 = 1$ (without loss of generality). Then the model is:

$$Y_n + \sum_{j=1}^{P} a_j Y_{n-j} = W_n \qquad (7.123)$$

and we will multiply this by $Y_{n-\tau}$ for $\tau = 0, 1, \ldots, P$ and then take expectations over the process $\boldsymbol{Y}$ to get a set of implicit equations for the MLE $\hat{\boldsymbol{a}}$:

$$E\left[Y_n + \sum_{j=0}^{P} \hat{a}_j Y_{n-j}\right] Y_{n-\tau} = E\left[W_n Y_{n-\tau}\right]$$

$$E\left[Y_n Y_{n-\tau}\right] + \sum_{j=0}^{P} \hat{a}_j E\left[Y_{n-j} Y_{n-\tau}\right] = E\left[W_n Y_{n-\tau}\right] \qquad (7.124)$$

which is:

$$r_{\boldsymbol{YY}}(\tau) + \sum_{j=0}^{P} \hat{a}_j r_{\boldsymbol{YY}}(\tau - j) = r_{\boldsymbol{WY}}(\tau) \qquad (7.125)$$

Now, using $Y_{n-\tau} = W_{n-\tau} - \sum_{j=1}^{P} a_j Y_{n-j-\tau}$ we get:

$$
\begin{aligned}
r_{WY}(\tau) &= E\left[W_n\left(W_{n-\tau} - \sum_{j=1}^{P} a_j Y_{n-j-\tau}\right)\right] \\
&= E[W_n W_{n-\tau}] - \sum_{j=1}^{P} a_j E[W_n Y_{n-j-\tau}] \quad (7.126) \\
&= \sigma^2 \delta[\tau] - \sum_{j=1}^{P} a_j r_{WY}(j+\tau)
\end{aligned}
$$

and since $j + \tau > 0$ it must be that $r_{YW}(j+\tau) = 0$ so that $r_{WY}(\tau) = \sigma^2 \delta[\tau]$. This gives us the following equations for the MLE IIR parameters:

$$
r_{YY}(\tau) + \sum_{j=1}^{P} \hat{a}_j r_{YY}(\tau - j) = \sigma^2 \delta[\tau] \quad (7.127)
$$

which, for $\tau \geq 1$ is conveniently written in matrix form:

$$
\begin{pmatrix}
r_{YY}(0) & r_{YY}(1) & \cdots & r_{YY}(P-1) \\
r_{YY}(1) & r_{YY}(0) & \cdots & r_{YY}(P-2) \\
\vdots & \vdots & \ddots & \vdots \\
r_{YY}(P-1) & r_{YY}(P-2) & \cdots & r_{YY}(0)
\end{pmatrix}
\begin{pmatrix}
\hat{a}_1 \\ \hat{a}_2 \\ \vdots \\ \hat{a}_P
\end{pmatrix}
= - \begin{pmatrix}
r_{YY}(1) \\ r_{YY}(2) \\ \vdots \\ r_{YY}(P)
\end{pmatrix}
\quad (7.128)
$$

along with $\hat{\sigma}^2 = r_{YY}(0) + \sum_{j=1}^{P} \hat{a}_j r_{YY}(j)$ for the $\tau = 0$ case. We can write this equation as $\mathbf{R}_{YY}\hat{a} = -r_{YY}$, then $\mathbf{R}_{YY}$ is known as an "autocorrelation matrix", and r_{YY} could be called a specific "autocorrelation vector".

It should be noted that since the above equation arises from a linear regression problem, this is just a special case of the normal equations (2.16) which we can write in the form $(\mathbf{Y}^T\mathbf{Y})\hat{a} = \mathbf{Y}^T T_1[y]$ where $\mathbf{Y}$ is the $N \times P$ matrix formed with columns containing the signal y time-delayed by $0, 1, 2, \ldots P - 1$ samples respectively. The matrix $\mathbf{Y}^T\mathbf{Y}$ is known as the (empirical) *autocorrelation matrix*. The time-delayed, autoregressive nature of the origins of this matrix makes it a Toeplitz matrix, the inversion of which can be computed using the $O(P^2)$ Levinson recursion (Proakis and Manolakis, 1996, section 11.3.1).

The above results can be derived as special cases of a fundamental idea in linear-Gaussian DSP known as linear minimum mean square filtering or Wiener filtering (after Norbert Wiener, an influential 20th century mathematician).

The IIR MLE estimates above go by the name of *linear prediction analysis* (LPA) and this has practical and widespread usage in DSP. For example, LPA is used as a fundamental component in *digital speech coding* for voice communications, representing the so-called *source-filter theory* of speech production, where the input to the filter W_n is considered as the "excitation source" (the major sources of sound energy) and the TF described by the model parameters $\boldsymbol{a}, \sigma^2$ is used to represent the "resonances" (controlled by the shape of the vocal tract which changes when articulating different speech sounds). See Section 8.3.

The ARMA case (for both $P, Q \geq 1$) is, unfortunately, nonlinear with no simple closed-form solution as in the AR or MA cases, thus requiring numerical optimization (Proakis and Manolakis, 1996, page 856).

Nonparametric power spectral density (PSD) estimation

We are very often presented with a signal x_n, $n = 1, 2, \ldots, N$ that we assume is a realization of a (zero-mean, weakly stationary) GP X_n with a specific PSD. Then the question arises of how to estimate the PSD of the signal. One approach is to compute an estimate of the autocorrelation and then compute the DFT (efficiently, using the FFT) of this estimator, exploiting the Wiener-Khintchine theorem. For this we will need the length $2N - 1$ empirical autocorrelation expectation values:

$$\hat{r}_{\boldsymbol{XX}}(\tau) = \frac{1}{Z(N,\tau)} \begin{cases} \sum_{n=\tau+1}^{N} x_{n-\tau} x_n & \tau \in 0, 1, \ldots N-1 \\ \sum_{n=1}^{N-|\tau|} x_{n-\tau} x_n & \tau \in -1, -2, \ldots, -(N-1) \end{cases}$$

$$(7.129)$$

where $Z(N, \tau)$ is a normalization term, for which there are two obvious choices:

(1) For $Z(N,\tau) = N - |\tau|$, the estimator is unbiased since $E[\hat{r}_{\boldsymbol{XX}}(\tau)] = r_{\boldsymbol{XX}}(\tau)$ while the variance of the estimator increases as $|\tau| \to N$ since there are only $N - |\tau|$ terms in each summation above,

(2) Alternatively, for $Z(N,\tau) = N$ there is a finite length bias $E[\hat{r}_{\boldsymbol{XX}}(\tau)] = \left(1 - \frac{|\tau|}{N}\right) r_{\boldsymbol{XX}}(\tau)$, however, the variance is smaller than for the above estimator.

The FFT to the unbiased autocorrelation estimator is not guaranteed to avoid negative PSD estimates, and so is generally avoided in practice.

In the limit as $N \to \infty$, the variance of both estimators goes to zero and both estimators are unbiased. So, in practice there is a trade-off – we can get an unbiased autocorrelation estimate but at the expense of it being more variable than a biased one. Nonetheless, since computing the autocorrelation is an $O(N^2)$ operation, it will be more efficient to estimate the PSD directly from the signal using a single $O(N \ln N)$ FFT:

$$\hat{P}_{\boldsymbol{XX}} = \frac{1}{N} \left| \sum_{n=1}^{N} x_n \exp\left(-2\pi i k \frac{n-1}{2N-1}\right) \right|^2 \qquad (7.130)$$

This is such a well-used estimator that it has its own name, the *periodogram*. It has the expected value:

$$E\left[\hat{P}_{\boldsymbol{XX}}\right] = \sum_{\tau=-(N-1)}^{N-1} \left(1 - \frac{|\tau|}{N}\right) r_{\boldsymbol{XX}}(\tau) \exp\left(-2\pi i k \frac{\tau}{2N-1}\right)$$

(7.131)

which is biased because the value is obtained by first multiplying the autocorrelation with a *triangular window* signal $u_n = 1 - |n|/N$. The source of this window bias is the DFT in (7.130) above which is implicitly based on the $Z(N,\tau) = N$ autocorrelation estimator. It is instructive to see this bias in the frequency domain:

$$E\left[\hat{P}_{\boldsymbol{XX}}\right] = \frac{1}{2\pi} \int_{-\pi}^{\pi} P_{\boldsymbol{XX}}(\phi) P_{\boldsymbol{UU}}(\omega - \phi)\, d\phi \qquad (7.132)$$

$$= P_{\boldsymbol{XX}} \star_{2N-1} P_{\boldsymbol{UU}} \qquad (7.133)$$

which means that the power spectrum of the triangular window $P_{\boldsymbol{UU}} = |F_k[\boldsymbol{u}]|^2$ is convolved with the true PSD of the signal $P_{\boldsymbol{XX}}$. So, for the periodogram to be a faithful (unbiased) estimate of the true PSD $r_{\boldsymbol{XX}}(\tau)$, we must have $P_{\boldsymbol{UU}}(\omega) = \delta[\omega]$. Otherwise, the window spectrum "spreads out" any frequency features. The window power spectrum is (DeFatta *et al.*, 1988, section 6.7.1):

$$P_{\boldsymbol{UU}}(\omega) = \frac{1}{N}\left(\frac{\sin(N\omega/2)}{\sin(\omega/2)}\right)^2 \qquad (7.134)$$

While this does converge on the delta function for $N \to \infty$, for any finite N it has finite width, and this width increases as N becomes smaller. Thus, small N periodograms suffer from non-negligible bias which manifests itself as frequency spreading (Figure 7.10). This is a somewhat counterintuitive but unavoidable feature of any kind of practical nonparametric PSD estimation based on the direct periodogram estimate.

This variance result is derived under the assumption that the signal is a GP.

We cannot eliminate the frequency spreading bias of the periodogram, but we can address the variance of the estimator which is (Proakis and Manolakis, 1996, page 906):

$$\mathrm{var}\left[\hat{P}_{\boldsymbol{XX}}\right] = P_{\boldsymbol{XX}}^2(\omega)\left[1 + \left(\frac{\sin(N\omega/2)}{N\sin(\omega/2)}\right)^2\right] \qquad (7.135)$$

So, even taking $N \to \infty$ the variance of the periodogram does not go to zero which means that there is a fundamental limit to the reliability of this estimator – estimated PSD components will always differ substantially across realizations.

One widely used approach to reducing this variance is by sample averaging: divide the signal into $M = N/L$ non-overlapping signal segments $\boldsymbol{x}^1, \boldsymbol{x}^2, \ldots, \boldsymbol{x}^M$ of length L each, compute the periodogram of each $\hat{P}_{\boldsymbol{XX}}^m$, then form the sample average $\hat{P}_{\boldsymbol{XX}} = \frac{1}{M}\sum_{m=1}^{M} \hat{P}_{\boldsymbol{XX}}^m$. The result, known as *Barlett's method*, is an estimator with $1/M$ times the variance of the periodogram (Proakis and Manolakis, 1996, section 12.2.1). Clearly, we would want M to be large but this means the L becomes small, which increases the frequency spreading bias of the periodogram. Computationally, this method requires $O\left(N\left(1+\ln L\right)\right)$ operations, which is an improvement over the periodogram provided $L < 0.38N$.

Since each segment above is non-overlapping and therefore uncorrelated with the others, the more segments available, the smaller the variance in the estimate. So, if we were able to somehow increase the number of segments, we could reduce the variance still further. One way to do this is allow overlap between segments: this idea is central to the *Welch method* (Proakis and Manolakis, 1996, section 12.2.2). With, for example, 50% overlap and segment size $L = N/M$ we will have $J = 2M - 1$ segments. Provided the overlap is sufficiently small, the cross-correlation between segments will be sufficiently small, thereby approximately halving the variance of the Bartlett estimator. The Welch method also attempts to mitigate the spreading bias of the periodogram by *pre-windowing* the data with the window signal $\boldsymbol{u}$:

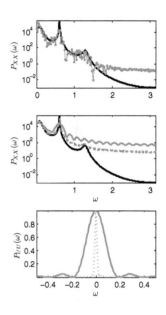

$$\hat{P}_{\boldsymbol{XX}}\left(k\right) = \frac{1}{JLU} \sum_{j=0}^{J-1} \left| \sum_{n=1}^{L} u_n x_{j\,L/2+n} \exp\left(-2\pi i k \frac{n-1}{2L-1}\right) \right|^2 \quad (7.136)$$

where $U = \frac{1}{L}\sum_{n=1}^{L} u_n^2$ is a normalization factor ensuring that the window does not bias the estimate. Aside from an additional normalization factor $1/U$, the expected value of the estimator is the same as (7.132) with $P_{\boldsymbol{UU}}$ the power spectrum of the window signal $\boldsymbol{u}$. The variance is harder to predict for arbitrary window signals but for the triangular window it is approximately $1/J$ times the periodogram variance (7.135), see (Proakis and Manolakis, 1996, page 913). The computational effort is $O\left(N\left(1+\ln L\right)\right)$, as with the Bartlett method.

Fig. 7.10: The top panel shows a given three-pole pair IIR filter driven by a Gaussian white noise source has PSD (black curve). The periodogram is used to estimate this PSD using direct DFT computation (size $N = 256$) from a realization of the IIR filter output signal x (top panel, grey dots). The finite-length signal bias manifests as spreading of the true PSD across the frequency scale (middle panel, grey curves), which is worsened as the DFT size is decreased (solid size $N = 32$, dotted size $N = 256$). The cause of this spreading is the width of the power spectrum of the implicit triangular windowing of the periodogram estimator (bottom panel, solid curve size $N = 32$, dotted $N = 256$) which is convolved with the true PSD in the frequency domain.

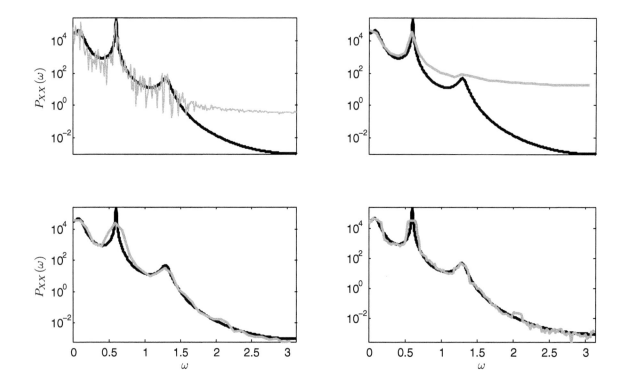

Fig. 7.11: Example behaviour of various nonparametric PSD estimators (grey curves), given the output of a known, three-pole pair IIR filter (analytically calculated power spectrum, black curves) applied to an $N = 512$ length realization of a zero-mean, Gaussian white noise signal. The periodogram (top left) with FFT size 512 has the highest variance of all estimators and substantial bias, particularly for high frequencies. By contrast Bartlett averaging (top right) with $M = 8$ segments of the signal with resulting FFT size 64, has much reduced variance but worse bias. The Welch method (bottom left) with 50% overlap and the same segment and FFT size addresses the high frequency bias using a Hann window, but the additional windowing smooths off the sharp response peak at $\omega = 0.6$. Finally, the prolate spheroid Thomson multi-taper estimate (bottom right) with $\omega_{\mathrm{c}} = 0.08$ (leading to $K = 9$ windows and FFT size 512) has little bias at high frequencies and sharper response peak, but with slightly increased variance by comparison to the Welch method.

If we are prepared to put up with computing the $O\left(N^2\right)$ autocorrelation estimate, then we can address the increased variance for large $|\tau|$ by windowing the autocorrelation $\hat{r}_{\boldsymbol{XX}}$. This is the basis of the *Blackman-Tukey* method which, when the window $\boldsymbol{u}$ is narrow by comparison to the $2N - 1$ autocorrelation function, has the effect of smoothing the periodogram estimate by convolving the true PSD of the signal with the Fourier transform of the window signal. To ensure positivity of the PSD estimate, the window function must be symmetric and its Fourier transform must be positive, which restricts the range of useful windows which can be used. The Blackman-Tukey method requires $O\left(N\left(N + \ln N\right)\right)$ operations. Thus, smoothing of the periodogram can arise naturally when mitigating poor autocorrelation function estimates. Extending this idea of periodogram smoothing in a different direction,

Wahba (1980) proposes spline smoothing the log periodogram function (in the process making use of the *cepstrum* which will be discussed in Chapter 9).

All the PSD estimators above suffer from unavoidable frequency spreading and do not address this issue in a systematic way. One point of view is that, while we cannot eliminate spreading altogether because these are finite-length signals, we can design windows to optimally "manipulate" the spreading. Measuring the *spectral concentration* of a signal in the frequency range $[-\omega_c, \omega_c]$ for $0 < \omega_c < \pi$ as $\int_{-\omega_c}^{\omega_c} P_{\boldsymbol{XX}}(\omega)\, d\omega$, we can specify the following optimization problem (Thomson, 1982):

$$
\text{maximize} \quad \frac{1}{2\pi} \int_{-\omega_c}^{\omega_c} P_{\boldsymbol{UU}}(\omega)\, d\omega \tag{7.137}
$$

$$
\text{subject to} \quad \frac{1}{2\pi} \int_{-\pi}^{\pi} P_{\boldsymbol{UU}}(\omega)\, d\omega = 1
$$

with respect to the window signal $\boldsymbol{u}$. In other words, find the window with largest spectral concentration in the region subject to a fixed total PSD of the entire window. The corresponding Lagrangian is:

$$
L(\boldsymbol{u}, \beta) = \frac{1}{2\pi} \int_{-\omega_c}^{\omega_c} P_{\boldsymbol{UU}}(\omega)\, d\omega + \beta \left(\frac{1}{2\pi} \int_{-\pi}^{\pi} P_{\boldsymbol{UU}}(\omega)\, d\omega - 1 \right) \tag{7.138}
$$

where β is the Lagrange multiplier. Assuming the window signals are real, we can express the power spectrum of the windows using the square magnitude of the DFT coefficients, leading to the following matrix-vector representation:

$$
L(\boldsymbol{u}, \lambda) = \boldsymbol{u}^T \boldsymbol{A} \boldsymbol{u} - \lambda \left(\boldsymbol{u}^T \boldsymbol{u} - 1 \right) \tag{7.139}
$$

where (without loss of generality) we rescale the multiplier $\lambda = -\beta/(2\pi)$ for convenience. The symmetric $N \times N$ matrix $\boldsymbol{A}$ has *Dirichlet kernel* elements:

$$
a_{nm} = \frac{\sin(\omega_c(n-m))}{\pi(n-m)} \tag{7.140}
$$

for $n, m \in 1, 2, \ldots N$. Differentiating the Lagrangian with respect to $\boldsymbol{u}$ we get the eigenvalue problem $\boldsymbol{A}\boldsymbol{u} = \lambda \boldsymbol{u}$. The N eigenvectors $\boldsymbol{u}^m$, $m = 1, 2, \ldots, N$ which solve this problem are known as *discrete prolate spheroidal* or *Slepian* functions. The corresponding eigenvalues have the property that $\lambda_l \approx 1$ for $l \in 1, 2, \ldots, \lfloor N\omega_c/\pi \rfloor$ and the rest are close to zero. Using these eigenvector windows, the *Thomson multi-taper* PSD estimate is formed (Thomson, 1982):

$$
\hat{P}_{\boldsymbol{XX}}(k) = \frac{1}{M} \sum_{m=1}^{M} \left| \sum_{n=1}^{N} u_n^m x_n \exp\left(-2\pi i k \frac{n-1}{N} \right) \right|^2 \tag{7.141}
$$

where $M = \lfloor N\omega_{\mathrm{c}}/\pi \rfloor$. The computational complexity is $O\left(M N \left(1 + \ln N\right)\right)$, which is less effort than Blackman-Tukey provided $M \ll N$ (which it usually will be in practice). In this estimate, each eigenvector acts as a window for the signal, and the average is formed over all M windows where $\lambda_k \approx 1$. Since the eigenvectors are all mutually orthogonal, then provided that the PSD does not vary rapidly over the interval $[\omega - \omega_{\mathrm{c}}, \omega + \omega_{\mathrm{c}}]$, each windowed estimate of the spectrum is orthogonal to every other estimate, so this method averages over M mutually independent PSD estimates (Thomson, 1982). This results in variance reduction by a factor of $1/M$ over the simple periodogram variance (7.135) as $N \to \infty$, but without the need to compromise on the frequency resolution due to the DFT size reduction in the Bartlett or Welch methods. The mean value of the estimator is (Thomson, 1982):

$$E\left[\hat{P}_{\boldsymbol{XX}}\right] = P_{\boldsymbol{XX}} \star_N \frac{1}{M} \sum_{m=1}^{M} |\boldsymbol{U}^m|^2 \qquad (7.142)$$

Often in practice M is chosen to be somewhat smaller than this because we want to use only the eigenvectors corresponding to the largest magnitude eigenvalues as part of the PSD estimate, as those comfortably satisfy the desired spectral concentration characteristic.

So, the true PSD is smoothed by the average power spectrum of the eigenvector windows, and the parameter ω_{c} controls the extent of this smoothing. Increasing this parameter reduces the variance but also the effective resolution. When M is chosen as above, then the average power spectrum approximates an ideal low-pass filter. Thus, the Thomson multi-taper PSD is, to a very close approximation, the estimate obtained by putting the true PSD through a frequency moving average filter of width $2\omega_{\mathrm{c}}$. This is not ideal for peaked PSDs where any sharp peaks will be smoothed away, but it is obviously well-suited to smooth PSDs. An alternative multi-taper approach, which minimize bias instead of spreading, uses sinusoids instead of Slepian functions and achieves similar performance but without the need to solve a potentially very large eigenvector/eigenvalue problem (Riedel and Sidorenko, 1995).

Parametric PSD estimation

The previous section developed several methods for nonparametric PSD estimation, the general finding is that at least one (smoothing) parameter is needed to provide a useful estimate. Thus, the primary value of nonparametric PSD estimation is that we can assume very little about the signal. However, there are situations where we can assume a parametric model for the signal and therefore make more accurate PSD estimates. The first such model is based on the all-pole AR (IIR) TF, and proposes the use of LPA and the corresponding MLE (7.128). This will work well for "peaked" spectra, and the order of the model P should be chosen to be twice the number of PSD maxima at non-zero frequencies (because the poles come in complex conjugate pairs for real filters), plus one additional pole each for any peaks at zero or Nyquist frequencies.

There are many distinct AR PSD estimation variants. In practice we have only a finite signal x_n for $n = 1, 2, \ldots, N$ so will need autocorrelation estimates for (7.128). The biased autocorrelation estimate, exploiting the symmetry $r_{\boldsymbol{XX}}(-\tau) = r_{\boldsymbol{XX}}(\tau)$ and the fact that the nor-

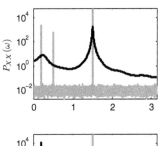

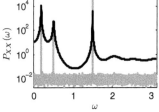

Fig. 7.12: Performance of two parametric PSD estimators (black curves) of a sample signal ($N = 128$) from three sinusoids (radian frequencies $\phi = 0.2$, 0.5 and 1.5, amplitudes $r = 1.0$, 0.5 and 5.0) with additive i.i.d. Gaussian noise (standard deviation $\sigma = 0.1$). For comparison, empirically estimated PSD using a high-resolution ($N = 10^5$) sinusoidal multitapered PSD is shown in each panel (grey curves). The Yule-Walker method (top) with $P = 12$ smears out the peaks, the covariance method substantially improves the peak resolution (bottom).

malization constant $Z(N, \tau) = N$ is the same across all autocorrelation values, can be computed using:

$$\hat{r}_{XX}(\tau) = \sum_{n=1}^{N-\tau} x_n x_{n+\tau} \tag{7.143}$$

for $\tau = 0, 1, \ldots P$. With these autocorrelations, the resulting PSD estimate is usually called the *Yule-Walker* PSD (Figure 7.12). While simple, it makes implicit use of the rectangular window, and other window functions u can be used. Windowed Yule-Walker methods usually have improved spectral resolution (Figure 7.12). Alternatively, explicitly minimizing the squared error over the finite length signal $E(a) = \sum_{n=P+1}^{N} \left(x_n + \sum_{j=1}^{P} a_j x_{n-j} \right)^2$ gives rise to the following autocorrelation estimates (Makhoul, 1975):

$$\hat{r}_{XX}(i, j) = \sum_{n=1}^{N-P} x_{n+j} x_{n+i} \tag{7.144}$$

for $i, j \in 0, 1, , \ldots, P$. Since $\hat{r}_{XX}(i, j) = \hat{r}_{XX}(j, i)$, the resulting autocorrelation matrix is symmetric but not Toeplitz. This is known as the *covariance* LPA or sometimes the *least-squares* method. Empirically, covariance LPA PSD estimates appear to be more accurate than Yule-Walker (Figure 7.12) even though Yule-Walker models are always stable since the poles always lie within the unit circle (Makhoul, 1975). Coming from a different direction, *maximum entropy* PSD *estimation*, that is, the problem of finding the most uniform PSD given the exact autocorrelation values $r_{XX}(\tau)$ for $\tau = 0, 1, \ldots, P$, can be shown to be solved by the linear system(7.128). Of course, in practice, we only have autocorrelation estimates which means that this method, known as *Burg* LPA, does not have obvious advantages over covariance LPA (Proakis and Manolakis, 1996, section 12.3.3).

So far, we have considered the LPA parameter P as fixed. However, as with all statistical modelling, we can always increase P and thereby overfit. AR PSD modelling requires some kind of complexity control. There are many possibilities, including cross-validation and Bayesian approaches (see Section 4.2), but regularization such as AIC (4.23) and BIC (4.24) are widely-used in this context. To compute these quantities we will need twice the NLL for the AR model:

$$
\begin{aligned}
2E &= N \ln\left(2\pi\hat{\sigma}^2\right) + \frac{1}{\hat{\sigma}^2} \sum_{n=P}^{N} \left(x_n - \sum_{j=1}^{P} \hat{a}_j x_{n-j} \right)^2 \\
&= N \ln\left(2\pi\hat{\sigma}^2\right) + N - P \tag{7.145}
\end{aligned}
$$

using the MLE Gaussian variance estimator $\hat{\sigma}^2$. The use of these model selection criteria does not guarantee that the perfect model can be found, indeed unless the signal is generated by an LPA model then there is no guarantee that the model order selected this way will not be much larger than a more parsimonious model which could conceivably be constructed for the signal (Figure 7.13). However, this is preferable to not using any kind of complexity control.

Roughly speaking, any spectral analysis method which posits a specific linear, statistical signal model might be described as a parametric method. Thus, we can find a very large array of parametric methods based on sums of sinusoids, for example the *Pisarenko* and MUSIC methods, discussed in the next section (Proakis and Manolakis, 1996, section 12.5).

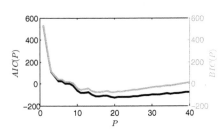

Fig. 7.13: Regularized LPA model selection, applied to PSD estimation for a sample signal (length $N = 100$) containing three sinusoids (radian frequencies $\phi = 0.2$, 0.5 and 1.5, amplitudes $r = 1.0$, 0.5 and 5.0) in additive white Gaussian noise (standard deviation $\sigma = 0.1$). The use of AIC (black) and BIC (grey) suggests similar LPA orders $P = 20$ (AIC) and $P = 19$ (BIC).

Subspace analysis: using PCA in DSP

Assume that we have a zero-mean GP signal Z_n, $n = 1, 2, \ldots, N$, with autocorrelation matrix r_{ZZ}. A typical DSP problem is recovering the signal from noisy observations:

$$X_n = Z_n + W_n \tag{7.146}$$

where the noise term W_n is an i.i.d., zero-mean GP with variance σ^2. We know, invoking statistical stability, that X_n is itself another GP. Estimating Z_n is a matter of solving a conjugate Gaussian Bayesian linear regression problem, where Z_n are the unknown regression parameters and the input variables are all ones (see Section 6.4). The MAP problem is:

$$\hat{Z} = \arg \min_{Z} \left[\frac{1}{\sigma^2} \| Z - X \|_2^2 + \left\| R_{XX}^{-1} Z \right\|_2^2 \right] \tag{7.147}$$

with the solution:

$$\hat{Z} = \frac{1}{\sigma^2} \left(\frac{1}{\sigma^2} I + \left(R_{ZZ}^{-1} \right)^T R_{ZZ}^{-1} \right)^{-1} X \tag{7.148}$$

known as the *Wiener filter* in DSP circles.

From the discussion on PCA (Section 6.6), diagonalizing the autocorrelation matrix, $R_{ZZ} = U \Lambda U^T$ will allow us to separate the signal into a sum of independent component signals, since the signal is a GP, and thus simplify the above expression:

$$
\begin{aligned}
\hat{Z} &= \frac{1}{\sigma^2} \left(\frac{1}{\sigma^2} I + U \Lambda^{-2} U^T \right)^{-1} X \\
&= U \frac{1}{\sigma^2} \left(\frac{1}{\sigma^2} I + \Lambda^{-2} \right)^{-1} U^T X \\
&= \sum_{n=1}^{N} u_n \frac{\lambda_n^2}{\lambda_n^2 + \sigma^2} u_n^T X
\end{aligned} \tag{7.149}
$$

where λ_n are the eigenvalues associated with the eigenvectors of R_{ZZ}, appearing on the diagonal of Λ. Let us explore what this equation does.

Firstly, the data is projected down onto the orthogonal eigenvectors of the autocorrelation matrix, which isolates the contribution of each eigenvector to the signal. The resulting projection coefficient is then multiplied by a "damping factor" $\lambda_n^2/\left(\lambda_n^2 + \sigma^2\right)$. Then, these weighted eigenvectors are added together to reconstruct the input signal.

The damping factor has the following effect: if the variance of the observation noise W_n is large by comparison to the square eigenvalue, then the corresponding eigenvector will be downweighted. As the observation noise $\sigma^2 \to 0$, the damping factor goes to 1, in other words, if the noise variance is small relative to the square eigenvalues, then the eigenvector signal components pass through the MAP estimation procedure unchanged. Thus, we can say that those components that have too small variance cannot be recovered, so these are effectively removed from the reconstruction. Following the PCA view, we could also just retain only those components above a certain damping factor threshold to "simplify" the signal to just its most prominent components.

The model above requires that we have a completely specified, $N \times N$ autocorrelation matrix for the prior. A much simpler model is where the prior GP is considered stationary, in which case the autocorrelation matrix is parameterized only by time difference. Then, the distribution of the signal is time-invariant, and since it is finite length, the autocorrelation matrix is circulant $\mathbf{R}_{ZZ}^N(\tau)$, so the eigenvectors are the DFT components. The eigenvalues are the DFT of the autocorrelation function $r_{ZZ}^N(\tau)$ which is the PSD of the prior autocorrelation. Thus (7.149) acts as an LTI filter: first, the signal is transformed into the frequency domain. Then, the amplitude of each frequency component is multiplied by a "damping" filter with frequency response $H(k) = \lambda_k^2/\left(\lambda_k^2 + \sigma^2\right)$. The result is then converted back into the time domain. Equivalently, the filtering can take place in the time domain, where the impulse response of the damping filter is convolved with the signal, to produce the MAP estimate for $\mathbf{Z}$. We can, of course, skip the reconstruction step, and use the frequency-domain representation as a MAP PSD estimate for $\mathbf{Z}$, so, we can view this kind of subspace filtering as a form of regularized, nonparametric PSD estimation (Figure 7.14).

The model above makes few assumptions about the form of the signal and the prior, and would therefore reasonably be described as nonparametric. Much more specific, parametric signal models have also been proposed in the context of subspace filtering, here we will look at *sinusoidal models* which are widely used in many DSP applications. Consider that the signal Z_n is a sum of M sinusoids with arbitrary frequencies ω_m, positive, real amplitudes a_m and random phases ϕ_m uniformly distributed on $[-\pi, \pi]$:

$$Z_n = \sum_{m=1}^{M} a_m \exp\left(i\omega_m n + \phi_m\right) \qquad (7.150)$$

The (unbiased) autocorrelation estimate of any one of these component signals is:

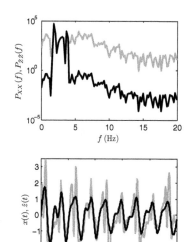

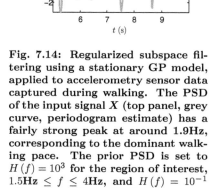

Fig. 7.14: Regularized subspace filtering using a stationary GP model, applied to accelerometry sensor data captured during walking. The PSD of the input signal X (top panel, grey curve, periodogram estimate) has a fairly strong peak at around 1.9Hz, corresponding to the dominant walking pace. The prior PSD is set to $H(f) = 10^3$ for the region of interest, $1.5\text{Hz} \leq f \leq 4\text{Hz}$, and $H(f) = 10^{-1}$ otherwise. The noise level is set at $\sigma^2 = 10$. The PSD of the MAP output signal $\hat{Z}$ is therefore strongly attenuated outside the region of interest (top panel, black curve). The effect on the input signal (bottom panel, grey curve) is to smooth out the more rapid fluctuations (bottom panel, black curve).

Such signals are not GPs in the classical sense, since they are entirely deterministic. However, these are "degenerate" GPs, for example, an AR model with poles lying exactly on the unit circle.

Because the signal here is complex, we need to use the complex-conjugate form of the autocorrelation.

$$
\begin{aligned}
\hat{r}_{\mathbf{ZZ}}(\tau) &= \frac{1}{N-\tau} \sum_{n=1}^{N-\tau} z_n \bar{z}_{n+\tau} & (7.151)\\
&= \frac{a^2}{N-\tau} \sum_{n=1}^{N-\tau} \exp\left(-i\omega n - \phi\right) \exp\left(i\omega\left(n+\tau\right)+\phi\right)\\
&= a^2 \exp\left(i\omega\tau\right)
\end{aligned}
$$

and using the fact that the autocorrelation is linear, it follows that:

$$
\hat{r}_{\mathbf{ZZ}}(\tau) = \sum_{m=1}^{M} a_m^2 \exp\left(i\omega_m\tau\right) \tag{7.152}
$$

Let us form the associated $N \times N$ autocorrelation matrix using delays $\tau \in 0,1\ldots,N-1$ from the autocorrelation function above:

$$
\hat{\mathbf{R}}_{\mathbf{ZZ}} =
\begin{bmatrix}
\hat{r}_{\mathbf{ZZ}}(0) & \hat{r}_{\mathbf{ZZ}}(1) & \cdots & \hat{r}_{\mathbf{ZZ}}(N-1)\\
\hat{r}_{\mathbf{ZZ}}(1) & \hat{r}_{\mathbf{ZZ}}(0) & \cdots & \hat{r}_{\mathbf{ZZ}}(N-2)\\
\vdots & \vdots & \ddots & \vdots\\
\hat{r}_{\mathbf{ZZ}}(N-1) & \hat{r}_{\mathbf{ZZ}}(N-2) & \cdots & \hat{r}_{\mathbf{ZZ}}(0)
\end{bmatrix} \tag{7.153}
$$

By construction, we can also write this in the following form:

$$
\hat{\mathbf{R}}_{\mathbf{ZZ}} = \sum_{m=1}^{M} \mathbf{s}_m a_m^2 \bar{\mathbf{s}}_m^T \tag{7.154}
$$

where $\mathbf{s}_m = \left[1, \exp\left(i\omega_m\right), \exp\left(i2\omega_m\right), \ldots, \exp\left(i\left[N-1\right]\omega_m\right)\right]^T$ are a set of M, linearly-independent, vectors. These vectors span the space of every N-dimensional *time-delayed vector* $\mathbf{z}_n = \left[z_n, z_{n+1}, \ldots, z_{n+N-1}\right]^T$. For this reason, this subspace of the signal $\mathbf{z}$ is called the *signal subspace*, $\mathcal{S}$. If $N > M$ then $\hat{\mathbf{R}}_{\mathbf{ZZ}}$ has rank M. This means that the ordered eigenvalues $\lambda_1 \geq \lambda_2 \geq \cdots \geq \lambda_N$ of $\hat{\mathbf{R}}_{\mathbf{ZZ}}$ are partitioned such that $\lambda_m > 0$ for $m = 1, 2, \ldots, M$ and the rest are zero. So, we can express the autocorrelation matrix using the reduced eigenvector reconstruction:

$$
\hat{\mathbf{R}}_{\mathbf{ZZ}} = \sum_{m=1}^{M} \mathbf{u}_m \lambda_m \bar{\mathbf{u}}_m^T \tag{7.155}
$$

where $\mathbf{u}_d$ are the corresponding eigenvectors for the eigenvalues λ_n, $n = 1, 2, \ldots, N$.

Now, going back to the linear signal model (7.146), the estimated autocorrelation for the observed signal is:

$$
\hat{r}_{\mathbf{XX}}(\tau) = \sum_{m=1}^{M} a_m^2 \exp\left(i\omega_m\tau\right) + \sigma^2 \delta\left[\tau\right] \tag{7.156}
$$

and it follows that the corresponding autocorrelation matrix has the simple form $\hat{\mathbf{R}}_{XX} = \hat{\mathbf{R}}_{ZZ} + \sigma^2 \mathbf{I}$. This matrix has full rank N, and the ordered eigenvectors of $\hat{\mathbf{R}}_{XX}$, $\boldsymbol{v}_n$ for $n = 1, 2, \ldots, N$ satisfy $\mu_n = \lambda_n + \sigma^2$. So, we can express the autocorrelation matrix of the data as:

$$\hat{\mathbf{R}}_{XX} = \sum_{m=1}^{M} \boldsymbol{v}_m \left(\lambda_m + \sigma^2 \right) \bar{\boldsymbol{v}}_m^T + \sum_{n=M+1}^{N} \boldsymbol{v}_n \sigma^2 \bar{\boldsymbol{v}}_n^T \qquad (7.157)$$

This decomposition into the two subspaces – $\mathcal{S}$ which spans the sinusoidal signal model, and $\mathcal{N}$ which spans the noise – coupled with the mutual orthogonality of the eigenvectors of the autocorrelation matrix of the observed signal, gives an intuitive understanding of how to separate out the signal from the noise. For example, following PCA, the first M eigenvectors (the first M principal components) can reconstruct the signal subspace while removing the noise (Figure 7.15):

$$\hat{\mathbf{Z}} = \sum_{m=1}^{M} \boldsymbol{v}_m \hat{\lambda}_m \bar{\boldsymbol{v}}_m^T \mathbf{X} \qquad (7.158)$$

where $\boldsymbol{v}_m, \hat{\lambda}_m$ are the eigenvectors and eigenvalues of the estimated autocorrelation matrix. In practice we may not know the noise level σ^2, but inspection of the eigenvalues of $\hat{\mathbf{R}}_{XX}$ may reveal a noise threshold which plausibly splits the subspaces.

Unlike the "generic" MAP filtering above, the sinusoidal model allows us to extract more specific information. Taking $D = M + 1$, then eigenvector $M + 1$ is contained in $\mathcal{N}$ so that it must be orthogonal to each of the signal vectors in $\mathcal{S}$:

$$\bar{\boldsymbol{s}}_m^T \boldsymbol{v}_{M+1} = 0, \quad m = 1, 2, \ldots M \qquad (7.159)$$

and by setting $z = \exp(i\omega_m)$ this can also be written as a polynomial in z with coefficients from the eigenvector $\boldsymbol{v}_{M+1}$. The roots of this polynomial all lie on the unit circle, and the arguments of these roots are the frequencies ω_m (Stoica and Moses, 2005, page 161). With this information, we can estimate the associated square amplitudes $\hat{a}_m^2$ (and $\hat{\sigma}^2$) using (7.156). This is known as *Pisarenko's* method.

While simple, Pisarenko's method depends upon the accuracy of a single eigenvector of the estimated autocorrelation matrix. Thus, it is prone to high variability. A way to improve the reliability of this estimate is to use many more noise subspace eigenvectors, a method known as *MUSIC (MUltiple SIgnal Classification)*. Consider any complex sinusoidal vector parametrized by the angular frequency ω, $\boldsymbol{s}(\omega) = [1, \exp(i\omega), \exp(i2\omega), \ldots, \exp(i[N-1]\omega)]$, then for $\omega = \omega_m$ for any $m = 1, 2, \ldots, M$ we must have $\bar{\boldsymbol{s}}(\omega)^T \boldsymbol{v}_n = 0$ for all $n = M + 1, M + 1, \ldots, L$ for some $L < N$. Therefore, with the weights

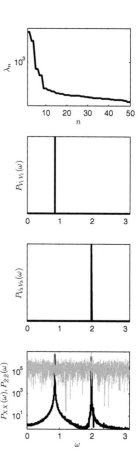

Fig. 7.15: **Subspace PCA filtering applied to a sinusoidal signal model (two real sinusoids, frequencies** $\omega_1 = 0.86$ **and** $\omega_2 = 1.99$**, both having unit amplitude, signal length** $N = 3,000$**), with additive white Gaussian observation noise of variance** $\sigma^2 = 100$**. The first few autocorrelation matrix eigenvalues** $\hat{\lambda}_n$ **are large (top panel), decreasing rapidly, which suggests a signal subspace of order** $M \leq 10$**. The first and third eigenvectors clearly capture the two sinusoidal components (middle two panels, PSDs estimated using the periodogram). The PCA reconstruction using the first four eigenvectors** $\hat{\mathbf{Z}}$ **(black curve, periodogram PSD estimate) is effective at removing the noise in the input signal** X **(grey curve, periodogram PSD estimate).**

$c_n > 0$, the function:

$$P(\omega) = \frac{1}{\sum_{n=M+1}^{L} c_n \left| \bar{\boldsymbol{s}}(\omega)^T \boldsymbol{v}_n \right|^2} \qquad (7.160)$$

will be infinite at each frequency ω_m in the sinusoidal model. This widely-used *pseudospectrum* behaves like (but is not!) a PSD estimate. Typically, the weights are either constant or set equal to the reciprocal of the noise subspace eigenvectors $c_n = \hat{\lambda}_n^{-1}$ (Figure 7.16).

Computationally speaking, the most intensive parts of subspace filtering are: estimating the autocorrelation matrix entries, $O\left(N L^2\right)$ where L is the number of autocorrelation time delays, and computing eigenvector decompositions, $O\left(L^2\right)$. Thus, PCA subspace filtering ($L = N$) is very difficult for long signals.

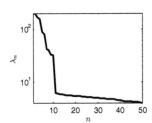

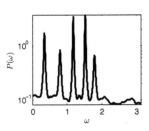

Fig. 7.16: Subspace MUSIC frequency estimation algorithm with eigenvalue weighting and $L = 50$ eigenvectors, applied to a sinusoidal signal model (five real sinusoids e.g. $M = 10$ complex-conjugate exponentials, frequencies $\omega = [0.34, 0.79, 1.17, 1.51, 1.78]$, amplitudes $r = [1.29, 0.88, 1.78, 1.93, 0.77]$, signal length $N = 3,000$), with additive white Gaussian observation noise of variance $\sigma^2 = 5$. The first ten autocorrelation matrix eigenvalues $\hat{\lambda}_n$ are large (top panel), the rest are decreasing rapidly, which confirms the model order. The pseudospectrum has clear peaks at the given frequencies.

7.5   The Kalman filter (KF)

In this section we will discuss in detail one of the most important linear-Gaussian DSP methods, the *Kalman filter* (KF). The probabilistic structure of this method is that of a two-level hierarchical PGM with a latent Markov state $\boldsymbol{Z}_{n-1} \to \boldsymbol{Z}_n$ with observations which depend upon that state, $\boldsymbol{Z}_n \to \boldsymbol{X}_n$ (Figure 5.3). We can write the description of this model in several forms, but the following "generative" form makes it easy to see how to sample from the model:

$$\begin{aligned} \boldsymbol{Z}_n &= \mathbf{A}\boldsymbol{Z}_{n-1} + \boldsymbol{U} \\ \boldsymbol{X}_n &= \mathbf{B}\boldsymbol{Z}_n + \boldsymbol{V} \end{aligned} \qquad (7.161)$$

where $\boldsymbol{Z}_n \in \mathbb{R}^M$ and $\boldsymbol{X}_n \in \mathbb{R}^D$. The random *state noise* $\boldsymbol{U} \in \mathbb{R}^M$ and *observation noise* $\boldsymbol{V} \in \mathbb{R}^D$ are multivariate Gaussian with zero mean, $\boldsymbol{U} \sim \mathcal{N}(\boldsymbol{0}, \mathbf{S})$ and $\boldsymbol{V} \sim \mathcal{N}(0, \boldsymbol{\Sigma})$. Together with the *state update* $\mathbf{A} \in \mathbb{R}^{M \times M}$ and *observation* (or *emission*) model parameters $\mathbf{B} \in \mathbb{R}^{D \times M}$, the Markov chain and observed signal are fully defined as discrete-time (multivariate) stochastic processes (Figure 7.17). Finally, to get the iteration started, we define $\boldsymbol{Z}_1 \sim \mathcal{N}(\boldsymbol{\mu}_1, \mathbf{S}_1)$. Given this definition, it is clear from the affine transformation properties of the multivariate Gaussian (Section 1.4) that $\boldsymbol{Z}_n, \boldsymbol{X}_n$ are multivariate Gaussian for all time indices, $n \in \mathbb{N}$.

Given this model, we typically need to solve one or more of the following problems:

(1) *Filtering*: given the KF model parameters and observed data up to index $1, 2, \ldots, n$, find the distribution of the latent Markov states up to that index. This can be solved using the so-called *forward recursion* algorithm,

(2) *Smoothing*: with the KF model parameters and observed data for all $n \in 1, 2, \ldots, N$, find the distribution of the latent Markov states for the whole observed signal. This is solved using the *backward*

recursion algorithm (or, in practice a combination of the forward and backward recursions),

(3) *Model fitting*: given observed data, estimate maximum likelihood values of the Markov, $\mathbf{A}, \mathbf{S}$ and observation parameters $\mathbf{B}, \boldsymbol{\Sigma}$ (also the initial parameters $\mathbf{S}_1, \boldsymbol{\mu}_1$). This can be performed using the *Baum-Welch* algorithm (an instance of E-M) which is uses forward-backward recursions to compute state and Markov probabilities at each time index, from which E-M model parameter estimates can be obtained. Iterative use of *Viterbi decoding* (see next item) and parameter re-estimation is also used.

We will also discuss *Viterbi decoding*: given observed data and KF model parameters, find the sequence of Markov states that maximizes the conditional probability of the complete Markov state sequence given the observed data. This is intimately related to the forward algorithm but using max-product or max-sum algebra in place of sum-product algebra.

Junction tree algorithm (JT) for KF computations

Given that the structure of the associated PGM is a tree, exact state inference is possible using JT message-passing (Algorithm 5.1). There are several ways in which the JT algorithm can be used here, but because we are interested in solving the above problems (filtering, smoothing and model fitting), we will use *time-slice* cliques $C_n = \{\boldsymbol{z}_{n-1}, \boldsymbol{z}_n, \boldsymbol{x}_n\}$ for $n > 1$ and $C_1 = \{\boldsymbol{z}_1, \boldsymbol{x}_1\}$ (here, for notational clarity, we use variable names rather than indices in defining variable sets) and corresponding clique factors $h_n(\boldsymbol{z}_{n-1}, \boldsymbol{z}_n, \boldsymbol{x}_n) = f(\boldsymbol{z}_n|\boldsymbol{z}_{n-1}) \otimes f(\boldsymbol{x}_n|\boldsymbol{z}_n)$. The observation density is $f(\boldsymbol{x}_n|\boldsymbol{z}_n) = \mathcal{N}(\boldsymbol{x}_n; \mathbf{B}\boldsymbol{z}_n, \boldsymbol{\Sigma})$, the Markov transition density $f(\boldsymbol{z}_n|\boldsymbol{z}_{n-1}) = \mathcal{N}(\boldsymbol{z}_n; \mathbf{A}\boldsymbol{z}_{n-1}, \mathbf{S})$, and the initial density is $f(\boldsymbol{z}_1) = \mathcal{N}(\boldsymbol{z}_1; \boldsymbol{\mu}_1, \mathbf{S}_1)$. Using this JT, and assuming realized observed data $\boldsymbol{x}_n$ (fixed values), we get the following *forward* and *backward* messages between cliques respectively:

$$\mu_{n \to n+1}(\boldsymbol{z}_n) = \bigoplus_{\boldsymbol{z}_{n-1}} h_n(\boldsymbol{z}_{n-1}, \boldsymbol{z}_n, \boldsymbol{x}_n)$$
$$\otimes \mu_{n-1 \to n}(\boldsymbol{z}_{n-1}) \qquad (7.162)$$
$$\mu_{n \to n-1}(\boldsymbol{z}_{n-1}) = \bigoplus_{\boldsymbol{z}_n} h_n(\boldsymbol{z}_{n-1}, \boldsymbol{z}_n, \boldsymbol{x}_n)$$
$$\otimes \mu_{n+1 \to n}(\boldsymbol{z}_n) \qquad (7.163)$$

To get the iterations started we also need the "boundary" messages $\mu_{N+1 \to N}(\boldsymbol{z}_N) = \otimes_{\mathrm{id}}$ (the identity element for the $\otimes$ operator, which will be 1 for the usual product) and $\mu_{1 \to 2}(\boldsymbol{z}_1) = f(\boldsymbol{z}_1) \otimes f(\boldsymbol{x}_1|\boldsymbol{z}_1)$. In the usual probability integral-product semiring for continuous variables (where $\otimes \mapsto \times$ and $\oplus \mapsto \int$), we can show by induction that the above

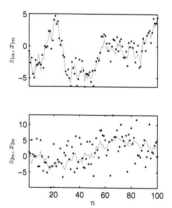

Fig. 7.17: **Typical 2D trajectory from a Kalman filter (KF) model (top panel: dimension 1 and bottom panel: dimension 2). The 2D Gaussian Markov variable z_n (grey lines) is unobserved, the observed data is x_n (black dots). Typical DSP operations include reconstructing the best estimate of the latent signal z from the observed signal x (filtering and smoothing).**

messages represent the following distributions:

$$
\begin{aligned}
\mu_{n \to n+1}(z_n) &= f(x_n|z_n) \int f(z_n|z_{n-1}) \mu_{n-1 \to n}(z_{n-1}) \, dz_{n-1} \\
&= f(x_1, x_2, \ldots, x_n, z_n) \tag{7.164} \\
\mu_{n \to n-1}(z_{n-1}) &= \int f(z_n|z_{n-1}) f(x_n|z_n) \mu_{n+1 \to n}(z_n) \, dz_n \\
&= f(x_n, x_{n+1}, \ldots, x_N|z_{n-1}) \tag{7.165}
\end{aligned}
$$

often denoted $\alpha(z_n)$ and $\beta(z_{n-1})$ in the literature, respectively.

Forward filtering

Let us first examine the forward messages. For several reasons it will be more useful to work with conditional messages $f(z_n|x_1, \ldots, x_n) = \hat{\alpha}(z_n)$ and multiply these using the *scale factors* $c_n = f(x_n|x_1, \ldots, x_{n-1})$. Because $f(x_1, \ldots, x_n) = \prod_{n'=1}^{n} c_{n'}$, we have $\mu_{n \to n+1}(z_n) = \hat{\alpha}(z_n) \times \prod_{n'=1}^{n} c_{n'}$. So, we can compute the scale factors recursively as part of forward message passing using the conditional messages:

$$
c_n \hat{\alpha}(z_n) = f(x_n|z_n) \int f(z_n|z_{n-1}) \hat{\alpha}(z_{n-1}) \, dz_{n-1} \tag{7.166}
$$

By Gaussian marginalization and conditioning properties, we know that the forward messages are multivariate normal and hence parameterized by mean and covariance alone, $f(z_n|x_1, \ldots, x_n) = \mathcal{N}(z_n; \mu_{n|n}, \Sigma_{n|n})$. We will use the notation $\mu_{n|n}$ to indicate the mean parameter $E_{Z_n|X_1, \ldots, X_n}[Z_n]$ of these conditional messages and $\Sigma_{n|n}$, the corresponding covariance parameter. To establish closed-form recursive formulae for these parameters, we first need to evaluate the following integral:

$$
\begin{aligned}
\int f(z_n|z_{n-1}) \hat{\alpha}(z_{n-1}) \, dz_{n-1} &\propto \int f(z_n, z_{n-1}|x_1, \ldots, x_{n-1}) \, dz_{n-1} \\
&= \int \mathcal{N}(z_n; A z_{n-1}, S) \times \tag{7.167} \\
&\quad \mathcal{N}(z_{n-1}; \mu_{n-1|n-1}, \Sigma_{n-1|n-1}) \times \\
&\quad dz_{n-1}
\end{aligned}
$$

The integrand here is a special case of the general Gaussian conditioning rule (Section 7.1), so the integral is simply obtained by inspection as $f(z_n|x_1, \ldots, x_{n-1}) = \mathcal{N}(z_n; \mu_{n|n-1}, \Sigma_{n|n-1})$ with $\mu_{n|n-1} = A \mu_{n-1|n-1}$ and $\Sigma_{n|n-1} = A \Sigma_{n-1|n-1} A^T + S$. This distribution can be seen as the *prediction* for Z_n obtained by propagating the previous message through the Markov chain and then marginalizing away the previous step.

Next, we must combine with the observation density to absorb the information from the next observation x_n, to obtain (using the general normal conditioning rule), the joint density $f(z_n, x_n | x_1, \ldots, x_{n-1})$ which is:

$$\mathcal{N}\left(\left(\begin{array}{c} z_n \\ x_n \end{array}\right); \left(\begin{array}{c} \mu_{n|n-1} \\ \mathbf{B}\mu_{n|n-1} \end{array}\right), \left(\begin{array}{cc} \Sigma_{n|n-1} & \Sigma_{n|n-1}\mathbf{B}^T \\ \mathbf{B}\Sigma_{n|n-1} & \mathbf{B}\Sigma_{n|n-1}\mathbf{B}^T + \Sigma \end{array}\right)\right)$$
$$(7.168)$$

Finally, in this form it is easy to compute the conditional $f(z_n | x_1, \ldots, x_n)$ to get the updated conditional forward messages:

$$
\begin{aligned}
\mu_{n|n} &= \mu_{n|n-1} + \mathbf{K}_n \left(x_n - \mathbf{B}\mu_{n|n-1}\right) \\
\Sigma_{n|n} &= \Sigma_{n|n-1} - \mathbf{K}_n \mathbf{B}\Sigma_{n|n-1} \\
\mathbf{K}_n &= \Sigma_{n|n-1}\mathbf{B}^T \left(\mathbf{B}\Sigma_{n|n-1}\mathbf{B}^T + \Sigma\right)^{-1}
\end{aligned}
\qquad (7.169)
$$

where we have defined the *Kalman gain* $\mathbf{K}_n$. Now, we have enough information to write the full forward message passing recursion compactly:

$$
\begin{aligned}
\mu_{n|n-1} &= \mathbf{A}\mu_{n-1|n-1} \\
\Sigma_{n|n-1} &= \mathbf{A}\Sigma_{n-1|n-1}\mathbf{A}^T + \mathbf{S} \\
\mu_{n|n} &= \mu_{n|n-1} + \mathbf{K}_n \left(x_n - \mathbf{B}\mu_{n|n-1}\right) \\
\Sigma_{n|n} &= \Sigma_{n|n-1} - \mathbf{K}_n \mathbf{B}\Sigma_{n|n-1} \\
\mathbf{K}_n &= \Sigma_{n|n-1}\mathbf{B}^T \left(\mathbf{B}\Sigma_{n|n-1}\mathbf{B}^T + \Sigma\right)^{-1}
\end{aligned}
\qquad (7.170)
$$

for $n = 2, 3, \ldots, N$.

Finally, to get the recursion started with the $n = 1$ case, we need $\mu_{1 \to 2}(z_1) = c_1 f(z_1 | x_1)$, for which we need $f(z_1) = \mathcal{N}(z_1; \mu_1, \mathbf{S}_1)$ and $f(x_1 | z_1) = \mathcal{N}(x_1; \mathbf{B}z_1, \Sigma)$. The general normal conditioning rule then gives us $f(z_1 | x_1) = \mathcal{N}\left(z_1; \mu_{1|1}, \Sigma_{1|1}\right)$ with:

$$
\begin{aligned}
\mathbf{K}_1 &= \mathbf{S}_1 \mathbf{B}^T \left(\mathbf{B}\mathbf{S}_1 \mathbf{B}^T + \Sigma\right)^{-1} \\
\mu_{1|1} &= \mu_1 + \mathbf{K}_1 \left(x_1 - \mathbf{B}\mu_1\right) \\
\Sigma_{1|1} &= \mathbf{S}_1 - \mathbf{K}_1 \mathbf{B}\mathbf{S}_1
\end{aligned}
\qquad (7.171)
$$

These updates can be given a *predictor* followed by *Bayesian corrector* interpretation. Given the previous latent state distribution conditional on the data observed so far, perform the Markov update which involves propagating this distribution through the chain. Then, use Bayes' rule to update using the distribution given the next observation. Then, the Kalman gain can be understood as the term which balances the "spread" of the prediction against the spread of the observation in the context of Bayesian inference for conjugate Gaussian models (see for instance, Section 6.4).

Backward smoothing

While it is possible in theory to compute backwards messages $f\left(\boldsymbol{x}_{n+1}, \ldots, \boldsymbol{x}_{N} \mid \boldsymbol{z}_{n}\right)$ using backwards message passing $C_{n} \rightarrow C_{n-1}$, the resulting linear-Gaussian messages are challenging to compute analytically. Given that we are generally only interested in the conditional $f\left(\boldsymbol{z}_{n} \mid \boldsymbol{x}_{1}, \ldots, \boldsymbol{x}_{N}\right)$, we will instead compute these quantities directly. The strategy to obtain this will be to reverse the Markov chain $\boldsymbol{Z}_{n+1} \rightarrow \boldsymbol{Z}_{n}$ and then obtain the desired conditional distribution of $\boldsymbol{Z}_{n}$ in terms of the distribution of $\boldsymbol{Z}_{n+1}$.

To reverse the Markov chain we can again use the general normal conditioning rule. The relevant joint distribution $f\left(\boldsymbol{z}_{n+1}, \boldsymbol{z}_{n} \mid \boldsymbol{x}_{1}, \ldots, \boldsymbol{x}_{n}\right)$ has density:

$$\mathcal{N}\left(\left(\begin{array}{c} \boldsymbol{z}_{n} \\ \boldsymbol{z}_{n+1} \end{array}\right) ;\left(\begin{array}{c} \boldsymbol{\mu}_{n \mid n} \\ \boldsymbol{\mu}_{n+1 \mid n} \end{array}\right),\left(\begin{array}{cc} \boldsymbol{\Sigma}_{n \mid n} & \boldsymbol{\Sigma}_{n \mid n} \mathbf{A}^{T} \\ \mathbf{A} \boldsymbol{\Sigma}_{n \mid n} & \boldsymbol{\Sigma}_{n+1 \mid n} \end{array}\right)\right) \qquad (7.172)$$

From this we can obtain the posterior (the *backwards* Markov chain transition density) $f\left(\boldsymbol{z}_{n} \mid \boldsymbol{z}_{n+1}, \boldsymbol{x}_{1}, \ldots, \boldsymbol{x}_{n}\right)$ which (using conditional independence properties of the PGM) is also $f\left(\boldsymbol{z}_{n} \mid \boldsymbol{z}_{n+1}, \boldsymbol{x}_{1}, \ldots, \boldsymbol{x}_{N}\right) = \mathcal{N}\left(\boldsymbol{z}_{n} ; \boldsymbol{\mu}_{n \mid N}, \boldsymbol{\Sigma}_{n \mid N}\right)$. The parameters are:

$$\begin{aligned} \boldsymbol{\mu}_{n \mid N}\left(\boldsymbol{z}_{n+1}\right) &= \boldsymbol{\mu}_{n \mid n} + \mathbf{L}_{n}\left(\boldsymbol{z}_{n+1} - \boldsymbol{\mu}_{n+1 \mid n}\right) \\ \boldsymbol{\Sigma}_{n \mid N} &= \boldsymbol{\Sigma}_{n \mid n} - \mathbf{L}_{n} \boldsymbol{\Sigma}_{n+1 \mid n} \mathbf{L}_{n}^{T} \\ \mathbf{L}_{n} &= \boldsymbol{\Sigma}_{n \mid n} \mathbf{A}^{T} \boldsymbol{\Sigma}_{n+1 \mid n}^{-1} \end{aligned} \qquad (7.173)$$

Now, from the forward recursions we have $\boldsymbol{\mu}_{n \mid n}$, $\boldsymbol{\Sigma}_{n \mid n}$ and $\boldsymbol{\Sigma}_{n+1 \mid n}$ but we do not know the value $\boldsymbol{z}_{n+1}$ since it is a latent variable. However, since we want $f\left(\boldsymbol{z}_{n} \mid \boldsymbol{x}_{1}, \ldots, \boldsymbol{x}_{N}\right)$ and we have the conditional $f\left(\boldsymbol{z}_{n} \mid \boldsymbol{z}_{n+1}, \boldsymbol{x}_{1}, \ldots, \boldsymbol{x}_{N}\right)$, we can use the *law of total expectation* to find the conditional expectations $\boldsymbol{\mu}_{n \mid N} = E_{\boldsymbol{Z}_{n} \mid \boldsymbol{X}}\left[\boldsymbol{Z}_{n}\right]$ and $\boldsymbol{\Sigma}_{n \mid N} = \operatorname{cov}_{\boldsymbol{Z}_{n} \mid \boldsymbol{X}}\left[\boldsymbol{Z}_{n}\right]$ directly. For the mean parameter, we have:

$$\begin{aligned} E_{\boldsymbol{Z}_{n} \mid \boldsymbol{X}}\left[\boldsymbol{Z}_{n}\right] &= E_{\boldsymbol{Z}_{n+1} \mid \boldsymbol{X}}\left[E_{\boldsymbol{Z}_{n} \mid \boldsymbol{Z}_{n+1}, \boldsymbol{X}}\left[\boldsymbol{Z}_{n}\right]\right] \\ &= E_{\boldsymbol{Z}_{n+1} \mid \boldsymbol{X}}\left[\boldsymbol{\mu}_{n \mid N}\left(\boldsymbol{Z}_{n+1}\right)\right] \\ &= E_{\boldsymbol{Z}_{n+1} \mid \boldsymbol{X}}\left[\boldsymbol{\mu}_{n \mid n} + \mathbf{L}_{n}\left(\boldsymbol{Z}_{n+1} - \boldsymbol{\mu}_{n+1 \mid n}\right)\right] \\ \boldsymbol{\mu}_{n \mid N} &= \boldsymbol{\mu}_{n \mid n} + \mathbf{L}_{n}\left(\boldsymbol{\mu}_{n+1 \mid N} - \boldsymbol{\mu}_{n+1 \mid n}\right) \end{aligned} \qquad (7.174)$$

Similarly, for the covariance, we have:

$$
\begin{aligned}
\operatorname{cov}_{\boldsymbol{Z}_n|\boldsymbol{X}}\left[\boldsymbol{Z}_n\right] &= E_{\boldsymbol{Z}_{n+1}|\boldsymbol{X}}\left[\operatorname{cov}_{\boldsymbol{Z}_n|\boldsymbol{Z}_{n+1},\boldsymbol{X}}\left[\boldsymbol{Z}_n\right]\right] \\
&+ \operatorname{cov}_{\boldsymbol{Z}_n|\boldsymbol{Z}_{n+1},\boldsymbol{X}}\left[E_{\boldsymbol{Z}_n|\boldsymbol{Z}_{n+1},\boldsymbol{X}}\left[\boldsymbol{Z}_n\right]\right] \\
&= E_{\boldsymbol{Z}_{n+1}|\boldsymbol{X}}\left[\boldsymbol{\Sigma}_{n|n} - \mathbf{L}_n\boldsymbol{\Sigma}_{n+1|n}\mathbf{L}_n^T\right] \\
&+ \operatorname{cov}_{\boldsymbol{Z}_n|\boldsymbol{Z}_{n+1},\boldsymbol{X}}\left[\boldsymbol{\mu}_{n|n} + \mathbf{L}_n\left(\boldsymbol{Z}_{n+1} - \boldsymbol{\mu}_{n+1|n}\right)\right]
\end{aligned} \tag{7.175}
$$

This gives us:

$$
\operatorname{cov}_{\boldsymbol{Z}_n|\boldsymbol{X}}\left[\boldsymbol{Z}_n\right] = \boldsymbol{\Sigma}_{n|n} - \mathbf{L}_n\boldsymbol{\Sigma}_{n+1|n}\mathbf{L}_n^T + \mathbf{L}_n\boldsymbol{\Sigma}_{n+1|N}\mathbf{L}_n^T \tag{7.176}
$$

Thus, we have the following simple recursion for $n = N-1, N-2 \ldots 1$:

$$
\begin{aligned}
\boldsymbol{\mu}_{n|N} &= \boldsymbol{\mu}_{n|n} + \mathbf{L}_n\left(\boldsymbol{\mu}_{n+1|N} - \boldsymbol{\mu}_{n+1|n}\right) \\
\boldsymbol{\Sigma}_{n|N} &= \boldsymbol{\Sigma}_{n|n} + \mathbf{L}_n\left(\boldsymbol{\Sigma}_{n+1|N} - \boldsymbol{\Sigma}_{n+1|n}\right)\mathbf{L}_n^T \\
\mathbf{L}_n &= \boldsymbol{\Sigma}_{n|n}\mathbf{A}^T\boldsymbol{\Sigma}_{n+1|n}^{-1}
\end{aligned} \tag{7.177}
$$

and the starting $n = N$ case is obtained at the end of the forward recursion, $\boldsymbol{\mu}_{N|N}$ and $\boldsymbol{\Sigma}_{N|N}$. This recursion is known as *Kalman* or *Rauch-Tung-Striebel* (RTS) *smoothing*. See Bishop (2006, page 641) and Shumway and Stoffer (2016, pages 297-298) for alternative derivations of this recursion that do not use iterated expectations.

Incomplete data likelihood

It will be important for various applications including parameter estimation to compute the incomplete NLL, $E = -\ln f(\boldsymbol{x};\boldsymbol{p})$ where $\boldsymbol{p} = (\mathbf{A},\mathbf{S},\mathbf{B},\boldsymbol{\Sigma},\boldsymbol{\mu}_1,\mathbf{S}_1)$ are the KF model parameters. Rather than attempting direct marginalization of $f(\boldsymbol{x},\boldsymbol{z};\boldsymbol{p})$, we can use the scale factors $f(\boldsymbol{x};\boldsymbol{p}) = \prod_{n=1}^N c_n$. The utility of this approach is that the NLL can then be computed recursively during forward message passing. These scale factors are easy to obtain by propagating the Markov state through the chain and then through the observation density to obtain the probably density values $c_n = \mathcal{N}(\boldsymbol{x}_n;\boldsymbol{\mu},\boldsymbol{\Sigma})$ with mean $\boldsymbol{\mu} = \mathbf{B}\mathbf{A}\boldsymbol{\mu}_{n-1|n-1}$ and covariance $\boldsymbol{\Sigma} = \mathbf{B}\left(\mathbf{A}\boldsymbol{\Sigma}_{n-1|n-1}\mathbf{A}^T + \mathbf{S}\right)\mathbf{B}^T + \boldsymbol{\Sigma}$. This relies on first having computed the forward message parameters for that time index. The initial scale factor will then be $c_1 = \mathcal{N}(\boldsymbol{x}_1;\mathbf{B}\boldsymbol{\mu}_1,\mathbf{B}\mathbf{S}_1\mathbf{B}^T + \boldsymbol{\Sigma})$ in an entirely analogous way. The NLL can be computed recursively as $E_n = E_{n-1} - \ln c_n$ starting with $E_0 = 0$.

Viterbi decoding

Using the above message structure, given observed data and with fixed transition and initial distribution, we can compute the probability of the most probable state sequence by forward message passing in the max-product semiring (mapping $\oplus \mapsto \max$ and $\otimes \mapsto \times$):

$$\mu_{n\to n+1}^{\max}(z_n) = \max_{z_{n-1}}\left[f(z_n|z_{n-1})f(x_n|z_n)\mu_{n-1\to n}^{\max}(z_{n-1})\right]$$
$$= f(x_n|z_n) \times \qquad (7.178)$$
$$\max_{z_{n-1}}\left[f(z_n|z_{n-1})\mu_{n-1\to n}^{\max}(z_{n-1})\right]$$

with initial message $\mu_{1\to2}^{\max}(z_1) = f(z_1)f(x_1|z_1)$. At the root index N of the message passing, we can obtain the most likely state which is $\hat{z}_N = \arg\max_{z_N}\mu_{N\to N+1}^{\max}(z_N)$. To find the most probable sequence of states as well, we also need to retain which state obtained the maximum at each iteration:

$$\Delta_{n\to n+1}(z_n) = \arg\max_{z_{n-1}} f(z_n|z_{n-1})\mu_{n-1\to n}^{\max}(z_{n-1}) \quad (7.179)$$

Now, using $\hat{z}_N$ we can *backtrack* along the forward message path to find $\hat{z}_{n-1} = \Delta_{n\to n+1}(\hat{z}_n)$ for $n = N, N-1, \ldots, 2$. This sequence of states is known as the *Viterbi path* or *sequence*.

For the KF, this most probable path has a simple relationship to the forward/backward messages described above. First, note that for the multivariate normal over the random variables $(X,Y)^T$, we have:

$$\max_y\left[\mathcal{N}\left(\begin{pmatrix}x\\y\end{pmatrix},\begin{pmatrix}\mu_X\\\mu_Y\end{pmatrix},\begin{pmatrix}\Sigma_{XX}&\Sigma_{XY}\\\Sigma_{XY}&\Sigma_{YY}\end{pmatrix}\right)\right] = \mathcal{N}(x;\mu_X,\Sigma_{XX}) \qquad (7.180)$$

and:

$$\hat{y}(x) = \arg\max_y\left[\mathcal{N}\left(\begin{pmatrix}x\\y\end{pmatrix},\begin{pmatrix}\mu_X\\\mu_Y\end{pmatrix},\begin{pmatrix}\Sigma_{XX}&\Sigma_{XY}\\\Sigma_{XY}&\Sigma_{YY}\end{pmatrix}\right)\right]$$
$$= \mu_Y + \Sigma_{XY}\Sigma_{XX}^{-1}(x-\mu_X) \qquad (7.181)$$

We can therefore see that maximization over one variable in the pair is equivalent to marginalizing out that variable. Applying this to the KF forward messages:

$$\max_{z_{n-1}} f(z_n,z_{n-1}|x_1,\ldots,x_{n-1}) = \qquad (7.182)$$
$$\mathcal{N}\left(z_n; A\mu_{n-1|n-1}, A\Sigma_{n-1|n-1}A^T + S\right)$$

So, we recover exactly the same result for the mean parameter as with forward message passing described above. Also:

$$
\begin{aligned}
\Delta_{n \to n+1}\left(\boldsymbol{z}_n\right) &= \boldsymbol{\mu}_{n-1|n-1} + \boldsymbol{\Sigma}_{n-1|n-1}\mathbf{A}^T\left(\mathbf{A}\boldsymbol{\Sigma}_{n-1|n-1}\mathbf{A}^T + \mathbf{S}\right)^{-1} \times \\
&\qquad \left(\boldsymbol{z}_n - \mathbf{A}\boldsymbol{\mu}_{n-1|n-1}\right) \qquad\qquad\qquad\qquad (7.183) \\
&= \boldsymbol{\mu}_{n-1|n-1} + \boldsymbol{\Sigma}_{n-1|n-1}\mathbf{A}^T\boldsymbol{\Sigma}_{n|n-1}^{-1}\left(\boldsymbol{z}_n - \boldsymbol{\mu}_{n|n-1}\right)
\end{aligned}
$$

Now, applying backtracking, we have $\hat{\boldsymbol{z}}_N = \max_{\boldsymbol{z}_N} f\left(\boldsymbol{z}_N|\boldsymbol{x}_1,\ldots,\boldsymbol{x}_N\right) = \boldsymbol{\mu}_{N|N}$. Thus, for $n = N-1, N-2, \ldots, 1$ we have:

$$
\begin{aligned}
\hat{\boldsymbol{z}}_n &= \Delta_{n+1 \to n+2}\left(\boldsymbol{\mu}_{n+1|N}\right) \qquad\qquad\qquad\qquad (7.184) \\
&= \boldsymbol{\mu}_{n|n} + \boldsymbol{\Sigma}_{n|n}\mathbf{A}^T\boldsymbol{\Sigma}_{n+1|n}^{-1}\left(\boldsymbol{\mu}_{n+1|N} - \boldsymbol{\mu}_{n+1|n}\right) \\
\boldsymbol{\mu}_{n|N} &= \boldsymbol{\mu}_{n|n} + \mathbf{L}_n\left(\boldsymbol{\mu}_{n+1|N} - \boldsymbol{\mu}_{n+1|n}\right) \qquad\qquad (7.185)
\end{aligned}
$$

This means that Viterbi decoding in the KF is the same as performing backwards recursions for the mean parameter alone.

Baum-Welch parameter estimation

In most practical situations we do not have parameter values $\boldsymbol{p}$ for the KF, that is, we know neither transition density, initial distribution, nor parameters for the observation distribution. These will need to be estimated from some given data, $\boldsymbol{x}$. The existence of latent variables means we cannot do maximum likelihood directly. As with mixture modelling (Section 6.2), this is an inference situation for which the E-M algorithm is ideally suited (Algorithm 5.3).

Let us first recall E-M expressed in one step which is $\boldsymbol{p}^{i+1} = \arg\min_{\boldsymbol{p}} E_{\boldsymbol{Z}|\boldsymbol{X}}\left[-\ln f\left(\boldsymbol{x}, \boldsymbol{Z}; \boldsymbol{p}\right)\right]$ with:

$$
E_{\boldsymbol{Z}|\boldsymbol{X}}\left[-\ln f\left(\boldsymbol{x}, \boldsymbol{Z}; \boldsymbol{p}\right)\right] = -\int f\left(\boldsymbol{z}|\boldsymbol{x}; \boldsymbol{p}^i\right)\ln f\left(\boldsymbol{z}, \boldsymbol{x}; \boldsymbol{p}\right)d\boldsymbol{z} \qquad (7.186)
$$

the expected NLL with respect to the latent states given the observed data. The model parameters are $\boldsymbol{p} = \left(\mathbf{A}, \mathbf{S}, \mathbf{B}, \boldsymbol{\Sigma}, \boldsymbol{\mu}_1, \mathbf{S}_1\right)$. Expanding out the negative log of the PGM likelihood:

$$
\begin{aligned}
-\ln f\left(\boldsymbol{z}, \boldsymbol{x}; \boldsymbol{p}\right) &= -\ln f\left(\boldsymbol{z}_1; \boldsymbol{\mu}_1, \mathbf{S}_1\right) - \sum_{n=2}^{N}\ln f\left(\boldsymbol{z}_n|\boldsymbol{z}_{n-1}; \mathbf{A}, \mathbf{S}\right) \\
&\qquad - \sum_{n=1}^{N}\ln f\left(\boldsymbol{x}_n|\boldsymbol{z}_n; \mathbf{B}, \boldsymbol{\Sigma}\right) \qquad\qquad\qquad (7.187)
\end{aligned}
$$

So, expected NLL is:

$$
\begin{aligned}
E_{\mathbf{Z}|\mathbf{X}}\left[-\ln f\left(\mathbf{x},\mathbf{Z};\mathbf{p}\right)\right] &= E_{\mathbf{Z}_1|\mathbf{X}}\left[-\ln f\left(\mathbf{Z}_1;\boldsymbol{\mu}_1,\mathbf{S}_1\right)\right] & (7.188) \\
&- \sum_{n=2}^{N} E_{\mathbf{Z}_n,\mathbf{Z}_{n-1}|\mathbf{X}}\left[\ln f\left(\mathbf{Z}_n|\mathbf{Z}_{n-1};\mathbf{A},\mathbf{S}\right)\right] \\
&- \sum_{n=1}^{N} E_{\mathbf{Z}_n|\mathbf{X}}\left[\ln f\left(\mathbf{x}_n|\mathbf{Z}_n;\mathbf{B},\boldsymbol{\Sigma}\right)\right]
\end{aligned}
$$

using $E_{\mathbf{Z}|\mathbf{X}}\left[-\ln f\left(\mathbf{Z}_i\right)\right] = E_{\mathbf{Z}_i|\mathbf{X}}\left[-\ln f\left(\mathbf{Z}_i\right)\right]$ for any latent variable $\mathbf{Z}_i$. So, we can mostly isolate terms in the expected NLL with respect each of the parameters in $\mathbf{p}$, thereby simplifying the optimization problem:

(1) *E-step*: Use the forward-backward messages to compute the posterior hidden state probabilities, whose expected sufficient statistics are $E_{\mathbf{Z}_n|\mathbf{X}}\left[\mathbf{Z}_n\right] = \boldsymbol{\mu}_{n|N}$ and $E_{\mathbf{Z}_n|\mathbf{X}}\left[\mathbf{Z}_n\mathbf{Z}_n^T\right] = \boldsymbol{\Sigma}_{n|N} + \boldsymbol{\mu}_{n|N}\boldsymbol{\mu}_{n|N}^T$. Similarly, the two-step Markov probability $f\left(\mathbf{z}_n,\mathbf{z}_{n-1}|\mathbf{x}_1,\ldots,\mathbf{x}_N\right)$ is jointly Gaussian with covariance $\mathrm{cov}_{\mathbf{Z}_n,\mathbf{Z}_{n-1}|\mathbf{X}}\left[\mathbf{Z}_n,\mathbf{Z}_{n-1}\right] = \boldsymbol{\Sigma}_{n-1|n-1}\mathbf{A}^T\boldsymbol{\Sigma}_{n|n-1}^{-1}\boldsymbol{\Sigma}_{n|N}$ (Bishop, 2006, page 641). From this, we get $E_{\mathbf{Z}_n,\mathbf{Z}_{n-1}|\mathbf{X}}\left[\mathbf{Z}_n\mathbf{Z}_{n-1}^T\right] = \boldsymbol{\Sigma}_{n-1|n-1}\mathbf{A}^T\boldsymbol{\Sigma}_{n|n-1}^{-1}\boldsymbol{\Sigma}_{n|N} + \boldsymbol{\mu}_{n|N}\boldsymbol{\mu}_{n-1|N}^T$.

(2) *M-step*: For the Markov chain parameter $\mathbf{A}$ we need to find:

$$
\hat{\mathbf{A}} = \arg\min_{\mathbf{A}}\left[-\sum_{n=2}^{N} E_{\mathbf{Z}_n,\mathbf{Z}_{n-1}|\mathbf{X}}\left[\ln f\left(\mathbf{Z}_n|\mathbf{Z}_{n-1};\mathbf{A},\mathbf{S}\right)\right]\right] \quad (7.189)
$$

with solution:

$$
\hat{\mathbf{A}} = \left(\sum_{n=2}^{N} E\left[\mathbf{Z}_n\mathbf{Z}_{n-1}^T\right]\right)\left(\sum_{n=2}^{N} E\left[\mathbf{Z}_{n-1}\mathbf{Z}_{n-1}^T\right]\right)^{-1} \quad (7.190)
$$

where have suppressed the expectation variables for clarity. Using similar matrix calculus arguments for the state error covariance S, the solution is:

$$
\begin{aligned}
\hat{\mathbf{S}} &= \frac{1}{N-1}\sum_{n=2}^{N}\left[E\left[\mathbf{Z}_n\mathbf{Z}_n^T\right] - \hat{\mathbf{A}}E\left[\mathbf{Z}_{n-1}\mathbf{Z}_n^T\right]\right. \\
&\left. - E\left[\mathbf{Z}_n\mathbf{Z}_{n-1}^T\right]\hat{\mathbf{A}}^T + \hat{\mathbf{A}}E\left[\mathbf{Z}_{n-1}\mathbf{Z}_{n-1}^T\right]\hat{\mathbf{A}}^T\right] & (7.191)
\end{aligned}
$$

For the emission distribution parameter $\mathbf{B}$, we need:

$$\hat{\mathbf{B}} = \underset{\mathbf{B}}{\arg\min}\left[-\sum_{n=1}^{N} E_{\mathbf{Z}_n|\mathbf{X}}\left[\ln f\left(\mathbf{x}_n|\mathbf{Z}_n;\mathbf{B},\mathbf{\Sigma}\right)\right]\right]$$

$$= \left(\sum_{n=1}^{N} \mathbf{x}_n E\left[\mathbf{Z}_n^T\right]\right)\left(\sum_{n=1}^{N} E\left[\mathbf{Z}_n \mathbf{Z}_n^T\right]\right)^{-1} \qquad (7.192)$$

For the observation distribution $\mathbf{\Sigma}$:

$$\hat{\mathbf{\Sigma}} = \underset{\mathbf{\Sigma}}{\arg\min}\left[-\sum_{n=1}^{N} E_{\mathbf{Z}_n|X}\left[\ln f\left(\mathbf{x}_n|\mathbf{Z}_n;\mathbf{B},\mathbf{\Sigma}\right)\right]\right] \qquad (7.193)$$

$$= \frac{1}{N}\sum_{n=1}^{N}\left[\mathbf{x}_n \mathbf{x}_n^T - \hat{\mathbf{B}}E\left[\mathbf{Z}_n\right]\mathbf{x}_n^T - \mathbf{x}_n E\left[\mathbf{Z}_n^T\right]\hat{\mathbf{B}} + \hat{\mathbf{B}}E\left[\mathbf{Z}_n \mathbf{Z}_n^T\right]\hat{\mathbf{B}}^T\right]$$

Finally, for the initial state distribution, the updates are analogous to the usual MLEs:

$$\hat{\mu}_1 = E\left[\mathbf{Z}_1\right] \qquad (7.194)$$

$$\hat{\mathbf{S}}_1 = E\left[\mathbf{Z}_1 \mathbf{Z}_1^T\right] - E\left[\mathbf{Z}_1\right]E\left[\mathbf{Z}_1^T\right]$$

To monitor convergence, we need the incomplete data NLL $-\ln f\left(\mathbf{x};\mathbf{p}\right)$ which can be computed from the scale factors discussed above.

Some observations on the reliability of E-M for the KF are necessary. There is an identifiability issue with the KF model described above, since the scaling of the latent state variables (determined by $\mathbf{A}$ and $\mathbf{S}$) can be compensated with a corresponding rescaling of $\mathbf{B}$ and $\mathbf{\Sigma}$. This increases the number of local minima in the NLL, making E-M for the KF somewhat unreliable in practice. In probabilistic PCA, this is addressed by restricting the latent state variables to unit multivariate normals and the observation density to be spherical (see Section 6.6). A similar restriction for the KF can be made to substantially reduce the number of degrees of freedom of the model thereby aiding the reliability of E-M parameter estimation.

Choosing "good" starting values for the E-M iteration is important. A trick that is often used is to start by assuming that the latent states are independent. This makes the KF model similar to probabilistic PCA and where values of the parameters $\mathbf{S},\mathbf{B},\mathbf{\Sigma}$ are more straightforward to estimate than in the full, dependent KF model.

Kalman filtering as signal subspace analysis

There is an intimate relationship between signal subspace analysis (see Section 7.4) and Kalman smoothing. To simplify the discussion, we will focus on the case of 1D latent variables with 1D observations. Indeed, this relationship may not come as a surprise since the latent signal is a discrete-time GP described by a Markov chain (whose autocorrelation matrix can be computed directly), observed through another, i.i.d. conditional GP. This is very similar to the subspace model (7.146) except that the KF has an additional transformation parameter b. We can thus find the MAP solution to this regularization problem analytically, just as we can with signal subspace analysis.

To do this, let us write down the KF likelihood from the PGM, ignoring the initial condition:

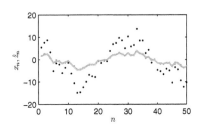

$$f\left(\boldsymbol{x}, \boldsymbol{z}\right) \propto \prod_{n=1}^{N} \mathcal{N}\left(x_n; b\, z_n, \sigma^2\right) \prod_{n=2}^{N} \mathcal{N}\left(z_n; a\, z_{n-1}, \tau^2\right) \tag{7.195}$$

so that the NLL is:

Fig. 7.18: Kalman filter (KF) smoothing of an observed 1D signal x (black dots) and the recovered latent 1D latent signal $\hat{z}$ (grey curve) using Tikhonov regularization.

$$-\ln f\left(\boldsymbol{x}, \boldsymbol{z}\right) \propto \frac{1}{2\sigma^2}\sum_{n=1}^{N}\left(x_n - b\,z_n\right)^2 + \frac{1}{2\tau^2}\sum_{n=2}^{N}\left(z_n - a\,z_{n-1}\right)^2 \tag{7.196}$$

where we have omitted constants that do not vary with any of the model parameters. This can be minimized analytically with respect to $\boldsymbol{z}$ by writing it as a "sum-of-norms" functional:

$$E\left(\boldsymbol{z}\right) = \|\boldsymbol{x} - \mathbf{Q}\boldsymbol{z}\|_{\mathbf{W}}^2 + \|\mathbf{R}\boldsymbol{z}\|_{\mathbf{V}}^2 \tag{7.197}$$

where $\mathbf{W} = \frac{1}{2\sigma^2}\mathbf{I}$ and $\mathbf{Q} = b\mathbf{I}$ are $N \times N$ matrices, and matrix $\mathbf{R}$ has 1's on the diagonal and $-a$'s on the first lower diagonal, and $\mathbf{V} = \frac{1}{2\tau^2}\mathbf{I}$. This is a Tikhonov regularization problem (see Section 2.2) with analytical solution:

$$\hat{\boldsymbol{z}} = \left(\mathbf{Q}^T\mathbf{W}\mathbf{Q} + \mathbf{R}^T\mathbf{V}\mathbf{R}\right)^{-1}\mathbf{Q}^T\mathbf{W}\boldsymbol{x} \tag{7.198}$$

See Figure 7.18. It is interesting to contrast the linear-Gaussian KF with certain kinds of nonlinear smoothing methods such as *total variation denoising*:

$$E\left(\boldsymbol{z}\right) = \frac{1}{2}\sum_{n=1}^{N}|x_n - z_n|^2 + \lambda\sum_{n=2}^{N}|z_n - z_{n-1}| \tag{7.199}$$

These ideas are explored in more details in Section 9.3). For more details on the MAP regularization interpretation of the KF, see Ohlsson *et al.* (2010). Also, as a linear-Gaussian subspace method, the KF has an interpretation in terms of Bayesian nonparametric Gaussian process regression (see Section 10.2).

7.6   Time-varying linear systems

Fourier transforms are fundamentally time-invariant operations. Thus, Fourier analysis makes the implicit assumption that frequency is important and time is not. In practice, this is rarely the case: most signals we encounter have some degree of time variance, for example, the amplitude or phase of frequency components in the signal change over time. In these circumstances Fourier analysis is inappropriate and we need to turn to time-varying linear methods instead.

Short-time Fourier transform (STFT) and perfect reconstruction

Perhaps the simplest time-varying linear analysis method is obtained by assuming local time-invariance and then applying Fourier analysis to a sliding time window. This leads to the following simple time-varying DTFT spectral estimator for the signal x_n with $n \in \mathbb{Z}$:

$$X^k(\omega) = \sum_{n \in \mathbb{Z}} x_n T_{kL}[w_n] \exp(-i\omega n) \qquad (7.200)$$

where $k \in \mathbb{Z}$ is the *window index* and and $L \in \mathbb{N}$ is the *overlap* between windows. The window function $w_n \in \mathbb{R}$ selects a set of samples in the input signal for DTFT analysis, which is a rectangular window of size $W \in \mathbb{N}$ centred on time index n:

$$w_n = \begin{cases} 1 & n \in \frac{1}{2}\left[-(W-1), W-1\right] \\ 0 & \text{otherwise} \end{cases} \qquad (7.201)$$

For an elementary sliding window we have $L = 1$ in which case k coincides with the sample index n. This is known as (discrete-time) *short-time Fourier transform* (STFT) *analysis* of the signal $\mathbf{x}$.

The square magnitude of the DTFT function $\left|X^k(\omega)\right|^2$ represents an estimate of the local power spectrum in window k. The size of the window W determines the trade-off between sensitivity to rapid local spectral changes (better when W is small) and frequency resolution (better when W is large), in accordance with the limitations of Heisenberg uncertainty (Section 7.2). The representation $X^k(\omega)$ is known as a *time-frequency analysis plane* with coordinates $(k, \omega) \in (\mathbb{Z}, \mathbb{R})$.

In practice however, this simple estimator will suffer from various problems since the window is rectangular and time-limited. A finite-length DFT is being performed in each window so that periodic continuation will cause spurious high frequency components (Gibbs phenomena) to appear in each window unless the signal values at the start and end of the window are close in value. There will also be *spectral leakage* because multiplication by the rectangular window causes convolution in the frequency domain with the sinc function, smearing out the spectrum of the signal. This suggests the use of non-rectangular windowing. For example, the simple triangular window:

$$w_n = \begin{cases} 1 - \frac{2|n|}{W} & n \in \frac{1}{2}\left[-(W-1), W-1\right] \\ 0 & \text{otherwise} \end{cases} \qquad (7.202)$$

mitigates discontinuities across window edges (both edges have values close to zero) and has a Fourier transform which is closer to a Dirac

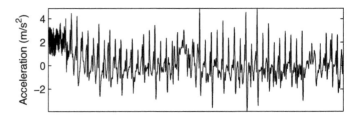

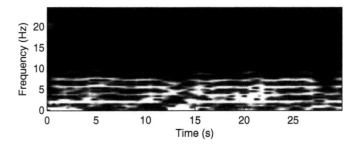

Fig. 7.19: **Sample-by-sample** ($L = 1$) **sliding discrete short-time Fourier transform (STFT) analysis of a recording of walking behaviour from a single axis of a hip-worn accelerometer (sample rate ~97Hz, STFT window size** $W = 256$, **triangular window). The signal in the time domain (top panel) shows clear periodicities whose cycle waveshape and frequency changes over time. These cycles are visible as large magnitude frequency components with harmonics in the STFT domain (bottom panel, brighter indicates larger magnitude components).**

delta function (the convolution identity) than that of the rectangular window (Figure 7.19).

Apart from analysis, we are often interested in filtering, that is modifying, the time-frequency space of the signal and then reconstructing the signal in the time domain. To do this, we will want to change each $X^k(\omega)$ in some way. If we add up each local DFT, we get:

$$
\begin{aligned}
\sum_{k \in \mathbb{Z}} X^k(\omega) &= \sum_{k \in \mathbb{Z}} \sum_{n \in \mathbb{Z}} x_n T_{kL}[w_n] \exp(-i\omega n) \\
&= \sum_{n \in \mathbb{Z}} x_n \exp(-i\omega n) \sum_{k \in \mathbb{Z}} T_{kL}[w_n]
\end{aligned}
\tag{7.203}
$$

Now, if for all $n \in \mathbb{Z}$:

$$
\sum_{k \in \mathbb{Z}} T_{kL}[w_n] = 1
\tag{7.204}
$$

then we get:

$$
\begin{aligned}
\sum_{k \in \mathbb{Z}} X^k(\omega) &= \sum_{n \in \mathbb{Z}} x_n \exp(-i\omega n) \\
&= X(\omega)
\end{aligned}
\tag{7.205}
$$

which is just the DTFT of the signal $\boldsymbol{x}$. It follows that we can reconstruct the signal using STFT *synthesis*, using the corresponding IDTFT:

$$\frac{1}{2\pi} \int_{-\pi}^{\pi} X\left(\omega\right) \exp\left(i\omega n\right) d\omega \;=\; \sum_{k\in\mathbb{Z}} \frac{1}{2\pi} \int_{-\pi}^{\pi} X^k\left(\omega\right) \exp\left(i\omega n\right) d\omega$$

$$= \sum_{k\in\mathbb{Z}} x^k = x_n \qquad (7.206)$$

where $\boldsymbol{x}^k = \boldsymbol{x} \circ T_{kL}\left[\boldsymbol{w}\right]$.

The requirement (7.204) that the sum of the window functions over k is unity with respect to n is known as the *constant overlap add* constraint that makes the STFT equivalent to the DTFT and allowing *perfect reconstruction*. Up to a constant specific to each window (which is straightforward to divide out) all windows have this property for all $L = 1$ but this sliding STFT is a highly redundant representation of the signal. We can construct much more parsimonious time-frequency representations with constant overlap add windows for larger values $L \leq W$. For example, the triangular window (7.202) and the *Hann* window:

$$w_n = \begin{cases} \cos^2\left(\frac{\pi n}{W}\right) & n \in \frac{1}{2}\left[-\left(W-1\right), W-1\right] \\ 0 & \text{otherwise} \end{cases} \qquad (7.207)$$

satisfy this constraint for $L = W/2$. Other windows can be constructed to have specific leakage with a given overlap (Borß and Martin, 2012).

In the real world, all signals are finite so the DTFT is replaced by the DFT with periodic continuation as a consequence. But this also means that each DFT/IDFT in the analysis and synthesis can be efficiently computed using FFTs. Each window is represented by a time and frequency discrete representation vector $\boldsymbol{X}^k$ for each $k \in 1, 2, \ldots, K$, and this is manipulated to produce $\tilde{\boldsymbol{X}}^k$ so that the modified signal can be resynthesized. The modifications might involve some kind of time-varying filtering, so that $\tilde{\boldsymbol{X}}^k = \boldsymbol{H}^k \circ \boldsymbol{X}^k$.

Continuous-time wavelet transforms (CWT)

The STFT is time-varying but we cannot escape Heisenberg uncertainty. If we consider the STFT in continuous-time, this would be based on windowed complex exponentials $u\left(t\right) = T_{t'}\left[w\left(t\right)\right]\exp\left(i\omega t\right)$ for $t, t' \in \mathbb{R}$. Since the window function $w\left(t\right)$ is symmetric about $t = 0$, the Heisenberg time spread around any t' is:

$$\sigma_t^2 = \int_{-\infty}^{\infty} \left(t - t'\right)^2 \left|u\left(t\right)\right|^2 dt = \int_{-\infty}^{\infty} t^2 \left|w\left(t\right)\right|^2 dt \qquad (7.208)$$

which is independent of the shift t' and frequency ω. Similarly, the FT of the window function is also real and symmetric about $\omega = 0$, so the frequency spread around any ω' is:

It is worth pointing out that with rectangular windows of length $W = L$ and time-invariant spectral modifications, we recover the overlap-add method of long convolution (Algorithm 7.2).

$$\sigma_\omega^2 = \int_{-\infty}^{\infty} (\omega - \omega')^2 |U(\omega)|^2 \, d\omega$$

$$= \int_{-\infty}^{\infty} \omega^2 |W(\omega)|^2 \, d\omega \qquad (7.209)$$

where $U(\omega) = F_\omega [u(t)]$ and $W(\omega) = F_\omega [w(t)]$.

Thus the frequency spread is independent of frequency shift ω'. We can visualize this as a "tiling" of the time-frequency plane with resolution that is independent of time and frequency (Figure 7.20). Heisenberg uncertainty means that the area of each tile remains fixed: increasing the window length W decreases the frequency spread, but reciprocally increases the time spread. There are alternatives to the STFT that have different ways of distributing the Heisenberg uncertainty across the time-frequency plane, perhaps the most widely-used of which is the *wavelet transform*. Wavelets are basis functions much like the windowed complex exponentials of the STFT that are localized in both time and frequency. They are much more flexible than the STFT bases as they can be constructed to have certain useful properties such as the compactness of their representation of signals of a given smoothness and control over their compactness in time.

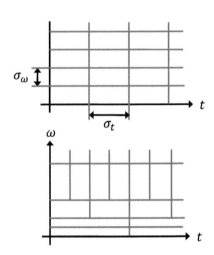

Prototypical wavelet analysis uses the *(forward) continuous wavelet transform* (CWT) of the real signal $x(t)$:

$$X(\tau, s) = W_{\tau,s} [x(t)] = \int_{-\infty}^{\infty} x(t) \frac{1}{\sqrt{s}} \psi\left(\frac{\tau - t}{s}\right) dt \qquad (7.210)$$

where $\tau \in \mathbb{R}$ is the *translation* or time shift parameter, and $s \in \mathbb{R}$ with $s > 0$ is the *scale* parameter. The function $\psi(t)$ is known as the *mother wavelet* function, chosen such that $\|\psi\| = 1$.

For any $v > 0$, the Fourier transform scaling property shows that expanding the scale of the time axis causes the frequency axis to be compressed:

$$F_\omega [\psi(vt)] = \frac{1}{v} \Psi\left(\frac{\omega}{v}\right) \qquad (7.211)$$

Fig. 7.20: Illustration contrasting time-frequency (t, ω) Heisenberg "tiling" properties of two widely-used time varying analysis methods: the short-time Fourier transform (STFT, left) and the discrete wavelet transform (DWT, right). The STFT has time-frequency resolution quantified by σ_t and σ_ω, respectively, which is independent of time and and frequency. Whereas, the DWT provides a different tiling of the time-frequency plane that depends upon time-frequency location.

So, the wavelet scale parameter s is analogous to the frequency parameter in the STFT. It therefore makes sense to think of the wavelet *translation-scale plane* (τ, s) as being analogous to the time-frequency plane of the STFT.

Mallat (2009, page 105) shows that if the FT $\Psi(\omega)$ of the mother wavelet satisfies the following *admissibility* condition:

$$C_\psi = \int_0^{\infty} \frac{|\Psi(\omega)|^2}{\omega} d\omega < \infty \qquad (7.212)$$

then the inverse $ICWT$ exists and we can recover the original signal from its CWT $X(\tau, s)$:

$$x(t) = \frac{1}{C_\psi} \int_0^{\infty} \int_{-\infty}^{\infty} X(\tau, s) \frac{1}{\sqrt{s}} \psi\left(\frac{t - \tau}{s}\right) d\tau \frac{ds}{s^2} \qquad (7.213)$$

If the mother wavelet is continuously differentiable and if $\Psi(0) = 0$, then the integral in (7.212) is finite and the wavelet is admissible. This means that the mother wavelet must have a spectrum with zero amplitude at zero frequency. In effect, the mother wavelet behaves like a frequency-selective bandpass filter (see Section 7.3). This condition on the FT in turn implies that $\int_{-\infty}^{\infty} \psi(t)\,dt = 0$ so the mother wavelet function is also "wave-like" with equal area above and below the horizontal axis.

Wavelet analysis imposes far less constraints on the mathematical structure of the problem than Fourier analysis, so we should expect to encounter many more complexities since more choices have to be made. Wavelet theory is therefore far more complex than classical Fourier analysis, and the following sections are only a brief overview of what is an active area of research. For further reference, the book by Mallat (2009) is an excellent and readable introduction, a particularly strong on the connections between wavelet analysis and time-varying linear DSP in general. To go even deeper the celebrated book by Daubechies (1992) is essential reading.

Discretization and the discrete wavelet transform (DWT)

The CWT is highly redundant: not all the time-scale transform information is required to reconstruct the original signal using the ICWT. Also, in practice we have finite-length, discrete-time signals so the CWT needs appropriate discretization. Ideally, we want this discrete transform to use an orthonormal basis, as this simplifies the analysis. These criteria, along with those imposed by the CWT above, lead to the *discrete wavelet transform* (DWT), which we describe next.

To develop the DWT, we restrict the continuous translation-scale plane to the discretized, *dyadic* plane $(m, k) \in \mathbb{Z}^2$. The goal is to find an appropriate choice of mother wavelet function such that the following set of orthonormal functions is a basis for any real signal $x(t)$:

$$\psi^{m,k}(t) = \frac{1}{\sqrt{2^k}} \psi\left(\frac{t - 2^k m}{2^k}\right) \qquad (7.214)$$

The signal must also be square integrable $\int |x(t)|^2\,dt < \infty$ (i.e. finite energy).

This dyadic tiling is organized such that every doubling of Heisenberg spread in frequency leads to a halving of spread in time (Figure 7.20). Fourier scaling (7.211) means that bandpass mother wavelets localized at frequency ω are shifted to $\omega_k = 2^{-k}\omega$ at scale k, and wavelets at scale $k+1$ have central frequency $\omega_{k+1} = 2^{-(k+1)}\omega = \frac{1}{2}\omega_k$, so, increasing k gives bandpass wavelets of reducing frequency coverage. Thus, provided the wavelet functions have frequency response behaviours that allow constant additive overlap in spectral amplitude (like the constant overlap property in the STFT above), then the full bandwidth of the signal can be covered by the (infinite) dyadic tiling without any loss of information: we can split the signal into separate frequency bands at each scale k, and then simply add them back together again to recover the original signal.

To be practical however, we need a finite representation. Discrete signals have finite maximum spectra (at the Nyquist frequency), which sets the smallest scale value k. At the largest scale value (corresponding to the lowest frequency, including zero), we can construct a function with low-pass frequency response which is complementary in that it has constant amplitude overlap in the frequency domain with the lowest frequency wavelet. This known as the *scaling* function $\phi(t)$. Now the set of functions $\psi^{m,k}(t)$ and $\phi^m(t) = \phi(t - m)$ together allow full tiling of the finite-duration translation-scale plane in a finite way.

The scaling function, as a low-pass filter, integrates over some temporal region of the signal and the output of that filtering operation is thus an approximate representation of the input signal. If we compress or expand the temporal scale of this function, we will create approximations at different resolutions. This can be formalized as a *multiresolution approximation system* of representations of the signal at different levels of detail (Mallat, 2009, pages 264-267). More exactly, a dyadic approximation at a given resolution k is an integration of the signal over a finite temporal region of the signal of size proportional to 2^k. These regions are nested inside each other: every region at scale $k+1$ contains 2 smaller, higher resolution regions at scale k. This integration is performed by the scaling function so that the translated scaling functions at resolution k:

$$\phi^{m,k}(t) = \frac{1}{\sqrt{2^k}}\phi\left(\frac{t - 2^k m}{2^k}\right) \qquad (7.215)$$

provide a basis for the approximation of the signal at that resolution. Furthermore, the nesting property means that the set $\phi^{m,k}(t)$ for $(m, k) \in \mathbb{Z}^2$ is a basis for the entire multiresolution approximation system. Therefore, the scaling functions at the coarser level $k = 1$ can be written as an expansion in terms of the more detailed level $k = 0$, giving the *two-scale equation* (Mallat, 2009, page 270):

$$\frac{1}{\sqrt{2}}\phi\left(\frac{t}{2}\right) = \sum_{m \in \mathbb{Z}} h_m \phi(t - m) \qquad (7.216)$$

The coefficients of this expansion:

$$h_m = \left\langle \frac{1}{\sqrt{2}}\phi\left(\frac{t}{2}\right), \phi(t - m) \right\rangle \qquad (7.217)$$

play a fundamental role in the DWT. Mallat (2009, Theorem 7.2) shows that they define an FIR filter (Section 7.3) called the *scaling* filter whose FT $H(\omega)$ must satisfy the *conjugate mirror* property:

$$|H(\omega)|^2 + |H(\omega + \pi)|^2 = 2 \qquad (7.218)$$

along with $H(0) = \sqrt{2}$. The filter $H(\omega + \pi)$ is what we get if we take $H(\omega)$ and mirror it about the Nyquist frequency. This means that the two filters, $H(\omega)$ and $H(\omega + \pi)$, split the signal into two complementary parts that, when added together, exactly reconstruct the input signal.

Given an approximation at scale k, we can subtract the approximation at the next coarser level $k+1$. What remains is the *detail signal* between e.g. levels k and $k+1$. Mallat (2009, Theorem 7.3) shows that for fixed k, the wavelet functions $\psi^{k,m}(t)$ are an orthonormal basis for this detail signal, and that the wavelet at scale $k=1$ can be written in the finer scaling function basis $k=0$:

$$\frac{1}{\sqrt{2}}\psi\left(\frac{t}{2}\right) = \sum_{m\in\mathbb{Z}} g_m \phi(t-m) \tag{7.219}$$

with expansion coefficients:

$$g_m = \left\langle \frac{1}{\sqrt{2}}\psi\left(\frac{t}{2}\right), \phi(t-m) \right\rangle \tag{7.220}$$

This defines another FIR filter known as the *detail filter*, which also must satisfy the conjugate mirror property (7.218). It is determined by the scaling filter $H(\omega)$ through the relationship $G(\omega) = \exp(-i\omega)\bar{H}(\omega+\pi)$. This tells us that the detail filter is obtained from the scaling filter by mirroring it around the Nyquist frequency. Taking the inverse FT gives us a direct relationship between the coefficients of the scaling and detail filters:

$$g_m = (-1)^{1-m} h_{1-m} \tag{7.221}$$

for $m \in \mathbb{Z}$. Furthermore, the mother wavelet function ψ must satisfy the constraint:

$$\Psi(\omega) = \frac{1}{\sqrt{2}} G\left(\frac{\omega}{2}\right) \Phi\left(\frac{\omega}{2}\right) \tag{7.222}$$

The scaling and detail filter coefficients are sufficient for us to now specify the DWT as an explicit algorithm. The overall idea is that applying the scaling filter to the signal computes the approximation at level k, which allows us to then compute the coarser approximation at level $k+1$ along with the wavelet coefficients using the detail filter at level $k+1$. The scaling functions $\phi^{m,k}(t)$ and wavelets $\psi^{m,k}(t)$ are orthonormal bases of the approximation and detail signals respectively, so their expansion coefficients at each level are:

$$\begin{aligned} a_n^k &= \langle x(t), \phi^{n,k}(t) \rangle \\ d_n^k &= \langle x(t), \psi^{n,k}(t) \rangle \end{aligned} \tag{7.223}$$

Typically, we will not be able to write down analytical expressions for ϕ, ψ directly. This would seem to be a major problem for computing these wavelet and scaling coefficients, but it turns out that we do not need these expressions because these coefficients can in fact be computed recursively in terms of each other. Mallat (2009, Theorem 7.10) shows that:

$$a_m^{k+1} = \sum_{n\in\mathbb{Z}} h_{n-2m} a_n^k = \left(R\left[\boldsymbol{h}\right] \star \boldsymbol{a}^k\right)_{2m}$$

$$d_m^{k+1} = \sum_{n\in\mathbb{Z}} g_{n-2m} a_n^k = \left(R\left[\boldsymbol{g}\right] \star \boldsymbol{a}^k\right)_{2m} \qquad (7.224)$$

Decimation is a subsampling operation widely used for sample rate conversion in multirate signal processing.

This tells us that the approximation and detail coefficients at level $k+1$ are obtained from those at level k, through filtering with the conjugate mirror FIRs. The filter convolution uses the reversed impulse response *decimated* by a factor of 2 at each stage, that is, by dropping every other coefficient from the convolution sequence. The detail coefficients are the coefficients of the DWT, along with the final, lowest frequency scale approximation. Thus, we can view the forward DWT as a recursive set of filtering and decimation operations.

We have to start the recursion somewhere. A simple solution is to consider the discretized signal x_n as an accurate approximation to the underlying continuous signal $x(t)$. In that case, we can identify the top-level (finest approximation) coefficients with the input signal, i.e. $a_n^0 = x_n$. This also sets the number of scales, since we will find that decimation reduces the number of coefficients a_m^k, d_m^k at each scale to the point where the convolution is no longer possible. More sophisticated approaches to discretization for initializing the recursion are given in Mallat (2009, page 301).

The final issue to raise is that for finite-duration signals x_n for $n = 1, 2, \ldots, N$ it will not be possible to compute the convolutions since in most cases the FIR coefficients $\boldsymbol{h}, \boldsymbol{g}$ are infinite in duration. The simplest solution might be to truncate the coefficients appropriately, but this causes a distortion in the filter's frequency response which has undesirable effects on the wavelet analysis. Another approach is assume periodicity, that is, replace the convolutions in (7.224) with circular convolutions. In which case, larger scale wavelets will effectively "wrap" around the time axis. Of course, this complexity must be taken into consideration when interpreting the DWT coefficients.

Inverting the DWT to recover the original signal can be explained in terms of running the analysis filter bank backwards, i.e. filtering the detail and approximation coefficients at the coarser scale $k + 1$, and combining these to reconstruct the approximation at the finer scale k. Because the scaling and wavelet functions at level $k+1$ are a basis for the scaling function at level k, it can be shown that (Mallat, 2009, Theorem 7.10):

$$a_m^k = \sum_{n\in\mathbb{Z}} h_{m-2n} a_n^{k+1} + \sum_{n\in\mathbb{Z}} g_{m-2n} d_n^{k+1} \qquad (7.225)$$

This defines the *inverse* DWT (IDWT), in which firstly, the coefficients $\boldsymbol{a}^{k+1}, \boldsymbol{d}^{k+1}$ are interpolated by inserting zeros between each value in each coefficient vector, then convolving with the scaling and detail filters, and then summing up the result. As with the DWT, for finite-length signals, the convolutions can be replaced by circular variants. In this way, the DWT/IDWT form a one-one pair of orthonormal transformations between the input signal and the wavelet coefficient sequences, analogous to the DFT/IDFT pair. We can also construct an orthonormal DWT $N \times N$ matrix $\mathbf{W}$ where the n-th column is a discretized wavelet obtained by applying the IDWT to the length N standard basis signal $\boldsymbol{e}_n$. Then the DWT of the signal $\boldsymbol{x}$ is just $W_{m,k}[\boldsymbol{x}] = \mathbf{W}^T \boldsymbol{x}$, and the IDWT $W_{m,k}^{-1}[\boldsymbol{X}] = (\mathbf{W}^T)^{-1} \boldsymbol{X} = \mathbf{W} \boldsymbol{X}$.

The decimation (and interpolation) of the filtering implementation of the DWT described above mean that the DWT/IDWT are both $O(N)$ operations. This contrasts with the fastest way of computing the DFT – the FFT – requiring $O(N \log N)$ computational effort. Thus, the increase in representational efficiency of the DWT comes with even more computational efficiency than the FFT, a remarkable fact.

Wavelet design

As stated above, wavelet transforms are unlike Fourier transforms in that there are an infinity of basis functions that can, for example, satisfy the time-frequency tiling properties of the multiresolution DWT. Wavelet bases are, therefore, usually specified indirectly through various design criteria, and then computed numerically. In this section we will describe these criteria and some of the most widely-used wavelets, see Figure 7.21 for an illustration of some of these wavelet functions.

Design criteria include their ability to efficiently represent particular classes of signals of varying smoothness. The *vanishing moment* criteria, as expressed by $\int_{-\infty}^{\infty} t^k \psi(t) \, dt = 0$ for $k \in 0, 1, \ldots, K$, tells us that the wavelet is orthogonal to any polynomial of degree K. The basic intuition here is that, as the number of vanishing moments increases, if the smoothness of the signal $x(t)$ is held constant, then the wavelet coefficients at the fine scales decrease in magnitude (Mallat, 2009, Theorem 6.3). Thus, the number of vanishing moments partly controls the extent to which any detail (high frequency components, edges, spikes etc.) in the signal requires large magnitude wavelet coefficients (Figure 7.22). This is important for capturing a *sparse representation* of the signal, where hopefully only a small fraction of the coefficients are large so that the signal is effectively summarized by its wavelet time-frequency representation. Another criteria is the *support* of the wavelet ψ. A wavelet with large support will cause high-frequency detail to spread among many of the wavelet coefficients, reducing the sparsity of the representation. So, to help with the interpretability of the transform, it is useful to keep the support of the wavelet small.

Vanishing moments and support are independent properties of wavelets, but orthogonality (for example in the DWT) forces the support size and the number of vanishing moments to become coupled. Precisely: if ψ has $K+1$ vanishing moments, then the support size is at least $2(K+1) - 1$ (Mallat, 2009, Theorem 7.7).

The simplest wavelet is the piecewise constant *Haar* wavelet, whose

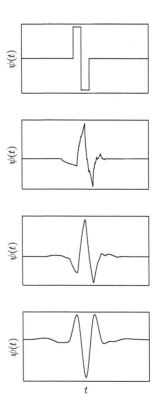

Fig. 7.21: A selection of wavelet functions $\psi(t)$ (top to bottom): piecewise constant Haar wavelet; order $K = 1$ (two vanishing moments) Daubechies wavelet; symmlet with $K = 3$ (four vanishing moments) and coiflet with $K = 7$ (eight vanishing moments).

compact wavelet function is given by:

$$\psi\left(t\right) = \begin{cases} -1 & 0 \le t < \frac{1}{2} \\ 1 & \frac{1}{2} \le t < 1 \\ 0 & \text{otherwise} \end{cases} \qquad (7.226)$$

This is ideally matched to representing piecewise constant functions. The scaling FIR filter coefficients are compact with only two elements, $\boldsymbol{h} = \frac{1}{\sqrt{2}}\begin{pmatrix} 1 & 1 \end{pmatrix}$. It has only one vanishing moment. This wavelet can be considered as the first in a hierarchy of *spline* wavelets including the *Battle-Lemarié* wavelets. For splines of degree Q, these wavelets have $Q+1$ vanishing moments. Explicit expressions for the FT of the scaling FIR filter are given in Mallat (2009, page 277) and by numerical Fourier inversion, a truncated set of FIR coefficient values are given for orders $Q = 1$ (linear) and $Q = 3$ (cubic) splines (Mallat, 2009, Table 7.1).

The *Daubechies* wavelets have compact support of minimum size for a fixed number of vanishing moments (the Haar wavelet is a special case with $K = 0$). This makes them widely used in practice. Explicit expressions for the wavelet function ψ are not known but computational procedures for generating scaling FIR coefficients can been derived to satisfy the moment and support constraints exactly (Sherlock and Kakad, 2002). A list of explicit FIR coefficient values for vanishing moment constraints $K = 1, 2, \ldots, 9$ is given in Mallat (2009, Table 7.2). Daubechies wavelet functions are highly asymmetric and this can be a problem for applying wavelet transforms in practice. Daubechies *symmlets* are designed to be much more symmetric at the same time as having compact support with a specified number of vanishing moments. Finally, *coiflets* are designed such that both the wavelet and scaling functions have a specified number of vanishing moments (with $\int_{-\infty}^{\infty} \phi\left(t\right) dt = 1$ since scaling functions act as low-pass filters).

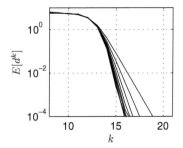

Fig. 7.22: **The number of vanishing moments for a wavelet function determines the expected magnitude of fine scale (large k) DWT coefficients. Here, the different curves correspond to changing the order K of the Daubechies wavelet applied to a random discrete Gaussian process, low-pass filtered to emphasize that the high frequency (fine scale) wavelet coefficients become sensitive to the choice of K.**

Applications of the DWT

Because of the simplicity of e.g. the DWT, wavelet analysis/synthesis finds widespread use in practice, for example, *wavelet shrinkage* estimation of a latent signal in noise where it is known that it is sparse in the wavelet basis. Consider the situation where we have i.i.d. Gaussian wavelet reconstruction errors so that the signal $\boldsymbol{X} \in \mathbb{R}^N$ is generated by $\boldsymbol{X} \sim \mathcal{N}\left(\mathbf{W}\boldsymbol{V}, \sigma^2 \mathbf{I}\right)$ where $\mathbf{W} \in \mathbb{R}^{N \times N}$ is an orthonormal DWT matrix and $\boldsymbol{V} \in \mathbb{R}^N$ are the unknown wavelet coefficients. Also, assume that these coefficients are *sparse* (only a few are large in magnitude). A good model for this situation is where the coefficients are i.i.d. Laplace distributed (4.43) with zero median and scale parameter b. The posterior distribution of the signal is then:

$$f\left(\boldsymbol{v}|\boldsymbol{x}; \mathbf{W}, \sigma^2, b\right) \quad \propto \quad \exp\left(-\frac{1}{2\sigma^2}\|\mathbf{W}\boldsymbol{v} - \boldsymbol{x}\|_2^2\right)$$
$$\times \quad \exp\left(-\frac{1}{b}\|\boldsymbol{v}\|_1\right) \tag{7.227}$$

The MAP solution is obtained by minimizing the following mixed L_2-L_1 objective (Section 2.2):

$$E\left(\boldsymbol{v}\right) = \frac{1}{2}\|\mathbf{W}\boldsymbol{v} - \boldsymbol{x}\|_2^2 + \lambda\|\boldsymbol{v}\|_1 \tag{7.228}$$

where we have used $\lambda = \sigma^2/b > 0$ as a single regularization parameter that controls the trade-off between sparsity of the coefficients (due to the Laplace prior) and the likelihood error. Since $\mathbf{W}$ is orthonormal, it follows that we can solve this simply by finding $\tilde{\boldsymbol{v}} = \mathbf{W}^T\boldsymbol{x}$ and then applying the shrinkage rule (2.19). But, since $\mathbf{W}^T\boldsymbol{x}$ is the forward DWT, we can compute this using the filter bank algorithm. This makes the entire shrinkage estimator $O\left(N\right)$.

Candès (2006) observes that the MAP orthonormal wavelet shrinkage discussed here and Fourier signal subspace analysis (7.149) are obviously examples of the same MAP procedure: first compute the forward orthonormal transform, then shrink the transform coefficients towards zero, and finally, reconstruct the signal using the inverse transform on the modified transform coefficients. The main difference is that Fourier signal subspace analysis assumes stationarity (time-invariance), whereas wavelet shrinkage is time-varying (Figure 7.23).

Wavelet analysis is ubiquitous in DSP and machine learning, so it is not possible to summarize everything here. Wavelet analysis, particularly through the computational simplicity of the DWT, has been applied to signal and image denoising, data compression, compressive sensing and "inverse problems" such as super-resolution signal estimation and missing sample recovery. For an excellent overview, see Mallat (2009). Similarly, wavelets have found important usage in statistical applications such as density estimation (Vidakovic, 1999).

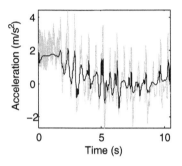

Fig. 7.23: DWT-based wavelet shrinkage applied to the accelerometer walking signal in Figure 7.19, showing that DWT offers a simple solution to simultaneously attenuate high-frequency interference signals (in the first 2 seconds) and yet retain the location of the peaks in the signal (coiflet wavelet basis with 10 vanishing moments and shrinkage parameter $\lambda = 2.5$ for a signal of length $N = 1024$).

Discrete signals: sampling, quantization and coding

8

Digital signal processing and machine learning require digital data which can be processed by algorithms on computer. However, most of the real-world signals that we observe are real numbers, occurring at real time values. This means that it is impossible in practice to store these signals on a computer and we must find some approximate signal representation which is amenable to finite, digital storage. This chapter describes the main methods which are used in practice to solve this representation problem.

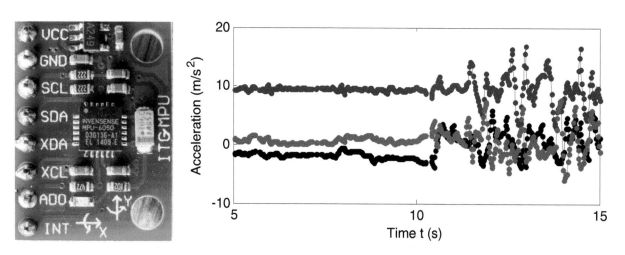

Fig. 8.1: Today's digital sensors are typically tiny and ubiquitous and can produce vast quantities of discrete-time, discrete-valued data. *Microelectromechanical systems* (MEMS) sensor chips such as this Invensense MPU-6050 (left), commonly found in smartphones, measures both 3-axis acceleration and orientation at typical sample rates of 50Hz and upwards, to a resolution of 16 bits. Typical 50Hz digital acceleration signals recorded by the sensor (right panel) in the X-axis (black), Y-axis (blue) and Z-axis (red). Each point is a single acceleration value. Sensor board photograph © Nevit Dilmen, CC BY-SA 3.0.

Digital signals can be found in all walks of life. For example, in 'consumer applications' one can find digital recordings of speech and music, photographs and video. Communications infrastructure such as the fixed land line and mobile telephone networks carry digital speech and video signals from one part of the world to the other. Digital sensors such as accelerometers or gyroscopes embedded in smartphones capture movements of the device (Figure 8.1). In civil engineering, structures such

Machine Learning for Signal Processing: Data Science, Algorithms, and Computational Statistics. Max A. Little.
© Max A. Little 2019. Published in 2019 by Oxford University Press. DOI: 10.1093/oso/9780198714934.001.0001

as bridges are monitored by digital strain gauges which record changing loads and stresses. Magnetometers in spacecraft record a stream of digital data about the flux in magnetic field strength encountered. In intensive care units in hospitals, patient's vital signs such are continuously monitored using digital electrocardiograms and photoplethysmographs which measure electrical and blood flow and oxygenation properties of the cardiovascular system. What all these digital signals have in common is that they are represented by a sequence of finite range, discrete-valued numbers which are captured at specific points in time.

8.1   Discrete-time sampling

To formalize the presentation, consider representing a real-world signal of interest by a real function f of a real time variable $f : \mathbb{R} \to \mathbb{R}$, we will usually write this $f(t)$ with $t \in \mathbb{R}$. In order to store this signal on a computer, we will have to approximate time as an integer value. There are a great many ways in which we can do this, but the difficulty will be that we cannot have access to f at all values of t, instead, we will only know the signal at a finite set of values of the time variable. *Sampling* is the process by which a selection of these values of f are captured. Typically, one obtains values at $f(t_n)$ at the sampling time points t_n, $n \in \mathbb{Z}$, which are generally increasing, $t_n < t_{n+1}$ for all n. We will denote this sampled sequence of values as f_n. When $t_{n+1} - t_n = \Delta t$ for all n, then we say that the sampling is *uniform*.

Ideally we want the sampling to introduce no error such that the signal can be perfectly recovered, or *reconstructed*, from the sampled values, but we have to expect that this is not generally possible. However, it turns out that by placing some constraints on the possible form of f, then perfect reconstruction is possible. The forms of these constraints will determine the nature of the sampling.

A unifying view we will take is that sampling is essentially a problem in *interpolation*, that is, finding a function which goes exactly through the sampled points and which coincides with f between these points (Figure 8.3). This has much in common with the problem of regression in statistics and machine learning, and indeed, the mathematical techniques in these two domains are intimately related. Where regression differs from interpolation is that in interpolation, the measurements are not noisy, that is, they have no observational error.

Typically, the process of sampling takes a real-world signal $f(t)$ captured by some kind of electronic measurement process (e.g. sensor) and applies *pre-filtering* (most carried out using some kind of analogue electronic device) to the signal to make it amenable to sampling. Then, an *analogue-to-digital converter* (ADC) samples values of the processed signal at uniform time intervals. These digital samples are stored and processed using some kind of digital hardware or software algorithm. The digital signal is sent to a *digital-to-analogue* converter (DAC) whose output is *post-filtered* (again, often using an analogue electronic device) to create a reconstruction of the original signal $\hat{f}(t)$ (Figure 8.2).

Bandlimited sampling

As suggested above, the perfect reconstruction of f from the discrete-time samples f_n obtained at time points t_n alone is an ill-posed problem. To make headway, we have to impose additional constraints. In particular, we will assume we are working with "nice" signals in the Hilbert space l_2. We will also impose the constraint that the reconstructed function $\hat{f}$ comes from the space of *bandlimited* functions. These are functions which have restricted support in the frequency domain such that their Fourier transforms satisfy $\hat{F}(\omega) = 0$ for $|\omega| > 2\pi B$. Here, $B > 0$ is the maximum frequency content in Hz (known as the *bandwidth* of the function in signal processing literature). This is a kind of smoothness constraint because it limits the speed of fluctuations in the function. Next, we will assume that the reconstruction function has minimum norm so that $\left\|\hat{f}\right\|^2$ is as small as possible; we can understand this as a *minimum energy* constraint. This gives us enough information to formulate the following constrained minimization problem with respect to $\hat{f}$:

$$\text{minimize} \quad \left\|\hat{f}\right\|^2 \tag{8.1}$$

$$\text{subject to} \quad \left\|f_n - \hat{f}_n\right\|^2 = 0$$
$$\hat{F}(\omega) = 0, \ |\omega| > 2\pi B$$

The solution to this problem can be obtained exactly in closed form (Yen, 1956):

$$\hat{f}(t) = \sum_{m=1}^{N} \sum_{n=1}^{N} f_m a_{nm} \text{sinc}\left(2B\left(t - t_n\right)\right) \tag{8.2}$$

The coefficients a_{nm} are those of a matrix $\mathbf{A} = \mathbf{B}^{-1}$, the inverse of the $N \times N$ *kernel matrix* $\mathbf{B}$ with entries:

$$b_{nm} = \text{sinc}\left(2B\left(t_m - t_n\right)\right) \tag{8.3}$$

In action (Figure 8.3), one can see the major limitation of this kind of nonuniform sampling: perfect reconstruction is not possible because there may be insufficient information between gaps in the signal where we were not able to get samples. In those gaps, the bandwidth and energy limitations have to substitute for the missing data.

Uniform bandlimited sampling: Shannon-Whittaker interpolation

This particularly simple form of interpolation is probably the oldest sampling theorem, and the one which is most commonly used in DSP

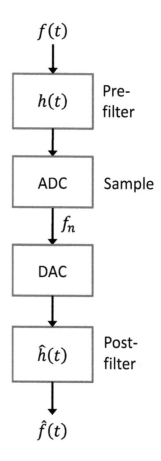

Fig. 8.2: **A block diagram of the process of discrete-time digital sampling hardware, which captures a continuous time signal** $f(t)$ **as a set of digital samples** f_n **using pre-filtering and analogue-to-digital conversion (ADC). These samples are then used to create a continuous-time signal using digital-to-analogue conversion (DAC), which is then post-filtered to create the reconstruction** $\hat{f}(t)$.

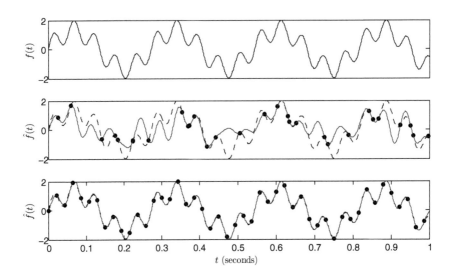

Fig. 8.3: Illustrating bandlimited sampling (interpolation). Top panel shows the original, bandlimited input signal $f(t)$. Middle panel shows non-uniform (i.e. irregular time interval, sample sequence f_n represented as black dots). The reconstruction $\hat{f}(t)$ obtained from these samples is shown in blue (dotted black line is $f(t)$). The bottom panel depicts uniform (Shannon-Whittaker) reconstruction (interpolation) at the optimal sampling rate.

Shannon was a brilliant mathematician who also invented information theory described in chapter 1 and thus played a critical role in developing the digital information revolution in the 20th century.

applications. It is usually closely associated with the mathematician, Claude Shannon although the essential mathematical concepts have a much longer history. To illustrate the idea, we will derive this as a special case of the bandlimited, nonuniform sampling above. If we set $t_n = n\,\Delta t$ and $\Delta t = 1/(2B)$, then $b_{nm} = \text{sinc}\,(2B\,(t_m - t_n)) = \delta\,[m - n]$, the Kronecker delta. This makes $\mathbf{B} = \mathbf{B}^{-1} = \mathbf{A} = \mathbf{I}$, the $N \times N$ identity matrix. Thus, $\sum_{n=1}^{N} a_{nm} \text{sinc}\,(2B\,(t - t_n))$ simplifies to $\text{sinc}\,(2B\,t - m)$, which in turn means that (8.2) becomes:

$$\hat{f}(t) = \sum_{m=1}^{N} f_m \text{sinc}\,(2B\,t - m) \qquad (8.4)$$

This is known as the *Shannon-Whittaker reconstruction formula*. Provided that the continuous-time signal f has bandwidth at most B, the reconstruction is perfect: not only does the interpolating function go through all the f_n, but the reconstruction error $\left\| f - \hat{f} \right\|$ is zero. This formula is also very useful in practice because (8.4) is an LTI convolution which can be (approximately) implemented as an *analogue electronic post-filter* (Figure 8.2).

The requirement that the sampling interval is at most $1/(2B)$ is known as the *Nyquist criterion* and it gives the longest time duration, relative to the bandwidth of the signal, which ensures perfect reconstruction. Alternatively, if we fix Δt, then we require a maximum bandwidth of $B = 1/(2\Delta t)$, or, $B = \frac{1}{2}S$ where $S = 1/\Delta t$ is the sampling frequency or *sampling rate* in Hz. The $\frac{1}{2}S$ bandwidth requirement is known as the *Nyquist frequency*.

It is nearly always the case that no real-world signal f is exactly bandlimited, therefore all signals violate the Nyquist criterion to some extent. To use Shannon sampling, in practice it is common for another analogue electronic filter to be applied to the signal before sampling, which acts to bandlimit the real signal to the bandwidth B. This is the pre-filtering step (Figure 8.2).

Another way to explain Shannon interpolation is to appreciate that (for an infinite number of samples) the sampling operation makes a signal *periodic* in the frequency domain, and, provided that the original signal is appropriately bandlimited, then these unwanted frequency-domain repetitions can be removed in post-conditioning using an appropriate linear filtering operation. The frequency response of the ideal filter which solves this problem is rectangular, in fact, it is the filter with response $H(\omega) = 0$ for $|\omega| > 2\pi B$ that has the sinc function $\hat{h}(t) = \mathrm{sinc}(2Bt)$ as impulse response. This is indeed the reconstruction function (8.4) (see Figure 8.4).

We next show why uniform sampling introduces frequency domain periodicity in the case of an infinite number of samples, by examining the *impulse train* which can be represented in the form of a Fourier series:

$$\sum_{n \in \mathbb{Z}} \delta[t - n\,\Delta t] \;=\; \frac{1}{\Delta t} \sum_{n \in \mathbb{Z}} \exp\left(i\,2\pi n \frac{t}{\Delta t}\right) \qquad (8.5)$$

Now, the sampled continuous-time signal can be written as the product of the input signal and the Fourier-series representation of the impulse train:

$$f_n(t) \;=\; f(t) \frac{1}{\Delta t} \sum_{n \in \mathbb{Z}} \exp\left(i\,2\pi n \frac{t}{\Delta t}\right) \qquad (8.6)$$

In the Fourier domain, the exponential function is transformed to the Dirac delta, so we get:

$$F_n(\omega) \;=\; \frac{2\pi}{\Delta t} F(\omega) \star \sum_{n \in \mathbb{Z}} \delta\left[\omega - \frac{2\pi n}{\Delta t}\right]$$

$$\;=\; \frac{2\pi}{\Delta t} \sum_{n \in \mathbb{Z}} F\left(\omega - \frac{2\pi n}{\Delta t}\right) \qquad (8.7)$$

where the last line uses the sifting property of the Dirac delta under convolution. In effect, the sampled signal $f_n(t)$, in the frequency domain, is the sum of an infinite sequence of copies of the input signal $f(t)$,

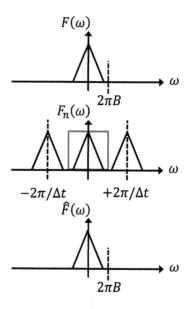

Fig. 8.4: *Shannon-Whittaker uniform sampling and (perfect) reconstruction in the frequency domain ω (rad/s).* **Assume that the spectrum $F(\omega)$ of the original, continuous time signal $f(t)$ (top panel) is bandlimited to $2\pi B$, where $B = 1/(2\Delta t)$ is the bandwidth (Hz) and Δt is the uniform sampling interval (secs). Then, although the sampled signal $f_n(t) = f(n\,\Delta t)$ is *periodized* by the sampling (that is, an infinite number of unwanted copies of the signal $F(\omega)$ are created at multiples of $2\pi/\Delta t$, middle panel), this periodization does not cause any overlap (*aliasing*) between the copies. Thus, applying a perfect *reconstruction* filter (blue box, middle panel) to the sampled signal (a post-conditioning step often performed in hardware using an analogue electronic filter) removes all copies but the desired, original spectrum (bottom panel). If $f(t)$ is not bandlimited to B, then a similar filter to the reconstruction filter can be applied prior to sampling, ensuring that periodization does not cause any frequency overlap.**

shifted by multiples of frequency $(\Delta t)^{-1} = S$ Hz. These copies can be removed by ideal low-pass filtering in post-filtering, but only if there is no overlap in frequency. Thus, perfect reconstruction will only be possible if the bandwidth B is sufficiently small to avoid overlap. This can be ensured by using similar low-pass pre-filtering. However, when this condition is violated an effect known as *aliasing* occurs: low-frequency components in the input signal are "folded" (aliased) to overlap with high-frequency components in the reconstruction, distorting the higher frequencies. This aliasing effect gets worse as the overlap increases.

Generalized uniform sampling

Despite the simplicity and ease of implementation of Shannon-Whittaker sampling, there are many kinds of signals which are poorly represented in the bandlimited model. These include signals with jumps, or signals that are continuous but with discontinuous derivatives. Such signals are abundant in practice. For example, gamma ray intensity signals recorded from *well logs* (drill holes used in exploration geophysics applications in the petroleum industry) often show abrupt jumps with changing depth as the drill hole passes through stacked stratigraphic layers (Figure 8.5). Digitizing these kinds of signals justifies more general sampling approaches. To do this, we will re-interpret uniform bandlimited sampling from the more abstract perspective of minimal-error projections onto orthonormal bases.

Firstly, we note that the uncountable space l_2 contains the space of bandlimited functions V which is countable with an orthonormal basis generated by the sinc function, $e_n(t) = \text{sinc}(t-n)$. For the sake of simplicity and without loss of generality, we will take the unit sampling rate $\Delta t = 1$ (we can always rescale time without changing the signal), in which case V is the set of l_2 functions such that $F(\omega) = 0$ for $|\omega| > \pi$.

The orthonomality of the sinc function can be seen from the fact that the inner product is a convolution $\langle \text{sinc}(t-n), \text{sinc}(t-m)\rangle = \int_{\mathbb{R}} \text{sinc}(t-n)\,\text{sinc}(t-m)\,dt$, and the sinc function has the rectangular pulse as Fourier transform. But, the rectangular pulse function is unchanged under multiplication with itself: $\text{rect}(\omega)\text{rect}(\omega)=\text{rect}(\omega)$. So, on application of the inverse Fourier transform, it follows that $\langle \text{sinc}(t-n), \text{sinc}(t-m)\rangle = \text{sinc}(n-m) = \delta[n-m]$.

Now, the bandlimited sampling and reconstruction of any signal in l_2 can be represented as a *projection* operator $P : l_2 \to V$:

$$
\begin{aligned}
\hat{f} = P[f] &= \sum_{n \in \mathbb{Z}} \langle f, e_n \rangle \, e_n \\
&= \sum_{n \in \mathbb{Z}} f_n e_n
\end{aligned}
\tag{8.8}
$$

In other words, sampling is obtained by projecting f down onto each sinc basis, which simply evaluates $f(n)$ since $\int_{\mathbb{R}} \text{sinc}(t-n)\,dt = 1$ (this is a consequence of the interpolating property). This is just a restatement of (8.4) with $N \to \infty$ and $B = \frac{1}{2}$, and $e(t)$ is just the impulse response of the post-filter $\hat{h}(t)$. Using the Hilbert projection theorem, $\hat{f}$ is the solution to the following constrained optimization problem:

$$\text{minimize} \quad \left\| f - \hat{f} \right\|^2 \tag{8.9}$$
$$\text{subject to} \quad \hat{F}(\omega) = 0, \ |\omega| > \pi$$

Provided that $f \in V$, then it must be that f is invariant under P, i.e. $f = P[f]$. In other words, Shannon-Whittaker sampling with sinc pre-filtering to restrict the input signal to the space of bandlimited functions, allows an exact reconstruction, as we have established above.

With sampling, one the issues we have to deal with in practice is that theoretically ideal pre-filtering/post-filtering cannot be performed in reality. In practice, we have to construct a filter that approximates e.g. the rectangular response of the sinc function. However, we can make life easier if we choose filters which are more manageable than the sinc filter. For example, the sinc filter has infinite impulse response and it decays in amplitude as $|t|^{-1}$ (see Figure 7.1). This means that any approximate (analogue, or recursive digital) filter will have to be close to unstable to match this response.

The more abstract presentation of this section allows us to consider bases constructed from something other than $e(t) = \text{sinc}(t)$. In other words, we can consider a more general space of functions W than the set of bandlimited functions:

$$W = \left\{ f(t) = \sum_{n \in \mathbb{Z}} c_n e(t - n) : c_n \in l_2 \right\} \tag{8.10}$$

The last condition requires that the infinite coefficient sequence is square summable, $\sum_{n \in \mathbb{Z}} |c_n|^2 < \infty$. Of course, many possible functions $e(t)$ could be used here, but to make the sampling algorithm tractable we consider only time-invariant, linearly independent basis functions with l_2 reconstructions. We also want it to be possible that any $f \in l_2$ can be approximated with as small a reconstruction error as we like by choice of the uniform sampling interval (note that such a basis may not be orthonormal). A very useful time-localized set of interpolants which satisfy these conditions are the *B-spline functions* $b_M(t)$. These are polynomial functions of degree $M \geq 0$ which can be used to form a basis $e_n(t) = b_M(t - n)$. With this basis, each function in W is piecewise polynomial on each interval $[n, n+1]$ (M odd) and $\left[n - \frac{1}{2}, n + \frac{1}{2}\right]$ (M even), such that the function is continuous and has derivatives of order $M - 1$. They have finite support on the interval $\left[-\frac{M+1}{2}, \frac{M+1}{2}\right]$.

The case $M = 0$ is just the rectangular function $\text{rect}(t)$, which is the only orthogonal B-spline. This is obviously the appropriate model for piecewise constant signals. Choosing $M = 1$ selects the linear *triangular function* $b_1(t) = t + 1$ for $-1 \leq t < 0$, $b_1(t) = -t + 1$ for $0 \leq t \leq 1$, and 0 otherwise. This function parameterizes all piecewise linear signals.

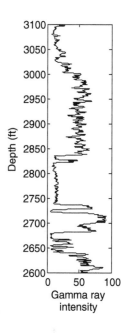

Fig. 8.5: Gamma ray intensity signal measured from a drill hole in Sherman County, Kansas, USA (source: Kansas Geological Survey, University of Kansas). The "sampling rate" is $\Delta t = 0.1$ feet/sample.

We can generate all B-spline functions by repeated convolution, $b_M = b_0 \star b_{M-1}$ for $M \geq 1$.

For any given input signal $f \in l_2$, we need to obtain the c_n to locate the best interpolating function in W. As with Shannon-Whittaker sampling, we find this by solving the minimum squared reconstruction error problem:

$$\text{minimize} \quad \left\| f - \hat{f} \right\|^2 \tag{8.11}$$

$$\text{subject to} \quad \hat{f} \in W$$

The solution is the projection:

$$P[f] = \sum_{n \in \mathbb{Z}} \langle f, e_n \rangle \, \hat{e}_n \tag{8.12}$$

such that $c_n = \langle f, e_n \rangle$. The functions $\hat{e}_n$ comprise the unique, translation-invariant *dual basis* to e_n in W, and, like e_n, they are not necessarily orthonormal. However, together these two sets of functions are always *biorthonormal*: $\langle e_n, \hat{e}_m \rangle = \delta[n - m]$. Note that in the case of the sinc function, then $\hat{e}_n = e_n$ so that $c_n = f_n$. This is also the case for the constant and linear B-splines.

For example, for the constant spline, $\hat{e}_n(t) = e_n(t) = 1$ for $t \in \left[n - \frac{1}{2}, n + \frac{1}{2}\right]$ and 0 otherwise. Thus, the analogue pre-filtering $h(t)$ is just a moving average filter or *integrator* of width 1, i.e. $c_n = \int_{n-1/2}^{n+1/2} f(t) \, dt$. We can see this as smoothing out the function f by replacing it with its average on each sampling interval. Actual ADC hardware which can perform this "block sampling" is relatively simple to implement in practice by comparison to a precision approximation to the ideal sinc pre-filter (Mallat, 2009, page 69). Finally, if f actually lives in W, then f is constant on each interval, so $c_n = f_n$ and we have the exact reconstruction $f = P[f]$. The entire piecewise constant spline sampling process samples the averaged signal on each interval, and the reconstruction post-filter averages this impulse train on each sampling interval.

For splines of order $M \geq 2$, things are not quite as simple. The constraint that the function matches the samples of the input signal at the sampling points exactly, $\left\| f_n - \hat{f}_n \right\|^2 = 0$, the c_n coefficients are obtained as an inverse discrete convolution of the sequence f_n with the B-spline sampled at discrete intervals. Using this coefficient expression, we can rearrange the computation of the c_n in terms of the *cardinal spline* basis $\hat{e}_n(t) = \hat{h}(t - n)$, which plays the same role as the sinc function in bandlimited pre-filtering. There is no simple closed form formula for this basis function, but it has the following frequency response (Unser, 1999):

$$\hat{H}(\omega) = \left(\text{sinc}\left(\frac{\omega}{2\pi}\right) \right)^{M+1} \frac{1}{B_M(\exp(i\omega))} \tag{8.13}$$

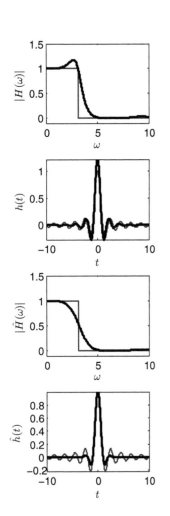

Fig. 8.6: Pre- and post-filtering for order quadratic B-spline uniform sampling. From top panel: pre-filter frequency response; pre-filter impulse response; post-filter frequency response and post-filter impulse response. Black lines are spline sampling, blue lines show the corresponding bandlimited (sinc) filtering for comparison.

where $B_M(z) = \sum_{n \in \mathbb{Z}} b_M(n) z^{-n}$ is the *z-transform* of the discrete sampled B-spline. The sinc term comes from the B-spline convolution used to construct b_M from b_0. Note that because the spline has finite support, the impulse response of this discrete sampled B-spline is also finite, but its inverse has infinite impulse response. Therefore, although the B-spline has finite support, the impulse response of the cardinal spline post-filter (8.13) does not, as with the sinc function. However, it goes to zero orders of magnitude faster than $\mathrm{sinc}(t)$, which is exactly what we require for a stable analogue filtering operation. Like the sinc function, the cardinal spline post-filter has the interpolating property, i.e. $\hat{h}(n) = \delta[n]$ for all $n \in \mathbb{Z}$.

Unlike in the bandlimited situation, the corresponding pre-filter $e_n(t) = h(t - n)$ is different, but orthonormal to, the post-filter, with frequency response (Unser, 1999):

$$H(\omega) = \left(\mathrm{sinc}\left(\frac{\omega}{2\pi}\right)\right)^{M+1} \frac{B_M(\exp(i\omega))}{B_{2M+1}(\exp(i\omega))} \qquad (8.14)$$

Figure 8.6 compares the frequency and impulse responses of the cardinal spline pre- and post-filter to the sinc function. Interestingly, both filters converge on the ideal bandlimited sinc function as $M \to \infty$, in this sense, Shannon-Whittaker sampling can be considered a special case of spline sampling (Unser, 1999).

8.2   Quantization

Sampling addresses the problem that continuous-time signals $f(t)$ cannot be stored in a finite amount of memory. If we have a discrete-time, discrete-valued signal, we can represent this with no loss of fidelity at all. If each measurement in an i.i.d. signal $\boldsymbol{x} = (x_1, x_2, \dots, x_N)$ takes $r = |\Omega_X|$ bits, then this signal requires exactly $N r$ total bits to represent it. However (as discussed in Section 1.5), if the signal is a random process and the distribution of each measurement X in the signal is non-uniform, we can use entropy coding to achieve a compressed representation with a smaller number of bits. The compressed number of bits lies between $N H_2[X]$ and $N H_2[X] + N$, and both are bounded above by $N r$. This compression is extremely useful in practice because signals can easily become unmanageably large.

Entropy coding is achieved by using algorithms such as Huffman coding, which creates a code for each discrete value in Ω_X. Each code is a sequence of bits that is not a *prefix* for any other sequence, that is, each binary code sequence is not embedded at the start of any other binary code sequence. Using this, any signal can be encoded as a sequence of bits without the need for any special marker between codes. This allows the codes to be uniquely deciphered and hence used to recover the original, uncompressed signal exactly.

By contrast, if these signals are continuous-valued, $f(t) \in \mathbb{R}$, we need to find a way to store each sampled value using a finite, discrete

Alternatives to Huffman coding include arithmetic coding *which can come arbitrarily close to encoding a signal at its empirical entropy rate.*

digital representation. This cannot be performed without introducing some error, called *loss* or *distortion* in the discipline. The most widely used technique for converting continuous-valued to discrete-valued signals is *quantization*, implemented using a *quantizer* Q which maps each sample X onto the discrete quantized version Y. This quantizer is described generally in terms of the conditional probabilistic relationship $f_Q(Y = y | X = x)$. A deterministic quantizer can be written as $y = Q(x)$ or $f_Q(y|x) = \delta(y - Q(x))$, with $Q : \Omega_X \to \Omega_Y$ The joint density $f(x, y) = f(x) f_Q(y|x)$ takes into account both source (input signal) and quantized output distributions. We will see examples of random quantizers when investigating *dithering* later. The practical electronic implementation of such a quantizer uses an ADC. In practice, the range Ω_Y is often an integer represented using a finite number of bits.

In the context of continuous-valued signals, the quantizer domain is a subset of the real line $\Omega_X \in \mathbb{R}$. A quantizer maps a partition of Ω_X – a set of K non-overlapping intervals which together make this domain – onto a finite set of values Ω_Y. Each interval is assigned a *code* $k \in 1, 2, \ldots, K$. The set of quantizer output values associated with each interval are known as quantization *levels*, $v_1, v_2 \ldots, v_K$. For deterministic quantizers, within each interval, each value of the domain is mapped onto the same level, so Q is by definition an irreversible, many-one mapping. The intervals can be defined by their $K + 1$ ordered edge values or *thresholds* $u_k \in \Omega_X$ with $u_0 < u_2 < \cdots < u_K$. The domain is then $\Omega_X = \cup_{k=1}^{K-1} U_k$ for the regions $U_k = [u_k, u_{k+1})$ or $U_k = (u_k, u_{k+1}]$. For unbounded domains we might have e.g. $u_1 = -\infty$ or $u_{K+1} = \infty$, or both. Accordingly a deterministic quantization function or rule can be written using indicator notation:

$$Q(x) = \sum_{k=1}^{K} v_k \mathbf{1}[x \in U_k] \tag{8.15}$$

That Q is many-one is exactly what we should expect as we cannot hope to exactly recover the underlying signal $f(t)$ from its discrete representation $Q(f(t))$. The inverse function $x = Q^{-1}(y)$ is not uniquely determined by Q, and we typically want to design a quantizer which in some sense to be defined minimizes the error entailed by reconstructing f from its quantized values. This reconstruction process is usually performed by an electronic DAC.

We will first look at the simplest of quantizer functions, the so-called *scalar uniform quantizers* (Figure 8.11). Here, the thresholds are equispaced, $u_k = u_1 + (k - 1)\Delta$ for $k = 1, 2, \ldots, K$ and $\Delta > 0$ is known as the quantization *width*. As an example, consider rounding the value x to the nearest integer. The quantizer is defined by the intervals $U_k = \left(k - \frac{1}{2}, k + \frac{1}{2}\right]$ and $v_k = k$, so that $\Delta = 1$. For general Δ, the uniform quantizer rule which rounds to the nearest integer multiple of Δ is:

$$Q\left(x\right) = \Delta \left\lfloor \frac{x}{\Delta} + \frac{1}{2} \right\rfloor \qquad (8.16)$$

For a bounded input domain Ω_X, the number of digital bits required to represent the code for each input value is some fixed finite value $r \in \mathbb{N}$. So, an r-bit uniform quantizer maps some standard, normalized domain $\Omega_X \in \mathbb{R}$, onto the full range which can be represented using this number of bits. It will have $K = 2^r$ levels. In so-called *two's-complement* number representation where the leading bit represents the sign of x, we would have a mapping such as $Q : [-1, 1) \to [-2^{r-1}, 2^{r-1} - 1]$, and $\Delta = 2^{1-r}$. For example, an $r = 4$ bit uniform rounding quantizer has $K = 2^4 = 16$ levels, which would map $[-1, 1)$ onto $[-8, 7]$ with $\Delta x = 2^{-3} = 0.125$.

This normalized domain is usually obtained by some kind of non-digital 'pre-scaling' of the signal f before the ADC.

The systematic analysis of the performance and properties of quantization functions such as Q has broadly proceeded along two tracks: *high resolution analysis* which focuses on quantizers with small quantization error, and *rate-distortion theory* which addresses the trade-off between quantization with a given error and the bit rate required to attain that error (Gray and Neuhoff, 1998).

Quantization can be viewed as a kind of classification (section 6.3) where the goal is to find the optimal class (quantization region U_k) to which each continuous-valued sample in the signal should belong. Likewise, the process of designing an optimal quantizer is extremely similar to the clustering algorithms explored in section 6.5, and we will see these same concepts arise in the course of the following sections.

Rate-distortion theory

As discussed above, continuous-valued signals cannot be represented without some loss of fidelity. There is always a trade-off: as we increase the number of bits required to represent each measurement in the signal, we reduce the error, and vice versa. What, therefore, is the optimal trade-off between the number of bits per sample and the quantization error? Alternatively, if we have a budget for a maximum number of bits per sample, what is the inevitable error we must tolerate? Addressing these questions is the subject of *rate-distortion theory*, which is intimately connected to the very practical concept of *lossy data compression* which we will discuss later.

Our first task is to be precise what we mean by distortion. We can choose a loss function $L\left(x, y\right)$, very typically the square loss $L\left(x, y\right) = \left(x - y\right)^2$. Distortion is then the expected loss given the quantizer and source distribution:

$$\begin{aligned} E\left(Q\right) &= E_{X,Y}\left[L\left(X, Y\right)\right] \\ &= \int \int L\left(x, y\right) f\left(x, y\right) dx \, dy \qquad (8.17) \\ &= \int \int L\left(x, y\right) f\left(x\right) f_Q\left(y|x\right) dx \, dy \end{aligned}$$

For deterministic quantizers this becomes:

$$
\begin{aligned}
E\left(Q\right) &= \int\int L\left(x,y\right)f\left(x\right)\delta\left(y-Q\left(x\right)\right)dy\,dx \\
&= \int L\left(x,Q\left(x\right)\right)f\left(x\right)dx \qquad\qquad (8.18) \\
&= E_X\left[L\left(X,Q\left(X\right)\right)\right]
\end{aligned}
$$

Next, we want to quantify the number of bits per sample in the quantized output, which can be measured as the mutual information (logarithm base 2) between the input and the quantized output:

$$
\begin{aligned}
I\left(Q\right) &= I_2\left[X,Y\right] \\
&= H_2\left[Y\right]-H_2\left[Y|X\right] \qquad\qquad (8.19)
\end{aligned}
$$

In the deterministic quantizer case, since Y is determined by X, there is no information gained by knowing X if we know Y, so that $H_2\left[Y|X\right]=0$. In this case, $I\left(Q\right)=H_2\left[Y\right]$. Using these two quantities $E\left(Q\right)$ and $I\left(Q\right)$, we can express the *rate-distortion* trade off as finding the quantizer that minimizes the bit rate $r=I\left(Q\right)$ for some given distortion e:

$$
\text{minimize}_Q \quad I\left(Q\right) \qquad\qquad (8.20)
$$
$$
\text{subject to} \quad E\left(Q\right)\le e
$$

The solution is written as a function $r\left(e\right)$. Similarly, we can stipulate a maximum bit rate r and minimize the distortion $e=E\left(Q\right)$. The *distortion-rate* function $e\left(r\right)$ is the solution to:

$$
\text{minimize}_Q \quad E\left(Q\right) \qquad\qquad (8.21)
$$
$$
\text{subject to} \quad I\left(Q\right)\le r
$$

The functions $r\left(e\right)$ and $e\left(r\right)$ are inverses of each other. It is sometimes awkward to deal with constrained optimization problems, so we can instead express the trade-off using the Lagrangian formulation for the trade-off parameter $\lambda>0$:

$$
\hat{Q}=\text{argmin}_Q\left[E\left(Q\right)+\lambda I\left(Q\right)\right] \qquad\qquad (8.22)
$$

Readers may recognize this expression as being similar to the idea of regularization. The analogue for distortion in Bayesian formulations would be the negative log likelihood, and the bit rate would be the negative log prior or model complexity.

Computing $r\left(e\right)$ or $e\left(r\right)$ analytically is difficult. From the properties of mutual information, $0\le r\left(e\right)=I_2\left[X,Y\right]\le H_2\left[X\right]$. Similarly, for loss functions such as the square loss, the distortion is non-negative $e\left(r\right)\ge0$. The function $r\left(e\right)$ is non-increasing with increasing e, and is convex; therefore $e\left(r\right)$ is non-decreasing with increasing r. This means that (8.22) is a convex problem. This can be used to show that for Gaussian i.i.d. sources with variance σ^2 (Gray and Neuhoff, 1998):

$$r_G(e) = \max\left(\frac{1}{2}\log_2\frac{\sigma^2}{e}, 0\right) \tag{8.23}$$

$$e_G(r) = \sigma^2 2^{-2r} \tag{8.24}$$

See Figure 8.7. Engineers often use $e_G(r)$ to justify the rule of thumb that the *signal-to-(quantization)-noise-ratio* (SQNR) of a fixed bit rate quantizer improves by approximately 6dB (decibels) per additional quantization bit:

$$\begin{aligned} \text{SNR}(r) &= 10\log_{10}\frac{\sigma^2}{e_G(r)} \\ &= r\,20\log_{10}2 \tag{8.25} \\ &\approx 6.02r \end{aligned}$$

Therefore an $r = 4$-bit quantizer with $K = 2^r = 16$ levels has a theoretical SQNR of approximately 24dB for Gaussian signals, and with 16 bits, $65,536$ levels, the approximate SNR is 96dB. The latter bit rate is very often used in applications such as audio coding.

While analytical calculation may be difficult, useful bounds are known for these functions (Makhoul *et al.*, 1985):

$$e_{\text{SB}}(r) \le e(r) \le e_G(r) \tag{8.26}$$

The Gaussian is the maximum entropy distribution on $\mathbb{R}$ with a given variance (see Section 4.5), therefore, all non-Gaussian sources will benefit from a better trade-off. This explains the upper bound. All distortion-rate functions are equal to or larger than the *Shannon lower bound* (Makhoul *et al.*, 1985):

$$e_{\text{SB}}(r) = \frac{1}{2\pi e}2^{2(H_2[X]-r)} \tag{8.27}$$

This is usually easy to compute as it relies only upon knowledge of the entropy of the univariate source distribution $H_2[X]$. For a given variance, these distortion-rate relationships ($e_{\text{SB}}(r)$, $e(r)$ and $e_G(r)$) can be written in the form $e(r) = s\sigma^2 2^{-2r}$, and are tabulated for a range of distributions in Table 8.1.

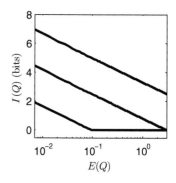

Fig. 8.7: Rate-distortion curves $r_G(e)$ where $e = E(Q)$ and $r = I(Q)$ for quantizing an i.i.d. Gaussian source, for different variances $\sigma^2 = 0.1$ (bottom line), $\sigma^2 = 3.16$ (middle line) and $\sigma^2 = 100$ (top line).

Source distribution $f(x)$	Shannon lower bound $e_{\mathrm{SB}}(r)$	Distortion-rate function $e(r)$	$e_{\mathrm{G}}(r)$	Variable-rate $e_{\mathrm{VR}}(r)$	Fixed-rate (Panter-Dite) $e_{\mathrm{LM}}(r)$
Uniform	$\frac{6}{\pi e} \approx 0.70$		1	1	1
Laplace	$\frac{e}{\pi} \approx 0.87$		1	$\frac{e^2}{6} \approx 1.23$	$\frac{9}{2} = 4.50$
Gaussian	1	1	1	$\frac{\pi e}{6} \approx 1.42$	$\frac{\sqrt{3}\pi}{2} \approx 2.72$

Table 8.1: Scale factors s for quantizer distortion-rate relationships $e(r) \approx s\,\sigma^2 2^{-2r}$ in scalar quantization, for various distributions with the same variance σ^2.

Lloyd-Max and entropy-constrained quantizer design

For the rate-distortion trade-off functions described in the previous section, analytical results are in general difficult to compute. One can instead turn to numerical methods. The *Lloyd-Max algorithm* (Gray and Neuhoff, 1998) is widely-used for this purpose. Given a source density function $f(x)$ and a fixed number of thresholds K, it estimates the levels $\boldsymbol{v}$ and thresholds $\boldsymbol{u}$ for the quantizer which optimizes the square loss distortion for a given fixed bit rate $r = \log_2 K$. For a deterministic quantizer, the square loss distortion (8.18) can be written as:

$$
\begin{aligned}
E(Q) &= E_X\left[(X - Q(X))^2\right] & (8.28) \\
&= \sum_{k=1}^{K} \int_{u_k}^{u_{k+1}} (x - v_k)^2 f(x)\,dx
\end{aligned}
$$

and the following holds at local minima for a given quantizer defined by $(\boldsymbol{u}, \boldsymbol{v})$:

$$
\frac{\partial E(Q)}{\partial u_k} = 0 \implies u_k = \frac{1}{2}(v_{k-1} + v_k), \quad k = 2, 3, \ldots K \quad (8.29)
$$

$$
\frac{\partial E(Q)}{\partial v_k} = 0 \implies v_k = \frac{\int_{u_k}^{u_{k+1}} x\,f(x)\,dx}{\int_{u_k}^{u_{k+1}} f(x)\,dx}, \quad k = 1, 2, \ldots K \quad (8.30)
$$

The expression (8.29) for u_k can be understood as the optimal classification decision boundary for each quantization region U_k for a one-dimensional LDA classifier with equal variances for each class, and this lies exactly between the two adjacent levels (Section 6.3). Similarly, (8.30) is just the expected value of each region which follows from the fact that the square loss function is minimized by the mean. Therefore, given a set of levels, it is possible to find the thresholds, and vice versa. The problem is that of course we do not know either initially. The Lloyd-Max algorithm involves starting with guess values and iteratively applying these two equations as updates whilst monitoring $E(Q)$ to detect convergence to a required tolerance (Gray and Neuhoff, 1998).

Since (8.29) minimizes $E(Q)$ given v, and the same for (8.30) given u, it must be that the sequence E^i, $i = 1, 2, \ldots$ is non-increasing so that the algorithm converges on a local minima for $E(Q)$. Furthermore, if the source density $f(x)$ is log-concave then $E(Q)$ is convex so that the local optima is in fact the global one (Makhoul *et al.*, 1985). This is true for the Gaussian and Laplace densities, for example, in which case, any set of initial values for v which are all distinct will lead to the optimal scalar quantizer for the given density and rate. Otherwise, multiple randomized restarts will be required to gain confidence in finding a good quantizer (see Section 2.6).

In theory at least! As with K-means, it is possible for numerically degenerate solutions to arise, see Bormin and Jing (2007) for examples.

Algorithm 8.1 The Lloyd-Max algorithm for fixed-rate quantizer design.

---

(1) *Initialization.* Choose initial estimates for v satisfying $v_1^0 < v_2^0 < \cdots < v_K^0$, choose convergence tolerance $\epsilon > 0$, set iteration number $i = 0$,

(2) *Update thresholds.* Set $u_k^{i+1} = \frac{1}{2}\left(v_{k-1}^i + v_k^i\right)$ for $k = 2, 3, \ldots, K$,

(3) *Update levels.* Set $v_k^{i+1} = \int_{u_0(k)}^{u_1(k)} x f(x)\, dx / \int_{u_0(k)}^{u_1(k)} f(x)\, dx$ where $u_0(k) = u_k^{i+1}$ and $u_1(k) = u_{k+1}^{i+1}$ for $k = 1, 2, \ldots K$,

(4) *Compute distortion.* Calculate $E^{i+1} = \sum_{k=1}^{K} \int_{u_0(k)}^{u_1(k)} \left(x - v_k^i\right)^2 f(x)\, dx$, $\Delta E = E^i - E^{i+1}$ and if $\Delta E < \epsilon$, exit with quantizer $\left(v^{i+1}, u^{i+1}\right)$,

(5) *Iteration.* Update $i \leftarrow i + 1$, go back to step 2.

For large K the distortion obtained at convergence of Algorithm 8.1 is approximated by the high-resolution *Panter-Dite* formula (Gray and Neuhoff, 1998):

$$e_{\text{LM}}(r) \approx \frac{1}{12} 2^{-2r} \left[\int f(x)^{\frac{1}{3}}\, dx\right]^3 \tag{8.31}$$

which for the univariate Gaussian is $\sigma^2 \frac{\sqrt{3}\pi}{2} 2^{-2r}$. Values for other distributions are given in Table 8.1. This is one of an extensive set of results from high-resolution quantization theory (**Gray and Neuhoff, 1998**).

Readers may recognize that the Lloyd-Max algorithm is in fact the same as K-means (see Section 6.5) but where the density function is known. Instead, we can use data x_n for $n = 1, 2, \ldots, N$ from a signal to (implicitly and approximately) estimate this distribution, which leads to K-means for quantization. This algorithm replaces the classification step (6.14) with quantization, and the levels are updated from the empirical means of all data points assigned to each quantization region U_k.

Because we do not know anything about the underlying distribution of the data, this algorithm, as with K-means, is only guaranteed to find a locally optimal solution and therefore would benefit from multiple random restarts to gain confidence that the result is close to optimal.

While the Lloyd-Max quantizer is useful we know that we can often improve the bit rate by performing quantization followed by entropy coding to an average rate of $H_2[Y]$.

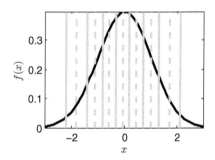

Algorithm 8.2 The K-means algorithm for fixed-rate quantizer design.

(1) *Initialization.* Choose initial estimates for v satisfying $v_1^0 < v_2^0 < \cdots < v_K^0$, set iteration number $i = 0$,

(2) *Update quantized data indicators.* Set $z_n^{i+1} = \{k : Q^i(x_n) = v_k\}$ for $n = 1, 2, \ldots, N$,

(3) *Update levels.* Set $v_k^{i+1} = \frac{1}{N_k}\sum_{n:z_n^{i+1}=k} x_n$ for $k = 1, 2, \ldots K$ where N_k is the number of data points assigned to quantizer region k,

(4) *Compute distortion.* Calculate empirical distortion $E^{i+1} = \frac{1}{N}\sum_{n=1}^{N}\left(x_n - Q^{i+1}(x_n)\right)^2$, $\Delta E = E^i - E^{i+1}$ and if $\Delta E < \epsilon$, exit with quantizer (v^{i+1}, u^{i+1}) where $u_k^{i+1} = \frac{1}{2}\left(v_{k-1}^i + v_k^i\right)$ for $k = 2, 3, \ldots, K$,

(5) *Iteration.* Update $i \leftarrow i + 1$, go back to step 2.

Fig. 8.8: The Lloyd-Max scalar quantizer for $K = 8$ quantization levels for a univariate Gaussian with variance $\sigma^2 = 1$. The algorithm converged to a solution to within $\epsilon = 10^{-5}$ in 70 iterations (top panel). The distortion $E(Q)$ (black line) rapidly converges on a solution close to the Panter-Dite high resolution approximation (grey line, $e_{\text{LM}}(r)$). In the bottom panel, superimposed over the Gaussian PDF (black curve) are the quantization levels v_k (grey vertical lines) which are more closely spaced towards the mode of the distribution at $x = 0$, and become further separated towards the tails. The thresholds u_k lie between the levels by construction (grey dashed vertical lines).

This raises the question of how to find, for a given $\lambda > 0$, the optimal (deterministic) quantizer for the Lagrangian cost function of (8.22), $F(Q) = E(Q) + \lambda I(Q) = E(Q) + \lambda H_2[Y]$, directly. By differentiating $F(Q)$, we obtain a modification of Lloyd-Max which stipulates an additional penalty term for the threshold updates:

$$u_k = \frac{1}{2}(v_{k-1} + v_k) - \frac{\lambda}{2(v_{k-1} - v_k)}\log_2\frac{p_{k-1}}{p_k} \qquad (8.32)$$

where $p_k = P(X \in U_k) = \int_{u_k}^{u_{k+1}} f(x)\, dx$ is the probability of the quantizer region U_k. Effectively, the penalty term adjusts the classifier decision boundary by treating the region probability as a prior term and modifying the quantizer to make the MAP decision for each region. This modification results in a useful, but greedy algorithm which is only guaranteed to find a local optima.

In practice, what we find is that the optimal quantizers produced by this algorithm are approximately uniform, even while they are an improvement over Lloyd-Max quantizers. For example, running this algorithm with $K = 8$ for a Gaussian with unit variance and $\lambda = 0.12$, produces a distortion $E(Q) \approx 0.087$ for a rate of $r = H_2[Y] \approx 2.0$ bits/sample. By contrast, a comparable Lloyd-Max quantizer with $K = 4$ levels and fixed bit rate of $r = \log_2 K = 2$ bits, has a distortion of $E(Q) \approx 0.118$ for the same Gaussian source. For large bit rates, the square error distortion for entropy-constrained quantizer has been shown as approximately:

Algorithm 8.3 Iterative variable-rate entropy-constrained quantizer design.

---

(1) *Initialization.* Choose initial estimates for $v_1^0 < v_2^0 < \cdots < v_K^0$, set $u_k^0 = \frac{1}{2}\left(v_{k-1}^0 + v_k^0\right)$, set $p_k^0 = \int_{u_k^0}^{u_{k+1}^0} f(x)\,dx$, and initialize iteration number $i = 0$,

(2) *Update thresholds.* Set $u_k^{i+1} = \frac{1}{2}\left(v_{k-1}^i + v_k^i\right) - \lambda\left(\log_2 p_{k-1}^i - \log_2 p_k^i\right) / \left(2\left(v_{k-1}^i - v_k^i\right)\right)$ for $k = 2, 3, \ldots, K$,

(3) *Update probabilities.* Set $p_k^{i+1} = \int_{u_0(k)}^{u_1(k)} f(x)\,dx$ where $u_0(k) = u_k^{i+1}$ and $u_1(k) = u_{k+1}^{i+1}$, for $k = 1, 2, \ldots K$,

(4) *Update levels.* Set $v_k^{i+1} = \int_{u_0(k)}^{u_1(k)} x f(x)\,dx / p_k^{i+1}$ for $k = 1, 2, \ldots K$,

(5) *Compute Lagrangian.* Calculate $F^{i+1} = \sum_{k=1}^{K} \int_{u_0(k)}^{u_1(k)} \left(x - v_k\right)^2 f(x)\,dx - \lambda \sum_{k=1}^{K} p_k^{i+1} \log_2 p_k^{i+1}$, $\Delta F = F^i - F^{i+1}$ and if $\Delta F < \epsilon$, exit with solution $\left(v^{i+1}, u^{i+1}\right)$,

(6) *Iteration.* Update $i \leftarrow i + 1$, go back to step 2.

$$e_{\mathrm{VR}}(r) \approx \frac{1}{12} 2^{2(H_2[X]-r)} \tag{8.33}$$

where $r = H_2[Y]$ is the variable bit rate. Values of this expression are shown in Table 8.1, demonstrating the tangible improvements possible. In fact, it is known that for square distortion and high rates, uniform quantizers followed by entropy coding are known to be optimal entropy-constrained quantizers (Gray and Neuhoff, 1998).

Finally we note that Algorithm 8.3 can be readily adapted to exploit an implicit empirical estimate of the distribution, much as K-means is the 'empirical density' counterpart to the Lloyd-Max algorithm. The quantized data indicators are computed using the MAP LDA discriminant rule (6.23):

$$z_n = \underset{k \in 1, 2, \ldots, K}{\arg\min} \left[\left(x_n - v_k\right)^2 - \lambda \log_2 p_k\right] \tag{8.34}$$

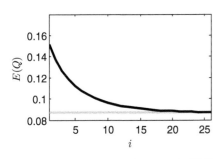

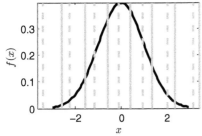

Fig. 8.9: Entropy-constrained scalar iterative quantizer, Algorithm 8.3, for $K = 8$ quantization levels, univariate Gaussian with variance $\sigma^2 = 1$, and trade-off parameter $\lambda = 0.12$. The algorithm converged to within $\epsilon = 10^{-6}$ in 26 iterations (top panel). The distortion $E(Q)$ (black line) converges to the high resolution approximation (grey line, $e_{\mathbf{VR}}(r)$). In the bottom panel, superimposed over the Gaussian PDF (black curve) are the quantization levels v_k (grey vertical lines) which are almost exactly uniformly spaced, $\Delta x \approx 1.0$. The thresholds u_k lie between the levels (grey dashed vertical lines) but are shifted by $\frac{\lambda}{2\Delta x}\log_2 \frac{p_{k-1}}{p_k}$.

We can put this expression into the same form as (6.23) if divide through both terms inside the minimization by $\lambda = 2\ln(2)\sigma^2$. Note that this σ is the common standard deviation across all Gaussian mixture components around each quantizer level in the LDA, not the standard deviation of the source variable X; so, if the quantizer is approximately uniform then $\sigma^2 \approx E(Q)$. Updates for the probabilities are replaced by their empirical estimates $p_k = \frac{N_k}{N}$, and for testing convergence the Lagrangian cost function becomes $F(Q) = \frac{1}{N}\sum_{n=1}^{N}(x_n - Q(x_n))^2 - \lambda\sum_{k=1}^{K} p_k \log_2 p_k$. The latter is related to the joint entropy $E_{X,Z}[-\log_2 f(Z)f(X|Z)]$ by $F(Q) = \lambda E_{X,Z}[-\log_2 \pi_Z \mathcal{N}(X, Q(X))] -\lambda Z(\sigma)$, where $Z(\sigma)$ is the Gaussian normalizing function which depends only upon σ.

Statistical quantization and dithering

Previously we have treated quantization as a (regularized) quantization error minimization problem where the goal is to minimize the overall error for each sample in the signal. If instead the statistical properties of the signal, the quantization error or the quantized signal, are of primary interest, then we can use characteristic functions to view quantization as a discrete-time sampling problem. That is, we wish to recover the distribution of the source $f(x)$ from quantized samples. Since the CF is the Fourier transform of the PDF, then we can pose the uniform quantization problem as a Shannon-Whittaker interpolation problem (Section 8.1).

A similar theory then arises: provided that the source density function is appropriately bandlimited, then perfect reconstruction of the PDF using sinc interpolation is always possible. This is *Widrow's quantizing theorem*: for a uniform quantizer with quantization width Δ and levels $v_k = k\Delta$ for $k \in \mathbb{Z}$, if the CF $\psi_X(s)$ of the density $f(x)$ satisfies $\psi_X(s) = 0$ for $|s| \geq \frac{\pi}{\Delta}$, then the CF of X can be recovered from the CF of $Y = Q(X)$. As a corollary, the PDF of X can be computed from the PDF of Y (Widrow and Kollar, 1996).

To understand this theorem, we need to model the quantization process using a sequence of convolutions. The quantization process leads to a discrete distribution which is an (infinite) series of Dirac delta functions placed at each quantization level v_k:

$$f_Y(y) = \sum_{k=-\infty}^{\infty}\int_{u_k}^{u_{k+1}} f(x)\,dx\,\delta(y - v_k) \qquad (8.35)$$

where the weights are the probabilities of each quantizer region U_k, $P(X \in U_k)$. We can, however, represent this as the product of the impulse train with the convolution of the density of X with a rectangular function $f_R(x) = \mathbf{1}\left[-\frac{\Delta}{2} \leq x \leq \frac{\Delta}{2}\right]$ of width Δ:

$$f_Y(y) = \sum_{k=-\infty}^{\infty}\delta(y - v_k)\int_{\mathbb{R}} f_X(y - x)f_R(x)\,dx \qquad (8.36)$$

In this form it is straightforward to compute the CF of Y:

$$\psi_Y(s) = \sum_{k=-\infty}^{\infty} \psi_X\left(s + k\frac{2\pi}{\Delta}\right) \text{sinc}\left[\frac{\Delta}{2}\left(s + k\frac{2\pi}{\Delta}\right)\right] \qquad (8.37)$$

where $\text{sinc}(x) = \sin(x)/x$. This is in the form of an infinite number of copies of the ψ_X (multiplied by sinc), each shifted by an integer multiple of $\frac{2\pi}{\Delta}$. As with bandlimited sampling, if we can isolate the central copy $k = 0$ using interpolation (which is equivalent to low-pass filtering in sampling), we can then invert the central copy to recover $f(x)$. Of course, this is only possible if the shifted copies of the CF do not overlap which will be true if the band limited condition given in the quantizing theorem is true.

Since the PDF of the sum of RVs is obtained by convolving their PDFs, their CFs are multiplied. This has an interesting consequence: if the CF of an RV is not bandlimited, we can make it so by adding an independent RV whose CF is bandlimited. Intuitively, since it is the fine details in the density function which contribute to the magnitude of the CF at large magnitude s values, these details are "smeared out" by randomly perturbing an RV with additional, independent random values.

Most PDFs are only approximately bandlimited, but in many cases the magnitude of their CFs is sufficiently small at $s = \pm\frac{\pi}{\Delta}$ that there is little effective overlap for an appropriate choice of quantization width Δ. For the Gaussian whose CF is $\exp\left(-\frac{1}{2}\sigma^2 s^2\right)$, this will be true provided σ is approximately equal to Δ or larger. By contrast, this is not true for the sine signal $x(t) = A\sin(t)$ with amplitude $A > 0$. The density is:

$$f(x) = \frac{1}{\pi\sqrt{1 - \left(\frac{x}{A}\right)^2}} \qquad (8.38)$$

on the sample space $\Omega_X = [-A, A]$. The corresponding CF is the *Bessel function of the first kind* $\psi_X(s) = J_0(As)$. The peaks of this function are proportional to $1/\sqrt{|x|}$ as $x \to \pm\infty$, therefore Δ must be very much smaller than A for reconstruction of this density to be at all feasible. The problem with this density is the existence of singularities at $x = \pm A$ which lead to severe and unavoidable Gibb's phenomena (see Figure 8.10).

If Widrow's quantizing theorem holds, then the quantization error $\epsilon = Q(x) - x$ is uniformly distributed RV on $\left[-\frac{\Delta}{2}, \frac{\Delta}{2}\right]$, and therefore has variance $\Delta^2/12$. But in fact the necessary and sufficient condition for the quantization error to be uniform is where $\psi_X\left(\frac{2\pi k}{\Delta}\right) = 0$ for all $k \in \mathbb{Z}_0$ where $\mathbb{Z}_0 = \mathbb{Z} - \{0\}$ (Wannamaker, 2003, Theorem 4.1). Furthermore, ϵ and X will be uncorrelated e.g. $E[\epsilon X] = 0$ (Widrow and Kollar, 1996), but of course, uncorrelated does not imply independent.

By extending this statistical model of quantization using *joint* CFs, we can address several very important practical questions, including those pertaining to the statistical relationship between ϵ and X. A

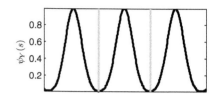

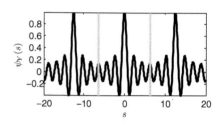

Fig. 8.10: **To allow reconstruction of the input density using Shannon-Whitaker interpolation after uniform quantization of width Δ, the characteristic function of the output, convolved with the sinc function of the quantizer, must be bandlimited so that $\psi_Y(s) = 0$ for $|s| > \frac{\pi}{\Delta}$. While the Gaussian is effectively bandlimited (top panel), this is not true for the sine wave (bottom panel). The grey vertical lines are at $\pm\frac{\pi}{\Delta}$, here $\Delta = \frac{1}{2}$ and both densities have the same standard deviation 0.469. The sine wave peak-to-peak amplitude is 3.**

deterministic quantizer Q always produces the same output from the same input, therefore the error is completely determined by the input. This deterministic relationship can be problematic in many practical applications. For example, in audio DSP, deterministic quantization noise introduces noticeable *harmonic* distortion.

The sinusoidal signal example presented earlier demonstrates that for many practical signals, Widrow's theorem is not satisfied. The primary limitation is the deterministic nature of the quantizer which passes information in the input through to the error signal. However, a *stochastic* quantizer can control this statistical relationship to a certain extent. We will look at a widely used and effective approach to stochastic quantization known as (*non-subtractive*) *dithering*. In this approach, a stochastic process called a *dither* signal W_n, usually independent of the input signal, is added to the input signal before quantization to control the statistical properties of the quantization process. The quantizer output is then $y = Q(x + w)$. The quantization error signal becomes $\epsilon = Q(x + w) - x$.

In subtractive dithering, the dither signal is subtracted from the quantizer output as well as being added to the quantizer input. Subtractive dithering is not often used in practice because it effectively requires transmission of the dither signal along with the quantized signal.

Any stochastic process could be used as a dither signal, but if we want certain statistical properties of the quantizer to hold then we need to apply particular restrictions on the CF of the dither. We will define a *dither of order* $M > 0$ as a process which satisfies $\psi^{(m)}\left(\frac{2\pi}{\Delta}k\right) = 0$ for all $k \in \mathbb{Z}_0$ and all $m = 0, 1, \ldots, M - 1$. As a particularly useful class of dither signals, (Wannamaker, 2003, page 27) explores the *M rectangular PDF dither* (*MRPDF*) obtained by summing M i.i.d. uniform random processes with mean zero having support Δ. It has the CF:

$$\psi(s) = \left[\frac{2}{\Delta s} \sin\left(\frac{\Delta s}{2}\right)\right]^M \tag{8.39}$$

Therefore, 1RPDF is uniform on $\left[-\frac{\Delta}{2}, \frac{\Delta}{2}\right]$ and 2RPDF is the *triangle distribution* on $[-\Delta, \Delta]$. It is straightforward to demonstrate that MRPDF processes are dithers of order M.

One of the primary values of dithering is that we can substantially reduce the statistical dependence between the input and the quantization error. More precisely: if the CF of the dither $\psi^{(i)}\left(\frac{2\pi}{\Delta}k\right) = 0$ for all $k \in \mathbb{Z}_0$ and all $i = 0, 1, 2, \ldots, M - 1$, then $E\left[\epsilon^m | X\right] = E\left[\epsilon^m\right]$ for all $m = 0, 1, 2, \ldots M$, and vice versa (Wannamaker, 2003, Theorem 4.8). This allows us to calculate moments of the error ϵ from moments of the dither W. For example, provided that the dither is at least second-order, then $E\left[\epsilon^2\right] = E\left[W^2\right] + \frac{\Delta^2}{12}$. Wannamaker (2003) then goes on to show that using non-subtractive MRPDF dither, the first $M > 0$ moments of the error signal are independent of the input signal. In particular for MRPDF dither of order $M \geq 2$, the error variance is $E\left[\epsilon^2\right] = (M + 1)\frac{\Delta^2}{12}$ (Wannamaker, 2003, Theorem 4.10).

Simple (non-subtractive) dithering has limitations. We cannot make the error signal completely independent of the input, and it is not possible to make the error uniformly distributed for arbitrary input distributions (Wannamaker, 2003, Theorem 4.6). Neither can we make it i.i.d. (Wannamaker, 2003, page 100). We can also never achieve error

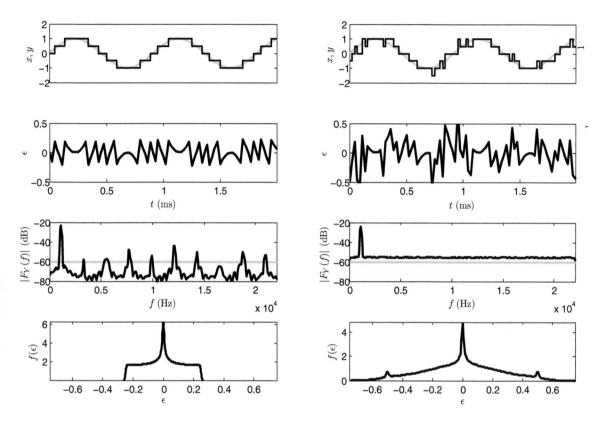

Fig. 8.11: **Quantization and dithering. A sinewave signal (grey curve, top row)** $x(t) = \sin(2\pi\phi t)$ **with** $\phi = 1,100$
and sample rate $S = 44.1\text{kHz}$ **is quantized using a deterministic uniform quantizer of width** $\Delta = \frac{1}{2}$, **to produce**
the quantized signal y **(black lines, top row left). This produces the error signal** ϵ **(second row, left), whose**
PSD has large peaks at harmonics of the input signal (third row, left). The PDF of the error signal is not
uniform (bottom row, left). Using the second-order i.i.d. 2RPDF triangular dither w **to create a stochastic**
quantizer, we obtain the results in the right column. Here, the PSD of the output signal y **has a flat noise**
floor, although the PDF of the error signal clearly has larger variance than in the deterministic case.

variance less than $\Delta^2/12$.

However, dither can control the *cross-moments* of the error. Provided
the joint CFs vanish at each pair of non-zero integer multiples of π/Δ
and the dither is independent of the input, then the autocorrelation of
the error is the same as the autocorrelation of the dither (Wannamaker,
2003, Theorem 4.9). So, if this holds then i.i.d. dither makes the PSD
of the error flat. An example is the MRPDF i.i.d. dither independent of
the input signal for $M \geq 2$ which has has flat error PSD (Wannamaker,
2003, page 74).

Furthermore, non-subtractive dither can allow moments of the input
X to be computed from moments of the output Y: provided some restrictions on the statistics of the dither are satisfied (Wannamaker, 2003,
Section 4.4.2). This includes computation of joint-moments, for example

the autocorrelation, so that:

$$F_Y(\omega) = F_X(\omega) + F_W(\omega) + \frac{\Delta^2}{6S} \tag{8.40}$$

where S is the sampling rate in Hz.

Finally, it can be shown that 2RPDF (triangular dither) has the smallest error variance of any dither for which the error mean and variance are independent of the input signal. Of course, the total error variance is smaller for 1RPDF (uniform) dither, but the error variance is dependent upon the input signal. Triangular dither is therefore a popular choice in applications such as audio DSP because input signal-dependent quantization noise is more perceptually obvious than flat noise. However, it also has important applications in scientific measurement since decoupling quantization noise from the signal gives makes it easier to model and predict the measurement error (see Figure 8.11).

So far we have considered it desirable to have stochastic quantization whose error is ideally i.i.d. or at least spectrally flat. There are, however, reasons to consider stochastic quantizers which produce non-i.i.d. quantization error. In particular, in audio DSP, quantization error in different parts of the spectrum is more noticeable than others because human auditory perception is much more sensitive to sound in the region 100Hz-1kHz than any other frequency range. So, it is possible to create dither signals that have particular spectra which can "shape" the quantization noise, a process known as *noise-shaping*. For example, by passing an MRPDF signal through an FIR filter with transfer function $H(z)$ we can create an autocorrelated dither signal W_n with PSD $F_W(\omega) = |H(\exp(i\omega))|^2$. With certain restricted coefficient sequences for $H(z)$, this creates a non-i.i.d. stochastic quantizer whose error signal has spectrum $F_\epsilon(\omega) = F_W(\omega) + \frac{\Delta^2}{6S}$ (Wannamaker, 2003, Corollary 5.1). More sophisticated approaches include a feedback arrangement whereby the quantization error is filtered with an additional FIR filter and then subtracted from the input signal, this arrangement gives much more flexibility to controlling the spectrum of the quantization noise, although the analysis becomes much more complex and useful results are only known for special classes of FIR filters (Craven and Gerzon, 1992).

Vector quantization

So far we have only discussed *scalar* quantization, that is the quantization of single samples at a time. It is often possible, however, to obtain a somewhat better rate-distortion trade off by quantizing many samples at a time. When the signal is not i.i.d., preceding the scalar quantization by some kind of transformation which makes the samples more independent is an improvement because the variance of the transformed signal is smaller than that of the original signal (see Section 8.3). However, quantizing a group of $D > 1$ samples simultaneously, a technique known as *vector quantization* (VQ), often improves on scalar quantization even

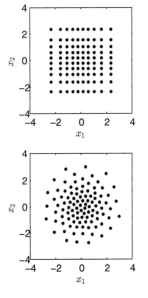

Fig. 8.12: **Comparing scalar with vector quantization (VQ), applied to standard ($\sigma = 1$, $\mu = 0$) Gaussian i.i.d. signals ($N = 100,000$ samples) with $K = 100$ quantizer output values. A scalar quantizer (parameters estimated using K-means with $K = 10$ for one co-ordinate only) produces a rectangular grid (top panel). The distortion is $e_S \approx 0.046$. By contrast, if the data are taken together in pairs, the resulting VQ (parameters estimated using K-means with $K = 100$) produces a more "circular" pattern, which outperforms the scalar quantizer with distortion $e_{VQ} \approx 0.039$, or a reduction in distortion of 16%.**

when the signal is i.i.d. The essential reason for this improvement is geometric: VQ can create a more efficient *tiling* of the D-dimensional space than scalar quantization in each dimension separately. That is, for the same bit rate, it is possible to devise a partition of the space which has lower distortion than scalar quantization.

As a simple illustration, let us examine two tilings in $D = 2$ dimensions. Consider the case of the uniform scalar quantizer with Δ quantization width, which in two dimensions leads to the regular square tiling of the plane. Assuming the quantizer output values are centered in each square cell, the square loss distortion for each cell is:

$$
\begin{aligned}
e_S(\Delta) &= \int_0^\Delta \int_0^\Delta \left[\left(x - \frac{1}{2}\Delta \right)^2 + \left(y - \frac{1}{2}\Delta \right)^2 \right] dy \, dx \\
&= \frac{1}{6}\Delta^4
\end{aligned}
\tag{8.41}
$$

By contrast, the *hexagonal* tiling of the plane with side length Δ has distortion:

$$
\begin{aligned}
e_H(\Delta) &= 12 \int_0^{\frac{1}{2}\Delta} \int_0^{\sqrt{3}x} \left[\left(x - \frac{1}{2}\Delta \right)^2 + \left(y - \frac{\sqrt{3}}{2}\Delta \right)^2 \right] dy \, dx \\
&= \frac{5\sqrt{3}}{8}\Delta^4
\end{aligned}
\tag{8.42}
$$

Now, two VQs will have the same bit rate if the number of quantization output values over the full range is the same for both. Solving for the hexagonal side length which makes the area of both the square and the hexagon equal gives us $\Delta_H = \frac{\sqrt{2}\sqrt[4]{3}}{3}\Delta_S$, so that:

$$
\frac{e_H(\Delta_H)}{e_S(\Delta_S)} = \frac{5\sqrt{3}}{9} \approx 0.96
\tag{8.43}
$$

So, for the same bit rate, the distortion of the hexagonal tiling, which can only be obtained using VQ, is a little smaller than the distortion of the square tiling obtained using uniform scalar quantization. Conversely, similar calculations show the corresponding improvement in bit rate for equal square loss distortion is 0.028 bits (Berger and Gibson, 1998).

Because the possible geometric configurations in two and greater dimensions are much more complex than for the scalar case, few useful theoretical results are known for VQ. For example, it is known that for i.i.d. Gaussian processes and the square loss distortion, VQ can reduce the bit rate by at most only 0.255 bits over a scalar quantizer with entropy coding (Berger and Gibson, 1998). However, as discussed earlier in this section, we can always use K-means (or the entropy-constrained version), to estimate the VQ parameters, and this is a simple way to increase quantizer performance (see Figure 8.12). That being said, although VQ is quite simple, it is much more bulky to implement in practice because the number of quantizer codes typically needs to grow as $O\left(\kappa^D\right)$ which

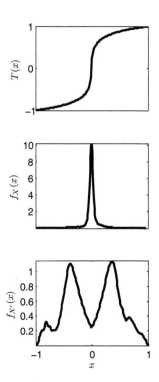

Fig. 8.13: Digital audio signals generally have densities which are sharply peaked at 0 with heavier-than-Gaussian tails (middle panel). Companding transformation functions $X' = T(X)$ such as μ-law (top panel) compress the region in the tails and expand the region near 0, making X' more uniform than X (bottom panel). As a result the quantization error for a uniform quantizer applied to X' is lower than that for X, and the quantization noise is perceptually "masked" by the louder parts of the signal. Here a 44.1kHz sample-rate audio signal is companded and quantized using a uniform $r = 8$ bit quantizer, giving an RMSE $\sqrt{E(Q)} = 1.8 \times 10^{-3}$ and bit-rate $I(Q) = 5.3$ whereas direct uniform quantization gives RMSE $\sqrt{E(Q)} = 2.3 \times 10^{-3}$ (bit-rate $I(Q) = 7.7$).

is exponential in dimension, where κ is the number of quantizer levels in each dimension. This means that to achieve useful compression using VQ, we need to transmit an exponentially-growing set of codes, and this overhead is often larger than the gain in sample bit rate achieved.

8.3   Lossy signal compression

In this section we will describe some examples of where discrete-time quantization and rate-distortion theory has been used to produce immensely practical signal compression algorithms, often known as *codecs* (short for *encoder-decoder*). These codec algorithms are mostly *lossy*, that is, unlike basic entropy coding, some information in the signal is lost. However, the loss can be tolerated in practice and these algorithms are embedded as part of e.g. international technology standards in communications infrastructure including GSM cellular telephony. Others are widely use in practical digital storage applications to reduce the amount of memory required to store large amounts of signal data.

Audio companding

One of the simplest techniques involves *companding*, which is a portmanteau word for *compression-expanding*. Here, a nonlinear transformation is applied to the signal before quantization. At the receiver, the inverse transformation is applied. An example is the *μ-law* telephony standard, whose transformation which is a signed logarithm:

$$T(x) = \text{sign}(x) \frac{\ln(1 + \mu |x|)}{\ln(1 + \mu)} \tag{8.44}$$

with inverse:

$$T^{-1}(x) = \frac{1}{\mu}\text{sign}(x)\left[(1 + \mu)^{|x|} - 1\right] \tag{8.45}$$

where $\mu = 2^r - 1$ and r is the bit-depth of the uniform quantizer. The distribution of typical audio signals which is highly peaked around $x = 0$, with heavy tails. This transformation "expands" the range of X around 0, and "compresses" this range in the tails near ± 1. The result is that the distribution of $T(X)$ is more uniform than that of X. This causes the quantization error of a uniform quantizer to be more evenly distributed across the whole range of X. Therefore, the quantization error is worse for large-amplitude signals and better for quieter signals. The perceptual effect is that the quantization error is "masked" by the signal itself to an extent, and the overall error $E(Q)$ is smaller. But since companding makes the signal more uniform before quantization, this causes the bit rate $I(Q)$ to be higher, as expected from rate-distortion theory (Figure 8.13).

Linear predictive coding (LPC)

More complex techniques can be used to exploit autocorrelation in the signal. For example, if the signal is generated by an LTI system driven by a Gaussian i.i.d. noise source, then linear prediction can be used to estimate the impulse response of the system; whereupon, the whitened residual can be encoded. This residual will have much smaller spread than the original signal. Therefore, for a given quantization error, the residual can be quantized using fewer bits than required for the original signal, providing compression.

Although most real-world signals are unlikely to be generated by exactly such a mechanism, there are practical examples which closely fit this structure. For example, voice and audio signals have non-trivial (short-time) spectra that can be fairly easily captured using linear prediction. In fact, *linear predictive coding* (LPC) can be effective in any situation where it is reasonable to describe the signal as the output of a driven LTI system. LPC initially assumes that the data is generated by an M-th order, recursive LTI system:

$$X_n = \sum_{m=1}^{M} a_m X_{n-m} + Z_n \tag{8.46}$$

with the coefficients $\boldsymbol{a} \in \mathbb{R}^M$ and Z_n is an i.i.d. Gaussian process. LPC uses LPA (Section 7.4) to estimate the coefficients $\hat{\boldsymbol{a}}$ from a given signal x_n, and we can use this to calculate the residual $z_n = x_n - \sum_{m=1}^{M} \hat{a}_m x_{n-m}$. Now, it is tempting to simply quantize the residual, $q_n = Q(z_n)$, but the decoding $\hat{x}_n$ must of course use q_n rather than z_n, i.e.:

$$\hat{x}_n = \sum_{m=1}^{M} \hat{a}_m \hat{x}_{n-m} + q_n \tag{8.47}$$

with $\hat{x}_m = x_m$ for $m = 1, 2, \ldots, M$ to get the iteration started. But then, on decoding, the quantization error $\epsilon_n = z_n - q_n$ will be amplified by the recursion, and $\hat{x}_n$ will quickly begin to diverge from x_n. If instead, we quantize the *prediction error* $e_n = x_n - \sum_{m=1}^{M} \hat{a}_m \hat{x}_{n-m}$ between the decoding and the input signal, $q_n = Q(e_n)$, then the decoding recursion (8.47) becomes:

$$\hat{x}_n = \sum_{m=1}^{M} \hat{a}_m \hat{x}_{n-m} + Q\left(x_n - \sum_{m=1}^{M} \hat{a}_m \hat{x}_{n-m} \right) \tag{8.48}$$

In effect, q_n now serves the dual purpose of quantizing the LPA residual, and compensating for the ongoing prediction error induced by the quantization. Note that if the quantization was perfect e.g. $Q(x) = x$, then we would have $\hat{x}_n = x_n$. Therefore, this arrangement prevents the recursion from diverging away from the input signal.

Compensating for the prediction error in this way is a useful trick, but now the recursion (8.48) is no longer linear: it has the highly nonlinear quantizer function in the loop. This means that we cannot use methods

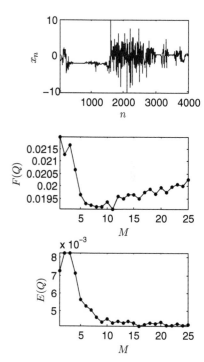

Fig. 8.14: Using linear predictive coding (LPC) to compress MEMs digital accelerometry signals (top panel, signal length $N = 4000$ samples). Distortion-rate trade-off as a function of the LPC model order M with regularization parameter $\lambda = 1.25 \times 10^{-6}$ and $r = 6$ bit uniform LPA residual quantizer with first order MRPDF dither (middle panel). This produces distortion $E(Q)$ (bottom panel). The optimal model order is $M = 11$, which gives RMSE distortion $\sqrt{E(Q)} = 0.07$, and compressed signal length $I(Q) = 11810$ bits. Direct uniform quantization of the input signal to $r = 16$ bits leads to RMSE $\sqrt{E(Q)} = 1.3 \times 10^{-4}$ and signal length of $R_X = 64000$, so that the compression ratio is 5.4 to 1, in other words LPC compresses this signal to less than 19% of the original size.

such as transfer function pole analysis to determine whether the recursion is stable: although by definition the input signal will be bounded, the quantized signal q_n will need to compensate for any instability. Also, transfer function pole analysis of the recursion is not possible. Furthermore, since the LPA analysis determines the optimal coefficients which minimize the residual error $E_Z = \sum_{n=1}^{N} z_n^2$, they do not exactly correspond to the optimal coefficients for the quantized residual error $E_Q = \sum_{n=1}^{N} q_n^2$. In fact, finding those coefficients will be extremely hard because E_Q is discontinuous and non-convex, and the parameters are continuous.

In summary, at the encoding end, first, M-th order LPA analysis is used to find the optimal coefficients $\hat{a}_m$ from the input signal x_n. The residual z_n is computed to determine the quantization width Δ, which depends upon the scale of the residual and the number of bits used in the quantizer. Dither is useful here to prevent deterministic quantization artifacts. Then, we use the recursion (8.48) to determine the quantized residual signal q_n. The quantized residual is often entropy coded to further reduce the number of bits per sample. The initial conditions x_m are packaged together with the coefficients and the entropy-coded residual, and sent to the decoder. The decoder recursion (8.47) is applied to this information to reconstruct the signal $\hat{x}_n$.

A simple and widely-used special case of LPC known as differential pulse coded modulation *(DPCM) assumes that $M = 1$ and $a_1 = 1$.*

In practice, LPC can often achieve fairly high signal compression at low distortion $E(Q) = \frac{1}{N} \sum_{n=1}^{N} (x_n - \hat{x}_n)^2$. However, determining the actual *compression ratio*, that is, the ratio between the number of bits in the original signal and the compressed signal, takes some care. Let us assume that the original digital signal is quantized to r bits, then the original signal uses $R_X = N r$ bits. The encoded (quantized, entropy-coded) signal takes up $(N - M) H[Q]$ bits, and we need the M coefficients and M initial condition values. To get best results the coefficients would ideally need to be floating-point, 32 bits at minimum, and the initial conditions have the same bit resolution as the input signal. Thus, in one practical implementation, the compression ratio is:

$$\frac{R_X}{I(Q)} = \frac{N r}{M \times (32 + r) + (N - M) H[Q]} \tag{8.49}$$

To achieve good compression, we want this ratio to be as large as possible which means making $I(Q)$ as small as we can. Now, $M \times (32 + r)$ is linearly increasing with M, so we want M to be small. Also, if $N \gg M$, then $(N - M) H[Q]$ can dominate over $M \times (32 + r)$. Thus, for longer signals the impact of the model order M on the compression ratio can be negligible, and it is the per-sample ratio $r/H[Q]$ that matters. However, as discussed above, LPC is, essentially, an LTI-system approach which assumes that the spectrum of the signal is time-invariant. This is very often not true in reality, and it is better in that case to break the signal up into smaller time windows which are approximately time-invariant, and apply LPC separately to each window. In which case the effect of the model order M cannot be ignored. Also, note that we can always entropy-code the original signal and this is certainly a simpler

procedure than using LPC, so it is reasonable to argue that a more realistic estimate of the achievable compression ratio would be computed
using $R_X = NH[X]$, where $H[X]$ is the entropy of the (quantized)
input signal. We need to set the model order M and the number of
quantizer bits. This requires optimizing the distortion-rate trade off
$F(Q) = E(Q) + \lambda I(Q)$ with respect to both parameters, a combinatorial optimization problem (see Section 2.6).

It is worth pointing out that the recursive framework of LPC is quite
flexible: for example, while classical LPC uses the purely linear predictor $\sum_{m=1}^{M} a_m X_{n-m}$, any arbitrary function of the previous samples,
$g(X_{n-1}, \dots X_{n-M})$, could be used instead. This invites ample opportunities to use various nonlinear regression methods (see for example
Section 6.4). Similarly, if the residual is expected to be non-Gaussian
then loss functions other than square loss might produce better results
(Section 4.4). See also Press (1992, page 571) for how to make LPC
completely *lossless* at the expense of compression ratio. The literature
on LPC is extensive, for more details see Gray and Neuhoff (1998) and
references therein.

Transform coding

It is apparent from the previous section that many signals are autocorre-
lated and can therefore be compressed more readily if these correlations
are taken into account. Using the duality between time and frequency
for LTI systems, we can look to perform compression in the frequency
rather than the time domain. This idea extends to time varying systems. In fact, wherever a transform exists which effectively decorrelates
the signal, quantizing the transform coefficients will obtain a reduction
in bitrate for a given distortion by comparison to quantizing the time
samples of the signal directly. Often, only a few transform coefficients
are significant: that is, most of them will have small variances. This
means for a given distortion, that only a few coefficients need to be
coded accurately and the rest can be coded in much smaller bit rates.

Let us consider a discrete-time signal x_n, $n = 1, 2, \dots, N$ which is
transformed using an orthonormal transform represented by the ma-
trix $\mathbf{A}$ to give the coefficient vector $\boldsymbol{\phi} = \mathbf{A}\boldsymbol{x}$ (for example, a DFT or
DWT, Sections 7.2 and 7.6). The coefficients will be subsequently quan-
tized. Since the transform is orthonormal, the (square loss) distortion in
the signal space is equal to the distortion in the coefficient space (Wiegand and Schwarz, 2011, page 186). This means that the distortion is
$E(Q) = \frac{1}{N} \sum_{n=1}^{N} \left(\phi_n - \hat{\phi}_n \right)^2$ where $\hat{\phi}_n = Q_n(\phi_n)$ is the quantized n-
th coefficient, and the problem of *transform coding* is to determine the
parameters of the N quantizers Q_n which optimize the rate-distortion
problem.

The distortion-rate trade-off function applies, but it is different for
each coefficient and we will denote this by $e_n(r_n)$ where r_n is the rate
for the n-th coefficient. Over the entire set of N coefficients, we can
specify the average bit rate $r = \frac{1}{N} \sum_{n=1}^{N} r_n$ and the corresponding av-

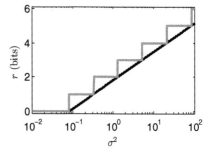

Fig. 8.15: In transform coding, under the fixed, high-rate quantization
approximation, the number of bits
(black line, next largest integer in
grey) required to quantize a particular coefficient depends upon the variance of the coefficient σ^2, the choice
of Lagrange multiplier λ (here $\lambda = 0, 5$), and the distribution of each coefficient (here, the Laplace distribu-
tion).

erage distortion will be $\frac{1}{N}\sum_{n=1}^{N} e_n(r_n)$. One way to pose the choice of quantization parameters which solve the transform coding problem is to optimize the average distortion:

$$\text{minimize}_{Q_n} \quad \frac{1}{N}\sum_{n=1}^{N} e_n(r_n) \tag{8.50}$$

$$\text{subject to} \quad r = \frac{1}{N}\sum_{n=1}^{N} r_n$$

The corresponding Lagrangian is $\frac{1}{N}\sum_{n=1}^{N} e_n(r_n)+\lambda\left(\frac{1}{N}\sum_{n=1}^{N} r_n - r\right)$ for the multiplier λ (see Section 2.1). Assuming that the distortion-rate functions are all non-negative, strictly convex with a continuous first derivative, the solution to the optimization problem is:

$$\frac{\partial e_n}{\partial r_n}(r_n) = -\lambda \tag{8.51}$$

for all $n = 1, 2, \ldots, N$. In words: the optimal transform coding quantizers all share the same rate of change of distortion with r_n, and each quantizer can be determined individually based on the statistics of each transform coefficient. Since (approximate, high-rate) distortion-rate functions (Section 8.2) can be written in the form $e(r) \approx s\sigma^2 2^{-2r}$ where s depends upon the distribution function, the solution to (8.51) is (Wiegand and Schwarz, 2011, page 192):

$$r_n = \max\left(\frac{1}{2}\log_2\left(\frac{s\,\sigma_n^2\,2\ln 2}{\lambda}\right), 0\right) \tag{8.52}$$

In practice we need to use $\lceil r_n \rceil$ bits. Often the coefficient distributions are heavy-tailed, e.g. Laplace distributed (to a first approximation anyway, see Figure 8.15).

As an example of how effective this approach can be, we will examine an application of this idea to digital musical audio signals (Figure 8.16). These are typically quantized at $r = 16$ bits. Using the *discrete cosine transform* (DCT) the distribution of the coefficients is much more heavy-tailed than the direct signal samples in the time domain. The consequence is that, for a choice of λ which would achieve a distortion-rate trade-off which is comparable to direct scalar quantization in the time domain of 16 bits, each r_n, computed using (8.52), is on average much smaller than 16 bits (often close to an order of magnitude smaller). With entropy coding this leads to a large compression ratio, a value of around 10 or more is not unusual (in the example of Figure 8.16, the per-sample compression ratio after entropy coding is approximately 16). This is an illustration of just how successful lossy compression using transform coding (with the right choice of orthonormal basis) can be.

It should be noted that in practice, this compression ratio will be reduced by the overhead required to transmit the model description.

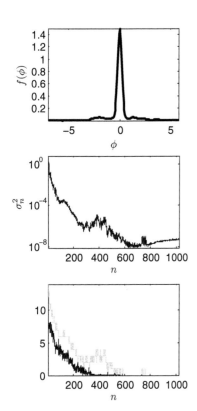

Fig. 8.16: Efficient transform coding of musical audio signals at distortion-rate trade off $\lambda = 5\times10^{-7}$. After applying the DCT ($M = 1024$) to the signal, the distribution of coefficients is very heavy-tailed (top, coefficient $n = 5$). The variance of the coefficients drops off rapidly with increasing n (middle), so that the number of quantizer bits r_n (fixed-rate Laplace model, $s = \frac{9}{2}$) also falls rapidly (bottom, grey curve). The entropy H_n of the coefficients (bottom, black curve) is significantly smaller than the quantizer bits (average $H = 0.91$ bits per sample). To achieve the same distortion ($E(Q) \approx 3\times10^{-7}$), direct scalar quantization requires $H = 7.3$ bits per sample.

Essentially, the codec is determined by the N variances σ_n^2 (which determine the rate variables r_n). A practical scheme would usually require no more than 32 (floating-point) bits for each variance, giving an overhead of $32M$ bits. An upper value of $M = 1024$ would be ample, so this overhead is essentially negligible for most digital music recordings of realistic durations. However, to take advantage of entropy coding, the *Huffman coding tree* for each quantizer needs to be transmitted as well. Several efficient data structures for representing the tree exist, for example, one that requires $\lceil \frac{3}{2} \cdot 2^r \rceil$ bits (Chowdhury *et al.*, 2002) leading to a total of $\sum_{m=1}^{M} \lceil \frac{3}{2} \cdot 2^{r_m} \rceil + 32M$ bits to send the entire model (in the example of Figure 8.16, the Huffman tree overhead is, on average, approximately 500 bits per coefficient). By contrast, with direct scalar quantization in the time domain, to perform entropy coding only the one Huffman tree needs to be transmitted. One could consider having a universal fixed Huffman tree e.g. based on a Laplace model for the coefficient distributions, but this would necessarily reduce the per-sample compression ratio. Thus, entropy transform coding is only really useful for signals which are of sufficient duration to make the overhead a negligible fraction of the final compressed bit length.

Transform coding has been pushed far beyond its original, simple formulation as discussed here, in the Motion Picture Experts Group (MPEG) Layer 3 digital audio codec, commonly known as *MP3 format* (Brandenburg, 1999). One of the limitations of classical rate-distortion theory is the dominant role played by the square loss function, which is a substantial limitation in practice. For example, human auditory perception is far more sensitive to distortion for certain frequencies than others. This means that the distortion-rate trade-offs for the different DCT coefficients discussed earlier are not equal. By allowing more distortion in low and high frequencies than mid ranges (e.g. 500Hz-4kHz), the MP3 algorithm can achieve much more perceptually acceptable distortion-rate trade offs in practice. Another notable advance has been to take *psychoacoustic masking* into account whereby quiet frequency components adjacent to loud components are not easily perceived: the quiet component is said to be "masked" by the louder component. Because of this, more distortion is perceptually acceptable for the quiet, masked component than the louder masking frequency. This and other advances mean that MP3 can achieve almost imperceptible distortion at compression ratios up to 10 for CD-quality audio, which means saving 90% of the memory required to store the digital audio recording. This has enabled a revolution in the accessibility of high-quality music since low-bandwidth storage and internet transmission then becomes feasible.

8.4   Compressive sensing (CS)

Discrete-time signal acquisition using bandlimited uniform sampling via Shannon-Whittaker interpolation, as described in Section 8.1, is ubiquitous and most practical ADC/DAC hardware uses this principle. Typically, signals are acquired this way, stored in their full-bandwidth uncompressed form, and then compressed using some kind of lossless or lossy technique (Section 8.3). This raises the question: if the signal can be compressed, is it actually necessary to acquire the full-bandwidth signal in the first place? The answer to this question is in fact "no" – we can indeed do much better than is implied by Shannon-Whittaker if we have more information about the signal.

In Shannon-Whittaker interpolation, we assume the signal comes from the space of bandlimited l_2 functions V. This is a particular, but actually quite non-specific, model. Consider a (particularly simple) example of such a function, a single sinusoid:

$$f(t) = \sin(2\pi\phi t) \tag{8.53}$$

where ϕ is the frequency in Hz. Now, for this particular signal, we can obtain the frequency parameter ϕ from the sampled digital signal at *any* time instant t_0 (taking into account that the $\sin^{-1}$ function usually gives us the principal value only):

$$\phi = \frac{1}{2\pi t_0} \sin^{-1}(f(t_0)) \tag{8.54}$$

This (overly simplified) example illustrates that many signals might be contained in V but contain a lot less (spectral) information than implied by their full bandwidth; actually there is *no information at all* in the frequency domain for this signal for frequencies in the range $(0, \phi)$. All we need is the value of ϕ to reconstruct it perfectly. On the other hand, in the natural domain in which we acquire the signal digitally (here, the time domain), (nearly) every uniformly-sampled digital value is non-zero. So, the frequency variable ϕ can be obtained using a *single sample* of this function – there is no need to sample the signal at anywhere other than at one arbitrary time instant, which might as well be chosen randomly.

Thus, if we are to use classical uniform sampling, this signal has a bandwidth of at least ϕ, so this requires a sampling frequency of $S \geq 2\phi$. So, we can view uniform sampling as the absolute worst-case required for perfect reconstruction if all we know is that the signal is in V, and many real signals do not require anything like this worst-case sampling effort, and this can lead to a gross waste of digital storage resources. The fast-growing area of *compressive sensing* (CS) has developed various theoretical tools which enable far more efficient sampling than classical uniform Shannon-Whittaker interpolation for specific and quite ubiquitous families of signals and sampling methods, to be defined next.

Sparsity and incoherence

As most of CS has been developed in the context of discrete-time signals $\boldsymbol{f} = (f_1, f_2, \ldots, f_N)$, we will restrict our attention to these objects. In this context, with reference to classical uniform sampling theory, the advantage of CS is that it demonstrates the successful use of *undersam-*

pling, that is, when we only have M measurements from the N uniform samples, where $M < N$ (Candès and Wakin, 2008). We will take a generalized view of sampling where we acquire signals digitally through some kind of correlation with *sensing functions* $\boldsymbol{u}_n$ through which we obtain digital samples $g_n = \langle \boldsymbol{u}_n, \boldsymbol{f} \rangle$. For example, with bandlimited sampling (Section 8.1) the acquisition corresponds to the use of discrete, standard basis functions $\boldsymbol{u}_n = \boldsymbol{e}_n$ and the samples are then just $f_n = \langle \boldsymbol{e}_n, \boldsymbol{f} \rangle$.

Signals like (8.53), which have very little spectral information inside their bandwidth, are called *sparse* (in the frequency domain). More formally we say that a function, represented in the orthonormal basis $\boldsymbol{a}_n$, $\boldsymbol{f} = \sum_{n=1}^{N} x_n \boldsymbol{a}_n$, is K-sparse if only $K \ll N$ of the coefficients $\boldsymbol{x} = (x_1, x_2, \ldots, x_N)$ are non-zero (Candès and Wakin, 2008). For example, if $\mathbf{A}$ is the Fourier matrix then sparsity refers to the discrete Fourier domain.

Similarly, the *coherence* between the sensing orthobasis $\mathbf{U}$ and the sparse representation basis $\mathbf{A}$ is defined as the maximum absolute dot product between each pair of basis functions (Candès and Wakin, 2008):

$$\mu(\mathbf{A}, \mathbf{U}) = \sqrt{N} \max_{j,k \in 1,2,\ldots,N} |\langle \boldsymbol{a}_k, \boldsymbol{u}_j \rangle| \tag{8.55}$$

which lies in the range $\left[1, \sqrt{N}\right]$. The main goal in CS, is to choose systems of bases which have low coherence, i.e. those which are *incoherent*, because then the undersampling can be large as we will see later. For example, the standard (delta) uniform sampling basis and the Fourier basis have $\mu(\mathbf{A}, \mathbf{U}) = 1$, therefore they are maximally incoherent. Because of this incoherence, $\boldsymbol{f}$ has a highly non-sparse (*dense*) representation in $\mathbf{U}$ (Candès and Wakin, 2008). This is true of our signal (8.53) for which we can take any sample of the signal in the time domain and use this to recover the signal. In other words, for a signal that is sparse in a domain whose bases are incoherent with the sensing basis, the information in the signal is well spread-out in the sensing domain. This is important because it means we do not need to acquire nearly as many measurements as suggested by Shannon-Whittaker interpolation.

Exact reconstruction by convex optimization

Given an undersampled signal with $M < N$ digital measurements g_m, $m = 1, 2, \ldots, M$, we want to reconstruct the true signal f_n. This, of course, is an ill-posed problem but as with bandlimited sampling, we can apply constraints to make this problem tractable. CS theorists have investigated the use of various constraints, the simplest and most well-studied of which is L_1-regularization (Candès and Wakin, 2008):

$$\begin{aligned} \text{minimize} \quad & \|\hat{\boldsymbol{x}}\|_1 \\ \text{subject to} \quad & \langle \mathbf{A}\hat{\boldsymbol{x}}, \boldsymbol{u}_m \rangle = g_m, \ m = 1, 2 \ldots M \end{aligned} \tag{8.56}$$

and the constraints simplify to $\sum_{n=1}^{N} \hat{x}_m a_{n,m} = f_m$, $m = 1, 2 \ldots M$ in

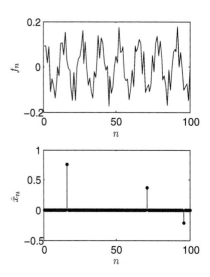

Fig. 8.17:   Compressive sensing illustration.   (Top panel) Input signal f_n of length $N = 100$, uniformly sampled.   This signal is sparse with $K = 3$ non-zero coefficients in the discrete cosine transform basis, which has coherence $\mu(\mathbf{A}, \mathbf{U}) = \sqrt{2}$ with the standard (delta) sensing basis. The recovery of the coefficients is exact (bottom panel) with $C = 1$, leading to a minimum of $M = 28$ measurements.

the case of the standard delta sensing basis e_n. In other words, we seek a set of coefficients $\hat{x}$ with minimum L_1-norm such that the M projections of the reconstruction onto the sensing basis agree with the M acquired samples. This reconstruction problem (8.56) is an LP (Boyd and Vandenberghe, 2004). It has been shown that for a K-sparse signal in the basis $\mathbf{A}$, provided (Candès and Wakin, 2008):

$$M \geq C \, \mu^2 \left(\mathbf{A}, \mathbf{U} \right) K \ln N \qquad (8.57)$$

where $C > 0$ is a constant, then with probability close to 1, $\hat{x} = x$. In other words, if (8.57) holds, the L_1 CS convex problem (8.56) (nearly always) recovers both the locations and the values of the non-zero coefficients exactly. The minimum number of measurements required, M, must increase with the square of the coherence, therefore, the smaller the coherence the better, as discussed earlier. Similarly, the number of measurements must linearly increase with the sparsity K, and logarithmically with the size of the signal N. Experiments show that constants as small as $C = 0.4$ or about 4-5 measurements per non-zero coefficient suffice in practice (Candès and Wakin, 2008).

Similar results have been shown to hold for the case where there is noise, e.g. where the measurements are corrupted by some error $g_m + \epsilon_m$. In this case, the reconstruction cannot be exact, but it comes close to finding the best solution possible as if the location of the K largest magnitude coefficients were known. These are remarkable results considering that discovering the number and locations of the non-zero coefficients is itself a difficult combinatorial optimization problem!

One of the other appealing aspects of CS is that the sensing basis can be random and will be largely incoherent with any basis $\mathbf{A}$. One simple way to construct such a random sensing basis is to choose N i.i.d. vectors uniformly distributed on the unit N-sphere and then orthonormalize them. Another random construction involves i.i.d. Bernoulli ± 1 entries (Figure 8.18). With probability close to 1, the coherence of such random sensing matrices with $\mathbf{A}$ is approximately $\sqrt{2 \ln N}$ (Candès and Wakin, 2008). These randomized constructions therefore give us a recipe for almost *universal* CS.

Compressive sensing in practice

Putting CS to work in practice requires some specific sensing hardware. The standard delta sampling basis implies the straightforward use of existing ADCs, only using randomized rather than uniform sampling time points. A slightly more complex method is known as *random demodulation* (Tropp *et al.*, 2010), and is extremely useful for very high-rate sparse signals where sampling at the full rate would be challenging (this is the situation for e.g. radar DSP). Here, the sparse input signal of bandwidth B is multiplied by a (pseudo) random ± 1 squarewave signal alternating around the Nyquist frequency $2B$. The product is then low-pass pre-filtered to a much lower bandwidth $\bar{B} < B$, and sampled at $2\bar{B}$ using a classical ADC. The act of multiplication by a randomized signal causes the spectrum of the (sparse) input signal to be "smeared" across the entire bandwidth B, since the spectrum of the input signal is convolved with the spectrum of the random squarewave. Consider an input sinusoid at much higher frequency than the low-rate sampling bandwidth $\bar{B}$. Even if this sinusoid has a much higher frequency than $\bar{B}$, the smearing leaves a 'trace' of the signal in the acquired, low-rate samples g_m. When there are multiple sinusoids, the acquired samples will contain a linear mixture of the frequency-shifted spectrum of the random squarewave, one for each input sinusoid, whose frequency, phase and amplitudes can be untangled using CS reconstruction. Tropp *et al.* (2010) show empirically that for a (sub) sampling rate obeying $\bar{S} \geq 1.7K \ln(B/K + 1)$, perfect reconstruction of sparse input signals of maximum bandwidth B containing K sinusoids with random frequencies, phases and amplitudes is possible with probability close to 1.

Having acquired the signal, we must then solve the reconstruction problem (8.56) to obtain the underlying signal f_n. This can be achieved using efficient interior-point methods (Section 2.4). However, note that this only needs to happen at the point of reconstruction so there is generally no need to store anything other than the compressed signal until the reconstructed signal is needed.

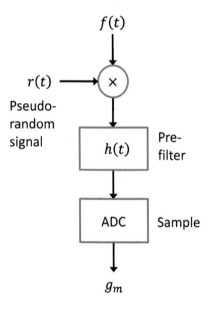

Fig. 8.18: Compressive sensing using random demodulation. The input signal $f(t)$ is multiplied by a pseudo-random ± 1 signal $r(t)$ generated at the Nyquist rate for the bandwidth of the input signal. The product is anti-aliased to a much lower rate, and sub-sampled at that low rate. Subsequent reconstruction is performed using e.g. L_1-regularized convex optimization.

Nonlinear and non-Gaussian signal processing

9

Linear, time-invariant (LTI) Gaussian DSP, as described in Chapter 7 has substantial mathematical conveniences that make it valuable in practical DSP applications and machine learning. When the signal really is generated by such an LTI-Gaussian model then this kind of processing is optimal from a statistical point of view. However, there are substantial limitations to the use of these techniques when we cannot guarantee that the assumptions of linearity, time-invariance and Gaussianity hold. In particular, signals that exhibit jumps or significant non-Gaussian outliers cause substantial adverse effects such as Gibb's phenomena in LTI filter outputs, and nonstationary signals cannot be compactly represented in the Fourier domain. In practice, many real signals show such phenomena to a greater or lesser degree, so it is important to have a 'toolkit' of DSP methods that are effective in many situations. This chapter is dedicated to exploring the use of the statistical machine learning concepts in DSP.

9.1 Running window filters

A running (sliding, or moving), window filter uses only information from a temporal neighborhood of each time step to determine the filter's output. Maximum likelihood filters do not have priors over the parameters. Running window filters are often extremely simple to implement and have very low computational complexity, and they can be surprisingly effective in the right circumstances.

In the following, the *window* function $I(n)$ returns the set of indices of the temporal neighborhood. The most common window function is the *centred rectangular window* of size W:

$$I(n) = \{n - W_2, \ldots, n-1, n, n+1, \ldots, n+W_2\} \qquad (9.1)$$

where $W_2 = \lfloor \frac{W-1}{2} \rfloor$ (usually W is odd). For finite length signals, there will be windows near the *boundaries* at $n=1$ and $n=N$ where there are elements $i \in I(n)$ for which $i < 1$ or $i > N$. One common approach to solving this is to *truncate* these windows by removing these elements.

Machine Learning for Signal Processing: Data Science, Algorithms, and Computational Statistics. Max A. Little.
© Max A. Little 2019. Published in 2019 by Oxford University Press. DOI: 10.1093/oso/9780198714934.001.0001

Another approach is to *extend* the signal instead: for example, by setting $x_i = 0$ for all $i < 1$ or $i > N$. This has the obvious disadvantage of introducing an artificial discontinuity at the boundary; a preferable approach is to set $x_i = x_1$ for $i < 1$, and $x_i = x_N$ for $i > N$.

Maximum likelihood filters

Consistent with the statistical machine learning approach, we can put a distribution over the samples in each window. If this distribution has a parameter, then the single parameter MLE is obtained by maximizing the joint probability over the samples in each window with respect to the parameter μ_n:

$$\hat{x}_n = \arg\max_{\mu_n} f\left(x_{I(n)}; \mu_n\right) \tag{9.2}$$

So, the filtered output signal is formed of the MLE parameter for each window.

We have already encountered a simple example of this: the moving average (MA) FIR filter (Section 7.3). This special case of (9.2) arises when we assume the signal inside each window is i.i.d. Gaussian, e.g.: $f\left(x_{I(n)}; \mu_n\right) = \prod_{i \in I(n)} \mathcal{N}\left(x_i; \mu, \sigma_n\right)$, then the MLE is just $\hat{x}_n = \frac{1}{W}\sum_{i \in I(n)} x_i$. Generally speaking, MLEs become more accurate as the sample size is increased, so it is useful to have the window size W as large as possible. However, increasing the window size also increases the timescale on which changes in the signal are expected to occur due to Heisenberg uncertainty (Section 7.2). This means that there is, generally, a trade-off between reducing the spread of the noise and retaining features on temporal scales smaller than the window size.

Of course, all the same issues with robustness to outliers arise as with any least-squares estimator, so it may be better to invoke a robust loss function. For example, a very useful range of filters can be constructed using the asymmetric check loss (4.88) with quantile parameter $q \in [0, 1]$. Assuming asymmetric Laplace distributed samples that (as above) are i.i.d. inside each window, the MLE is the q-th quantile, which, for a set of K data samples is the value in position $k = \lfloor qK \rfloor$ when the samples are sorted into ascending order.

When $q = \frac{1}{2}$ we recover the standard symmetric Laplace distribution, for which the MLE is just the median. The resulting *running median filter* is a standard method in nonlinear signal processing; its extension to two dimensions applied to images is used extensively (Arce, 2005, section 5.1). The median filter has certain properties that make it useful for noise removal from piecewise constant signals. In particular, piecewise constant signals pass through the filter unaffected, which means that for noise with small spread, they are able to remove most of the noise whilst keeping the jumps in the original signal relatively unaffected. This contrasts with MA filters that always smooth away edges. Running quantile filters are very useful for skewed noise which occurs in some kinds of instrumentation, such as the Kepler transit light curves, an astronomical application in exoplanetary discovery (see Figure 9.1). MLE running filters represent a very broad class of nonlinear, non-Gaussian filters, the book by Arce (2005) is a detailed resource on this topic.

Change point detection

Another kind of running window 'filter' that gets considerable use in practice are *change point detectors*. These signal processing methods look for abrupt changes in the level (jumps) of the signal, for example, that cannot be explained as noise. Assume that the left and right halves of the window have i.i.d. Gaussian samples with different means but the same variance:

$$f\left(x_{I_L(n)}; \mu_{Ln}\right) = \prod_{i \in I_L(n)} \mathcal{N}\left(x_i; \mu_{Ln}, \sigma\right) \tag{9.3}$$

$$f\left(x_{I_R(n)}; \mu_{Rn}\right) = \prod_{i \in I_R(n)} \mathcal{N}\left(x_i; \mu_{Rn}, \sigma\right) \tag{9.4}$$

where $I_L(n) = \{n - W_2, \ldots, n - 1, n\}$ and $I_R(n) = \{n + 1, n + 2, \ldots, n + W_2\}$ then the sample means in the left and half windows are the MLEs of the Gaussian mean parameters μ_{Ln} and μ_{Rn} respectively. Similarly, the MLE for the variance over the entire window (known as the *pooled sample variance*) is:

$$v_n = \frac{\sum_{i \in I_L(n)} (x_i - \mu_{Ln})^2 + \sum_{i \in I_R(n)} (x_i - \mu_{Rn})^2}{2W_2 - 3} \left(\frac{1}{W_2} + \frac{1}{W_2 - 1}\right) \tag{9.5}$$

The test statistic:

$$t_n = \frac{\hat{\mu}_{Ln} - \hat{\mu}_{Rn}}{\sqrt{v_n}} \tag{9.6}$$

can be shown to be distributed as a *Student's t-distribution* with $d = 2W_2 - 3$ degrees of freedom, with CDF:

$$F(t) = 1 - B\left(\frac{d}{d + t^2}, \frac{d}{2}, \frac{1}{2},\right) \tag{9.7}$$

Here, $B(x, a, b)$ is the *regularized incomplete beta function*. We would reject the explanation for a jump at n simply being due to Gaussian noise (the null hypothesis), if for example, $1 - F(t) < 0.05$. This would correspond to rejection at the 95% significance level.

In practice, running window jump detection techniques must trade off statistical power against the risk of complete failure: increasing the window size allows better detection power but there is the risk that there could be more than one jump in each window. This would violate the main assumption that the left and right halves of the window are drawn from one distribution only.

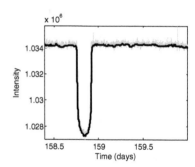

Fig. 9.1: **Application of the quantile running filter with $q = 0.25$ and window size $W = 50$. The signal is light intensity from an exoplanetary light curve from the Kepler satellite (original signal, light grey). The quantile filtered signal (black curve, each window representing approximately 50 minutes) effectively smooths away the large, positive and thus asymmetric, outliers in the original data.**

9.2　Recursive filtering

Running window filters above are perhaps the simplest kind of nonlinear/ non-Gaussian filters. Somewhat more complex filters involve *recursion*, that is, where the output from one time step forms (part of) the input for the next. These filters can be described as *semi-Markov chains*, in the form of the PGM $Y_{n-1} \rightarrow Y_n \leftarrow X_n$, where Y_n is the output of the filter at time index n and X_n is the (observed) input. The first order, linear IIR filter $y_n = a\,y_{n-1} + b\,x_n$ (Section 7.3) is in this form. The case $b = 1 - a$ (so that the TF magnitude at 0Hz is unity) is given the name *exponential smoothing* in time series forecasting applications.

This recursion is the linear IIR filter applied to the exponent of the variables, if the variables were Gaussian.

The probabilistic formulation of (one-step) recursive filtering is expressed in terms of the semi-Markov conditional distribution $f(y_n|y_{n-1}, x_n)$. This is true in general but we need to choose a particular form for this distribution, and the results will depend upon this choice. If the distribution of Y_n and X_n are assumed to be from the log-normal distribution (with sample space $\Omega_X, \Omega_Y > 0$), then the recursion $y_n = y_{n-1}^a x_n^b$ remains log-normal, for $a, b \in \mathbb{R}$ (this is a direct approach to expressing the recursive filter conditional distribution in terms of transformations of the input signal). The resulting filter has interesting behaviour. The output is always positive, and for signals that change slowly, with $a > 0$ and $b \geq 1$, the filter produces very well-localized outputs that are co-located with the peaks in the input signal. Rapid changes are filtered away. Changing to $a < 0$ forces rapid changes to produce large filter outputs instead (Figure 9.2).

Nonlinear median filtering (described in the previous section) can be made recursive, e.g. $y_n = \mathrm{median}\,(y_{n-1}, x_n)$; by contrast the recursive maximum (minimum) filters represent efficient, $O(N)$ solutions to the problem of computing the maximum (minimum) of a signal. For further details of these and other recursive filters, see Arce (2005, Section 6.4).

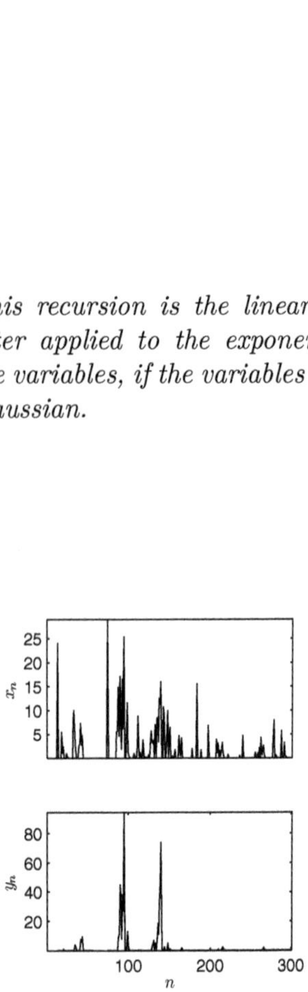

Fig. 9.2: **The first-order log-normal recursive filter applied to daily rainfall observations (mm rain/day, top panel) from a gauging station in Abingdon, UK. The recursion output is $y_n = y_{n-1}^{0.4} x_n$ with $y_1 = 10^{-6}$ (bottom panel). The small quantity 10^{-6} is added to the input to ensure that the filter does not immediately fall into an infinite sequence of zero outputs.**

9.3　Global nonlinear filtering

Both running and recursive nonlinear filtering methods above take a "local" view on the signal, that is, they process small parts of the sequence in turn. We can, instead, take a "global" view on the signal and this approach leads to a set of filters with uniquely useful properties. We will consider a broad class of nonlinear filters that can be expressed in terms of minimizing an appropriate (discrete) functional of the input data $\boldsymbol{x} \in \mathbb{R}^N$ with respect to the output data $\boldsymbol{y} \in \mathbb{R}^N$. We will investigate a few important special cases, including a group of *nonlinear diffusion* methods defined by the following functional:

$$E(\boldsymbol{y}) = \frac{1}{q}\sum_{n=1}^{N}|x_n - y_n|^q + \frac{\gamma}{p}\sum_{n=1}^{N-1}|y_{n+1} - y_n|^p \qquad (9.8)$$

with $p, q > 0$ and $\gamma > 0$ a continuous regularization parameter. Total variation denoising (2.47) (Figure 9.3) is a special case with $p = 1$ and $q = 2$, and can be minimized using quadratic programming or path-following methods (Section 2.4). This and other special cases (including $p = 1$, $q = 1$ and $p = 0$ with $q = 1$ or $q = 2$) generate *piecewise constant* outputs. This is because, essentially, the second term in the functional above is zero when the signal is constant (it is the first discrete derivative). Different algorithms are required for these different special cases (Little and Jones, 2011*a,b*) and for any $p, q < 1$, the functional is non-convex.

Yet another group are defined by minimizing the functional:

$$E\left(\boldsymbol{y}\right) = \frac{1}{q} \sum_{n=1}^{N} |x_n - y_n|^q + \frac{\gamma}{p} \sum_{n=2}^{N-1} |y_{n+1} - 2y_n + y_{n-1}|^p \tag{9.9}$$

The main difference between this and the nonlinear filter (9.8) is that the second (regularization) term is the discrete *second* rather than *first* difference of the signal. Thus, signals that are *piecewise linear* minimize this term. The convex case $q = 2, p = 1$ is known as L_1-*trend filtering* and can be solved as a quadratic program, but Kim *et al.* (2009) describe a specialized primal-dual interior-point implementation which has $O\left(N\right)$ complexity (Section 2.4).

Another specific global method is the *bilateral filter* that minimizes the following functional:

$$E\left(\boldsymbol{y}\right) = \sum_{n=1}^{N} \sum_{m=1}^{N} \left[1 - \exp\left(-\beta\left(y_n - y_m\right)^2\right)\right] \mathbf{1}\left[|n - m| \le W\right] \tag{9.10}$$

where $\beta > 0$ and $W \in \mathbb{N}$ are smoothing parameters. Using a particular, adaptive step-size gradient descent, it can be shown that this (non-convex) functional is approximately minimized using the following iteration (Little and Jones, 2011*a*):

$$y_n^{i+1} = \frac{\sum_{m=1}^{N} \exp\left(-\beta\left(y_n^i - y_m^i\right)^2\right) \mathbf{1}\left[|n - m| \le W\right] y_m^i}{\sum_{m=1}^{N} \exp\left(-\beta\left(y_n^i - y_m^i\right)^2\right) \mathbf{1}\left[|n - m| \le W\right]} \tag{9.11}$$

where $i = 1, 2, \ldots$ is the iteration number and starting with $y_n^1 = x_n$. The iteration is repeated until convergence is detected (when the output does not change significantly). The behaviour of this filter can be understood as follows: for $W = N$ then $\mathbf{1}\left[|n - m| \le W\right] = 1$ independent of n, m. In this case, the filter collapses down to soft mean shift clustering (Section 6.5) that eventually converges on an output that clusters the signal into a finite number of distinct levels. However, when $W < N$, this clustering only affects the signal on the temporal scale W. The result is a *local* clustering that adapts to the number and value of levels in each time region (Figure 9.4). This is practical because most sig-

The two-dimensional version of this filter is widely used in image processing.

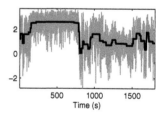

Fig. 9.3: Total variation denoising ($\gamma = 15$) applied to a fixed frequency bin values over time of the log power spectrum of raw accelerometry data. The original input signal x_n (grey curve) has noticeable and large fluctuations that are smoothed out by the filtering y_n (black curve). The smoothing results in a piecewise constant curve (at a certain time scale) unlike linear smoothing that always produces a smooth curve.

nals with defined peaks in their sample distribution are not statistically stationary.

Computationally, the bilateral filter requires $O\left(W^2\right)$ operations per iteration, convergence is usually obtained in around 20 iterations. So, provided W is small, implementation of this filter in DSP applications is straightforward.

9.4 Hidden Markov models (HMMs)

All the filters investigated in this chapter have unknown variables that are continuous, and the filtering goal is to recover the value of the continuous, latent variable. There are, however, many practical situations in which the latent variable is discrete while the observed signal is continuous. An example of this is a signal generated by a mixture model, whereby the latent variable is the indicator sequence (see Section 6.2). Filtering in this situation is a matter of using techniques such as E-M and is tractable in practical DSP applications. However, far more challenging is the situation where the indicator sequence is a Markov chain – a structure known as the *hidden Markov model* (HMM) – then it is no longer possible to independently estimate each indicator value since this depends upon previous indicator values in the sequence (Figure 5.3). For mixture models, computing the E-step (E-M) or maximizing over the latent state (ICM) requires $O\left(N K\right)$ operations where K is the number of possible states and N the length of the signal, which is tractable when K is small, but for the HMM the corresponding computation requires estimating all possible sequences of states, requiring exponential $O\left(K^N\right)$ operations. This is not tractable. However, the form of the HMM PGM is such that a corresponding JT can be defined on that tree to allow tractable computation using the junction tree algorithm (Algorithm 5.1). HMMs have exactly the same probabilistic structure as the Kalman filter (Section 7.5), except that the message passing involves discrete categorical distributions rather than Gaussians.

In using HMMs for DSP and machine learning, we typically need to solve one or more of the following problems:

(1) *Evaluation*: given a Markov transition density, observation model and observed data indices $1, 2, \ldots, n$, find the probability of the observed data under the model. This can be solved using the forward recursion algorithm (as with KF filtering),

(2) *Model fitting*: given observed data, estimate maximum likelihood values of the parameters of the transition and emission probability densities (and the posterior Markov state probabilities). This can be performed using discrete Baum-Welch which is uses forward-backward recursions to compute state and Markov probabilities at each time point, from which E-M model parameter estimates can be obtained. Iterative use of Viterbi decoding and parameter re-estimation is also used,

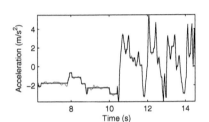

Fig. 9.4: Global nonlinear filtering using the bilateral filter ($\beta = 120$, $W = 10$) applied to raw accelerometry data sampled at ~48Hz. The raw input signal x_n (grey curve) has two apparently distinct regions: resting behaviour up to ~10.5s, and walking behaviour thereafter. The filter output y_n (black curve) is effective at maintaining almost constant output in the rest region, and performs almost no smoothing in the walking region.

(3) *Decoding*: given a Markov transition density, observation model and observed data, find the sequence of Markov states that maximizes the conditional probability of the complete Markov state sequence given the observed data. As with the KF, this can be solved by Viterbi decoding using max-product or max-sum algebra in place of max-sum algebra.

A naive approach to all the problem above would demand exponential computational complexity unless we use the JT algorithm, as we describe below. We will use the following notation: the HMM latent Markov random state sequence is $Z_n \in \{1, 2, \ldots, K\}$ for $n = 1, 2, \ldots, N$; we write $\mathbf{Z}$ for the full latent sequence. This stationary Markov chain has transition distribution represented in the matrix $\mathbf{P}$ with entries $f(z_n = k | z_{n-1} = k') = p_{kk'}$. The initial probability vector is $f(z_1 = k) = \pi_k$. The observed random (generally vector-valued) signal $\mathbf{X}_n$ has observation (*emission*) distribution $f(\mathbf{x}_n | z_n)$ and we write $\mathbf{X}$ for the full, joint data or $\mathbf{x}$ for a particular realization. Below, for notational clarity, we will usually omit the functional dependence of these distributions on the parameters.

Junction tree (JT) for efficient HMM computations

As with the KF, the nature of the problems we wish to solve strongly implicates the use of time-slice cliques $C_n = \{z_{n-1}, z_n, \mathbf{x}_n\}$ for $n > 1$ and $C_1 = \{z_1, \mathbf{x}_1\}$ and corresponding clique factors $h_n(z_{n-1}, z_n, \mathbf{x}_n) = f(z_n | z_{n-1}) \otimes f(\mathbf{x}_n | z_n)$. For realized observed data $\mathbf{x}_n$, the corresponding following forward and backward messages (see Algorithm 5.1) are:

$$\mu_{n \to n+1}(z_n) = \bigoplus_{z_{n-1}} h_n(z_{n-1}, z_n, \mathbf{x}_n) \otimes \mu_{n-1 \to n}(z_{n-1}) \quad (9.12)$$

$$\mu_{n \to n-1}(z_{n-1}) = \bigoplus_{z_n} h_n(z_{n-1}, z_n, \mathbf{x}_n) \otimes \mu_{n+1 \to n}(z_n) \quad (9.13)$$

The "boundary" message is $\mu_{N+1 \to N}(z_N) = \otimes_{\mathrm{id}}$ and $\mu_{1 \to 2}(z_1) = f(z_1) \otimes f(\mathbf{x}_1 | z_1)$. In the usual probability sum-product semiring (where $\otimes \mapsto \times$ and $\oplus \mapsto +$), the above messages represent the following distributions:

$$\mu_{n \to n+1}(z_n) = \sum_{k=1}^{K} f(z_n | z_{n-1} = k) f(\mathbf{x}_n | z_n) \mu_{n-1 \to n}(k) \quad (9.14)$$

$$\mu_{n \to n-1}(z_{n-1}) = \sum_{k=1}^{K} f(z_n = k | z_{n-1}) f(\mathbf{x}_n | z_n = k) \mu_{n+1 \to n}(k)$$

denoted $\alpha_n(z_n)$ and $\beta_{n-1}(z_{n-1})$, respectively. From this we can deduce the following very useful fact:

$$f(z_n, \mathbf{x}) = \mu_{n \to n+1}(z_n) \mu_{n+1 \to n}(z_n) \quad (9.15)$$

In words, the joint distribution of all the observed data with the Markov state at index n is the product of the forward and backward messages coinciding at that index. Furthermore, the marginal distribution of the observed data is just:

$$f\left(\boldsymbol{x}\right)=\sum_{k=1}^{K}f\left(z_{n}=k,\boldsymbol{x}\right)=\sum_{k=1}^{K}\mu_{n\to n+1}\left(k\right)\mu_{n+1\to n}\left(k\right) \qquad (9.16)$$

We will also need the two-step probability:

$$f\left(z_{n-1},z_{n},\boldsymbol{x}\right)=\mu_{n-1\to n}\left(z_{n-1}\right)f\left(z_{n}|z_{n-1}\right)f\left(\boldsymbol{x}_{n}|z_{n}\right)\mu_{n+1\to n}\left(z_{n}\right) \qquad (9.17)$$

Viterbi decoding

As with the KF, the most probable state sequence can be computed using forward message passing in the max-product semiring ($\oplus \mapsto$ max and $\otimes \mapsto \times$) over the discrete latent states:

$$\mu_{n\to n+1}^{\max}\left(z_{n}\right) \;=\; f\left(\boldsymbol{x}_{n}|z_{n}\right)\times \qquad (9.18)$$
$$\max_{k\in 1,2,\ldots,K}f\left(z_{n}|z_{n-1}=k\right)\mu_{n-1\to n}^{\max}\left(k\right)$$

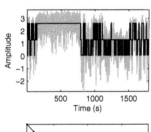

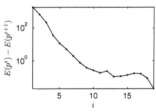

with initial message $\mu_{1\to 2}^{\max}\left(z_{1}\right)=f\left(z_{1}\right)f\left(\boldsymbol{x}_{1}|z_{1}\right)$. At the message passing root we obtain the most likely state $\hat{z}_{N}=\arg\max_{k\in 1,2,\ldots,K}\left[\mu_{N\to N+1}^{\max}\left(k\right)\right]$. The most probable sequence of states is then retained:

$$\Delta_{n\to n+1}\left(z_{n}\right) \;=\; \arg\max_{k\in 1,2,\ldots,K}f\left(z_{n}|z_{n-1}=k\right)f\left(\boldsymbol{x}_{n}|z_{n}\right)\mu_{n-1\to n}^{\max}\left(k\right)$$
$$=\; \arg\max_{k\in 1,2,\ldots,K}f\left(z_{n}|z_{n-1}=k\right)\mu_{n-1\to n}^{\max}\left(k\right) \qquad (9.19)$$

Backtracking along the forward message then finds $\hat{z}_{n-1}=\Delta_{n\to n+1}\left(\hat{z}_{n}\right)$ for $n=N,N-1,\ldots,2$.

Fig. 9.5: Hidden Markov modelling ($K = 3$, with Laplace observation distributions) applied to the same data in Figure 9.4. Trained using Baum-Welch (convergence in lower panel), the Viterbi decoded output results in a simple, piecewise constant curve of exactly three possible values, one per Markov state, unlike TVD (top panel). Also, unlike mixture models, the time dependence causes states to "persist" over time which is a realistic feature of many real signals.

Baum-Welch parameter estimation

Typically, we do not know the HMM model parameter values, i.e. the transition matrix $\mathbf{P}$, initial probabilities $\boldsymbol{\pi}$, and parameters for the observation distributions $\boldsymbol{\theta}$, collectively $\boldsymbol{p}=(\boldsymbol{\pi},\mathbf{P},\boldsymbol{\theta})$. We will demonstrate the use of E-M (Algorithm 5.3) for this estimation problem. The outline of E-M for the HMM is essentially the same as that for the KF, except with discrete rather than continuous state expectations.

For the E-step, we need the posterior distributions of the latent Markov states. These can be efficiently computed using (9.15), (9.16) and (9.17). For a single state:

$$\gamma_n (k) = f(z_n = k | \boldsymbol{x}) \tag{9.20}$$

$$= \frac{f(z_n = k, \boldsymbol{x})}{\sum_{k'=1}^{K} f(z_n = k', \boldsymbol{x})}$$

and for temporally adjacent states:

$$\xi_n (k, k') = f(z_n = k, z_{n-1} = k' | \boldsymbol{x}) \tag{9.21}$$

$$= \frac{f(z_{n-1} = k', z_n = k, \boldsymbol{x})}{\sum_{i=1}^{K} \sum_{i'=1}^{K} f(z_{n-1} = i, z_n = i', \boldsymbol{x})}$$

For the M-step, we will need negative log of the PGM likelihood $f(\boldsymbol{z}, \boldsymbol{x}; \boldsymbol{p})$:

$$-\ln f(\boldsymbol{z}, \boldsymbol{x}; \boldsymbol{p}) = -\ln f(z_1; \boldsymbol{\pi}) - \ln f(\boldsymbol{x}_1 | z_1; \boldsymbol{\theta})$$

$$- \sum_{n=2}^{N} [\ln f(z_n | z_{n-1}; \mathbf{P}) + \ln f(\boldsymbol{x}_n | z_n)] \tag{9.22}$$

$$= -\ln f(z_1; \boldsymbol{\pi}) - \sum_{n=2}^{N} \ln f(z_n | z_{n-1}; \mathbf{P})$$

$$- \sum_{n=1}^{N} \ln f(\boldsymbol{x}_n | z_n; \boldsymbol{\theta})$$

so that the expected NLL is:

$$E_{\boldsymbol{Z} | \boldsymbol{X}} [-\ln f(\boldsymbol{x}, \boldsymbol{Z}; \boldsymbol{p})] = E_{\boldsymbol{Z} | \boldsymbol{X}} [-\ln f(Z_1; \boldsymbol{\pi})]$$

$$+ E_{\boldsymbol{Z} | \boldsymbol{X}} \left[-\sum_{n=2}^{N} \ln f(Z_n | Z_{n-1}; \mathbf{P}) \right]$$

$$+ E_{\boldsymbol{Z} | \boldsymbol{X}} \left[-\sum_{n=1}^{N} \ln f(\boldsymbol{x}_n | Z_n; \boldsymbol{\theta}) \right] \tag{9.23}$$

$$= E_{Z_1 | \boldsymbol{X}} [-\ln f(Z_1; \boldsymbol{\pi})]$$

$$+ E_{Z_n, Z_{n-1} | \boldsymbol{X}} \left[-\sum_{n=2}^{N} \ln f(Z_n | Z_{n-1}; \mathbf{P}) \right]$$

$$+ E_{Z_n | \boldsymbol{X}} \left[-\sum_{n=1}^{N} \ln f(\boldsymbol{x}_n | Z_n; \boldsymbol{\theta}) \right]$$

So, for the M-step, we can optimize the expected NLL above with respect to each parameter $\boldsymbol{\pi}, \mathbf{P}, \boldsymbol{\theta}$ by optimizing over each of the terms in the above sum separately.

For the initial probability vector, solving $\hat{\boldsymbol{\pi}} = \arg\min_{\boldsymbol{\pi}} E_{Z_1 | \boldsymbol{X}} [-\ln f(Z_1; \boldsymbol{\pi})]$ leads to:

$$\hat{\pi}_k = f(z_1 = k|\boldsymbol{x}) = \gamma_1(k) \tag{9.24}$$

Similarly, for the transition density, $\hat{\mathbf{P}} = \arg\min_{\mathbf{P}} E_{Z_n, Z_{n-1}|\mathbf{X}}\left[-\sum_{n=2}^N \ln f(Z_n|Z_{n-1}; \mathbf{P})\right]$ is solved by:

$$
\begin{aligned}
\hat{p}_{k,k'} &= f(z_n = k|z_{n-1} = k', \boldsymbol{x}) \\
&= \frac{\sum_{n=2}^N \xi_n(k, k')}{\sum_{n=2}^N \gamma_{n-1}(k')}
\end{aligned}
\tag{9.25}
$$

Finally, for the observation distribution parameters, we have the same situation as for the mixture model e.g. for exponential family distributions, the parameter updates are in the form of (6.12). As a special case, for the multivariate Gaussian, the updates are:

$$
\begin{aligned}
\hat{\boldsymbol{\mu}}_k &= \frac{\sum_{n=1}^N \gamma_n(k)\boldsymbol{x}_n}{\sum_{n=1}^N \gamma_n(k)} \\
\hat{\boldsymbol{\Sigma}}_k &= \frac{\sum_{n=1}^N \gamma_n(k)(\boldsymbol{x}_n - \hat{\boldsymbol{\mu}}_k)(\boldsymbol{x}_n - \hat{\boldsymbol{\mu}}_k)^T}{\sum_{n=1}^N \gamma_n(k)}
\end{aligned}
\tag{9.26}
$$

Putting this all together we get the Baum-Welch algorithm for the HMM, as described below.

Algorithm 9.1 *Baum-Welch* expectation-maximization (E-M) for hidden Markov models (HMMs).

(1) *Initialization.* Start with a randomized set of parameters $\boldsymbol{p}^0 = (\boldsymbol{\pi}^0, \mathbf{P}^0, \boldsymbol{\theta}^0)$, set iteration number $i = 0$, choose a small convergence tolerance $\epsilon > 0$,

(2) *E-step.* Compute the conditional probabilities $\gamma_n(k) = f(z_n = k|\boldsymbol{x}; \boldsymbol{p}^i)$ and $\xi_n(k, k') = f(z_n = k, z_{n-1} = k'|\boldsymbol{x}; \boldsymbol{p}^i)$ using message passing (JT algorithm),

(3) *M-step.* Update parameters: initial probability $\pi_k^{i+1} = \gamma_1(k)$, transition density $p_{k,k'}^{i+1} = \sum_{n=2}^N \xi_n(k, k')\big/\sum_{n=2}^N \gamma_{n-1}(k')$, and emission distribution parameters $\boldsymbol{\theta}_k^{i+1} = \arg\min_{\boldsymbol{\theta}_k}\left[-\sum_{n=1}^N \sum_{k=1}^K f(z_n = k|\boldsymbol{x}; \boldsymbol{p}^i) \ln f(\boldsymbol{x}_n|z_n = k; \boldsymbol{\theta}_k)\right]$, $k = 1, 2, \ldots, K$,

(4) *Convergence check.* Calculate the value of the observed variable marginal NLL $E(\boldsymbol{p}^{i+1}) = -\ln f(\boldsymbol{x}; \boldsymbol{p}^{i+1})$ and the improvement $\Delta E = E(\boldsymbol{p}^i) - E(\boldsymbol{p}^{i+1})$ and if $\Delta E < \epsilon$, exit with solution $\boldsymbol{p}^{i+1}$,

(5) *Iteration.* Update $i \leftarrow i+1$, go back to step 2.

Baum-Welch is very similar to E-M for mixture models and the KF, but the more complex latent space means it is natural to expect that convergence will be slower. The temporal dependence of HMMs is a much more realistic model than a mixture model for many signals encountered in practical DSP applications, and the $O(N)$ message passing updates are relatively straightforward to implement in DSP hardware (Figure 9.5).

Model evaluation and structured data classification

Given an HMM with parameters $p = (\pi, \mathbf{P}, \theta)$, it is simple to use one complete forward message pass to compute $\mu_{N \to N+1}(z_N) = f(z_N, x; p)$ to find $f(x; p) = \sum_{k=1}^{K} \mu_{N \to N+1}(k)$. Thus, for a signal x (not necessarily one used to estimate parameters, and not necessarily of the same length as the training signals) we can evaluate the NLL $E(p) = -\ln f(x; p)$. This can be used to compare HMM models for any given signal. For example, we might have multiple models p^j for $j = 1, 2, \ldots, M$; then the ML decision about which HMM is most likely to have generated a new signal $\tilde{x}$ can be efficiently solved using $\hat{j} = \arg\min_j \left[-\ln f(\tilde{x}; p^j) \right]$. This can be viewed as a form of classification in which the data is *structured*, that is, not just a single observation, or a collection of i.i.d. observations. An example of this kind of classification occurs in *automatic speech recognition* (ASR) where multiple HMMs are trained to represent different sounds. Then, for each segment of speech, evaluating multiple sound HMMs allows the optimal sound category to be chosen for each segment.

Viterbi parameter estimation

Much as for mixture modelling, we can use maximization instead of expectation to simplify HMM model fitting (see Section 6.5) to derive a procedure known as *Viterbi training* (Algorithm 9.2). Consider that we have the decoded most probable sequence $\hat{z}$, then we can use this information to do direct maximum likelihood update of the HMM model parameters. For the initial probability and transition densities, these are just (conditional) categorical random variables, so we can use ML parameter estimates obtained by normalized counts:

$$
\begin{aligned}
\hat{\pi}_k &= \mathbf{1}\left[\hat{z}_1 = k\right] &\text{(9.27)}\\
\hat{p}_{k,k'} &= \frac{\sum_{n=2}^{N} \mathbf{1}\left[\hat{z}_{n-1} = k' \wedge \hat{z}_n = k\right]}{\sum_{n=2}^{N} \mathbf{1}\left[\hat{z}_{n-1} = k'\right]}
\end{aligned}
$$

Similarly, the observation parameters can now be updated using direct ML estimates, e.g. $\hat{\theta}_k = \arg\min_{\theta_k} \left[-\sum_{n=1}^{N} \ln f(x_n | \hat{z}_n; \theta_k) \right]$. For exponential family distributions, this becomes $a'(\hat{p}_k) = \frac{1}{N_k} \sum_{n:\hat{z}_n=k} g(x_n)$.

As with K-means clustering (Algorithm 6.2), since there are only a finite number of possible configurations and the NLL cannot increase,

Viterbi training will eventually converge on a fixed point. Technically then, there is no need for a convergence tolerance: checking that the NLL has not changed from the previous iteration would suffice. However, we may only need a solution at a certain level of accuracy and in that case we can save computational effort by curtailing the iterations early.

Avoiding numerical underflow in message passing

As we have seen above, message passing in the HMM JT is critical for computational scalability. However, for large N, the messages, which, for probability sum-product algebra are computed using cumulative products, become vanishingly small as $n \to N$ (forward) or $n \to 1$ (backwards). Thus, for practical numerical implementations it is necessary to find some way to avoid the numerical underflow that inevitably results.

Algorithm 9.2 *Viterbi training* for hidden Markov models (HMMs).

(1) *Initialization.* Start with a randomized set of parameters $p^0 = (\pi^0, \mathbf{P}^0, \theta^0)$, set iteration number $i = 0$, choose a small convergence tolerance $\epsilon > 0$,

(2) *Decoding.* Use Viterbi decoding to obtain the most probable sequence $\hat{z}^i$ given model parameters p^i,

(3) *Estimation.* Update parameters: initial probability $\pi_k^{i+1} = \mathbf{1}[\hat{z}_1 = k]$, transition density $p_{k,k'}^{i+1} = \sum_{n=2}^{N} \mathbf{1}[\hat{z}_{n-1} = k' \wedge \hat{z}_n = k] / \sum_{n=2}^{N} \mathbf{1}[\hat{z}_{n-1} = k']$, and emission distribution parameters $\hat{\theta}_k = \arg\min_{\theta_k}\left[-\sum_{n=1}^{N} \ln f(x_n|\hat{z}_n; \theta_k)\right]$, for $k = 1, 2, \ldots, K$,

(4) *Convergence check.* Calculate the value of the model NLL $E(p^{i+1}) = -\ln f(\hat{z}^i, x; p^{i+1})$ and the improvement $\Delta E = E(p^i) - E(p^{i+1})$ and if $\Delta E < \epsilon$, exit with solution p^{i+1},

(5) *Iteration.* Update $i \leftarrow i + 1$, go back to step 2.

One simple, widely applicable approach is to convert the probabilities involved into information instead i.e. apply the information map (log or negative log). The corresponding semiring algebra must change as a result. Let us look at Viterbi decoding first. Here, converting to log probabilities e.g. $a \mapsto \ln a$ requires switching to the max-sum semiring (mapping $\oplus \mapsto \max$ and $\otimes \mapsto +$). Use of negative logs $a \mapsto -\ln a$ requires the min-sum semiring instead (mapping $\oplus \mapsto \min$ and $\otimes \mapsto +$). Under this map, the multiplicative identity $1 \mapsto 0$, and additive identity $0 \mapsto \infty$ (and $0 \mapsto -\infty$ for negative logs). In the max-sum semiring, the forward messages become:

$$
\begin{aligned}
\mu_{n \to n+1}^{\ln \max}(z_n) &= \max_{k \in 1,2,\ldots,K} \left[\ln f\left(z_n | z_{n-1} = k\right) + \mu_{n-1 \to n}^{\ln \max}(k) \right] \\
&+ \ln f\left(\boldsymbol{x}_n | z_n\right) \quad\quad (9.28) \\
\Delta_{n \to n+1}(z_n) &= \arg\max_{k \in 1,2,\ldots,K} \left[\ln f\left(z_n | z_{n-1} = k\right) + \mu_{n-1 \to n}^{\ln \max}(k) \right]
\end{aligned}
$$

with initial message $\mu_{1 \to 2}^{\ln \max}(z_1) = \ln f(z_1) + \ln f(\boldsymbol{x}_1 | z_1)$. Backtracking remains unchanged.

To use forward-backward message passing to compute Markov posterior distributions, under the log probability map $a \mapsto \ln a$, then we will to map the product to the sum of logs, $a \otimes b \mapsto \ln a + \ln b$, and the sum to the "log-sum-exp" operator: $a \oplus b \mapsto \ln\left(e^{\ln a} + e^{\ln b}\right)$. For negative logs we map $a \otimes b \mapsto -\ln a - \ln b$ and $a \oplus b \mapsto -\ln\left(e^{-\ln a} + e^{-\ln b}\right)$. In the positive log algebra, the forward and backwards messages become:

$$
\begin{aligned}
\mu_{n \to n+1}^{\ln}(z_n) &= \ln \sum_{k=1}^{K} \exp l_f(k) \\
\mu_{n \to n-1}^{\ln}(z_{n-1}) &= \ln \sum_{k=1}^{K} \exp l_b(k) \quad\quad (9.29)
\end{aligned}
$$

where:

$$
\begin{aligned}
l_f(k) &= \ln f\left(z_n | z_{n-1} = k\right) + \ln f\left(\boldsymbol{x}_n | z_n\right) + \mu_{n-1 \to n}^{\ln}(k) \\
l_b(k) &= \ln f\left(z_n = k | z_{n-1}\right) + \ln f\left(\boldsymbol{x}_n | z_n = k\right) + \mu_{n+1 \to n}^{\ln}(k) \quad (9.30)
\end{aligned}
$$

To convert back to probabilities, all that is necessary is to take the exponent, e.g. $f(z_n, \boldsymbol{x}) = \exp\left(\mu_{n \to n+1}^{\ln}(z_n) + \mu_{n+1 \to n}^{\ln}(z_n)\right)$. However, for very large magnitude (positive/negative) log probabilities, we still risk numerical overflow in using the exponent. Therefore, we should only compute exponents of normalized log probabilities. Performing this normalization can be done using the "log-sum-exp trick" where $\ln \sum_k e^{x_k} = x^\star + \ln \sum_k e^{x_k - x^\star}$ for $x^\star = \max_k x_k$. This avoids the need to compute exponents of very large magnitude quantities since the exponent will be shifted down to small magnitude values.

9.5 Homomorphic signal processing

Given two discrete-time signals x_n, y_n for $n \in \mathbb{Z}$, find their convolution, $\boldsymbol{z} = \boldsymbol{x} \star \boldsymbol{y}$, and apply the DTFT to both sides to get $Z(\omega) = X(\omega) Y(\omega)$. It follows that $\ln Z(\omega) = \ln X(\omega) + \ln Y(\omega)$. In other words, the log-spectra is additive under convolutions. It also follows that:

$$
\begin{aligned}
F_n^{-1}\left[\ln Z(\omega)\right] &= F_n^{-1}\left[\ln X(\omega)\right] + F_n^{-1}\left[\ln Y(\omega)\right] \\
&= \hat{x}_n + \hat{y}_n \quad\quad (9.31)
\end{aligned}
$$

so that the inverse DTFT of the log-spectra of the convolved signal, $\hat{z}_n$, is the sum the inverse DTFTs of the log-spectra of the two signals. The signal $\hat{x}_n = F_n^{-1} [\ln X(\omega)]$ is called the *cepstra* of the signal x_n (a rearrangement of the word spectra). The utility of this representation is that, if the two cepstra do not overlap (i.e. $\hat{x}_n$ is zero or very small for values of n where $\hat{y}_n$ is large and vice versa), then the signal $\mathbf{z}$ can be easily separated into the two components by "masking". For example, if there is a cut-point N_c at which the cepstra can be separated, then we can form the two estimates:

$$\tilde{x}_n = \begin{cases} \hat{x}_n & n \le N_c \\ 0 & n > N_c \end{cases} \tag{9.32}$$

$$\tilde{y}_n = \begin{cases} \hat{y}_n & n > N_c \\ 0 & n \le N_c \end{cases} \tag{9.33}$$

from which we can reconstruct $\tilde{\mathbf{x}} = F_n^{-1} \left[\exp \left(F_\omega^{-1} [\tilde{\mathbf{x}}] \right) \right]$ and $\tilde{\mathbf{y}} = F_n^{-1} \left[\exp \left(F_\omega^{-1} [\tilde{\mathbf{y}}] \right) \right]$. This is known as *deconvolution*. Cepstral signal processing is thus particularly useful for problems where deconvolution is needed. One can also simply apply some kind of manipulation to the cepstra of a signal to remove the effect of some kind of convolution, such as undoing a specific linear filtering operation (see Section 7.3). This is called *liftering* (another rearrangement).

Cepstral processing is widely used in digital speech analysis (Rabiner and Schafer, 2007, Section 5). The primary reason for this is that speech production has a natural interpretation in terms of a source of acoustic excitation energy (the larynx) and a resonance that modifies this source of energy (the vocal tract, including the oral and nasal cavities, shaped by the position off the tongue, lips and jaw). Thus, there is a natural convolution operation in action: the source signal $\mathbf{u}$ is convolved with the vocal tract impulse response $\mathbf{h}$ to obtain the acoustic signal $\mathbf{x} = \mathbf{h} \star \mathbf{u}$. Often the goal of speech analysis is to separate out the vocal tract from the source, a task for which cepstral deconvolution is well suited. For example, during vowel sounds, the vocal source oscillates at the vocal pitch generating energy across the spectrum, and the vocal tract filters that excitation source to produce the particular vowel. Cepstral analysis allows fairly ready identification of the primary vocal source oscillation period from the second largest peak in the cepstra. The vocal tract resonances (on the order of 800Hz to 3kHz) occupy the first few milliseconds on the cepstral vector (Figure 9.6). Thus, the first few cepstral coefficients are highly responsive to the vocal tract resonances, and the rest are more relevant to the vocal excitation source.

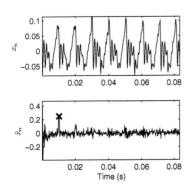

Fig. 9.6: Cepstral analysis (bottom panel: cepstra) for voiced speech signals (top panel: original input signal sampled at 6kHz) allows separation (deconvolution) of voiced excitation signal (pitch period of 10.3ms or 96.7Hz identified by black cross, bottom panel) from the vocal tract resonances (first few milliseconds).

Nonparametric Bayesian machine learning and signal processing

10

We have seen that stochastic processes play an important foundational role in a wide range of methods in DSP. For example, in Section 7.4 we treat a discrete-time signal as a Gaussian process, and thereby obtain many mathematically simplified algorithms, particularly based on the power spectral density. At the same time, in machine learning, it has generally been observed that nonparametric methods (see, for example, nonlinear regression in Section 6.4) outperform parametric methods in terms of predictive accuracy since they can adapt to data with arbitrary complexity. However, these techniques are not Bayesian so we are unable to do important inferential procedures such as draw samples from the underlying probabilistic model or compute posterior confidence intervals. But, Bayesian models are often only mathematically tractable if parametric, with the corresponding loss of predictive accuracy. An alternative, discussed in this section, is to extend the mathematical tractability of stochastic processes to Bayesian methods. This leads to so-called *Bayesian nonparametrics* exemplified by techniques such as Gaussian process regression and Dirichlet process mixture modelling that have been shown to be extremely useful in practical DSP and machine learning applications.

10.1 Preliminaries

Engineers, statisticians and other analysts who use mathematics for data analysis want to make life easy by making assumptions that simplify their mathematical models. For example, it is very common to assume that some data is i.i.d., and in doing so we end up with very tractable likelihoods that can be readily optimized, often analytically (see Section 4.2). However, sometimes this strategy is just not plausible because the i.i.d. assumption is a very strong one (it is false for most non-trivial signals, for instance). A weaker assumption is to invoke some kind of *invariance principle*, which usually means that the distribution is unchanged under the action of a group applied to the random variables (Section 1.1). This mathematical construction often leads to considerable algorithmic simplifications. To illustrate, consider *stationarity*

Machine Learning for Signal Processing: Data Science, Algorithms, and Computational Statistics. Max A. Little.
© Max A. Little 2019. Published in 2019 by Oxford University Press. DOI: 10.1093/oso/9780198714934.001.0001

For much more detail on the underlying algebraic structure of LTI DSP, see Puschel and Moura (2008)

whereby the distribution is invariant to (discrete) time shifts. Then, for (zero mean) Gaussian processes, the distribution is entirely characterized by the autocorrelation function, which is symmetric and whose corresponding autocorrelation matrix (for finite-duration signals) is circulant and therefore diagonalizable in the Fourier basis. Thus, assuming time shift invariance (in addition to Gaussianity and linearity) implies that there is a decomposition of the autocorrelation matrix in terms of the Fourier basis. The coefficients of the signal in the basis are the PSD (Section 7.4).

Exchangeability and de Finetti's theorem

While i.i.d. models are simple, they are not expressive enough for practical DSP and machine learning applications. A weaker assumption, that of *exchangeability*, has much more practical value but also deep implications for statistical machine learning, as we shall see next. An exchangeable set of N random variables X_n, $n = 1, 2, \ldots, N$ is one for which any *permutation* $\pi : \{1, 2, \ldots, N\} \to \{1, 2, \ldots, N\}$ leaves their joint distribution unchanged:

$$f(x_1, x_2, \ldots, x_N) = f\left(x_{\pi(1)}, x_{\pi(2)}, \ldots, x_{\pi(N)}\right) \qquad (10.1)$$

The meaning of this definition, for a discrete time signal, is that the variables do not depend upon their specific index number n, but they may be mutually dependent. Note that this does *not* imply that the samples are i.i.d., yet for any set of N i.i.d. random variables it is true. When $N \to \infty$ the infinite collection of random variables is said to be *infinitely exchangeable*.

A typical Bayesian model assumes that the data $x_n \in \Omega_X$ have some conditional distribution $g(x_n|p)$ (likelihood) where $p \in \Omega_P$ is a parameter with (prior) distribution $\mu(p)$. This makes the data items X_n *conditionally independent* of each other given the parameter. The joint distribution over the data is (infinitely) exchangeable:

$$f(x_1, x_2, \ldots) = \int_{\Omega_P} \prod_{n=1}^{\infty} g(x_n|q)\,\mu(q)\,dq \qquad (10.2)$$

This much is obvious, but the remarkable *de Finetti's* theorem states that, in the case where $\Omega_p = [0,1]$ and $\Omega_X = \{0,1\}$ are Bernoulli random variables, the implication also works the other way around: for infinitely exchangeable random variables, the representation (10.2) applies (Bernardo and Smith, 2000, prop. 4.1). In other words, the assumption of infinite exchangeability implies the existence of a random parameter upon which all the observations depend, e.g. $X_n \sim g(x|p)$. Furthermore, $P \sim \mu(p)$ and this random parameter value is also the limiting empirical frequency of 1's in the data:

$$p = \lim_{N \to \infty} \frac{1}{N} \sum_{n=1}^{N} x_n \qquad (10.3)$$

where the convergence is almost sure. This theorem can be generalized to when the $x_n \in \mathbb{R}$. Then, the prior parameter becomes a *distribution* G, one member of the space of all distributions over $\mathbb{R}$, call it $\mathcal{F}(\mathbb{R})$, which itself is drawn from the distribution $G \sim \mu(g)$, with each $X_n \sim g(x)$. The *Hewitt-Savage* theorem states that (Bernardo and Smith, 2000, prop. 4.3):

$$f(x_1, x_2, \ldots) = \int_{\mathcal{F}(\mathbb{R})} \prod_{n=1}^{\infty} g(x_n) \, \mu(g) \, dg \qquad (10.4)$$

where the distribution g (called the *de Finetti* or *mixing measure*) is the limiting EPDF, (1.45), of the data:

$$g(x) = \lim_{N \to \infty} \frac{1}{N} \sum_{n=1}^{N} \delta[x_n - x] \qquad (10.5)$$

Thus, merely by assuming infinite exchangeability over real-valued data, we get the following:

(1) A measure g from which the data is drawn,

(2) This measure g is itself random,

(3) A prior distribution μ for this measure,

(4) The random measure g is also the limiting EPDF of the data,

(5) A hierarchical Bayesian model, $G \sim \mu(g)$ and $X_n \sim g(x)$.

The distribution g is an example of a *random measure* which is an object of central importance in nonparametric Bayesian machine learning. This will allow us to do, for example, fully Bayesian nonparametric density estimation. We will explore models of such objects in detail later in this chapter.

We can illuminate some quite deep observations about the (somewhat fraught) relationship between Bayesian and frequentist modelling from this presentation on exchangeability. The frequentist assumption is that the parameter p (or distribution g) is an unknown constant (or fixed but unknown distribution) and the data are i.i.d. But this is a special case of the above situation where the prior μ is *degenerate* (for example, a Dirac delta measure), concentrated on some particular, fixed value (or particular distribution), g_F. In the real-valued data setting, this would be $\mu(g) = \delta[g - g_F]$. Then the joint distribution of the data becomes:

$$
\begin{aligned}
f(x_1, x_2, \ldots) &= \int_{\mathcal{F}(\mathbb{R})} \prod_{n=1}^{\infty} g(x_n) \, \delta[g - g_F] \, dg \\
&= \prod_{n=1}^{\infty} g_F(x_n) \qquad (10.6)
\end{aligned}
$$

and $\hat{g}_F(x) = \lim_{N \to \infty} \frac{1}{N} \sum_{n=1}^{N} \delta[x - x_n]$. So, we recover the limiting EPDF as the estimate for the fixed distribution from which the data is drawn. Thus, in fact, i.i.d. frequentist modelling can be derived from the

Fig. 10.1: **Infinite exchangeability, described in PGM form. Assuming the infinite number of observed random quantities x_n are exchangeable, then there exists a random measure g which has prior μ, from which these random observations x_n are drawn. Thus, assuming exchangeability implies the existence of a hierarchical Bayesian model for the observations. This is the structural content of de Finetti's theorem and generalizations (Hewitt-Savage and others).**

For now, we are glossing over the technical details of defining a degenerate prior over a distribution and focusing on the algebraic content.

infinite exchangeability assumption with all of the implications above; the primary point of divergence is that frequentists do not assume any *uncertainty* in the parameter value, but this does not mean that the distribution μ does not exist (O'Neill, 2009), as de Finetti's theorem shows.

Representations of stochastic processes

For any set of random variables with corresponding distribution, stochastic processes included, we need to find a way to uniquely describe their properties, a representation. While every stochastic processes is unique and there is no entirely "automatic" way to characterize all of them, there are common patterns that have been found to be useful. Certain representations, such as defining "self-marginal" f.d.d.s on countable index sets and invoking Kolmogorov extension, establish existence and uniqueness. Constructive algorithms for drawing sample paths of the process are useful in practice. For model inference purposes with combinatorial processes, sequential conditional distributions allow draws from corresponding combinatorial objects such as partitions or binary matrices which implicitly define the process. Finally, there are infinite limits of familiar, finite models such as the Gaussian mixture (Section 6.2) or linear Bayesian regression with Gaussian priors (Section 6.4).

In discrete, combinatorial Bayesian nonparametrics, the random measures that appear in de Finetti's theorem are often characterized in a particular way that requires some explanation. Such measures are generally *point processes*, represented as an infinite sum of Dirac delta functions or *atoms*, for illustration, considering the case of the real sample space:

$$g(x) = \sum_{i=1}^{\infty} w_i \delta [x - u_i] \tag{10.7}$$

with an infinite number of real, random weights $w_i > 0$ and random atom values $u_i \in \mathbb{R}$. This may seem highly counterintuitive and indeed much unlike the smooth, parametric distributions found in the rest of the book. But such models are essentially discrete, in the following way. Consider drawing a sample x_n from $g(x)$. All the probability mass is located on an infinite number of point masses that are contained in the real line. The probability of drawing a particular value is proportional to the weight of the associated mass on that atom value. Thus, all the samples have a finite probability of being drawn more than once. This is, of course, a property of discrete random variables that distinguishes them from continuous ones. Indeed, all discrete PMFs can be represented in the form of (10.7), whereupon the u_i are the elements of the sample space Ω_X of X and $g(x)$ is a PDF representation of the PMF.

Furthermore, the fact that samples are repeated means that the infinite set of samples x_n, $n = 1, 2, \ldots$ is *partitioned* by the atom values u_i they take on. In other words, the samples can be grouped together according to their shared values, the groups do not overlap, and together they make up the entire set of values u_i. Indeed, we can just as well represent the sequence of random draws x_n in terms of the cluster labels, to give the *indicator sequence* $z_n \in \mathbb{N}$ so that $x_n = u_{z_n}$. So, the indicator gives the atom index for that particular sample.

This partitioning is indeed central to combinatorial problems in machine learning such as clustering explored in Section 6.5. If we set the random atom values in (10.7) to the location parameter for each component $k = 1, 2, \ldots, K$ and weights of a mixture model (6.1), then we get the representation $g(x) = \sum_{k=1}^{K} \pi_k \delta[x - p_k]$, and draws from $g(x)$, a sequence of component location parameters, are indeed clustered, with frequency proportional to the weights. The corresponding sequence $z_n \in 1, 2, \ldots K$ identifies the cluster labels of these draws. This point process representation allows us to do useful things such as create a fully Bayesian treatment of nonparametric density estimation (Section 10.3 below).

Partitions and equivalence classes

Due to the discreteness of many nonparametric Bayesian models, we often find that they encode distributions over various combinatorial structures. Of particular relevance are *set partitions*, that is, ways of dividing a set A of $|A| = N$ elements up into K subsets $A_1, A_2, \ldots, A_K$ (groups or partitions), such that the subsets are exhaustive ($A_1 \cup A_2 \cup \cdots \cup A_K = A$) and mutually exclusive ($A_i \cap A_j = \emptyset$ for all $1 \leq i, j \leq K$ and $i \neq j$). For example, taking a set of three elements $A = \{1, 2, 3\}$, the partitions are: $\{\{1\}, \{2\}, \{3\}\}, \{\{1, 2\}, \{3\}\}, \{\{1, 3\}, \{2\}\}, \{\{1\}, \{2, 3\}\}$ and $\{\{1, 2, 3\}\}$, five different partitions. The number of such partitions is determined by the size of the set and is given by the *Bell numbers*, defined recursively as $B_{N+1} = \sum_{n=0}^{N} \binom{N}{n} B_n$ starting with $B_0 = 1$. The first few numbers in the sequence are 1, 1, 2, 5, 15, 52, 203, 877, 4140, 21147 and so on. The number B_N grows faster than e^N but slightly slower than $N!$.

Typically, Bayesian nonparametric models are exchangeable, for which the specific labels for the partitions are irrelevant. So, two partitions are the same under permutations of their labels, e.g. if we swap $\{1, 2, 3\} \rightarrow \{2, 3, 1\}$ then $\{\{1, 2\}, \{3\}\} \rightarrow \{\{2, 3\}, \{1\}\}$ so that the latter two partitions are not distinct, reducing the number of unique partitions from 5 down to 3. In this case, the only thing that distinguishes partitions is the size of the subsets. The number of such *permutation-invariant* partitions C_N has the sequence 1, 1, 2, 3, 5, 7, 11, 15, 22, 30 and so on. This is the same as the number of integer partitions (that is, the number of ways of adding together whole numbers less than or equal to N such that their sum is equal to N), which, as $N \rightarrow \infty$ tends to:

$$C_N \rightarrow \frac{1}{4N\sqrt{3}} \exp\left(\pi\sqrt{\frac{2N}{3}}\right) \tag{10.8}$$

Therefore, the growth of permutation-invariant partitions is, asymptotically, exponential in the number of elements, and there are clearly much fewer such partitions for a given N than if the labelling matters. Nonetheless, this explosive growth of the number of partitions with N means that brute-force clustering solutions are generally impractical (see combinatorial optimization, Section 2.6).

Structures other than partitions also naturally arise in the context of Bayesian nonparametrics. These include *equivalence relations* which group together members of a set according to some shared property. For example, each member of the set of all natural numbers $\mathbb{N}$ is divisible by 2 or not; so, the equivalence relation $x \sim y \iff x \equiv y \mod 2$. Then two numbers in $x, y \in \mathbb{N}$ are equivalent, $x \sim y$, if they are both either even or odd. Clearly, this relation partitions the integers into even or odd numbers. This is a general idea: an equivalence relation on a set partitions the set into subsets. Every member of a given subset is said to be in the same *equivalence class*, so they are all in the same partition element.

10.2   Gaussian processes (GP)

Regression is a ubiquitous tool in machine learning and DSP. Section 6.4 discusses connections between several kinds of regression. Linear regression, including linear-in-parameters regression, is simple with a straightforward, analytical Bayesian counterpart. Kernel regression methods and their discrete-time DSP convolution/filtering counterparts (Section 7.3) are nonlinear, but they are also not Bayesian. In this section we will see that there is a natural, Bayesian counterpart to nonlinear regression which makes use of the elegant mathematical properties of the (continuous-index) Gaussian process (GP).

From basis regression to kernel regression

Consider the linear-in-parameters regression model for $Y \in \mathbb{R}$ with M basis functions $h_i : \mathbb{R}^D \to \mathbb{R}$, $i = 1, 2, \ldots, M$ on the input vector $\boldsymbol{x} \in \mathbb{R}^D$:

$$E\left[Y|\boldsymbol{X}\right] = y\left(\boldsymbol{x}\right) = \sum_{i=1}^{M} w_i h_i\left(\boldsymbol{x}\right) = \boldsymbol{w}^T \boldsymbol{h}\left(\boldsymbol{x}\right) \tag{10.9}$$

with weight vector $\boldsymbol{w} \in \mathbb{R}^M$. Given regression input data $\boldsymbol{x}_n$ and output data y_n, for $n \in 1, 2, \ldots, N$, the L_2-norm regularized (ridge regression) MAP solution with regularization parameter $\lambda > 0$ is obtained by minimizing the following error functional with respect to the weights:

$$E\left(\boldsymbol{w}\right) = \frac{1}{2} \sum_{n=1}^{N} \left(\boldsymbol{w}^T \boldsymbol{h}\left(\boldsymbol{x}_n\right) - y_n\right)^2 + \frac{\lambda}{2} \left\|\boldsymbol{w}\right\|_2^2 \tag{10.10}$$

Using this regularized solution $\hat{\boldsymbol{w}}$, the prediction function can now be written entirely in terms of the *kernel function* $\kappa\left(\boldsymbol{x};\boldsymbol{x}'\right) = \boldsymbol{h}\left(\boldsymbol{x}\right)^T\boldsymbol{h}\left(\boldsymbol{x}'\right)$:

$$
\begin{aligned}
\hat{y}\left(\boldsymbol{x}\right) &= \hat{\boldsymbol{w}}^T\boldsymbol{h}\left(\boldsymbol{x}\right) \qquad\qquad (10.11)\\
&= \boldsymbol{k}\left(\boldsymbol{x}\right)^T\left(\mathbf{K}+\lambda\mathbf{I}\right)^{-1}\boldsymbol{y}
\end{aligned}
$$

with the *kernel matrix* $\mathbf{K}$ defined by $k_{nn'} = \kappa\left(\boldsymbol{x}_n;\boldsymbol{x}_{n'}\right)$, $n, n' \in 1, 2, \ldots, N$ and the vector $k_n\left(\boldsymbol{x}\right) = \kappa\left(\boldsymbol{x};\boldsymbol{x}_n\right)$ and the output data vector $\boldsymbol{y} = \left(y_1, y_2, \ldots, y_N\right)^T$. Now, defining the *weight function* $\boldsymbol{g}\left(\boldsymbol{x}\right) = \left(\mathbf{K}+\lambda\mathbf{I}\right)^{-1}\boldsymbol{k}\left(\boldsymbol{x}\right)$, the predictor is just $\hat{y}\left(\boldsymbol{x}\right) = \boldsymbol{g}\left(\boldsymbol{x}\right)^T\boldsymbol{y}$. So, this makes an important connection between regularized linear basis function regression and kernel regression (6.56). The regularization parameter λ controls the degree of smoothing in the regression.

Idealizing the situation to regression occurring across the continuum of the input space, this weight function can be computed analytically using Fourier analysis (Rasmussen and Williams, 2006, Chapter 7, pp 151-155). For example, it can be shown that the *Laplacian basis* $h\left(x\right) = \exp\left(-c\left|x\right|\right)$ with bandwidth $c > 0$ is its own idealized weight function, and the weight function for the *Gaussian basis* $h\left(x\right) = \exp\left(-\frac{c}{2}x^2\right)$ is well approximated by the sinc function (Figure 10.2).

If the input data x is univariate and the signal is uniformly sampled, this would just be describing a specific kind of linear FIR filter, see Section 7.3.

Distributions over function spaces: GPs

The derivation in the previous section does not make any explicit assumptions about the distribution of the parameter vector $\boldsymbol{w} \in \mathbb{R}^M$ in the regression. In this section, we will go further by assuming the $\boldsymbol{w}$ are spherical multivariate Gaussian with unit variance, $\boldsymbol{W} \sim \mathcal{N}\left(0, \mathbf{I}\right)$. Each realization $\boldsymbol{w}$ therefore determines a different prediction function $y\left(\boldsymbol{x}\right) = \boldsymbol{w}^T\boldsymbol{h}\left(\boldsymbol{x}\right)$ each from the space of prediction functions (Figure 10.3).

Now, using statistical stability, we know that the distribution $f\left(y;\boldsymbol{x}\right)$ must also be Gaussian because $y\left(\boldsymbol{x}\right) = \boldsymbol{w}^T\boldsymbol{h}\left(\boldsymbol{x}\right)$ is just a linear weighted combination of the $\boldsymbol{w}$'s. From this, we also know that the vector of values $\boldsymbol{y} = \left(y\left(\boldsymbol{x}_1\right), y\left(\boldsymbol{x}_2\right), \ldots, y\left(\boldsymbol{x}_N\right)\right)^T$ for any set of N points $\boldsymbol{x}$ must be multivariate Gaussian, and we can compute the mean vector and covariance matrix of the joint random variables $\boldsymbol{Y} = \left(Y_1, Y_2, \ldots, Y_N\right)^T$. To simplify notation, let us denote by $\mathbf{H}$ the M-by-N *design matrix* of with entries $h_{in} = h_i\left(\boldsymbol{x}_n\right)$. The mean vector is zero since:

$$
\begin{aligned}
\boldsymbol{m} &= E\left[\boldsymbol{Y}\right] = E\left[\boldsymbol{W}^T\mathbf{H}\right]\\
&= E\left[\boldsymbol{W}^T\right]\mathbf{H} = \mathbf{0} \qquad\qquad (10.12)
\end{aligned}
$$

Using this, we can then compute the covariance:

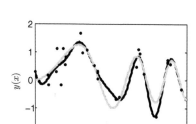

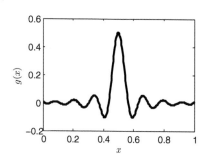

Fig. 10.2: **Linking kernel and regularized basis function regression.** Given $N = 30$ samples (top panel, black dots) from the model $y\left(x\right) = \frac{\sin\left(17.3x^2\right)}{x+0.5} + \epsilon$ where $\epsilon \sim \mathcal{N}\left(0, 0.3\right)$, the Gaussian kernel function $\kappa\left(x;x'\right) = \exp\left(-c\left(x-x'\right)^2\right)$ with $c = 50$ and $\lambda = 10^{-3}$ leads to a smoothed regression estimator (top panel, black curve). The equivalent weight function $g\left(x\right)$ behaves very much like the sinc function (bottom panel, black curve)

$$\begin{aligned}
\operatorname{cov}[Y_n, Y_{n'}] &= E\left[\boldsymbol{W}^T \boldsymbol{h}(\boldsymbol{x}_n) \boldsymbol{W}^T \boldsymbol{h}(\boldsymbol{x}_{n'})\right] \\
&= E\left[\boldsymbol{h}(\boldsymbol{x}_n)^T \boldsymbol{W}\boldsymbol{W}^T \boldsymbol{h}(\boldsymbol{x}_{n'})\right] \\
&= \boldsymbol{h}(\boldsymbol{x}_n)^T E\left[\boldsymbol{W}\boldsymbol{W}^T\right] \boldsymbol{h}(\boldsymbol{x}_{n'}) \\
&= \boldsymbol{h}(\boldsymbol{x}_n)^T \boldsymbol{h}(\boldsymbol{x}_{n'}) = k_{nn'}
\end{aligned} \tag{10.13}$$

We can recognize the last line as entries from the kernel matrix $\mathbf{K}$ from the previous section. What this is telling us is that the kernel matrix, which is defined in terms of the basis functions, directly determines the covariance of the multivariate Gaussian predictions from the regression model. So, we can write $f(\boldsymbol{y}; \boldsymbol{x}) = \mathcal{N}(\boldsymbol{0}, \mathbf{K})$.

Next, we will shift focus from a finite number of input data points N to the entirety of the input space. First, consider a subset $M < N$ of the data. From Section 1.4 we know that this subset of M samples are also multivariate Gaussian. We will consider these "self-marginal" distributions on subsets of $\mathbb{R}^D$ as f.d.d.s of some stochastic process. Also, these f.d.d.s are exchangeable, so we can apply the *Kolmogorov extension theorem* to show that a distribution for the process indexed by the continuous space $\mathbb{R}^D$ does indeed exist (Grimmett and Stirzaker, 2001, page 372). It also shows that the process, being Gaussian, is uniquely determined by appropriate mean and covariances; these are maps from $\mathbb{R}^D \to \mathbb{R}$, in other words, they are *functions*: we call them the *mean* $\mu(\boldsymbol{x})$ and *covariance functions* $\kappa(\boldsymbol{x}; \boldsymbol{x}')$. Furthermore, any one realized sample path of this process is also a function, $y(\boldsymbol{x})$. So, we can write $f(y(\boldsymbol{x})) = \operatorname{GP}(\mu(\boldsymbol{x}), \kappa(\boldsymbol{x}; \boldsymbol{x}'))$. Thus, the GP is a distribution over the space of general, nonlinear functions. All the size N subsets of the sample paths of the GP are multivariate Gaussian vectors with the distribution $\boldsymbol{Y} \sim \mathcal{N}(\mu(\boldsymbol{x}_n), \kappa(\boldsymbol{x}_n; \boldsymbol{x}_{n'}))$ for $n, n' \in 1, 2, \ldots, N$.

It is important to note that this does not constrain other properties of the sample paths (functions drawn from the GP), for instance, continuity or differentiability. These properties depend upon the specific parameters of the GP.

As draws from GPs can be arbitrary functions, machine learning practitioners often rely on a 'toolkit' of covariance functions which lead to functions with particular behaviours. Here are some specific examples (see Figure 10.4), starting with *spherically-symmetric* kernels in the form $\kappa(\boldsymbol{x}; \boldsymbol{x}') = g(\|\boldsymbol{x} - \boldsymbol{x}'\|)$. These kernels specify stationary GPs. The case $g(r) = b \exp\left(-\frac{1}{2}(r/\sigma)^2\right)$ with $b > 0$ and $\sigma > 0$ is known as the Gaussian or *squared exponential* kernel. Samples from a GP with this kernel are infinitely differentiable (Rasmussen and Williams, 2006, page 15) which is a desirable property in some situations. Parameter σ controls the "effective width" of the kernel; if this is large (small) then draws from the process are smooth (rough) and correlations extend across large (small) distances in the space of $\boldsymbol{x}$. The scale factor b controls the overall magnitude of the correlations. More generally, any function in the form $g(r) = b \exp(-(r/\sigma)^p)$ for $0 < p \le 2$ is a valid, stationary GP kernel. An important special case is the *Cauchy* kernel $p = 1$. In one dimension, this is the *Ornstein-Uhlenbeck* process which has non-differentiable sample paths (Rasmussen and Williams, 2006, page 86).

Bayesian linear regression is a special case with the kernel:

$$\kappa\left(\boldsymbol{x};\boldsymbol{x}'\right) = \sigma^2 + \left(\boldsymbol{x} - \boldsymbol{\mu}\right)^T\left(\boldsymbol{x}' - \boldsymbol{\mu}\right) \tag{10.14}$$

The parameter $\boldsymbol{\mu} \in \mathbb{R}^D$ determines the origin in the space of $\boldsymbol{x}$. Kernels can also encode *periodicities*, which are of substantial value in DSP applications. The following periodic kernel has an effective width $\sigma > 0$ and period $p > 0$:

$$\kappa\left(\boldsymbol{x};\boldsymbol{x}'\right) = \exp\left(-\frac{2}{\sigma^2}\sin^2\left[\frac{\pi}{p}\left\|\boldsymbol{x} - \boldsymbol{x}'\right\|\right]\right) \tag{10.15}$$

The periodic sample paths of this kernel are best understood in the Fourier domain: the effective width controls the frequency content. If it is large (small), then higher frequency components will be suppressed (amplified). A final kernel which is interesting from DSP applications is the *autoregressive* (AR/IIR) kernel over discrete-time indices $n, n' \in \mathbb{Z}$:

$$\kappa\left(n; n'\right) = a^{|n-n'|} \tag{10.16}$$

for $-1 < a < 1$. Sample paths from a GP with this kernel are realizations from a first order IIR filter with the single parameter a and i.i.d. Gaussian noise input. This can be extended to arbitrary order $P > 1$ IIR filtering using the general form of the autocovariance for IIR filters which is $\kappa\left(n; n'\right) = \sum_{j=1}^{P} b_j p_j^{|n-n'|}$ where the p_j are the poles of the TF (Section 7.4) and $b_j \in \mathbb{R}$ are arbitrary IIR initial conditions.

Using the properties of Mercer kernels from Section 6.1 we can take any kernel $\kappa'\left(\boldsymbol{x};\boldsymbol{x}'\right)$ and include a *jump discontinuity* at a chosen location $\boldsymbol{\mu}$, $\kappa\left(\boldsymbol{x};\boldsymbol{x}'\right) = \kappa'\left(\boldsymbol{x};\boldsymbol{x}'\right) + c\,\delta\left[\left(\boldsymbol{x} \geq \boldsymbol{\mu}\right) \wedge \left(\boldsymbol{x}' \geq \boldsymbol{\mu}\right)\right]$ where δ is the Kronecker delta (Figure 10.4).

We should point out the kernels are not restricted to real data: they can be constructed for many different data types, including sequences of symbols (*strings*), sets and graphs (Shawe-Taylor and Christianini, 2004, Chapter 11). For more kernel recipes, see Rasmussen and Williams (2006).

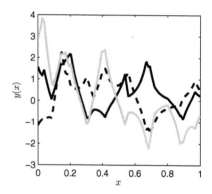

Fig. 10.3: Placing a spherical multivariate Gaussian over the weights w in a linear basis regression model $y(x) = w^T h(x)$ (here there are $M = 30$ basis functions $h_i(x) = \exp\left(-c\left|x - x_i\right|\right)$ where the x_i are drawn uniformly at random from $[0, 1]$). Drawing realizations from this distribution allows exploration of the implied space of prediction functions.

Bayesian GP kernel regression

From Section 4.5, we know that the multivariate Gaussian is an exponential family distribution, and when the covariance is fixed, it has a conjugate prior for the mean vector, and that prior is also multivariate Gaussian. Let us consider then that the prior over the function we wish to estimate from the data is a GP with zero mean function and covariance kernel $\kappa_0\left(\boldsymbol{x};\boldsymbol{x}'\right)$, we can write this as $f\left(y\left(\boldsymbol{x}\right);\kappa_0\left(\boldsymbol{x};\boldsymbol{x}'\right)\right) = \mathrm{GP}\left(y\left(\boldsymbol{x}\right);0,\kappa_0\left(\boldsymbol{x};\boldsymbol{x}'\right)\right)$. With analogy to the multivariate Gaussian situation, the function we wish to estimate is the mean function of the GP. To exploit conjugacy, we will assume that the likelihood for the function is also a GP with white noise covariance function, e.g. $f\left(\tilde{y}\left(\boldsymbol{x}\right)|y\left(\boldsymbol{x}\right)\right) = \mathrm{GP}\left(\tilde{y}\left(\boldsymbol{x}\right);y\left(\boldsymbol{x}\right),\sigma^2\delta\left[\boldsymbol{x} = \boldsymbol{x}'\right]\right)$ where $\tilde{y}\left(\boldsymbol{x}\right)$ is the observed function (the data). The implication is that the posterior function also has a GP distribution: $f\left(y\left(\boldsymbol{x}\right)|\tilde{y}\left(\boldsymbol{x}\right)\right) = \mathrm{GP}\left(y\left(\boldsymbol{x}\right);\mu\left(\boldsymbol{x}\right),\kappa\left(\boldsymbol{x};\boldsymbol{x}'\right)\right)$

We assume zero mean for simplicity in the following derivations, but it is straightforward to modify them to use an arbitrary mean function instead.

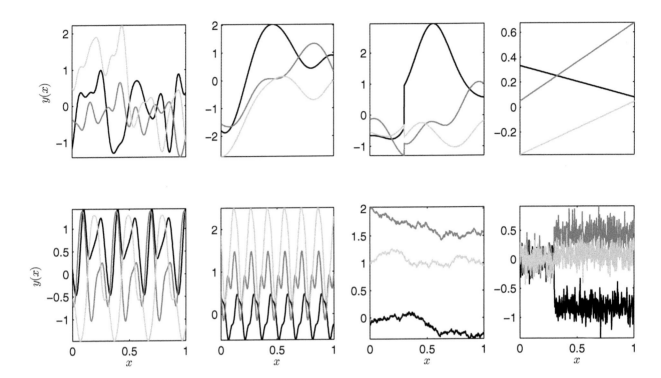

Fig. 10.4: Draws from a zero-mean Gaussian process (GP) with a range of covariance functions. From top left: Gaussian kernel, length scale $\sigma = 0.05$ and scale factor $b = 1$; Gaussian kernel with $\sigma = 0.2$, $b = 1$; Gaussian kernel, $\sigma = 0.2$ with jump discontinuity at $x = 0.3$; the linear regression kernel with $\sigma = 0.5$; periodic (harmonic) kernel with $\sigma = 1$ and $p = 0.3$; periodic kernel $\sigma = 1.5$, $p = 0.15$; AR(1) kernel with $a = 1 - 10^{-4}$; and finally the i.i.d. (white) noise kernel with jump discontinuity of 0.5 at $x = 0.3$, with $\sigma = 0.15$.

where $\mu\left(\boldsymbol{x}\right)$ and $\kappa\left(\boldsymbol{x};\boldsymbol{x}'\right)$ are the posterior mean and covariance function, respectively. These facts will form the basis of a method for tractable Bayesian nonlinear regression, which hinges on finding the parameters for the posterior and posterior predictive GPs.

In practice, we only ever have a finite set of samples, $\boldsymbol{x}_n, y_n$ for $n = 1, 2, \ldots, N$. Then, the likelihood GP marginalized over the data becomes the finite multivariate Gaussian $f\left(\tilde{\boldsymbol{y}}\,|\boldsymbol{x}_1, \boldsymbol{x}_2, \ldots, \boldsymbol{x}_N\,; \sigma^2\right) = \mathcal{N}\left(\tilde{\boldsymbol{y}}; \boldsymbol{y}\left(\boldsymbol{x}\right), \sigma^2\mathbf{I}\right)$, where $\tilde{\boldsymbol{y}} = \left(y_1, y_2, \ldots, y_N\right)^T$ and $\boldsymbol{y}\left(\boldsymbol{x}\right) = \left(y\left(\boldsymbol{x}_1\right), y\left(\boldsymbol{x}_2\right), \ldots, y\left(\boldsymbol{x}_N\right)\right)^T$. We can now use normal conjugacy to compute the posterior GP which has parameters:

$$\mu\left(\boldsymbol{x}\right) = \tilde{\boldsymbol{\kappa}}\left(\boldsymbol{x}\right)^T \left(\tilde{\mathbf{K}} + \sigma^2\mathbf{I}\right)^{-1} \tilde{\boldsymbol{y}} \tag{10.17}$$

$$\kappa\left(\boldsymbol{x};\boldsymbol{x}'\right) = \kappa_0\left(\boldsymbol{x};\boldsymbol{x}'\right) - \tilde{\boldsymbol{\kappa}}\left(\boldsymbol{x}\right)^T \left(\tilde{\mathbf{K}} + \sigma^2\mathbf{I}\right)^{-1} \tilde{\boldsymbol{\kappa}}\left(\boldsymbol{x}'\right) \tag{10.18}$$

Here, the $N \times N$ data kernel matrix $\tilde{\mathbf{K}}$ has entries $\tilde{k}_{nn'} = \kappa_0\left(\boldsymbol{x}_n; \boldsymbol{x}_{n'}\right)$ and the length N data kernel vector $\tilde{\boldsymbol{\kappa}}\left(\boldsymbol{x}\right)$ has elements $\tilde{\kappa}_n\left(\boldsymbol{x}\right) = \kappa_0\left(\boldsymbol{x}; \boldsymbol{x}_n\right)$.

Since the posterior GP mean coincides with its mode for any value of

x, the MAP posterior value of the function is just (10.17), and in fact this is the same as the regularized kernel regression MAP formula (10.11) with $\lambda = \sigma^2$. So, it is entirely justified to call GP regression *Bayesian kernel regression*, where the prior covariance kernel function determines the kernel in the equivalent, non-Bayesian kernel regression (Figure 10.5).

Similarly, the posterior predictive distribution is also a Gaussian, $f\left(\bar{y}|\boldsymbol{y};\sigma^2\right) = \mathcal{N}\left(\bar{y};\bar{\mu}\left(\bar{\boldsymbol{x}}\right),\bar{\sigma}^2\left(\bar{\boldsymbol{x}}\right)\right)$ with mean and variance:

$$\bar{\mu}\left(\bar{\boldsymbol{x}}\right) = \bar{\boldsymbol{\kappa}}\left(\bar{\boldsymbol{x}}\right)^T\left(\tilde{\mathbf{K}}+\sigma^2\mathbf{I}\right)^{-1}\tilde{\boldsymbol{y}} \tag{10.19}$$

$$\bar{\sigma}^2\left(\bar{\boldsymbol{x}}\right) = \kappa_0\left(\bar{\boldsymbol{x}};\bar{\boldsymbol{x}}\right) - \bar{\boldsymbol{\kappa}}\left(\bar{\boldsymbol{x}}\right)^T\left(\tilde{\mathbf{K}}+\sigma^2\mathbf{I}\right)^{-1}\bar{\boldsymbol{\kappa}}\left(\bar{\boldsymbol{x}}\right) + \sigma^2 \tag{10.20}$$

where $\bar{\kappa}_n\left(\bar{\boldsymbol{x}}\right) = \kappa_0\left(\bar{\boldsymbol{x}};\boldsymbol{x}_n\right)$. Once the $O\left(N^3\right)$ computational cost of inverting $\tilde{\mathbf{K}}+\sigma^2\mathbf{I}$ has been absorbed, making individual predictions requires $O\left(N\right)$ computations for the mean and $O\left(N^2\right)$ for the variance, per prediction.

GP regression gains several advantages over frequentist kernel regression. For example: we can compute the confidence in our regression estimates of the unknown function, for any value of $\boldsymbol{x}$. The spread of the posterior around the regression mode naturally increases away from the observed data, and how rapidly the spread increases away from the observed data depends upon the implicit "length scale" of the prior kernel. For example, the Gaussian kernel's length scale is determined by the σ parameter. Prior kernels with short length scales will therefore provide little certainty about the global shape of the regression curve, while giving high confidence to the regression predictions close the observed data (Figure 10.5).

We can use this confidence interval in a many applications. In DSP, with uniformly sampled data, we can effectively perform FIR filtering and estimate the posterior distribution of this filtering in one conceptual framework.Roberts *et al.* (2012) present several examples of GP regression for DSP, such as tidal height prediction and forecasting in oceanography, exoplanet light curve smoothing and inference in astronomy, and market indices in financial engineering. In sequential sampling situations, we can find regions of high posterior uncertainty and target sampling in those regions to give maximal reduction in uncertainty overall. GP regression has also been used in this sequential way as a tool for solving nonlinear optimization problems (Osborne *et al.*, 2009).

Being nonparametric, GP regression can very often outperform parametric regression in empirical machine learning studies, much as kernel regression often outperforms parametric methods such as linear regression. Both methods require access to the entire set of training data to make predictions. This raises the question of how to manage the model complexity of GP regression, which depends upon the smoothness of the prior mean and covariance functions. Smoother priors will generally lead to smoother (less complex) posterior functions, therefore, the usefulness of the predictions out-of-sample will depend upon these prior

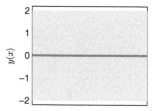

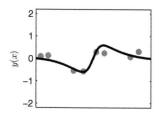

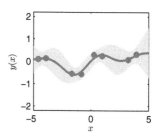

Fig. 10.5: **Bayesian GP regression requires choosing a prior mean function (top panel, dark grey line) and kernel covariance function (top panel, light grey filled area shows 95% confidence interval for each value of x). $N = 8$ noisy observations with standard deviation $\sigma = 0.2$ (middle panel, dark grey dots) are obtained from the underlying function to estimate, $y\left(x\right) = 2\sin\left(\frac{1}{5}\pi x\right)/\left(x^2+1\right)$ (middle panel, black curve). These observations from the likelihood GP are then used to find the posterior GP mean function (bottom panel, dark grey curve) and kernel covariance (bottom panel, light grey filled area, 95% interval).**

choices and whether these match the true length scale of the underlying function that generated the data (see also the next section).

The primary disadvantage of GP regression is computational complexity. Finding the matrix inverse $\left(\tilde{\mathbf{K}} + \sigma^2\mathbf{I}\right)^{-1}$ requires $O\left(N^3\right)$ operations in general which is prohibitive for very long signals. For this reason, a substantial amount of research effort has been devoted to improving the computational complexity of GP regression. Many of these techniques are based on the idea of finding a subset of the data with size $M \ll N$ from which a good approximation of the regression posterior can be estimated in only $O\left(M^3\right)$ operations. We will look at one of these methods known as the *informative vector machine* (Rasmussen and Williams, 2006, Section 8.3.3). Considering all subsets of size M of the data is a combinatorial problem with exponential complexity, which is obviously infeasible. Instead, we can use a greedy optimization heuristic (Section 2.6). A reasonable greedy criteria is to select a member of the data set that maximizes the difference in entropy of the posterior GP at a specific location in the input space of $\boldsymbol{x}_i$ before and after including an observed data item at that location in the subset:

$$\Delta H = H\left[f\left(y\left(\boldsymbol{x}_i\right)\middle|\boldsymbol{y}_{S\setminus i}\right)\right] - H\left[f\left(y\left(\boldsymbol{x}_i\right)\middle|\boldsymbol{y}_S\right)\right] \tag{10.21}$$

where S is a subset of data indices $1, 2, \ldots, N$ and the vector $\boldsymbol{y}_S = (y_i)_{i \in S}$ is the data in the selected subset. By this criteria, observations that most reduce the posterior uncertainty are included in the subset in a greedy, stepwise fashion. For the GP, the marginal at any location is univariate Gaussian with variance τ^2, which has entropy $\frac{1}{2}\ln\left(2\pi e\tau^2\right)$. The posterior variance of a univariate Gaussian after including a single observation is $\sigma_P^2 = \left(1/\tau^2 + 1/\sigma^2\right)^{-1}$, so that:

$$\begin{aligned} \Delta H &= \frac{1}{2}\ln\left(\frac{2\pi e\,\tau^2}{2\pi e\,\sigma_P^2}\right) \\ &= \frac{1}{2}\ln\left(\frac{\sigma^2 + \tau^2}{\sigma^2}\right) \end{aligned} \tag{10.22}$$

This expression is used in Algorithm 10.1. This can be understood as a form of greedy combinatorial neighbourhood search, where each neighbourhood is drawn from the entire space of $\boldsymbol{x}$ (c.f. Algorithm 2.8).

Algorithm 10.1 Gaussian process (GP) *informative vector machine* (IVM) regression.

(1) *Initialization.* Set $\Omega = \{1, 2, \ldots, N\}$ and $S = \emptyset$, and choose the final regression subset size M ,

(2) *Sample random subset.* Pick a subset R of indices uniformly at random from the data of size $|R| = T$ from Ω,

(3) *Minimize entropy scores.* Compute $\Delta H_i = \frac{1}{2} \ln \left((\sigma^2 + \tau_i^2)/\sigma^2 \right)$ for each $i \in 1, 2, \ldots T$ where τ_i^2 is the variance of the posterior for $y(\boldsymbol{x}_{R_i})$ given the data y_R and $\boldsymbol{x}_R$,

(4) *Select next item.* Solve $j = \arg\max_i \Delta H_i$, incorporate this into the working subset $S \leftarrow S \cup R_j$ and remove $\Omega \leftarrow \Omega \backslash R_j$,

(5) *Iteration.* Set $m \leftarrow m+1$, if $m < M$ go back to step 2 otherwise perform GP regression with subset of data with indices in S.

For a comprehensive discussion of other methods for reduced-complexity GP regression, see Rasmussen and Williams (2006, Chapter 8).

GP regression and Wiener filtering

Wiener filtering (see Section 7.4) is a special case of GP regression as we show next. Consider functions $y(x)$ sampled on $x_n = \Delta n$ where Δ is the sampling interval. For an observed digital signal $\boldsymbol{y}$ with samples $y_n = y(x_n)$, $n \in 1, 2, \ldots, N$, the length-N posterior GP mean function can be written as (Rasmussen and Williams, 2006, page 25):

$$
\begin{aligned}
\boldsymbol{\mu} &= \mathbf{K}_0 \left(\mathbf{K}_0 + \sigma^2 \mathbf{I} \right)^{-1} \boldsymbol{y} \\
&= \sum_{n=1}^{N} \boldsymbol{u}_n \frac{\alpha_n}{\alpha_n + \sigma^2} \boldsymbol{u}_n^T \boldsymbol{y}
\end{aligned}
\tag{10.23}
$$

where $\boldsymbol{u}_n$ and α_n are the eigenvectors and eigenvalues of the GP prior autocovariance matrix $\mathbf{K}_0$ with entries $k_{nn'} = \kappa_0 (\Delta n; \Delta n')$ and the posterior mean signal $\boldsymbol{\mu}$ has elements $\mu_n = \mu(x_n)$. This is exactly the MAP predictions given by the Wiener filter (7.149). The primary advantage of GP regression over Wiener filtering is the computation of the posterior covariance, in other words, we can perform Wiener filtering and also get an estimate of the uncertainty in that filtering operation.

For DSP applications, GP regression requires $O(N^3)$ computations to invert $\mathbf{K}_0 + \sigma^2 \mathbf{I}$ so this is not very practical. However, for special classes of signals, we can do the computations much more efficiently. In particular, consider translation-invariant GP prior autocovariance functions $\kappa_0(x; x') = g_0(x - x')$. Then $\mathbf{K}_0 + \sigma^2\mathbf{I}$ is Toeplitz (has constant diagonals), so it can be inverted using the Levinson recursion in $O(N^2)$ operations. This is a substantial improvement, but we can do better than that. If in addition we consider that the signal is periodic so that $y_{n+N} = y_n$ for all $n \in \mathbb{Z}$ and that the prior autocovariance is symmetric, then $\mathbf{K}_0 + \sigma^2\mathbf{I}$ is circulant, which means it can be diagonalized in the Fourier basis. We can exploit this to make an even more efficient implementation. Let us write out the posterior mean signal as $\boldsymbol{\mu} = \mathbf{K}_0\boldsymbol{h}$ where $\boldsymbol{h} = (\mathbf{K}_0 + \sigma^2\mathbf{I})^{-1}\boldsymbol{y}$. Rewriting the linear problem for $\boldsymbol{h}$ as $(\mathbf{K}_0 + \sigma^2\mathbf{I})\boldsymbol{h} = \boldsymbol{y}$, this is in the form of a circular convolution $(\boldsymbol{k}_0 + \sigma^2\boldsymbol{\delta}) \star_N \boldsymbol{h} = \boldsymbol{y}$ where $\boldsymbol{k}_0 + \sigma^2\boldsymbol{\delta}$ is the length N vector in the first row (or first column) of the circulant matrix $\mathbf{K}_0 + \sigma^2\mathbf{I}$. Then, we can use the convolution property of the DFT to show (see Section 7.2):

$$F_k\left[(\boldsymbol{k}_0 + \sigma^2\boldsymbol{\delta}) \star_N \boldsymbol{h}\right] = F_k\left[\boldsymbol{k}_0 + \sigma^2\boldsymbol{\delta}\right]F_k\left[\boldsymbol{h}\right] = F_k\left[\boldsymbol{y}\right] \quad (10.24)$$

and therefore the solution is:

$$\boldsymbol{h} = F_k^{-1}\left[\frac{F_k\left[\boldsymbol{y}\right]}{F_k\left[\boldsymbol{k}_0 + \sigma^2\boldsymbol{\delta}\right]}\right] \quad (10.25)$$

In words, we can effectively compute the GP matrix inversion problem using the DFT. The advantage here is that we can implement the DFT using an FFT requiring $O(N \ln N)$ operations. This is a *very* substantial saving over the general GP regression problem. We can also compute the posterior mean with similar efficiency, since it is another circular convolution, $\boldsymbol{\mu} = \boldsymbol{k}_0 \star_N \boldsymbol{h}$ where $\boldsymbol{k}_0$ is the first row or column of $\mathbf{K}_0$:

$$\boldsymbol{\mu} = F_k^{-1}\left[\frac{F_k\left[\boldsymbol{k}_0\right]}{F_k\left[\boldsymbol{k}_0 + \sigma^2\boldsymbol{\delta}\right]}F_k\left[\boldsymbol{y}\right]\right] \quad (10.26)$$

This is just a special case of (10.23) where DFT diagonalizes the prior GP autocovariance.

Other GP-related topics

GP regression is sufficiently versatile that it can be co-opted to solve other machine learning or DSP problems, for instance, various uses of GPs in classification (see Section 6.3) have been explored extensively. Here is a very simple example. Consider coding a two-class classification problem with class membership labels $Y \in \{0, 1\}$ given input data $\boldsymbol{X} \in \mathbb{R}^D$. We can use GP regression to directly model the decision boundary, by regressing on the training data $(\boldsymbol{x}_n, y_n)_{n=1}^N$, to find the posterior distribution for the function $y(\boldsymbol{x})$. Since the marginal posterior at each $\boldsymbol{x}$ is univariate normal, we can find the posterior classification probability as $p(\boldsymbol{x}) = 1 - \Phi\left(\frac{1}{2}; \mu(\boldsymbol{x}), \kappa(\boldsymbol{x}; \boldsymbol{x})\right)$ where $\Phi(x; \mu, \sigma^2)$ is the CDF of the univariate normal with parameters μ, σ^2. We can similarly find the posterior predictive probability using $\bar{\mu}(\boldsymbol{x}), \bar{\kappa}(\boldsymbol{x}; \boldsymbol{x})$ instead. The MAP classification would be $p(\boldsymbol{x}) > \frac{1}{2}$ to assign $Y = 1$ and $p(\boldsymbol{x}) \le \frac{1}{2}$ to

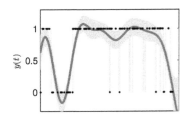

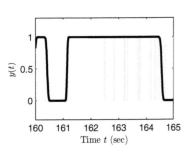

Fig. 10.6: Simple use of GP regression for classification. Thresholded instantaneous (10ms) log-energy of a digital audio signal (top panel, light grey curve) is randomly sampled (black dots). GP regression using Gaussian prior kernel with length scale 0.5 is used to predict the posterior distribution of the decision function (mean, dark grey curve; two standard deviation confidence interval, light grey shading). This is then used to predict the posterior $Y = 1$ classification probability (bottom panel, black curve).

assign $Y = 0$. For appropriate choices of (smooth) kernel function, the classification is regularized across the space of $\boldsymbol{x}$, which has practical applications in DSP (Figure 10.6).

Much more elaborate approaches to classification using GPs have been proposed. Typically these are discriminative, for example, replacing the linear decision function $b_0 + \boldsymbol{b}^T \boldsymbol{x}$ in logistic regression (6.26) with a non-linear function with GP prior (Rasmussen and Williams, 2006, Section 3.3). Parameter and posterior estimation quickly get complex however, as the simple Gaussian conjugacy which makes posterior inference so straightforward, is lost and we need to turn to numerical approximations.

GPs have also been applied to dimensionality reduction. By modelling the mapping in PPCA (Section 6.6) between latent $\boldsymbol{Z}$ and observed variables $\boldsymbol{X}$ as a GP indexed by the latent variables $\boldsymbol{Z}$, it is possible to form a general nonlinear PPCA known as the *Gaussian process latent variable model* (Lawrence, 2004).

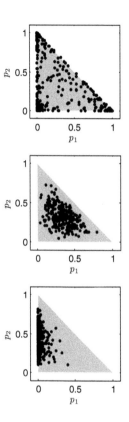

10.3 Dirichlet processes (DP)

In Section 4.8 we discussed the KDE, while in Section 6.2 the mixture model was explored. We can view both of these models as frequentist density estimators. For illustration, let us consider the simple Gaussian case for which $\Omega_X = \mathbb{R}$ with fixed variance σ^2, then both the kernel κ and the mixture component distributions f are Gaussian, and the only parameters are the means of each component. For the KDE, the means coincide with each of the N samples $\mu_n = x_n$ in the given data so there are N components, whereas for the mixture, there are K mean parameters μ_k. Which statistical model is best? It is possible to compare likelihoods for a certain dataset and pick the model with the larger of the two. Nonetheless it is clear that the KDE has "maximal" complexity in the sense that it requires N parameters (the values in the data). By contrast, the mixture model requires K parameters. All other things being equal, this makes the KDE much more likely to overfit.

Of course, we have freedom to vary the complexity of the model with the single bandwidth parameter σ^2, but this assumes that the complexity of the PDF curve is locally the same everywhere on the real line and so does not offer much control. On the other hand, if the data has intermediate complexity, we can always fit it with a mixture density with a large number of components. The question of exactly how many components, much like the choice of the KDE bandwidth, has no single, entirely satisfactory answer, however. Regularization methods (AIC, BIC, MDL etc. see Section 4.2) are not fully Bayesian so we cannot quantify the uncertainty in the density estimate. As an alternative, the *Dirichlet process* (DP) model offers a simple, fully Bayesian method for density estimation, addressing the problem of selecting the appropriate number of components for a certain dataset in an elegant way.

Fig. 10.7: Samples from a Dirichlet distribution. The sample space for the distribution is $p_k \in [0,1]$ and $\sum_{k=1}^{K} p_k = 1$ for $k = 1, 2, \ldots, K$. This is the $K - 1$-dimensional simplex, here shown for $K = 3$ (grey area). Small parameter values concentrate the samples around the edge of the simplex (top panel, parameter vector $a = (0.3, 0.3, 0.3)$), whereas large values concentrate samples around the simplex interior (middle panel, $a = (3.8, 3.8, 3.8)$). The concentration can be heterogeneous, favouring one edge over others (bottom panel, $a = (0.3, 3.8, 5.1)$).

The Dirichlet distribution: canonical prior for the categorical distribution

Our starting point for understanding the DP begins with a Bayesian treatment of the *Dirichlet distribution*. Consider a general, discrete random variable on a finite, discrete sample space $\Omega_X = \{1, 2, \ldots K\}$. Given no other information, we can say that it is categorically-distributed. The PMF is $f_X(x) = p_x$ with parameter vector $p \in [0,1]^K$ such that $\sum_{k=1}^K p_k = 1$. It is straightforward to show that the ML parameters for N i.i.d. draws from this random variable are just the normalized counts $\hat{p}_k = N_k/N$, which we can write using Kronecker delta form, $N_k = \sum_{n=1}^N \delta[x_n - k]$. Taking a simple Bayesian approach, we can turn to the exponential family, for which the conjugate prior to the categorical is the Dirichlet distribution with parameters $a_k > 0$ (Figure 10.7). We will write this as Dir $(p; a)$, and the random vector $p \sim$ Dir (a). The associated PMF is:

$$f(p; a) = C(a) \prod_{k=1}^K p_k^{a_k - 1} \tag{10.27}$$

using the normalization factor $C(a) = \Gamma\left(\sum_{k=1}^K a_k\right) \Big/ \prod_{k=1}^K \Gamma(a_k)$. The categorical likelihood for the data is:

$$f(x\,|p) = \prod_{n=1}^N p_{x_n} = \prod_{k=1}^K p_k^{N_k} \tag{10.28}$$

giving the joint distribution:

$$\begin{aligned}
f(p, x; a) &= C(a) \prod_{k=1}^K p_k^{a_k + N_k - 1} \\
&= \frac{C(a)}{C(a+N)} C(a+N) \prod_{k=1}^K p_k^{a_k + N_k - 1} \tag{10.29} \\
&= \frac{C(a)}{C(a+N)} \text{Dir}(p; a+N)
\end{aligned}$$

where $N = (N_1, N_2, \ldots, N_K)$. The evidence probability is:

$$\begin{aligned}
f(x; a) &= \int_{\Omega_P} f(p, x; a)\, dp \\
&= \frac{C(a)}{C(a+N)} \int_{\Omega_P} \text{Dir}(p; a+N)\, dp \tag{10.30} \\
&= \frac{C(a)}{C(a+N)}
\end{aligned}$$

Using these, it is simple to show the conjugacy of the posterior distribution:

$$f(p \,|\, x \,; a) \;=\; f(p, x; a)/f(x; a)$$
$$\;=\; \mathrm{Dir}(p; a + N) \tag{10.31}$$

and to derive a simple formula for the posterior predictive distribution, a conditional categorical:

$$
\begin{aligned}
f(x_{N+1} = x \,|\, x_N, x_{N-1}, \ldots, x_1 \,; a) \;&=\; f(x_{N+1} = x, x; a)/f(x; a) \\
&=\; \frac{C(a)}{C(a + N + \delta_x)} \Big/ \frac{C(a)}{C(a + N)} \\
&=\; \frac{C(a + N)}{C(a + N + \delta_x)} \\
&=\; \frac{a_x + N_x}{\sum_{k=1}^{K} a_k + N} \tag{10.32}
\end{aligned}
$$

where $\delta_x = e_x$, the standard basis vector in K dimensions. The proof of this relies on the fact that $\Gamma(x+1) = x\Gamma(x)$.

A key fact about the Dirichlet distribution is the *agglomeration property* of the marginals. So, if $p \sim \mathrm{Dir}(a)$ and then $p' = (p_1, \ldots, p_i + p_j, \ldots, p_K) \sim \mathrm{Dir}(a_1, \ldots, a_i + a_j, \ldots, a_K)$ for $1 \leq i, j \leq K$ and $i \neq j$. This means that forming new sample spaces by combining together parts of the sample space (which is $[0,1]^K$, $\sum_{k=1}^{K} p_k = 1$) results in another Dirichlet distribution over this smaller sample space.

The parameter vector p being a normalized probability vector means that it is a general discrete PMF. Of fundamental importance to Bayesian nonparametric machine learning is that this is a random vector which means that it is also a *random probability mass function*. Thus, the Dirichlet prior is a *distribution over distributions*. Furthermore, the Dirichlet distribution is conjugate to the categorical distribution, so it follows that the posterior is also Dirichlet. We will define the Dirichlet parameters using a set of K *base measure* parameters $b_k > 0$ with $\sum_{k=1}^{K} b_k = 1$, and a scaling parameter $\alpha > 0$. The Dirichlet parameters are then $a = \alpha b$. Here we will write this as $\mathrm{Dir}(p; \alpha, b)$. In this parameter arrangement, the conjugate Dirichlet-categorical model can be brought into alignment with the notion of *Bayesian updating* of random PMFs using observations. It is informative to look at the behaviour of this formulation. The mean and variance of each random variable p_k are:

The parameter α is also called a precision or inverse spread parameter since it is approximately inversely proportional to the variance.

$$
\begin{aligned}
\mu_k \;&=\; E[p_k] = b_k \tag{10.33} \\
\sigma_k^2 \;&=\; E\left[(p_k - b_k)^2\right] = \frac{b_k(1 - b_k)}{\alpha + 1}
\end{aligned}
$$

So, as $\alpha \to \infty$, then $\sigma_k^2 \to 0$ which implies that the p_k become highly concentrated around the mean b_k. For this reason, α is often known as the *concentration* parameter. The posterior Dirichlet distribution can then be written as:

$$f\left(\boldsymbol{p}\,|\,\boldsymbol{x}\,;\alpha,\boldsymbol{b}\right) = \text{Dir}\left(\boldsymbol{p};\alpha + N,\frac{1}{\alpha + N}\left[\alpha\boldsymbol{b} + \sum_{n=1}^{N}\boldsymbol{\delta}_{x_n}\right]\right) \qquad (10.34)$$

Thus, we can view the Bayesian updating procedure as accumulating samples into the base measure.

In order to draw samples x_n from this Bayesian model, we can use ancestral sampling in which we first draw a categorical parameter vector $\boldsymbol{p}\sim\text{Dir}\left(\alpha,\boldsymbol{b}\right)$, and then draw samples from the categorical distribution with parameter vector $\boldsymbol{p}$. This involves sampling from a Dirichlet distribution, but since the CDF is not available in closed form it is difficult to use the generic techniques described in Chapter 3. Instead we will look at two special algorithms. The first, most direct approach is the *normalized gamma* method: first, generate K random numbers y_k from the gamma distribution with PDF $y_k^{\alpha b_k - 1}\exp\left(-y_k\right)\big/\Gamma\left(\alpha b_k\right)$. Then, the normalized $p_k = y_k\big/\sum_{k=1}^{K}y_k$ have the required distribution (Devroye, 1986, Theorem 4.1). The second approach, the *stick-breaking* method, exploits the fact that the Dirichlet has marginal distributions which are Beta-distributed. First, draw $K-1$ random numbers $v_k\sim\text{Beta}\left(\alpha b_k,\alpha\sum_{j=k+1}^{K}b_j\right)$ for $k = 1, 2,\ldots, K-1$. Then the variables $p_k = v_k\prod_{j=1}^{k-1}\left(1 - v_j\right)$ and $p_K = 1 - \sum_{k=1}^{K-1}p_k$ have the desired distribution (Devroye, 1986, Theorem 4.2). This method is quite easily generalized and is used extensively in Bayesian nonparametrics, as we will discuss below.

Alternatively, using the posterior predictive distribution (10.32), we can describe how to draw samples x_n directly without first having to draw the distribution $\boldsymbol{p}$:

$$f\left(x_{N+1} = x\,|\,\boldsymbol{x};\alpha,\boldsymbol{b}\right) = \frac{\alpha b_x}{\alpha + N} + \frac{1}{\alpha + N}\sum_{n=1}^{N}\delta\left[x - x_n\right] \qquad (10.35)$$

Thus, we can formulate the following algorithm, known as a *Polyá urn* method, for drawing samples x_n from this Bayesian model: first, we draw a sample x_1 from the discrete categorical distribution with parameters $\boldsymbol{b}$, the base measure. For the next sample, we either draw from the base measure with probability $\alpha/(\alpha + 1)$, or with probability $1/(\alpha + 1)$ we draw the same value as x_1. The next draw comes either from the base measure (probability $\alpha/(\alpha + 2)$) or from one of either x_1 or x_2 (probability $1/(\alpha + 2)$) and so on. The concentration parameter determines the balance between prior and likelihood: as $\alpha \to \infty$, so the prior base measure dominates and the accumulating sequence of samples does not influence future draws. Conversely, for $\alpha \to 0$, the accumulating samples dominate the choice of future samples entirely.

Since samples will be repeated in this process, the second term in (10.35) is proportional to $\sum_{k=1}^{K}N_k\delta\left[x - k\right]$, which reveals a key property of the Polyá urn: as more samples are drawn, their probability of being selected on subsequent draws depends upon how many previous samples have taken on the same value. Thus, certain values of $x \in 1, 2, \ldots, K$ come to dominate over others, which is known as the *rich-get-richer* property of the Dirichlet-categorical model. Furthermore, the Polyá urn distribution is clearly exchangeable, so by de Finetti's theorem there exists a random parameter vector which by construction we know to be $\boldsymbol{p} \sim \mathrm{Dir}\,(\alpha, \boldsymbol{b})$.

Defining the Dirichlet and related processes

Having described the structure of the conjugate Dirichlet Bayesian model above, an informal way of describing the Dirichlet process (DP) is that it is the infinite-dimensional version, in a way which we make rigorous in this section. By analogy to the Dirichlet distribution being a distribution over distributions on finite, discrete sample spaces, so the DP is a distribution over distributions defined over uncountably infinite spaces Ω. Choose any finite measurable partition $A_1, A_2, \ldots, A_K$ on Ω. Then g is a DP-distributed measure if:

$$(g\,(A_1), g\,(A_2), \ldots, g\,(A_K)) \sim \mathrm{Dir}\,(\alpha, (b\,(A_1), b\,(A_2), \ldots, b\,(A_K)))$$
(10.36)

where b is the base measure, a distribution on the sample space Ω. We write that $g \sim \mathrm{DP}\,(\alpha, b)$. For a technical construction of this process which allow us to create useful probability measures, see Orbanz (2011). This definition states that random vector of elements of the finite partition are distributed according to a Dirichlet distribution, with parameters being the prior base measure of the corresponding element.

Much as with the Dirichlet distribution, it is conjugate to a general distribution, which in uncountable spaces can be defined using the EPDF (1.45) (the uncountable analogue to the categorical PMF). Draws g are distributions from which we can in turn draw samples $x_n \in \Omega$. Assuming the DP prior $f\,(g; \alpha, b) = \mathrm{DP}\,(g; \alpha, b)$ and the EPDF as likelihood, then the posterior must be a DP because (10.36) holds for all possible finite partitions, and the posterior then mirrors the finite case due to the conjugacy of the Dirichlet prior for the categorical distribution (Ferguson, 1973):

$$f\,(g\,|\,\boldsymbol{x}\,;\alpha, b) = \mathrm{DP}\left(g; \alpha + N, \frac{1}{\alpha + N}\left[\alpha b + \sum_{n=1}^{N} \delta_{x_n}\right]\right)$$
(10.37)

where $\delta_{x_n}\,(x) = \delta\,[x - x_n]$ is the Dirac delta function. The posterior measure $b' = \frac{1}{\alpha+N}\left[\alpha b + \sum_{n=1}^{N} \delta_{x_n}\right]$ is an interesting mixture of the prior measure $b\,(x)$ which may be smooth, and a sequence of atoms located at the value of each sample in the data.

As with the discrete case, there is a stick-breaking method for drawing sample paths $g \sim \mathrm{DP}\,(\alpha, b)$ directly, which is $v_k \sim \mathrm{Beta}\,(1, \alpha)$ for $k =$

The 'measurable' stipulation just means that each element of the partition is a subset of the underlying σ-algebra over Ω because we need to be able to define random variables over them.

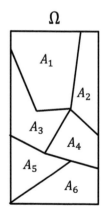

Fig. 10.8: The Dirichlet process (DP) is a stochastic process on a sample space Ω, and draws from the process g are random distributions. It has the property that the measure of any finite partition of the space is Dirichlet-distributed, in the sense that $(g\,(A_1), g\,(A_2), \ldots, g\,(A_K)) \sim \mathrm{Dir}\,(\alpha, (b\,(A_1), b\,(A_2), \ldots, b\,(A_K)))$. The diagram shows a $K = 6$ example.

$1, 2, \ldots$. Then the variables $w_k = v_k \prod_{j=1}^{k-1} (1 - v_j)$ and $x_k \sim b$ define the weights and atom locations of the point process representation of DP draws:

$$g(x) = \sum_{k=1}^{\infty} w_k \delta [x - x_k] \tag{10.38}$$

See Figure 10.9. In practice, this has to be truncated. We can also sample directly using the DP Polyá urn scheme:

$$f(x_{N+1} | \boldsymbol{x}; \alpha, b) = \frac{\alpha}{\alpha + N} b(x_{N+1}) + \frac{1}{\alpha + N} \sum_{n=1}^{N} \delta [x_{N+1} - x_n] \tag{10.39}$$

This discrete Polyá urn distribution invites a purely combinatorial analysis. Drawing samples from (10.39) in turn, we will initially begin with a sample drawn from b. Subsequent samples will either be drawn from b (with probability $\alpha/(\alpha + N)$), or take on a previously sampled value (with probability $1/(\alpha + N)$). Thus, the DP Polyá urn produces samples which are partitioned into K^+ distinct values, x_k for $k = 1, 2, \ldots K^+$. Note that K^+ and each N_k are random variables. So we can describe the urn with the following PDF:

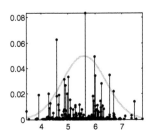

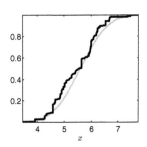

$$f(x_{N+1} | \boldsymbol{x}; \alpha, b) = \frac{\alpha}{\alpha + N} b(x_{N+1}) + \sum_{k=1}^{K^+} \frac{N_k}{\alpha + N} \delta [x_{N+1} - x_k] \tag{10.40}$$

starting with $K^+ = 0$ for the first draw from b. As samples are drawn in sequence, we may draw a new sample from b, then K^+ will increase by one and this new sample becomes available to be chosen again in subsequent draws. The N_k will increase according to which sample has just been drawn.

But given this discrete partitioned description, we can concentrate entirely on the partition element $z_n \in 1, 2, \ldots K^+$ (cluster) to which each draw x_n belongs. This approach leads to the so-called *Chinese restaurant process* (CRP):

$$f(z_{N+1} = k | \boldsymbol{z}; \alpha) = \begin{cases} \frac{N_k}{\alpha + N} & k \in 1, 2, \ldots, K^+ \\ \frac{\alpha}{\alpha + N} & k = K^+ + 1 \end{cases} \tag{10.41}$$

and we write that $\boldsymbol{z} \sim \text{CRP}(\alpha)$. Since the value of K^+ is incremented by one with probability $\alpha/(\alpha + N - 1)$, then the expected number of

Fig. 10.9: Draws from a DP, $g(x)$ (top panel, black spikes) are discrete distributions composed of an infinite, countable number of Dirac deltas drawn from the base measure $b(x)$ (top panel, grey curve, here, a Gaussian with $\mu = 5.6$, $\sigma = 0.8$). The top panel vertical axis represents probability values of the weights w_k. The corresponding sample ECDF, $G_N(x)$ has jumps at each sample value x_n drawn from $g(x)$ (bottom panel, black curve), which, with an infinite number of draws, converges on the base measure CDF $G(x)$ (bottom panel, grey curve).

partition elements is:

$$
\begin{aligned}
E\left[K^+ \mid N, \alpha\right] &= \sum_{n=0}^{N-1} \frac{\alpha}{\alpha + n} \\
&= \alpha\left(\psi\left(\alpha + N\right) - \psi\left(\alpha\right)\right) \qquad (10.42) \\
&\approx \alpha \ln\left(1 + \frac{N}{\alpha}\right) \\
&\approx \alpha \ln N
\end{aligned}
$$

where ψ is the *digamma* function. The approximation on the last line, which holds for large N and α shows that, unlike parametric Bayesian models, the complexity (in terms of the number of parameters) of the DP model is unbounded, growing with the size of the data, but falling short of nonparametric frequentist models (such as kernel regression or kernel density estimation, see next section) whose number of parameters is usually proportional to the size of the data itself. It also demonstrates that the concentration parameter α controls the rate of growth of partition elements. The number of parameters is not the only measure of complexity, of course, but it allows a direct comparison between parametric and nonparametric models here (see also Section *1.5*).

Another important observation is that as (10.41) is exchangeable over the z_n, it depends solely on the sizes of the partitions and the number of samples, not the ordering or specific labelling. Therefore, this is a distribution over permutation-invariant partitions of a set with N elements. Thus, the partition element counts N_k of z_n's form a random partition. It is useful to compute the distribution over this partition. This contains the same information as the joint distribution over the vector of indicator values z, which we can calculate using the CRP distribution (10.41). For N samples, we have terms $\alpha, \alpha+1, \alpha+2$ up to $\alpha+N-1$ in the denominator. Similarly, each sample for which K^+ is incremented contributes a term α to the numerator. Alternatively, for each of the remaining values of $z \in 1, 2, \ldots, K^+$, at the point when the k-th element of the partition is selected, there will be a term $N_k = 0$, and the second time selected, a term $N_k = 1$ and so on. Thus, the joint distribution is:

$$
\begin{aligned}
f\left(z \mid \alpha, N\right) &= f\left(N_1, N_2, \ldots, N_{K^+} \mid \alpha, N\right) = \frac{\alpha^{K^+}}{[\alpha]^N} \prod_{k=1}^{K^+} \left(N_k - 1\right)! \\
&= \frac{\Gamma\left(\alpha\right)}{\Gamma\left(N + \alpha\right)} \alpha^{K^+} \prod_{k=1}^{K^+} \Gamma\left(N_k\right) \qquad (10.43)
\end{aligned}
$$

where we use the rising factorial notation $[a]^n = a\left(a+1\right)\left(a+2\right)\cdots\left(a+n-1\right)$. The last line gives us a way to express the same distribution in terms of gamma functions, which is useful in certain computations. This distribution is an example of an *exchangeable partition probability function* (EPPF).

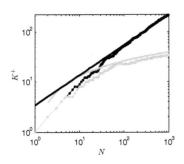

Fig. 10.10: Comparing the number of partition elements K^+ (components) generated by a DP (grey) versus PYP (black) process as a function of the number of samples, N. Whereas the PYP generates a power-law number of partition elements, for the DP, this number grows logarithmically with increasing number of draws. Continuous curves are approximations of the expected number of components, dots are simulated numbers.

While DPs are canonical priors for arbitrary distributions, the number of partition elements grows only logarithmically with N, and there is only one parameter to control this growth. In many situations, the partitioning may need to be much more aggressive, because we need to capture more detail to produce an accurate distributional model, and we may need more degrees of freedom to capture specific situations. The *Pitman-Yor process* (PYP) generalizes the DP with an additional degree of freedom, a *discount factor* β, which allows more control over the rich-get-richer effect of the DP. This controls the extent to which a small number of partition elements can dominate over the rest.

The PYP stick-breaking method for drawing PYP-distributed sample paths $g \sim \mathrm{PYP}(\alpha, \beta, b)$ generalizes the DP method: $v_k \sim \mathrm{Beta}(1 - \beta, \alpha + k\beta)$ for $k = 1, 2, \ldots$ and $0 \le \beta < 1$ and $\alpha > -\beta$. The variables w_k and $x_k \sim b$ define the weights and atom locations of the point process representation of PYP draws are obtained in the same way as for the DP. There is an analogous predictive indicator distribution for the PYP:

$$f\left(z_{N+1} = k \,|\, \boldsymbol{z}; \alpha, \beta\right) = \begin{cases} \frac{N_k - \beta}{\alpha + N} & k \in 1, 2, \ldots, K^+ \\ \frac{\alpha + \beta K^+}{\alpha + N} & k = K^+ + 1 \end{cases} \tag{10.44}$$

and a corresponding EPPF (Lijoi and Prüster, 2010):

$$f\left(\boldsymbol{z} \,|\, \alpha, \beta, N\right) = \frac{\prod_{k=1}^{K^+ - 1}(\alpha + k\beta)}{[\alpha + 1]^{N-1}} \prod_{k=1}^{K^+} [1 - \beta]^{N_k - 1} \tag{10.45}$$

Several special cases can be identified. For $\beta = 0$ we recover the DP. The case $\alpha = 0$ is known as a *stable process*. Finally, the special case $\alpha = 0$ and $\beta = \frac{1}{2}$ is known as the *normalized inverse-Gaussian process* (NIGP). Unfortunately, there are no known closed-form expressions for the f.d.d.s of the PYP except for the NIGP case. As with the CRP, it is possible to compute the expected number of partition elements as a function of N and the PYP parameters (Pitman, 2002):

$$E\left[K^+ \,|\, N, \alpha, \beta\right] \approx \frac{\Gamma(1 + \alpha)}{\beta \Gamma(\beta + \alpha)} N^\beta \tag{10.46}$$

for $N \to \infty$. This shows that the PYP has *power-law* growth in the number of elements with N, which is useful in situations where we expect many more, smaller partition elements than for corresponding DP parameter values (Figure 10.10).

Infinite mixture models (DPMMs)

One of the most valuable applications of the DP (and PYP) described above is *Bayesian nonparametric density estimation* through mixture modelling with an infinite number of components – so called *infinite mixture models*. DP-distributed density functions are discrete with the point process representation (10.7), but in practice density estimation of continuous random variables usually requires the density function to be (at least piecewise) smooth. If we convolve the point density with a *kernel* density function $h(\boldsymbol{x}; \boldsymbol{p})$, we get the infinite mixture model representation:

$$f(\boldsymbol{x}) = \sum_{k=1}^{K} w_k h(\boldsymbol{x}; \boldsymbol{x}_k) \tag{10.47}$$

This representation unifies several distinct kinds of density estimator. For finite K, the weights w_k can simply be identified with those in a finite mixture model, and h are just the mixture component densities. Taking $K \to \infty$, the weights can be obtained using stick-breaking from a DP prior, in which case we obtain a *DP (infinite) mixture model* (DPMM). If $h = \kappa$ is a kernel function and we set $K = N$, the number of data points in a data set with data items $\boldsymbol{x}_n$, then identifying $w_k = 1/N$ we obtain the KDE (4.96). By contrast, and if $K = N$ is finite and but $h(\boldsymbol{x}; \boldsymbol{p}) = \delta[\boldsymbol{x} - \boldsymbol{p}]$ the Dirac delta function, then we get the EPDF (1.45).

We can compare the suitability of these density estimators. The finite mixture model obviously suffers from the problem that we may not know K, the number of mixture components. Similarly, with the KDE, not all data points may have the same importance in the estimator. The EPDF is not continuous. None of these models are fully Bayesian and so prior information about the density cannot be taken into consideration. DPMMs, which we describe in this section, address these limitations.

The primary difficulty with developing a method to infer the DPMM parameters is that the number of parameters is effectively infinite. So, to simplify things, we will start with a modified version of the Gibbs sampler for the finite mixture model (Section 6.2), and then extend this to the infinite case. To do this, we need to provide a prior over the mixture weights $\boldsymbol{\pi} = (\pi_1, \pi_2, \ldots, \pi_K)$, the parameters of the categorical distribution from which the indicator variables z_n are drawn, which we will then integrate out (Figure 10.13). Placing a *symmetric* conjugate Dirichlet over these parameters, $\mathrm{Dir}\left(\frac{\alpha}{K}, \ldots, \frac{\alpha}{K}\right)$ obtains the posterior distribution $f(\boldsymbol{\pi}|\boldsymbol{z}; \alpha) = \mathrm{Dir}\left(\boldsymbol{\pi}; N_1 + \frac{\alpha}{K}, \ldots, N_K + \frac{\alpha}{K}\right)$ where N_k is the number indicators such that $z_n = k$. Marginalizing over the categorical parameters gives us the posterior predictive distribution which is a *Dirichlet-multinomial*:

$$f(\boldsymbol{z}|\alpha) = \frac{\Gamma(\alpha)}{\Gamma(\alpha+N)} \prod_{k=1}^{K} \frac{\Gamma(N_k + \alpha/K)}{\Gamma(\alpha/K)} \tag{10.48}$$

Using the above, we can find the conditional probability of each indicator variable upon all the others:

$$f(z_n = k|\boldsymbol{z}_{-n}, \alpha) = \frac{N_{k,-n} + \alpha/K}{N + \alpha - 1} \tag{10.49}$$

where $\boldsymbol{z}_{-n} = (z_1, \ldots, z_{n-1}, z_{n+1}, \ldots, z_N)$ and $N_{k,-n} = \sum_{n \neq m} \delta[z_n - k]$

for $n, m = 1, 2, \ldots, N$ (this is the number of data points assigned to cluster k, excluding the current data point $\boldsymbol{x}_n$). So, the conditional probability of the indicator variables given the component parameters is:

$$f\left(z_n = k \,|\, \boldsymbol{x}, \boldsymbol{p}, \boldsymbol{z}_{-n}; \alpha\right) = \frac{\left(N_{k,-n} + \alpha/K\right) f\left(\boldsymbol{x}_n; \boldsymbol{p}_k\right)}{\sum_{j=1}^{K} \left(N_{j,-n} + \alpha/K\right) f\left(\boldsymbol{x}_n; \boldsymbol{p}_j\right)} \qquad (10.50)$$

This leads to the two-step (block) Gibbs sampler:

(1) Sample indicators z_n from $f\left(z_n \,|\, \boldsymbol{x}, \boldsymbol{p}, \boldsymbol{z}_{-n}; \alpha\right)$ for each $n = 1, 2, \ldots, N$, and,

(2) Sample component parameters $\boldsymbol{p}_k$ from $f\left(\boldsymbol{p}_k \,|\, \boldsymbol{z}, \boldsymbol{x}; \boldsymbol{q}\right)$ for $k = 1, 2, \ldots, K$.

Other sequential sampling schemes are of course possible. For example, we can run through the indicators z_n in any (randomized) order, and the parameters in any (randomized) order. We could also "interleave" sampling of indicators and component parameters in some way.

The next step is to take $K \to \infty$. The process of doing so converts the mixture model into an (infinite) DPMM, in the following way. The posterior predictive assignment probability (10.49) becomes:

$$f\left(z_n = k \,|\, \boldsymbol{z}_{-n}; \alpha\right) = \frac{N_{k,-n}}{N + \alpha - 1} \qquad (10.51)$$

Consider the corresponding marginal probability of components which have no data points assigned, that is:

$$
\begin{aligned}
f\left(\tilde{\boldsymbol{z}}_{-n} \,|\, \boldsymbol{z}_{-n}; \alpha\right) &= 1 - \sum_{j : N_{j,-n} > 0} \frac{N_{j,-n}}{N + \alpha - 1} \\
&= 1 - \frac{1}{N + \alpha - 1} \sum_{j : N_{j,-n} > 0} N_{j,-n} \qquad (10.52) \\
&= 1 - \frac{N - 1}{N + \alpha - 1} \\
&= \frac{\alpha}{N + \alpha - 1}
\end{aligned}
$$

where $\tilde{\boldsymbol{z}}_{-n}$ denotes all indicator variables which point to cluster numbers not equal to z_n. We can identify (10.52) with the probability of generating a *new* cluster not assigned to any existing data considered, whereas (10.51) is the probability for any existing cluster.

We can now use these posterior predictive probabilities in a CRP-based Gibbs sampler. Assume that, given some finite number N of data points, there are K^+ clusters. Then the probability of the next point $\boldsymbol{x}_{N+1}$ belonging to an existing mixture component is:

$$f\left(z_{N+1} = k \,|\, \boldsymbol{x}, \boldsymbol{p}, \boldsymbol{z}; \alpha\right) \propto N_{k,-n} f\left(\boldsymbol{x}_{N+1} \,|\, \boldsymbol{p}_k\right) \qquad (10.53)$$

for $k \in 1, 2, \ldots, K^+$. We also need the probability of being assigned to a new cluster, but in that situation there are an infinite number of corresponding mixture component parameters, from which we cannot sample. So, we deal with this in the same was as with the indicators: by integrating them out. The marginal probability we need is:

$$f\left(z_{N+1} = K^+ + 1 \,|\, \boldsymbol{x}; \alpha\right) \propto \alpha \int f\left(\boldsymbol{x}_{N+1} \,|\, \boldsymbol{p}\right) f\left(\boldsymbol{p}; \boldsymbol{q}\right) d\boldsymbol{p} \qquad (10.54)$$

where $f\left(\boldsymbol{p}; \boldsymbol{q}\right)$ is the prior over the mixture component parameter $\boldsymbol{p}$. In the simple case that $f\left(\boldsymbol{x}_{N+1} \,|\, \boldsymbol{p}\right)$ and $f\left(\boldsymbol{p}; \boldsymbol{q}\right)$ are conjugate exponential family distributions (see Section 4.5) then the posterior predictive distribution $f\left(\boldsymbol{x}; \boldsymbol{q}\right)$ will have a simple closed form which makes computation of the indicator probabilities straightforward. Similarly, the posterior component parameter distributions are conjugate posteriors; Algorithm 10.2 brings these ingredients together.

Algorithm 10.2 Dirichlet process mixture model (DPMM) Gibbs sampler.

---

(1) *Initialization.* Set number of components $K^+ = 1$, set $z_n = 1$ for all $n = 1, 2, \ldots N$, and set iteration number $i = 0$,

(2) *Start data scan.* Set $n = 1$,

(3) *Indicator sampling.* Sample a new categorical $z_n \in 1, 2, \ldots, K^+ + 1$ using parameters $f\left(z_n = k \,|\, \boldsymbol{x}, \boldsymbol{p}, \boldsymbol{z}; \alpha\right)$ for $k \in 1, 2, \ldots, K^+$ and $f\left(z_n = K^+ + 1 \,|\, \boldsymbol{x}; \alpha\right)$,

(4) *Create new components.* If $z_n = K^+ + 1$, set $K^+ \leftarrow K^+ + 1$, then sample new component parameter $\boldsymbol{p}_{K^+}$ from the single-point posterior $f\left(\boldsymbol{p} \,|\, \boldsymbol{x}_n; \boldsymbol{q}\right) \propto f\left(\boldsymbol{x}_n \,|\, \boldsymbol{p}\right) f\left(\boldsymbol{p}; \boldsymbol{q}\right)$,

(5) *Indicator iteration.* Update $n \leftarrow n + 1$, and if $n \leq N$, return to step 3,

(6) *Component parameter sampling.* Given indicator values above, sample mixture component parameters $\boldsymbol{p}_k$ from component posteriors $f\left(\boldsymbol{p} \,|\, \boldsymbol{x}, \boldsymbol{z}; \boldsymbol{q}\right) \propto \prod_{n:z_n=k} f\left(\boldsymbol{x}_n \,|\, \boldsymbol{p}\right) f\left(\boldsymbol{p}; \boldsymbol{q}\right)$, for $k = 1, 2, \ldots, K^+$,

(7) *Gibbs iteration.* While i sufficiently small, update $i \leftarrow i + 1$, go back to step 2, otherwise exit.

It is important to note that Algorithm 10.2 can obtain $N_{k,-n} = 0$ so that any component k can no longer be assigned any data points. Thus, a reasonable adaptation of the algorithm removes any components for which that occurs.

Comparisons to the GMM Gibbs sampler (Section 6.2) are instructive, which we will do in the univariate Gaussian case for clarity. The existing and new cluster indicator posterior probabilities are:

$$f\left(z_n = k \,|\, \boldsymbol{\mu}, \boldsymbol{z}, \boldsymbol{x}; \mu_0, \sigma_0, \alpha\right) \propto \begin{cases} N_{k,-n}\mathcal{N}\left(x_n; \mu_k, \sigma^2\right) & k \in 1, 2, \ldots, K^+ \\ \alpha\mathcal{N}\left(x_n; \mu_0, \sigma^2 + \sigma_0^2\right) & k = K^+ + 1 \end{cases} \tag{10.55}$$

and the posterior mean parameter distributions are:

$$f\left(\mu_k | \boldsymbol{z}, \boldsymbol{x}; \mu_0, \sigma_0\right) = \mathcal{N}\left(\mu_k; \bar{\mu}_k, \bar{\sigma}_k\right) \tag{10.56}$$

$$f\left(\mu_{K^+} | x_n; \mu_0, \sigma_0\right) = \mathcal{N}\left(\mu_{K^+}; \frac{\sigma^2\mu_0 + \sigma_0^2 x_n}{\sigma^2 + \sigma_0^2},\right.$$
$$\left.\frac{\sigma^2\sigma_0^2}{\sigma^2 + \sigma_0^2}\right) \tag{10.57}$$

with:

$$\bar{\mu}_k = \frac{1}{\sigma^2 + \sigma_0^2 N_k}\left(\sigma^2\mu_0 + \sigma_0^2 \sum_{n:z_n=k} x_n\right)$$

$$\bar{\sigma}_k = \frac{\sigma^2\sigma_0^2}{\sigma^2 + \sigma_0^2 N_k} \tag{10.58}$$

DPMMs can be used pretty much anywhere density estimation is required. For illustration, let us see an example of LPC signal compression (Section 8.3). Hydrological (river flow) data is concatenated, and then segmented into non-overlapping windows of 50 observations. A first order $M = 1$ LPC model is fitted to each window, and this produces a set of coefficients. It would be useful to find a simple model for the distribution of these coefficients, so that a single coefficient index per window could be used in the data compression, replacing the need for storing e.g. a full, 4-byte floating-point representation of each coefficient. With $K^+ = K = 5$ clusters, DPMM is able to find a much more faithful match to the KDE (which requires all the data) than the GMM (see Figure 10.11) This means that at most 3 bits (instead of 32 bits) suffices for each window to perform effective (but lossy) compression of the coefficient representation.

DPMMs produce compact distribution estimates where the number of components grows slowly with N. For situations where more "aggressive" growth in complexity is expected, it is straightforward to adapt Algorithm 10.2 by replacing $N_{k,-n}$ with $N_{k,-n} - \beta$ in (10.53) and α with $\alpha + \beta K^+$ in (10.54) to obtain the *Pitman-Yor mixture model* (PYMM).

DPMM Gibbs samplers are simple to implement, yet they will often require a large number of iterations to converge, so they do not lend themselves to embedded DSP applications. A simpler, more tractable algorithm is known as *DP-means*, by analogy with K-means (Kulis and Jordan, 2012), applicable in the multivariate Gaussian case (Algorithm 10.3). The derivation of this algorithm follows that of obtaining K-means from the GMM Gibbs sampler (Section 6.5). First, assume that the prior over the mean is spherical, zero-mean Gaussian, $f(\boldsymbol{p}; \boldsymbol{q}) = f(\boldsymbol{\mu}; \rho) = \mathcal{N}(\mathbf{0}, \rho\mathbf{I})$ where $\rho > 0$ is a prior scaling parameter. Also, assume that the component likelihoods are spherical Gaussian, $f(\boldsymbol{x}|\boldsymbol{p}) = f(\boldsymbol{x}|\boldsymbol{\mu}) = \mathcal{N}(\boldsymbol{\mu}, \sigma\mathbf{I})$, with $\sigma > 0$ another scaling parameter. The assignment probabilities become:

$$f(z_n = k | \boldsymbol{x}, \boldsymbol{\mu}, \boldsymbol{z}; \sigma, \alpha) \propto \frac{N_{k,-n}}{(2\pi\sigma^2)^{D/2}} \exp\left(-\frac{1}{2\sigma^2}\|\boldsymbol{x}_n - \boldsymbol{\mu}_k\|_2^2\right) \tag{10.59}$$

for existing components and:

$$f(z_n = K^+ + 1 | \boldsymbol{x}; \sigma, \rho, \alpha) \quad \propto \quad \frac{\alpha}{(2\pi(\rho^2 + \sigma^2))^{D/2}} \times \tag{10.60}$$
$$\exp\left(-\frac{1}{2(\rho^2 + \sigma^2)}\|\boldsymbol{x}_n\|_2^2\right)$$

for new components. We will also write α as a function of both σ and ρ:

$$\alpha = \left(1 + \frac{\rho^2}{\sigma^2}\right)^{D/2} \exp\left(-\frac{\lambda}{2\sigma^2}\right) \tag{10.61}$$

where $\lambda > 0$ is a third parameter. This expression may seem rather contrived – indeed it is – but in this form, the new component assignment probability can be written as:

$$f(z_n = K^+ + 1 | \boldsymbol{x}; \sigma, \rho, \alpha) \propto \exp\left(-\frac{1}{2\sigma^2}\left[\lambda + \frac{\sigma^2}{\rho^2 + \sigma^2}\|\boldsymbol{x}_n\|_2^2\right]\right) \tag{10.62}$$

Now, as with K-means, we will take $\sigma^2 \to 0$, the consequence of which is that, much like K-means, Gibbs samples for z_n become direct assignments of observations $\boldsymbol{x}_n$ to their current closest components, *unless* λ is smaller than all of these distances, in which case $z_n = K^+ + 1$. As with K-means, the component mean posteriors collapse onto the mean for all the data assigned to each cluster, and the posterior for the new component mean is just the observation $\boldsymbol{x}_n$.

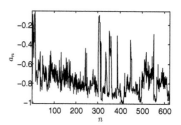

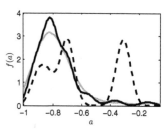

Fig. 10.11: Parametric and nonparametric density estimation of LPC coefficients a_n (top panel) for a first-order ($M = 1$) of 50-sample windows of hydrological (river flow rate) data (windows indexed $n = 1, 2, \ldots, N$). The Gaussian DP mixture for these coefficients (bottom panel, black curve, $\alpha = 10^{-2}$, $\sigma = 0.05$, $\sigma_0 = 0.1$ and μ_0 set to the median of the coefficient data, fit using Gibbs) more closely approximates the full kernel density estimate (using all the data) with only $K^+ = 5$ components. By contrast, a finite $K = 5$ Gaussian mixture does not produce nearly as good fit (bottom panel, dashed black curve).

Algorithm 10.3 Dirichlet process means (DP-means) algorithm.

(1) *Initialization.* Set number of components $K^+ = 1$, set $z_n^0 = 1$ for all $n = 1, 2, \ldots N$, set $\boldsymbol{\mu}_1^0 = \frac{1}{N} \sum_{n=1}^N \boldsymbol{x}_n$, and set iteration number $i = 0$,

(2) *Start data scan.* Set $n = 1$,

(3) *Update indicators.* Set the *distance vector* $\boldsymbol{d} = \left(\left\| \boldsymbol{x}_n - \boldsymbol{\mu}_1^i \right\|_2^2, \left\| \boldsymbol{x}_n - \boldsymbol{\mu}_2^i \right\|_2^2, \ldots, \left\| \boldsymbol{x}_n - \boldsymbol{\mu}_{K^+}^i \right\|_2^2, \lambda \right)$, then let $z_n^{i+1} = \arg\min_{j \in 1,2,\ldots,K^++1} d_j$,

(4) *Create new clusters.* If $z_n^{i+1} = K^+ + 1$, set $K^+ \leftarrow K^+ + 1$, then set new cluster parameter $\boldsymbol{\mu}_{K^+}^i = \boldsymbol{x}_n$,

(5) *Indicator iteration.* Update $n \leftarrow n + 1$, and if $n \leq N$, return to step 3,

(6) *Update component means.* Calculate $\boldsymbol{\mu}_k^{i+1} = \frac{1}{N_k} \sum_{n:z_n^{i+1}=k} \boldsymbol{x}_n$ for $k = 1, 2, \ldots, K^+$,

(7) *Convergence check, iteration.* If K^+ has increased from the previous iteration, or if any $z_n^i \neq z_n^{i+1}$, update $i \leftarrow i + 1$, go back to step 2, otherwise exit.

Kulis and Jordan (2012) show that the DP-means algorithm cannot increase the following error function:

$$E = \sum_{k=1}^{K^+} \sum_{n:z_n=k} \left\| \boldsymbol{x}_n - \boldsymbol{\mu}_k \right\|_2^2 + \lambda K^+ \tag{10.63}$$

which is just the K-means objective (6.63) but with an additional term λK^+ which acts as a regularizer, penalizing overly complex models (this is very similar in construction to AIC, MDL or BIC, see Section 4.2). Combine the lack of increase of (10.63) with the finite (but extremely large!) number of possible clustering arrangements, to find that the iteration must reach a fixed point. A limitation is that this fixed point depends entirely upon the ordering of the data, by contrast, K-means also depends upon the choice of the initial values of the cluster centroids $\boldsymbol{\mu}_k$. DP-means is quite fast: as with K-means, we require $O\left(N K^+\right)$ operations per iteration, and since $K^+ \approx \alpha \ln N$ it follows that this tends towards $O\left(N \ln N\right)$ operations per iteration (for large α and N). DP-means lends itself to large scale clustering problems in DSP (see Figure 10.12 for an example signal processing application from astronomy) but it is often outperformed by somewhat more complex algorithms (Raykov *et al.*, 2016a) due, for example, to the limitations of the choice of Gaussian mixture components.

DP-means is best understood as a computationally tractable approach to "infinite clustering", but infinite mixture modelling needs a similarly tractable algorithm. In these cases, *maximum a-posteriori DP* (MAP-DP) is a practical alternative to Gibbs sampling. The central concept is the use of iterative conditional modes (ICM) (Algorithm 5.2), that is, rather than sample from conditional distributions in the graphical model, ICM uses the MAP value of each distribution. This is guaranteed not to decrease the joint likelihood for the graphical model, and thus leads to a simple, deterministic approach to inference.

To develop this algorithm, we start with the DP Gibbs sampler (Algorithm 10.2) and collapse out the component parameters p, to obtain the prior and posterior predictive distributions for an observation x (Figure 10.13). This variable collapsing has several advantages. Firstly, it simplifies the modelling: we only need to figure out the predictive density function. Secondly, it simplifies the MAP estimation, since maximization is only required over the conditional probabilities of the indicators. This maximization is solved by brute-force computation over all possible values of $k \in 1, 2, \ldots, K^+ + 1$, and we do not need an analytical expression for the posterior mode of the component parameters. Thirdly, as we will see, the typical number of iterations to reach convergence of the algorithm is often substantially reduced.

Since we are using exponential family distributions, the prior and posterior distributions over the observation parameters p will have the same form. All the observations assigned to component k but excluding observation x_n is denoted $x_{k,-n}$. After this collapsing, only the posterior distributions for the indicators need to be sampled:

$$f(z_n = k \,|\, x, z \; \alpha, q) \propto \begin{cases} N_{k,-n} f(x_n \,|\, x_{k,-n} \; ; q) & k \in 1, 2, \ldots, K^+ \\ \alpha f(x_n; q) & k = K^+ + 1 \end{cases}$$

(10.64)

Replacing the Gibbs sampling steps with MAP steps, and replacing the MAP steps with minimized negative log posterior steps, we get MAP-DP, Algorithm 10.4.

When the component distributions are chosen appropriately for the particular problem, MAP-DP tends to perform very well (Raykov *et al.*, 2016*a*) by comparison to DP-means (which is less accurate) and Gibbs sampling (which is orders of magnitude more computationally intensive), see Figure 10.14. There are some simple improvements to make to the MAP-DP implementation given here, in particular, the count of cluster sizes $N_{k,-n}$ and the Bayesian hyperparameter updates $q_{k,-n}$ only need to change when an assignment for a particular component changes. The negative log posterior predictive distributions have a particular simple and elegant form using only the log normalizer function when written in sufficient statistic form (see Section 4.5), and this can be exploited to simplify the algorithm implementation even further (Raykov *et al.*, 2016*b*).

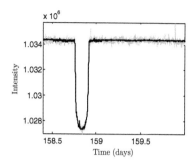

Fig. 10.12: Using Dirichlet process means (DP-means) to perform noise reduction in an astronomical problem: Kepler satellite exoplanet transit light (intensity) curves (grey). The input light curve is broken up into non-overlapping windows of 1/4 the transit period (2.21 days, one transit period only shown here for clarity). Each of the resulting $N = 60$ windows forms a set of $D = 807$ dimensional observation vectors x_n and these are clustered using DP-means with $\lambda = 2 \times 10^9$, which converges in two iterations with the final $K^+ = 2$. Rearranging the means μ_k in the sequence μ_{z_n} allows reconstruction of the de-noised light curves (black).

Algorithm 10.4 Maximum a-posteriori Dirichlet process mixture collapsed (MAP-DP) algorithm for conjugate exponential family distributions.

(1) *Initialization.* Set number of components $K^+ = 1$, set $z_n^0 = 1$ for all $n = 1, 2, \ldots N$, and set iteration number $i = 0$,

(2) *Start data scan.* Set $n = 1$,

(3) *Component conjugate hyperparameter Bayesian updates.* For $k \in 1, 2, \ldots, K^+$, given all observations assigned to cluster k excluding observation $\boldsymbol{x}_n$,

(4) *Update indicators.* Set the *distance vector* $d_k = -\ln f(\boldsymbol{x}_n \,|\, \boldsymbol{x}_{k,-n}\,; \boldsymbol{q}) - \ln N_{k,-n}$ for $k \in 1, 2, \ldots, K^+$ and $d_{K^++1} = -\ln f(\boldsymbol{x}_n; \boldsymbol{q}) - \ln \alpha$, then let $z_n^{i+1} = \arg\min_{k \in 1, 2, \ldots, K^++1} d_k$,

(5) *Create new clusters.* If $z_n^{i+1} = K^+ + 1$, set $K^+ \leftarrow K^+ + 1$,

(6) *Indicator iteration.* Update $n \leftarrow n + 1$, and if $n \leq N$, return to step 3,

(7) *Convergence check, iteration.* If K^+ has increased from the previous iteration, or if any $z_n^i \neq z_n^{i+1}$, update $i \leftarrow i + 1$, go back to step 2, otherwise exit.

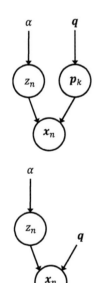

Fig. 10.13: The PGM for the generic Bayesian mixture model (finite and infinite DP cases) after integrating out the mixture (or DP stick-breaking) weights (top) and then also after integrating out the component parameters (bottom).

As with DP-means, we can write down an objective function, this is the NLL for the model with the component parameters integrated out (Raykov *et al.*, 2016a):

$$E = -\sum_{k=1}^{K^+} \sum_{n:z_n=k} \ln f(\boldsymbol{x}_n \,|\, \boldsymbol{x}_{k,-n}\,; \boldsymbol{q}) - K^+ \ln \alpha - \sum_{k=1}^{K^+} \ln \Gamma(N_k) + C(\alpha, N)$$
(10.65)

The second and third terms are just the negative log of the particular clustering configuration (10.43), and the final term can be ignored as it does not vary with any adjustable parameters during iteration.

Any model which collapses out the component parameters (e.g. MAP-DP) does not provide estimates of these at convergence, but computing these is straightforward. All that is needed is to solve the MLE for these parameters given the assignments at convergence for each cluster separately. For example, in the Gaussian case, the mean vector $\hat{\boldsymbol{\mu}}_k$ and covariance $\hat{\boldsymbol{\Sigma}}_k$ of each cluster $k \in 1, 2, \ldots, K^+$ can be estimated from the data partitioned according to the converged indicators $\boldsymbol{z}$ (see for example Figure 10.14).

Can DP-based models actually infer the number of components?

The short, simple answer is "no". The reason is that infinite DP mixtures have $\alpha \ln N$ components which is always increasing with N. So, for some data generated using a finite K component generating model, clearly the DPMM cannot consistently estimate K as N increases. Thus, DPMMs and other models based on the DP, are not suitable as estimators for K, in, for example, fixed clustering applications. Instead, it is better to think of DPMMs as "simple" models for a given N, where the number of components grows *sub-linearly* with N. By contrast KDEs are much more complex, since they grow in size as $O(N)$. At the same time, finite mixtures do not grow in complexity as N increases, and this is only appropriate if the structure of the data stays fixed as N increases. For general data where the underlying complexity is not known in advance, DPMMs tend to outperform finite mixtures and, unlike KDEs, represent a compression of the input data which is sensitive to the underlying, but unknown detail in the data. This is a common feature of most Bayesian nonparametric models: they have infinite dimensionality growing only slowly with the data observed and for any finite set of observations, this makes them very useful and efficient techniques for practical applications.

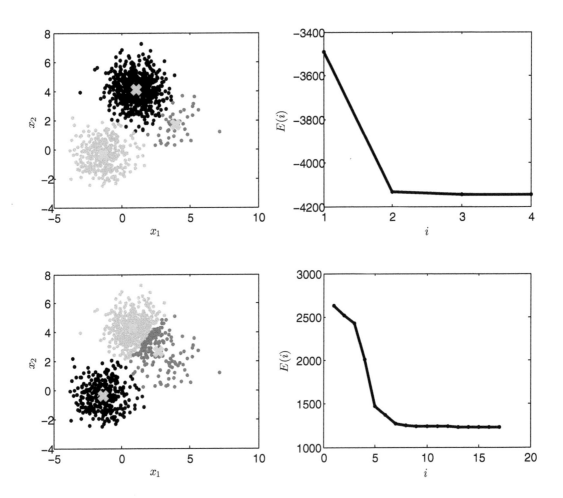

Fig. 10.14: MAP-DP (top panels) versus DP-means (bottom panels) for nonparametric DP clustering. Despite the $K = 3$ synthetic clusters here being spherical Gaussian (exactly matching the implicit assumptions of DP-means), DP-means cannot accurately identify the smallest cluster (grey) whereas MAP-DP, which has more flexible cluster models, can. MAP-DP converges much more quickly than DP-means, mostly because the component parameters are collapsed out. $E(i)$ refers to the objective function for each at iteration i.

Bibliography

Aji, S.M. and McEliece, R.J. (2000). The Generalized Distributive Law. *IEEE Transactions on Information Theory*, **46**(2).

Akaike, H. (1974). A new look at the statistical model identification. *IEEE Transactions on Automatic Control*, **19**(6), 716–723.

Arce, Gonzalo R. (2005). *Nonlinear signal processing: a statistical approach*. Wiley-Interscience, Hoboken, N.J.

Arlot, Sylvain and Celisse, Alain (2010). A survey of cross-validation procedures for model selecion. *Statistics Surveys*, **4**, 40–79.

Berger, Toby and Gibson, Jerry D. (1998). Lossy source coding. *IEEE Transactions on Information Theory*, **44**(6), 2693–2723.

Bernardo, J.M. and Smith, A.F.M. (2000). *Bayesian Theory*. John Wiley & Sons, Chichester.

Bertsekas, Dimitri P. (1995). *Nonlinear programming*. Athena Scientific, Belmont, Mass.

Bishop, Christopher M. (2006). *Pattern recognition and machine learning*. Information Science and Statistics. Springer, New York.

Bollobás, Béla (1998). *Modern graph theory*. Graduate Texts in Mathematics; 184. Springer, New York.

Bormin, Huang and Jing, Ma (2007). On asymptotic solutions of the Lloyd-Max scalar quantization. In *Information, Communications and Signal Processing, 2007 6th International Conference on*, pp. 1–6.

Borß, C. and Martin, R. (2012). On the construction of window functions with constant-overlap-add constraint for arbitrary window shifts. In *2012 IEEE International Conference on Acoustics, Speech and Signal Processing (ICASSP)*, Kyoto, Japan. IEEE.

Boyd, Stephen P. and Vandenberghe, Lieven (2004). *Convex optimization*. Cambridge University Press, Cambridge, UK; New York.

Brandenburg, Karlheinz (1999). MP3 and AAC Explained. In *Audio Engineering Society Conference: 17th International Conference: High-Quality Audio Coding*, Florence, Italy. Audio Engineering Society.

Byrd, R. H. and Payne, D. A. (1979). Convergence of the interatively reweighted least squares algorithm for robust regression.

Candès, E.J. (2006). Modern statistical estimation via oracle inequalities. *Acta Numerica*, **15**, 1–69.

Candès, E.J. and Wakin, M.B. (2008). An introduction to compressive sampling. *IEEE Signal Processing Magazine*, **25**(2), 21–30.

Casella, George (1985). An introduction to empirical Bayes data analysis. *The American Statistician*, **39**(2), 83–87.

Cavanaugh, J.E. and Neath, A.A. (1999). Generalizing the derivation of the Schwarz information criterion. *Communications in Statistics - Theory and Methods*, **28**(1), 49–66.

Cheng, Y.Z. (1995). Mean shift, mode seeking, and clustering. *IEEE Transactions on Pattern Analysis and Machine Intelligence*, **17**, 790–799.

Chib, Siddhartha and Greenberg, Edward (1995). Understanding the Metropolis-Hastings algorithm. *The American Statistician*, **49**(4), 327–335.

Chowdhury, R.A., Kaykobad, M., and King, I. (2002). An efficient decoding technique for Huffman codes. *Information Processing Letters*, **81**, 305–308.

Cortes, C. and Vapnik, V. (1995). Support-vector networks. *Machine Learning*, **20**(3), 273–297.

Cover, T. M. and Thomas, Joy A. (2006). *Elements of Information Theory* (2nd edn). Wiley-Interscience, Hoboken, N.J.

Cowles, M.K. and Carlin, B.P. (1996). Markov Chain Monte Carlo convergence diagnostics: A comparative review. *Journal of the American Statistical Association*, **91**(434), 883–904.

Craven, P.G. and Gerzon, M.A. (1992). Compatible improvement of 16 bit systems using subtractive dither. *Audio Engineering Society Conference Proceedings*.

Dasgupta, Sanjoy, Papadimitriou, Christos H., and Vazirani, Umesh Virkumar (2008). *Algorithms*. McGraw-Hill Higher Education, Boston.

Daubechies, I. (1992). *Ten lectures on wavelets*. Society for Industrial and Applied Mathematics, Philadelphia, USA.

DeFatta, D.J., Lucas, J.G., and Hodgkiss, W.S. (1988). *Digital signal processing: a system design approach*. Wiley.

Dempster, A.P., Laird, N.M., and Rubin, D.B. (1977). Maximum likelihood from incomplete data via the EM algorithm. *Journal of the Royal Statistical Society: Series B*, **39**, 1–38.

Devroye, Luc (1986). *Non-uniform random variate generation*. Springer-Verlag, New York. Luc Devroye. 25 cm. Includes index.

Dummit, David Steven and Foote, Richard M. (2004). *Abstract algebra* (3rd edn). Wiley, Hoboken, NJ.

Dunn, J.C. (1973). A fuzzy relative of the ISODATA process and its use in detecting compact well-separated clusters. *Journal of Cybernetics*, **3**(3), 32–57.

Ferguson, T.S. (1973). A Bayesian analysis of some nonparametric problems. *Annals of Statistics*, **1**(2), 209–230.

Gelman, Andrew (2004). *Bayesian data analysis* (2nd edn). Texts in statistical science. Chapman & Hall/CRC, Boca Raton, Fla.

Geweke, J. (1991). Evaluating the accuracy of sampling-based approaches to the calculation of posterior moments. Technical Report 148, Federal Reserve Bank of Minneapolis.

Gilchrist, Warren (2000). *Statistical modelling with quantile functions*. Chapman and Hall/CRC, Boca Raton. Warren G. Gilchrist. ill. ; 24 cm.

Gondzio, J. (2012). Interior point methods 25 years later. *European Journal of Operational Research*, **218**, 587–601.

Görür, D. and Teh, Y.W. (2011). Concave-convex adaptive rejection sampling. *Journal of Computational and Graphical Statistics*, **20**(3), 670–691.

Gray, R.M. and Neuhoff, D.L. (1998). Quantization. *IEEE Transactions on Information Theory*, **44**(6), 1–63.

Grimmett, Geoffrey and Stirzaker, David (2001). *Probability and random processes* (3rd edn). Oxford University Press, Oxford; New York.

Hairer, E., Lubich, Christian, and Wanner, Gerhard (2006). *Geometric numerical integration: structure-preserving algorithms for ordinary differential equations* (2nd edn). Springer, Berlin ; New York. Ernst Hairer, Christian Lubich, Gerhard Wanner. ill. ; 25 cm. Springer series in computational mathematics, 31.

Hand, David J. (2006). Classifier technology and the illusion of progress. *Statistical Science*, **21**(1), 1–14.

Hansen, Mark H. and Yu, Bin (2001). Model selection and the principle of minimum description length. *Journal of the American Statistical Association*, **96**(454), 746–774.

Hastie, Trevor, Tibshirani, Robert, and Friedman, J. H. (2009). *The elements of statistical learning : data mining, inference, and prediction* (2nd edn). Springer, New York, NY. Trevor Hastie, Robert Tibshirani, Jerome Friedman. ill. (some col.) ; 25 cm. Springer series in statistics,.

Henle, Michael (1994). *A combinatorial introduction to topology*. Dover, New York.

Hoffman, Matthew D. and Gelman, Andrew (2014). The No-U-Turn sampler: Adaptively setting path lengths in Hamiltonian Monte Carlo. *The Journal of Machine Learning Research*, **15**.

Horst, Reiner, Pardalos, P. M., and Thoai, Nguyen V. (2000). *Introduction to global optimization* (2nd edn). Nonconvex optimization and its applications. Kluwer Academic Publishers, Dordrecht ; Boston.

Humphreys, J. F. (1996). *A course in group theory*. Oxford Science Publications. Oxford University Press, Oxford; New York.

Jain, A.K. (2010). Data clustering: 50 years beyond K-means. *Pattern Recognition Letters*, **31**(8), 651–666.

Jiang, K., Kulis B. and Jordan, M.I. (2012). Small-variance asymptotics for exponential family Dirichlet process mixture models. In *Advances in Neural Information Processing Systems*, pp. 3158–3166.

Kaye, Richard and Wilson, Robert (1998). *Linear algebra*. Oxford University Press, Oxford; New York.

Kemeny, John G. and Snell, J. Laurie (1976). *Finite Markov chains*. Springer-Verlag, New York.

Kim, S.-J., Koh, K., Boyd, S., and Gorinevsky, D. (2009). L1 trend filtering. *SIAM Review*, **51**(2), 339–360.

Koh, K., Kim, S.-J., and Boyd, S. (2007). An interior-point method for large-scale L1-regularized logistic regression. *Journal of Machine Learning Research*, **8**, 1519–1555.

Kulis, B. and Jordan, M.I. (2012). Revisiting K-means: New algorithms via Bayesian nonparametrics. In *ICML 2012: Proceedings of the 29th International Conferencce on Machine Learning*, Edinburgh, Scotland, pp. 1131–1138. Omnipress.

Kurihara, K. and Welling, M. (2009). Bayesian K-means as a 'maximization-expectation' algorithm. *Neural Computation*, **21**(4), 1145–1172.

Landsman, Zinoviy M. and Valdez, Emiliano A. (2003). Tail conditional expectations for elliptical distributions. *North American Actuarial Journal*, **7**(4), 55–71.

Lawrence, N.D. (2004). Gaussian process latent variable models for visualisation of high dimensional data. In *Advances in Neural Information Processing Systems (NIPS)*, pp. 329–336.

Lighthill, M.J. (1958). *An Introduction to Fourier Analysis and Generalised Functions*. Cambridge University Press.

Lijoi, A. and Prüster, I. (2010). Models beyond the Dirichlet process. In *Bayesian Nonparametrics* (ed. N. Hjort, C. Holmes, P. Müller, and S. Walker), pp. 80–136. Cambridge University Press, Cambridge.

Little, M.A. and Jones, N.S. (2011*a*). Generalized methods and solvers for noise removal from piecewise constant signals. I. Background theory. *Proceedings of the Royal Society A: Mathematical, Engineering and Physical Sciences*, **467**(2135).

Little, M.A. and Jones, N.S. (2011*b*). Generalized methods and solvers for noise removal from piecewise constant signals. II. New methods. *Proceedings of the Royal Society A: Mathematical, Engineering and Physical Sciences*, **467**(2135).

Lloyd, Stuart (1982). Least squares quantization in PCM. *IEEE Transactions on Information Theory*, **28**(2), 129–137.

Makhoul, J. (1975). Linear prediction: a tutorial review. *Proceedings of the IEEE*, **63**(4), 561–580.

Makhoul, J., Roucos, S., and Gish, H. (1985). Vector quantization in speech coding. *Proceedings of the IEEE*, **73**(11), 1551–1588.

Mallat, S. G. (2009). *A wavelet tour of signal processing: the sparse way* (3rd edn). Elsevier/Academic Press, Amsterdam ; Boston.

Mckay, David J.C. (2003). *Information Theory, Inference and Learning Algorithms* (4th edn). Cambridge University Press.

Murphy, K.P. (2012). *Machine Learning: A Probabilistic Perspective.* MIT Press, Cambridge, MA.

Neal, Radford M. (2003). Slice sampling. *Annals of statistics*, 705–741.

Nesterov, I. U. E. (2004). *Introductory lectures on convex optimization: a basic course.* Applied optimization; v. 87. Kluwer Academic Publishers, Boston.

Nocedal, Jorge and Wright, Stephen J. (2006). *Numerical optimization* (2nd edn). Springer series in operations research. Springer, New York.

Ohlsson, H., Gustafsson, F., Ljung, L., and Boyd, S. (2010). State smoothing by sum-of-norms regularization. In *49th IEEE Conference on Decision and Control (CDC)*.

O'Neill, B. (2009). Exchangeability, correlation, and Bayes' effect. *International Statistical Review*, **77**(2), 241–250.

Orbanz, Peter (2009). Functional conjugacy in parametric Bayesian models. Technical report, Cambridge University.

Orbanz, Peter (2011). Projective limit random probabilities on Polish spaces. *Electronic Journal of Statistics*, **5**, 1354–1373.

Osborne, M.A., Garnett, R., and Roberts, S.J. (2009). Gaussian processes for global optimization. In *3rd International Conference on Learning and Intelligent Optimization (LION3)*, pp. 1–15.

Pei, Soo-Chang and Chang, Kuo-Wei (2016). Optimal discrete Gaussian function: the closed-form functions satisfying Tao's and Donoho's uncertainty principle with Nyquist bandwidth. *IEEE Transactions on Signal Processing*, **64**(12).

Pei, Soo-Chang and Tseng, Chien-Cheng (1998). A comb filter design using fractional-sample delay. *IEEE Transactions on Circuits and Systems II: Analog and Digital Signal Processing*, **45**(5), 649–653.

Pelleg, D. and Moore, A.W. (2000). X-means: Extending K-means with efficient estimation of the number of clusters. In *ICML '00: Proceedings of the Seventeenth International Conference on Machine Learning*, Volume 1, Stanford, California, USA.

Petersen, K.B. and Pedersen, M.S. (2008). The matrix cookbook. Technical report, Technical University of Denmark.

Pitman, J. (2002). Combinatorial stochastic processes. Technical Report 621, University of California.

Platt, J.C. (1999). Fast training of support vector machines using sequential minimal optimization. *Advances in Kernel Methods*, 185–208.

Powell, M. J. D. (1976). Some global convergence properties of a variable metric algorithm for minimization without exact line search. In *Nonlinear Programming: Proceedings of a Symposium in Applied Mathematics of the AMS and SIAM*, Volume 9, New York.

Press, William H. (1992). *Numerical recipes in C : the art of scientific computing* (2nd edn). Cambridge University Press, Cambridge ; New York.

Priestley, H.A. (2003). *Introduction to Complex Analysis.* Oxford University Press, Oxford, UK.

Proakis, John G. and Manolakis, Dimitris G. (1996). *Digital signal processing: Principles, algorithms and applications.* Prentice-Hall, Upper Saddle River, NJ, US.

Puschel, M. (2003). Cooley-tukey FFT like algorithms for the DCT. In *2003 IEEE International Conference on Acoustics, Speech and SIgnal Processing*, Hong Kong, China. IEEE.

Puschel, M. and Moura, J.M.F. (2008). Algebraic signal processing theory: Cooley-Tukey type algorithms for DCTs and DSTs. *IEEE Transactions on Signal Processing*, **56**(4), 1502–1521.

Rabiner, L.R. and Schafer, R.W. (2007). Introduction to digital speech processing. *Foundations and Trends in Signal Processing*, **1**(1), 1–194.

Rasmussen, C.E. and Williams, C.K.I. (2006). *Gaussian Processes for Machine Learning.* MIT Press.

Raykov, Y.P., Boukouvalas, A., Baig, F., and Little, M.A. (2016a). What to do when K-means clustering fails: a simple yet principled alternative algorithm. *PLoS One*, **11**(9), e0162259.

Raykov, Y.P., Boukouvalas, A., and Little, M.A. (2016b). Simple approximate MAP inference for Dirichlet processes mixtures. *Electronic Journal of Statistics*, **10**, 3548–3578.

Riedel, K.S. and Sidorenko, A. (1995). Minimum bias multiple taper spectral estimation. *IEEE Transactions on Signal Processing*, **43**(1), 188–195.

Robert, Christian P. and Casella, George (1999). *Monte Carlo statistical methods.* Springer texts in statistics. Springer, New York. Christian P. Robert, George Casella. ill. ; 24 cm.

Roberts, G.O. and Smith, A.F.M. (1994). Simple conditions for the convergence of the Gibbs sampler and Metropolis-Hastings algorithms. *Stochastic Processes and their Applications*, **49**, 207–216.

Roberts, S., Osborne, M., Ebden, M., Reece, S., Gibson, N., and Aigrain, S. (2012). Gaussian processes for time-series modelling. *Philosophical Transactions of the Royal Society A: Mathematical, Physical and Engineering Sciences*, **371**(1984).

Rosset, S. and Zhu, J. (2007). Piecewise linear regularized solution paths. *The Annals of Statistics*, **35**(3), 1012–1030.

Rotman, Joseph J. (2000). *A first course in abstract algebra* (2nd edn). Prentice Hall, Upper Saddle River, N.J.

Roweis, S.T. and Saul, L.K. (2000). Nonlinear dimensionality reduction by locally linear embedding. *Science*, **290**(5500), 2323–2326.

Schwarz, G. (1978). Estimating the dimension of a model. *Annals of Statistics*, **6**(2), 461–464.

Scott, David W. (1992). *Multivariate density estimation: theory, practice and visualization.* Wiley.

Shawe-Taylor, J. and Christianini, N. (2004). *Kernel Methods for Pattern Analysis*. Cambridge University Press, New York.

Sherlock, B.G. and Kakad, Y.P. (2002). MATLAB programs for generating orthonormal wavelets. In *Advances in Multimedia, Video and Signal Processing Systems*, pp. 204–208. World Scientific and Engineering Society Press.

Shumway, Robert H. and Stoffer, David S. (2016). *Time Series Analysis and Its Applications* (4th edn). Springer.

Silverman, B.W. (1998). *Density estimation for statistics and data analysis*. Chapman and Hall/CRC, New York.

Smith, J.O. (2010). *Physical audio signal processing*. W3K Publishing.

Stoica, P. and Moses, R. (2005). *Spectral Analysis of Signals*. Prentice-Hall, Upper Saddle River, NJ.

Stone, M. (1977). An asymptotic equivalence of choice of model by cross-validation and Akaike's criterion. *Journal of the Royal Statistical Society: Series B*, **39**(1), 44–47.

Sugiyama, Masashi, Krauledat, Matthias, and Müller, Klaus-Robert (2007). Covariate shift adaptation by importance weighted cross validation. *Journal of Machine Learning Research*, 985–1005.

Sutherland, W. A. (2009). *Introduction to metric and topological spaces* (2nd edn). Oxford University Press, Oxford.

Thomson, D.J. (1982). Spectrum estimation and harmonic analysis. *Proceedings of the IEEE*, **70**(9), 1055–1096.

Tibshirani, R.J. and Taylor, J. (2011). The solution path of the generalized lasso. *The Annals of Statistics*, **39**(3), 1335–1826.

Tipping, M.E. and Bishop, C.M. (1999). Probabilistic principal component analysis. *Journal of the Royal Statistical Society: Series B*, **61**(3), 611–622.

Tropp, Joel A., Laska, Jason N., Duarte, Marco F., and Romberg, Justin K. (2010). Beyond Nyquist: Efficient sampling of sparse bandlimited signals. *IEEE Transactions on Information Theory*, **56**(1), 520–544.

Tsochantaridis, I., Joachims, T., Hofmann, T., and Altun, Y. (2005). Large margin methods for structured and interdependent output variables. *Journal of Machine Learning Research*, 1453–1484.

Unser, Michael (1999). Splines: A perfect fit for signal and image processing. *IEEE Signal Processing Magazine*, **16**(6), 22–38.

Vaidyanathan, P.P. (2007). *The Theory of Linear Prediction*. Morgan and Claypool.

van der Maaten, L.J.P., Postma, E.O., and van den Herik, H.J. (2009). Dimensionality reduction: A comparative review. Technical Report TR2009-005, Tilburg University.

Vedaldi, A. and Soatto, S. (2008). Quick shift and kernel methods for mode seeking. In *European Conference on Computer Vision*, pp. 705–718.

Vidakovic, B. (1999). *Statistical Modelling by Wavelets*. John Wiley and Sons, New York.

Wahba, G. (1980). Automatic smoothing of the log periodogram. *Journal of the American Statistical Association*, **75**(369), 122–132.

Wannamaker, R.A. (2003). *The theory of dithered quantization*. Ph.D. thesis, University of Waterloo.

Widrow, B. and Kollar, I. (1996). Statistical theory of quantization. *IEEE Transactions on Instrumentation and Measurement*, **45**(2).

Wiegand, Thomas and Schwarz, Heiko (2011). Source coding: Part I of fundamentals of source and video coding. *Foundations and Trends in Signal Processing*, **4**(1-2), 1–222.

Wornell, G. and Willsky, A. (2004). 6.432 stochastic processes, detection, and estimation: Course notes.

Wu, C. F. J. (1983). On the convergence properties of the EM algorithm. *The Annals of Statistics*, **11**(1), 95–103.

Yen, J.L. (1956). On nonuniform sampling of bandwidth-limited signals. *IRE Transactions on Circuit Theory*, **3**(4), 251–257.

Index

acceptance
  probability, 85
  region, 75
agglomeration property, 329
Akaike information criteria (AIC), 102, 177
algebra, 1
  Borel, 13
  fundamental theorem, 10, 207
  matrix, 8
  max-product, 243
  max-sum, 243
  σ-algebra, 12
algorithm
  K-means, 171, 177
  K-medians, 173
  K-medoids, 174
  backward recursion, 242
  Baum-Welch, 243, 304, 308
  Box-Muller, 78
  deterministic, 152
  Dirichlet process means (DP-means), 338
  expectation-maximization (E-M), 146, 153, 175, 249
  forward recursion, 242, 304
  generalized E-M, 147
  greedy, 42
  iterative, 211
  junction tree (JT), 141, 304
  Lloyd-Max, 173, 278
  maximum a-posteriori Dirichlet process mixture (MAP-DP), 341
  message passing, 141, 243
  overlap-add, 223
  soft K-means, 175
  variational Bayes' (VB), 147
aliasing, 219, 270
analogue-to-digital converter (ADC), 266
Armijo bound, 54, 57, 58
atom, 316
autocorrelation, 197, 226
autocovariance, 24
autoregressive (AR), 212
auxiliary variable, 89

backtracking, 54, 248, 249, 306
bandwidth, 267
basis, 5, 181
  coherence, 295

incoherent, 295
  orthogonal, 6, 198
  orthonormal, 6, 257, 259, 270
  standard, 5
Bayes' rule, 16
Bayesian
  information criteria (BIC), 102
  nonparametrics, 313
Bayesian information criteria (BIC), 177
bilinear transform, 218
binary search inversion, 81
biorthonormal, 272
bit, 30, 273, 275
  prefix, 273
  reversal, 211
bootstrap, 107
burn-in period, 131

canonical joint density, 88
centroid, 174
cepstra, 312
chain rule, 134
chaotic sequence, 71
Cholesky decomposition, 79
classification, 154, 326
  error, 161
  linear discriminant analysis (LDA), 278
  structured, 309
cluster splitting, 177
clustering, 171, 317
  K-means, 279
  mean shift, 303
  nonparametric, 176
  parametric, 177
code, 274
  binary, 273
  Huffman, 273
codec, 288, 293
coding
  entropy, 273, 280
  transform, 291
collapsing, 108
common
  cause, 137
  effect, 138
companding, 288
  μ-law, 288
complementary slackness, 44
complex
  argument, 189

conjugate, 189
conjugate pairs, 236
exponential, 190
magnitude, 189
plane, 189
complexity
  computational, 38
  Kolmogorov, 28, 30, 31, 116
    conditional, 30
  model, 106, 276
compressed representation, 104
compressive sensing (CS), 294
concave, 36
conic section, 156
conjugate pair
  Bernoulli-Beta, 122
  Normal-Gamma, 123
  Poisson-Gamma, 123
connected, 84
constant
  overlap add, 255
  rejection, 75
  signal, 198
continuous-time Fourier transform, 202
convergence, 45, 84, 131, 145, 152
  diagnostics, 132
  linear, 46, 54, 146
  order, 45
  quadratic, 46, 56, 74
  rate, 45
  subgradient method, 61
  sublinear, 46
  superlinear, 46
convex, 7, 36, 42, 117, 128
  first-order condition, 36
  polytope, 60
  relaxation, 161
  second-order condition, 37
  upper bound, 161
convolution, 191, 195, 268
  circular, 199, 260
  discrete, 11
  operator, 191
correlation
  coefficient, 113
covariance, 18, 21
covariate, 162
critical points, 41
cross-correlation, 196, 226
curse of dimensionality, 133, 168

cutoff frequency, 215, 216, 220, 225
cycle, 34

d-connection, 138
d-separation, 138
data
  count, 167
  partially labelled, 176
  test, 105
  training, 105, 134, 155
  unlabelled, 176
data compression, 28, 103
  linear predictive coding (LPC), 338
  lossy, 275, 288
  MP3 format, 293
  ratio, 290, 292
decibel, 277
decimation, 260
  in frequency (DIF), 211
  in time (DIT), 211
decision boundary, 155
deconvolution, 312
decorrelation, 227
degrees of freedom, 102, 159, 178
  effective, 103
delta
  Dirac, 188
  Kronecker, 187
descent
  coordinate, 145
  gradient, 42
  steepest, 42
detailed balance, 27, 85, 89
deterministic, 274
digital-to-analogue converter (DAC),
      266, 274
dimensionality reduction, 178, 327
discrete
  Fourier matrix, 49
  Fourier transform (DFT), 49, 200
  time Fourier transform (DTFT), 192
discriminant, 155, 156
distortion, 274
distribution
  asymmetric Laplace, 127
  Bernoulli, 13, 120
  Beta, 121
  Beta-Binomial, 123
  binomial, 19
  Boltzman, 69
  Cauchy, 124
  χ-squared, 21
  conditional, 134
  Dirichlet, 328
  Dirichlet-multinomial, 335
  elliptical, 23, 114
  equilibrium, 26
  exponential, 73, 116, 124
  exponential family, 32, 109, 153, 165
  extreme value, 23

gamma, 77
Gaussian, 14, 118, 204
Gumbel, 73
joint, 82
Laplace, 73, 75, 112, 124, 173
location-scale family, 21
location-scale-shape family, 74
log-normal, 302
logistic, 73, 124
maximum entropy, 116
mixture, 129, 150
multivariate Gaussian, 21, 155, 162,
      163, 179, 182
  standard, 22
Negative-Binomial, 123
normal-Gamma, 121
normal-Wishart, 164
Poisson, 167
posterior predictive, 108
prior predictive, 108
proposal, 75
proposal transition, 85
stationary, 26
Student-t, 123
symmetric Dirichlet, 335
transition, 25, 84
truncated exponential, 77
uniform, 75, 124
univariate Gaussian, 151
Weibull, 73, 124
dithering, 274
domain
  frequency, 49, 193, 269
  time, 49, 193
duality gap, 45

edge, 34, 134
  directed, 34
  undirected, 34
eigenfunction, 9, 149
eigenvalue, 9, 183
eigenvector, 9, 183
element
  identity, 1
  inverse, 1
empirical
  Bayes, 109
  risk, 98, 113
  risk estimator, 98
energy, 196
entropy, 31
  cross, 33, 97
  differential, 29
  maximum, 115
  relative, 33
  Shannon, 29
equation
  difference, 197, 207
  eigenvalue, 193
  normal, 49, 230

ordinary differential (ODE), 206
  two-scale, 258
equiprobable hyper-ellipsoid, 181
estimator
  James-Stein, 109
  max-norm, 112
  ridge, 164
  robust, 112
event, 12
evidence, 16, 108
exchangeability, 95
exchangeable, 314, 317
  infinitely, 314
expectation, 17
  law of total, 246
  sample, 20
expectation maximization (E-M), 304
expected
  loss, 98
  value, 17
exponential smoothing, 302
extrapolate, 165

factorization, 135
fast Fourier transform (FFT), 11, 208
  Cooley-Tukey, 210
  radix, 210
feasible region, 42
feature, 149, 184
  space, 149
filter, 195, 239, 254
  L_1-trend, 303
  (Type I) Chebyshev, 221
  all-pass, 221
  bandpass (BPF), 220, 257
  bilateral, 303
  Butterworth, 220
  Butterworth low-pass, 217
  comb, 221
  conjugate mirror, 258
  detail, 259
  differentiator, 217, 218
  digital, 169
  elliptic, 221
  finite impulse response (FIR), 212,
      215, 258
  high-pass (HPF), 206
  infinite impulse response (IIR), 212
  inverse, 227
  Kalman, 304
  low-pass (LPF), 208, 213, 258, 270
  median, 302
  rational design, 212
  recursive, 302
  scaling, 258
  unstable, 212
  Wiener, 238
filtering
  forward-backward, 221
  Wiener, 325

finite-dimensional distribution (f.d.d.), 24
fixed point, 144, 146
formula
   de Moivre's, 189
   Euler's, 189
   Panter-Dite, 279
   Parseval's, 196
   Shannon-Whittaker reconstruction, 268
forward stagewise selection, 170
forward-backwards message, 243, 244, 246, 250
Fourier transform (FT), 192, 252, 282
function
   L_2-norm objective, 49
   absolute loss, 98
   bandlimited, 267
   basis, 165, 271
   Bessel, 283
   characteristic (CF), 19, 282
   covariance, 320
   cumulative distribution (CDF), 13
   delta, 187
   digamma, 333
   Dirac delta, 269
   discrete prolate spheroidal, 235
   dual objective, 44
   empirical cumulative distribution (ECDF), 20
   equality constraint, 42
   exchangeable partition probability (EPPF), 333
   Gaussian basis, 165
   global, 140
   Heaviside step, 188, 205
   hinge loss, 128
   inequality constraint, 44
   kernel, 184, 319
   Lagrangian, 43
   link, 166
   local, 140
   logistic, 157
   logistic error, 52
   logit, 166
   loss, 51, 98, 110, 161
   many-one, 274
   mean, 320
   measure, 12
   message, 141
   mixed L_2-L_1-norm objective, 50
   moment generating (MGF), 18
   nonlinear objective, 51
   objective, 36, 41, 152
   piecewise constant, 261
   piecewise linear, 76, 90, 126, 303
   piecewise linear objective, 59, 62
   polynomial basis, 165
   primal objective, 44
   probability density (PDF), 13
   probability mass (PMF), 13

quantile, 72, 77, 90
random PMF, 329
rectangular pulse, 189
rectangular window, 215
scaling, 258
semi-local, 140, 141
sensing, 295
sinc, 188, 267, 270
spline, 271
square loss, 98
transfer (TF), 195, 206
   analysis, 206
transfer (TF), 207
unit step, 188
weight, 319

Gaussian process (GP), 226, 318
Gaussian process latent variable model (GPLVM), 327
generalization, 101
generalized linear model (GLM), 53
geometric sequence, 198
Gibb's phenomena, 205, 215, 222, 225, 253, 283
Gibbs
   block, 84, 151
   random scan, 84
   systematic scan, 84
graph, 34, 134
   bipartite, 35
   clique, 243, 305
   connected, 35
   diameter, 35
   directed acyclic (DAG), 35, 134
   distance, 34
   factor, 143
   fully-connected, 35
   moral, 141
   unweighted, 34
   weighted, 34
group, 1
   Abelian, 1
   homomorphism, 2
      exponential map, 2
      negative logarithmic map, 3
   isomorphism, 2
   symmetry, 2

heavy tails, 112
Heisenberg uncertainty, 204, 253
   relation, 204
heuristic, 66, 177
high resolution analysis, 275
histogram, 129
homomorphic DSP, 311
hyperparameter, 95, 108
hyperplane, 179

identifiability, 251
ill-conditioned, 101

impulse response, 192, 222, 225, 289
   infinite, 271
   truncated, 215
impulse train, 269
independence, 16
   conditional, 134, 136
independent and identically distributed (i.i.d.), 24
inequality
   Jensen's, 36
inference
   approximate, 143
   exact, 140
infinite impulse signal, 187
information
   gain, 33
   map, 310
   mutual, 32, 276
informative vector machine, 324
initial conditions, 207
integrator, 272
interpolation, 261, 266
interquartile range, 125
invariance principle, 313
invariant, 271
inverse
   discrete Fourier transform (IDFT), 200
   discrete wavelet transform (IDWT), 260
   discrete-time Fourier transform (IDTFT), 193
iteration
   fixed-point, 73, 172
iterative, 45
   conditional modes (ICM), 144
   improvement, 67
   reweighted least squares (IRLS), 52, 166

joint
   CDF, 14
   distribution function, 93
   expectation, 18
   likelihood, 150
   moment, 18
   PDF, 15
   PMF, 15
   random variables, 14, 133

Kalman filter (KF), 24, 136, 143, 242, 249
Kalman gain, 245
Karush-Kuhn-Tucker (KKT) conditions, 45
kernel
   K-nearest neighbour, 168
   autoregressive (AR), 321
   box, 168
   density estimate (KDE), 94, 129, 167, 335

Dirichlet, 235
  function, 149
  Mercer, 149, 321
  periodic, 321
  squared exponential, 320
  trick, 160
Kolmogorov extension, 316
Kullback-Leibler divergence (KL), 33, 96,
    116, 145
kurtosis, 17

Lagrange multiplier, 42, 43, 155
least squares, 111
  linear, 49
  weighted, 50
level curves, 42
liftering, 312
likelihood, 16, 95, 162
  complete, 145
  complete data, 152
  incomplete, 153
  marginal, 108, 145
  negative log (NLL), 96
  out-of-sample, 105
  ratio, 97
linear
  congruential generator, 72
  discriminant analysis (LDA), 156
  hyperplane, 159
  orthogonality principle, 228
  phase, 214
  prediction analysis (LPA), 230, 236,
      289
  predictive coding (LPC), 289
  separability, 159
  stochastic DSP, 226
  system, 191
  systems theory, 187
  time-invariant (LTI), 289
  time-invariant (LTI) system, 191
local
  average, 168
  error, 217
  linear embedding (LLE), 185
  power spectrum, 253
log-concave, 77
log-spectra, 311
log-sum-exp trick, 311
loop, 34
loss
  0-1, 161
  hinge, 159
  logistic, 161

manifold, 185
marginalization, 15, 108
  of concave functions, 37
Markov
  blanket, 139, 151
  chain, 25, 134, 136, 141

aperiodic, 26
  Hamiltonian MCMC (HMC), 88
  higher-order, 136
  irreducible, 26, 84
  Monte Carlo (MCMC), 83, 130
  reversible, 27
  semi, 302
  property, 25
matrix, 8
  addition, 8
  adjacency, 34
  autocorrelation, 230, 238
  circulant, 11
  covariance, 10, 162, 179, 184
  determinant, 9
  diagonal, 9
  diagonalizable, 10
  discrete convolution, 199
  doubly stochastic, 27
  Gram, 149
  Hessian, 56
  identity, 9
  invertible, 9
  kernel, 149, 267, 319
  kernel design, 184
  orthonormal, 11
  permutation, 209
  positive-definite, 11, 50
  sample covariance, 181, 183
  singular, 9
  sparse, 209
  square, 8
  stochastic, 26
  symmetric, 10
  Toeplitz, 11
  trace, 9
  transition, 26
  transpose, 8
  triangular, 11
maximal clique, 141
maximum
  a-posteriori (MAP), 37, 99, 252
  likelihood (ML), 96
  likelihood estimator (MLE), 96
  local, 42
  margin, 159
  of convex functions, 37
  probability, 144
mean shift, 175
measure
  base, 117, 118, 329
  mixing, 315
  random, 315
median, 125, 126
medoid shift, 176
method
  backward Euler, 217
  BFGS, 58
  Conjugate gradient, 58
  greedy, 146

Marsaglia polar, 78
  Newton's, 53, 56, 73
  normalized gamma, 330
  Polyá urn, 330
  Quasi-Newton, 58
  simplex, 60
  stick-breaking, 330, 331
  subgradient, 61
  trapezoidal, 217
metric, 4
  absolute distance, 111
  city-block, 4, 111
  discrete, 4
  Euclidean, 4, 173
  Hamming distance, 66
  Mahalanobis distance, 4
  space, 4
  square Euclidean distance, 111
Metropolis-Hastings (MH), 85, 144
  random-walk, 86
minimum
  description length (MDL), 104
  energy, 267
  global, 52, 53, 155
  local, 41
  norm, 267
  square loss, 181
misclassification rate, 159
mixture
  component, 150
  density, 82
  finite, 123
  weights, 150, 153
mode, 150
  local, 176
model
  autoregressive (AR), 136
  autoregressive HMM, 136
  Bayesian, 94, 133
  complexity, 177
  Dirichlet process mixture (DPMM),
      335
  discriminative, 158
  Gaussian mixture, 152
  generalized linear, 158, 165
  hidden Markov (HMM), 136, 304
  hidden Markov filtering, 242
  hidden Markov fitting, 243
  hidden Markov smoothing, 242, 251
  hierarchical Bayesian, 136
  hierarchical probabilistic graphical, 242
  infinite mixture, 334
  linear expectation, 162
  linear-Gaussian, 143
  local linear, 185
  mixture, 136, 171, 304
  non-Bayesian, 95
  non-nested, 97
  nonparametric, 93, 167
  parametric, 93

Pitman-Yor mixture (PYMM), 338
probabilistic graphical (PGM), 249
sinusoidal, 239
statistical, 93
moment, 17, 125
central, 17
second raw, 204
vanishing, 261
moving average (MA), 212, 225, 228
multimodal, 150
multiresolution approximation system,
258
MUSIC, 241

nearest neighbours, 185
neighbourhood, 66, 170
nonlinear diffusion, 302
nonparametric
density estimation, 129
norm, 5
L_p, 5
Euclidean, 5
Frobenius, 163
squared L_2 weighted, 7
weighted L_q, 113
normalizer, 98
log, 117, 118
normed space, 5
notation
big-O, 38
NP-complete, 39, 66
number
Bell, 317
complex, 189
imaginary, 189
integer, 266
pseudo-random, 71
real, 1
transcendental, 71
two's-complement, 275
numerical overflow, 311
numerical underflow, 310
Nyquist
criterion, 268
frequency, 258, 269

Occam's razor, 101, 169
operator, 1
associative, 1
commutative, 1
distributive, 3, 140
linear, 7
reversal, 195
optimization, 36
combinatorial, 41, 170, 174, 177, 296
constrained, 42, 59
continuous, 41
global, 65
optimum
local, 36

predictions, 97
probability, 95
orthonormal, 50
out-of-sample, 105
outcome, 12
overfit, 102, 105

parameter, 41
bandwidth, 94, 129
canonical, 117
concentration, 329
location, 127
natural, 121
regularization, 263
partition, 316, 317
permutation invariant, 317
random, 333
path, 34
chain, 137
undirected, 137
perfect reconstruction, 255
periodic, 193, 198, 260, 269
extension, 198
permutation, 314
phase delay, 214, 216
piecewise exponential, 76
Pitman-Koopman lemma, 117
point estimates, 108
polar form, 189
pole, 207, 212, 217, 227
complex, 220
complex-conjugate pair, 220
polynomial
characteristic, 10
posterior, 16
power spectral density (PSD), 226
prediction, 97
primal-dual
interior-point, 60, 62
optimal point, 45
principal component (PC), 180
principal components analysis (PCA),
179, 183, 239
kernel (KPCA), 185
probabilistic, 136, 251
prior, 16
conjugate, 109, 120
updating, 108
probabilistic graphical model (PGM), 35,
82, 134, 150
probability, 12
acceptance, 86
conditional, 15
product
dot, 6
elementwise, 7
inner, 6
matrix, 8
matrix Kronecker, 208
non-commutative, 8

sample space, 14
scalar, 5
program
integer linear (ILP), 48
linear (LP), 59
quadratic (QP), 60
PSD estimator
Barlett's, 232
Blackman-Tukey, 234
Burg, 237
covariance LPA, 237
nonparametric, 231
periodogram, 231
Pisarenko, 241
regularized Wiener, 239
Thomson multi-taper, 235
Welch, 233
Yule-Walker, 237
pseudospectrum, 242

quadratic
discriminant analysis (QDA), 156
form, 112, 156
quantile matching, 126
quantization, 173, 273, 274
levels, 274
vector (VQ), 286
quantizer, 274
dithering, 284
optimal scalar, 279
scalar uniform, 274
stochastic, 284

random
demodulation, 296
restarting, 66, 69, 147
variable, 13, 93, 134
rate-distortion, 276
Rauch-Tung-Striebel smoothing, 247
reconstruction, 266
recurrence relation
linear, 26
nonlinear, 71
recursion
Levinson, 326
regression, 51, 136, 162
L_1-norm, 59, 62
Bayesian kernel, 323
Bayesian linear, 238
Gaussian process, 252
kernel, 168, 224
Lasso, 50, 60, 164, 170
least-angle (LAR), 170
linear, 162, 163, 170, 182, 230
linear-in-parameters, 164
locally linear kernel, 168
logistic, 157, 158, 166
polynomial, 102, 103
quantile, 127
ridge, 170, 318

support vector, 127
regularization, 37, 49, 97, 100, 159, 161, 170, 184, 237, 252
 L_1, 167
 L_2-norm, 49
 parameter, 64
 path, 64, 128
 Tikhonov, 101, 103, 252
residual, 59
resonator, 220
ring, 3
roots of unity, 189
rule
 classification, 158
 Sturge's, 129

saddle-point, 42
sample
 mean, 20, 111
 median, 111
 proposal, 75
 space, 12
 variance, 112
sampling, 266
 adaptive rejection (ARS), 75
 ancestral, 82
 bandlimited, 283, 295
 Gibbs, 83, 86, 143, 152
 independence, 86
 inverse transform, 73, 90
 nonuniform, 267
 rate, 269
 rejection, 75, 80
 Shannon-Whittaker, 271
 slice, 90
 uniform, 266
search
 exhaustive, 47
 golden section, 55
 greedy, 170, 174
 line, 54
 local, 36
 tabu, 67
semi-supervised, 176
semifield, 3
semiring, 3, 140
 max-product, 247, 306
sequential search inversion, 80
set, 1
Shannon, 268
 lower bound, 277
Shannon information map, 29, 31
Shannon-Whittaker interpolation, 293
shrinkage, 50
 ridge, 100
sift property, 188, 269
signal
 digital, 265

signal subspace analysis, 238, 251
signal-to-(quantization)-noise-ratio (SQNR), 277
simulated annealing, 68, 87
skewness, 17
small-variance asymptotics, 173
soft Gaussian assignment, 174
sparse, 263, 295, 297
 K-sparse, 295
sparse representation, 261
sparsity, 112, 127
spectral concentration, 235
spectral leakage, 253
spline, 262
 cardinal basis, 272
standard deviation, 17
stationarity, 263
 strong, 24
 weak, 24
stationary
 weakly, 226
statistic, 116
 order, 112, 126
 sufficient, 116, 118, 154
statistical
 inference, 95
 stability, 19, 238
step
 BFGS, 58
 Newton, 56, 58
step size
 constant, 53
 optimal, 54
stepwise
 backward, 170
 forward, 170
stochastic process, 23, 320
 autoregressive (AR), 28
 Chinese restaurant (CRP), 332
 continuous-time, 23
 Dirichlet (DP), 24, 327
 discrete-time, 23
 Gaussian (GP), 24
 normalized inverse-Gaussian (NIGP), 334
 Ornstein-Uhlenbeck, 320
 Pitman-Yor (PYP), 334
 point, 316
strong duality, 45
sub-graph, 135
subdifferential, 46
subgradient, 47, 128
subspace
 noise, 241
 signal, 240
sum of losses, 110
support, 84
 vector machine (SVM), 60, 128, 159

test

Geweke, 131
 Kolmogorov-Smirnov, 131
 Student's t, 131
theorem
 central limit, 21
 cross-correlation, 197
 de Finetti's, 314
 Hewitt-Savage, 315
 Hilbert projection, 271
 Kolmogorov extension, 320
 Mercer's, 149, 184
 Neyman factorization, 116
 Plancherel's, 196, 203
 source-coding, 30
 Widrow's quantization, 282
 Wiener-Khintchine, 197, 226, 231
theory
 information, 29
 rate-distortion, 275
thinning, 131
threshold, 274
time
 invariance, 252
 invariant, 191, 226
time and space efficiency, 38
time index, 193
time-frequency plane
 analysis, 253
 tiling, 256, 261
tolerance, 45
total variation (TV), 206
 denoising (TVD), 65, 129, 252, 303
transform
 location-scale-shape, 127
 translation-scale plane, 256
traveling salesman problem, 48
tree, 35, 243
 minimum spanning, 141
 width, 143

underfit, 102
undersampling, 295
unimodal, 150, 155
unit circle, 5, 219
unsupervised, 171

validation, 105
 M-fold, 106
 cross, 105, 106, 237
 leave-one-out, 106
variable
 elimination, 140
 frequency, 193
 indicator, 150, 153, 171
 latent, 136, 150
 parent, 134, 139
 selection, 47, 169
variance, 180
vector, 5
 orthogonal, 6

space, 5, 187
vertex, 34, 134
  child, 141
  clique, 141
  degree, 34
Viterbi
  decoding, 243, 247, 305, 306, 310
  path, 248
  training, 309

wavelet
  admissibility, 256
  Battle-Lemarié, 262
  coiflets, 262

continuous transform (CWT), 256
Daubechies, 262
detail signal, 259
discrete transform (DWT), 257
dyadic, 257
Haar, 261
mother, 256
shrinkage, 262
spline, 262
support, 261
symmlets, 262
transform, 256
weighted

mean, 113
median, 113
white noise, 227, 228
whitening, 227
window
  rectangular, 253
  sliding, 253
  triangular, 232, 253
Wolfe conditions, 58

zero, 227
  padding, 210, 222
zeros, 207